Second Edition

Handbook for

CRITICAL CLEANING

APPLICATIONS, PROCESSES, AND CONTROLS

Second Edition

Handbook for
CRITICAL
CLEANING
APPLICATIONS, PROCESSES, AND CONTROLS

Edited by Barbara Kanegsberg
Edward Kanegsberg

CRC Press
Taylor & Francis Group
Boca Raton London New York

CRC Press is an imprint of the
Taylor & Francis Group, an **informa** business

Cover design by David Kanegsberg.

CRC Press
Taylor & Francis Group
6000 Broken Sound Parkway NW, Suite 300
Boca Raton, FL 33487-2742

First issued in paperback 2017

© 2011 by Taylor and Francis Group, LLC
CRC Press is an imprint of Taylor & Francis Group, an Informa business

No claim to original U.S. Government works

ISBN 13: 978-1-138-07732-4 (pbk)
ISBN 13: 978-1-4398-2829-8 (hbk)

Visit the Taylor & Francis Web site at
http://www.taylorandfrancis.com

and the CRC Press Web site at
http://www.crcpress.com

To a beautiful, safe, productive world for the next generation

Noa Raeli Kanegsberg

To our most valuable collaborative efforts

Deborah Joan Kanegsberg and David Jule Kanegsberg

And to the memory and positive influence of

Israel Feinsilber

Jule Kanegsberg

Murray Steigman

Dr. Jacob J. Berman

Contents

Preface to the Second Edition xi

Preface to the First Edition xv

About the Second Edition xix

Editors.. xxv

Contributors' Bios xxvii

PART I   Process Implementation and Control

1   Evaluating, Choosing, and Implementing the Process: How to Get
Vendors to Work with You... 3
Barbara Kanegsberg

2   Cleaning Agent Balancing Act.. 15
Barbara Kanegsberg

3   Blunders, Disasters, Horror Stories, and Mistakes You Can Avoid 29
Arthur Gillman

4   Cleaning Practices and Pollution Prevention..................................39
Mike Callahan

5   Basis of Design for Life Science Cleanroom Facilities.....................61
Scott E. Mackler

6   Validating and Monitoring the Cleanroom.....................................89
Kevina O'Donoghue

7   Cleanroom Management, Validation of a Cleanroom Garment System,
and Gowning Procedure 103
Jan Eudy

8   Principles of Wiping and Cleaning Validation 119
Karen F. Bonnell and Howard Siegerman

9   Overview to Analytical and Monitoring Techniques 129
Ed Kanegsberg

10   Practical Aspects of Analyzing Surfaces 135
     Ben Schiefelbein

11   How Clean Is Clean? Measuring Surface Cleanliness and Defining
     Acceptable Levels of Cleanliness 145
     Mantosh K. Chawla

12   Cleaning Validations Using Extraction Techniques 161
     Kierstan Andrascik

13   Biomedical Applications: Testing Methods for Verifying Medical
     Device Cleanliness 173
     David E. Albert

14   Material Compatibility 183
     Eric Eichinger

PART II   Applications

15   Clean Critically: An Overview of Cleaning Applications 197
     Barbara Kanegsberg

16   Cleaning Validation of Reusable Medical Devices: An Overview of
     Issues in Designing, Testing, and Labeling of Reusable Devices 229
     John J. Broad and David A.B. Smith

17   Critical Cleaning for Pharmaceutical Applications 255
     Paul Lopolito

18   Cleaning in the Food Processing Industry 271
     Hein A. Timmerman

19   Electronic Assembly Cleaning Process Considerations 283
     Mike Bixenman

20   Precision Cleaning in the Electronics Industry: Surfactant-Free
     Aqueous Chemistries 319
     Harald Wack

21   Contamination-Induced Failure of Electronic Assemblies 333
     Helmut Schweigart

22   Surface Cleaning: Particle Removal 345
     Ahmed A. Busnaina

23   Cleaning Processes for Semiconductor Wafer Manufacturing
     (Aluminum Interconnect) 359
     Shawn Sahbari, Mahmood Toofan, and John Chu

24   Advanced Cleaning Processes for Electronic Device Fabrication
     (Copper Interconnect and Particle Cleaning) 379
     Shawn Sahbari and Mahmood Toofan

25 The Cleaning of Paintings 399
    Richard C. Wolbers and Chris Stavroudis

26 Road Map for Cleaning Product Selection for Pollution Prevention 411
    Jason Marshall

27 Wax Removal in the Aerospace Industry 429
    Bill Breault, Jay Soma, and Christine Fouts

28 Implementation of Environmentally Preferable Cleaning Processes for
    Military Applications.. 439
    Wayne Ziegler and Tom Torres

PART III   Safety and Regulations

29 Worker Protection and the Environment: Current Editorial
    Observations 455
    Barbara Kanegsberg

30 Health and Safety... 465
    James L. Unmack

31 Critical Cleaning and Working with Regulators: From a Regulator's
    Viewpoint 483
    Mohan Balagopalan

32 Momentum from the Phaseout of Ozone-Depleting Solvents Drives
    Continuous Environmental Improvement 491
    Stephen O. Andersen and Margaret Sheppard

33 Screening Techniques for Environmental Impact of Cleaning Agents 501
    Donald J. Wuebbles

Glossary of Terms and Acronyms.. 521

Index.. 525

Preface to the Second Edition

Why a Second Edition?

In the last few years, challenges to the manufacturing community have increased and so have performance expectations. With the ever-decreasing size of components, these expectations are becoming more difficult to meet. The second edition of the *Handbook for Critical Cleaning* hopes to help you meet these expectations and produce high-quality products in a cost-effective manner. Although cleaning is a process and not a chemical, increased awareness of the consequences of chemical use to workers, to the general public, and to the environment has prompted more stringent regulatory measures worldwide. Environmental and worker safety regulations are imperative to maintaining a decent quality of life on this planet and, perhaps, to our very survival. However, the goal of manufacturing is not to jump through regulatory hoops, but to produce efficient products. Compromising on the efficacy of cleaning and thereby producing a suboptimal product can affect public safety and can compromise on the quality of life. Manufacturers face the challenge of doing it all.

What Is Critical Cleaning?

Identify, then qualify/validate and monitor the critical cleaning steps. The terms "critical cleaning" and "precision cleaning" are often used interchangeably. However, we prefer the former term. Precision cleaning suggests cleaning in a highly restricted clean room, where each individual component is perhaps cleaned separately by a highly trained technician, where there are perhaps wet benches with automated product handling, and where there may be a multichamber-automated spray system that feeds directly into the clean room. This is a limited view of the important cleaning step. Perhaps the best way to define a critical cleaning step is to consider the negative consequences that arise if that step is not performed or is performed inadequately. Based on our experience, the important cleaning step, the *critical cleaning step*, may occur in a machine shop or in a job shop (e.g., a coating facility) in what appears, at first glance, like an automotive repair facility. If the soil (matter out of place) is not adequately removed at that step, subsequent processing and cleaning steps may not resolve the problem but may actually exacerbate contamination by inadvertent chemical reaction of the soil, drying of the soil, or by embedding the soil on the surface of the product. Contamination happens long before the product enters the clean room. A clean room can only minimize recontamination, and even the most sophisticated clean room or controlled environment cannot correct a contaminated product (Kanegsberg and Kanegsberg, 2010).

Lean Cleaning and Supply Chains

Economic pressures have led to the implementation of such concepts as lean cleaning and six-sigma; thus, we have to clean smarter. In fact, it is imperative that we clean smarter. Cleaning must be value added. Assess your own processes and understand the role that cleaning plays in those processes. Sometimes the value added is only appreciated by factoring in the costs of not cleaning or under-cleaning at a particular step. Critical cleaning is not just about *what* is done or *how* it is done. It is also about *when* it is done. This becomes even more important when we realize that most products are not built from scratch within one facility. There usually is a complex supply chain of autonomous or semiautonomous facilities, which may sometimes be separate divisions or departments of the same company. Regardless of whether the supply chain involves inter- or intracompany processes, it is crucial that communication take place and that process understanding and process integration occur. The most critical cleaning step may be one that needs to take place at a supplier, before the part reaches your facility.

Critical Cleaning and Surfaces

Cleaning is removing undesired materials from surfaces without changing the surface in an unacceptable manner. As products get smaller, the surface becomes a greater percentage of the product. When products are at the nanoscale, it can be said that the product *is* the surface.

New and Useful

Cleaning Is a Process

The economic and regulatory hurdles involved in introducing new cleaning agents have increased considerably (see "A balancing act" in book 2). Therefore, chemicals that have been developed for markets other than cleaning but have been adapted for the cleaning sector and complex blends have become increasingly prevalent. Therefore, the newer cleaning products are covered extensively in the second edition. Cleaning equipment has also evolved during the past decade, and meshing the appropriate cleaning agent with the right equipment requires a working understanding of chemistry, physics, and engineering. We have added discussion of ultrasonic techniques and monitoring. Partially spurred by regulatory pressures, an increased use of so-called nonchemical approaches is included.

Process Implementation

All of the knowledge in the world about cleaning agents, cleaning equipment, and process flow is of no use if you do not improve the cleaning process. We provide guidance to actually do something: to select, validate, implement, and monitor the cleaning process. We also cover new approaches to definitive, lean, analytical testing and provide discussions related to clean rooms, including construction and working in a clean room.

Applications

The application portion of the *Handbook for Critical Cleaning* has been expanded to include critical cleaning processes for high-value product such as for medical devices, pharmaceutical, food processing, aerospace, and military. Electronics cleaning, which had been considered to be "solved" a decade ago, has resurfaced as a critical issue due to such developments as miniaturization, increased component density, and replacement of lead solders with lead-free, higher temperature solders. Conservation of fine art may not immediately be thought of along with manufacturing, but this involves critical cleaning and the requirements are in some ways similar. Two art conservators outline the thought processes and trial-and-error determinations to match cleaning agents to the soil when cleaning or restoring paintings.

Safety/Environmental Considerations

Safety and environmental considerations are not only global issues but are also important concerns at the national and local levels, and they do not always coordinate or mesh well. You cannot ignore them, and you should not ignore them. We have not attempted to outline all regulations. Dealing with such a moving target would be frustrating and futile, and most engineers would develop glazed eyeballs. Instead, we have attempted to discuss a few topics that are important to the critical cleaning community and to provide strategies for working constructively with the regulatory world. Some of this guidance comes from members of the regulatory community.

Resource conservation is becoming an important topic in the twenty-first century. Efforts to minimize or recycle water, chemicals, and energy will increasingly become a factor in keeping process costs competitive. Green cleaning, which considers both safety and environmental impact, is discussed throughout the two books. The definition of green is not set in stone; it will continue to evolve.

Web-Based Material

Some of the authors have submitted non-print media (color illustrations, animations, film clips, etc.) to augment their chapters. These can be accessed via the "Downloads & Updates" tab on the web pages for these books at CRCPress.com.

The Lady in the Saffron Sari

Barbara Kanegsberg

"You must run, you must flee," implored the earnest gentleman as he ran toward us.

Puzzled and slightly alarmed, our daughter Deborah and I peered down a corridor of immense, multicolored marble slabs while balancing a finished wood cabinet door, a celadon green tile, and some decorative hardware. It was a brilliant, Southern California morning. The silhouette of the plaster Disneyland Matterhorn broke through a cloudless blue sky. The only obvious danger was the trauma of remodeling the kitchen.

"Why do we need to run?" I asked.

"You must hide, my wife must not see you," he replied.

"Why can't your wife see us?" our daughter asked.

"Because, you see, I told her, first we will select the marble, then the cabinets, then the tile, then the door pulls. You are coordinating. If she sees you, she will want to coordinate."

At that very moment, an elegant woman wearing a luminous, saffron-yellow sari came gliding across the marble yard.

"You see, dear," she said, putting her arm around the gentleman and steering him purposefully toward the exit, "they are coordinating. Let's go, we must coordinate too."

Coordinate, Extrapolate, Optimize

The lady in the saffron sari had the right idea. You, too, must coordinate. Achieving a high-quality manufactured product in a cost-competitive manner requires coordination of critical cleaning and contamination control within your company and perhaps coordination with the efforts of a complex supply chain. If you are in charge of selecting cleaning equipment, please read over the chapters on cleaning agents and coordinate the two efforts (and vice versa). Coordinating cleaning efforts with regulatory requirements, including safety, environmental, and validation requirements, is also time well spent. Whether you are a job shop, an initial fabricator, a final assembler, or you have a repair facility, understanding the importance of critical cleaning is a must to achieve a cost-competitive advantage.

It is reasonably safe to say that your manufacturing situation is unique. We suggest that you consider, blend, and extrapolate from the information and advice provided in both books, even perusing those chapters that seem outside of your field. We often combine the approaches of what, at first glance, seem to be unrelated fields. As you read the chapters, think about how approaches might apply to your application and where cleaning is really necessary. Always clean critically.

Acknowledgments

We want to express our profound gratitude to all contributors. Many of you composed your chapters during a time of professional and/or personal challenges; we thank all of you for your wonderful, useful, practical chapters. The information, expertise, and guidance provided in these chapters are invaluable. We would also like to thank Cindy Carelli, Jessica Vakili, Jennifer Smith, and the staff at CRC Press for supporting us throughout the process. A special thanks to Dr. Vinithan Sedumadhavan, the production project manager, for careful attention to detail and to turning the manuscripts into printed pages.

Our thanks also to our children, Deborah Kanegsberg and David Kanegsberg, daughter-in-law, Sandra Hart, and parents, Ruth Feinsilber and Mimi Steigman, for standing by us during the writing and editing process. Our granddaughter, Noa Raeli Kanegsberg, was a very special inspiration for creating this second edition.

Reference

Kanegsberg, B. and Kanegsberg, E. Contamination detection basics, *Controlled Environments Magazine*, June 2010.

Ed Kanegsberg
Barbara Kanegsberg

Preface to the First Edition

Adapted from: What is critical cleaning?, First Edition, *Handbook for Critical Cleaning.*

Critical cleaning is required for the physical manifestation of technology.

We are in the information age, an age of thought, ideas, communication. However, this technology is based on physical objects, parts, or components. Many of these objects require precision cleaning or critical cleaning because they are either intrinsically valuable, or they become valuable in the overall system or process in which they are used. Some parts or components require critical cleaning not because of the inherent value of the part itself but instead due to their place in the overall system. For example, inadequate cleaning of a small inexpensive gasket can potentially lead to catastrophic failure in an aerospace system.

Nearly all companies which manufacture or fabricate high-value physical objects (components, parts, assemblies) perform critical cleaning at one or more stages. These range from the giants of the semiconductor, aerospace, and biomedical world to a host of small to medium to large companies producing a dizzying array of components.

Soil

The concepts of contamination, cleaning, and efficacy of cleaning are open to debate and are intertwined with the overall manufacturing process and with the ultimate end-use of the assembled product.

Contamination or soil can be thought of as matter out of place (Petrulio and Kanegsberg, 1998). During manufacture, parts or components inevitably become contaminated. Contamination can come from the environment (dust, smog, skin particles, bacteria), from materials used as part of fabrication (oils, fluxes, polishing compounds), as a by-product of manufacturing, and as from residue of cleaning agent ostensibly meant to clean the component.

Cleaning

Cleaning processes are performed because some sort of soil must be removed. In a general sense, we can consider cleaning to be the removal of sufficient amounts of soil to allow adequate performance of the product, to obtain acceptable visual appearance as required, and to achieve the desired surface properties. You may notice that surface properties are included because most cleaning operations probably result in at least a subtle modification of the surface. If a change in the cleaning process removes additional soil and if as a result the surface acquires some undesirable characteristic (e.g., oxidation), then the cleaning process is not acceptable. Therefore, surface preparation and surface quality can be an inherent part of cleaning.

Identifying the Cleaning Operation

Cleaning processes and the need for cleaning would seem to be trivial to identify. If you had a child who appeared in the doorway covered with mud, you would do a visual assessment of the need for cleaning, perform site-directed immersion or spray cleaning in an aqueous/saponifier mixture with hand-drying. However, people perform critical cleaning operations without knowing it. This lack of understanding can detrimental to process control and product improvement.

Recognizing a cleaning step when it occurs is probably one of the major challenges in the components manufacturing community. Cleaning is often enmeshed as a step in the overall process rather than being recognized as a concept in itself. It may be considered as something that occurs before or after another process, but not as a process to be optimized on its own (Dorothy Rosa, personal communication). A cleaning process often not called a cleaning process. For example, optics deblocking (removing pitches and waxes), defluxing, degreasing, photoresist stripping and edge bead removal in wafer fabrication, and surface preparation prior to adhesion, coating, or heat treatment can all be thought of in terms of soil removal (cleaning). Sometimes the cleaning process is identified only by the name of the engineer who first introduced it.

The sociological and psychological bases for this aversion to discussing cleaning are no doubt fascinating, but are beyond the scope of this book. The important thing is for you to recognize a cleaning process when you see it.

There are several reasons. One obvious reason is process control. A second is trouble-shooting or failure analysis. If the product fails and you need to fix the process, it is crucial to identify not only where soil might be introduced but also what steps are currently being taken in soil removal. If the chemical being used comes under regulatory scrutiny, identifying cleaning is even more important. If a supplier provides the component and a problem arises, it is important to be able to recognize where the cleaning steps occur. Finally, identifying the cleaning steps allows you to apply technologies developed in other industries to your own process.

Critical Cleaning

Defining critical cleaning or precision cleaning is a matter of ongoing debate among chemists, engineers, production managers, and those in the regulatory community. Certainly the perceived value or end-use of the product is a factor as are the consequences of remaining soil. The level of allowable soil remaining after cleaning is a consideration. Precision cleaning has been defined as the removal of soil from objects that already appear to be clean in the first place (Carole LeBlanc, personal communication). In some instances, however, high levels of adherent soil are involved in the processing of critical devices. Precision cleaning was once euphemistically said to be YOUR cleaning process for YOUR critical application, whereas everyone else's process could be considered as general cleaning (Kanegsberg, 1993). In one sense, there is some truth that the manufacturer is often the one best able to understand process criticality. At the same time, recognizing general cleaning and critical cleaning as parts of other operations can lead to overall industrial process improvement.

Why Should You Be Concerned about Critical Cleaning?

Critical cleaning issues are becoming increasingly important. Competitive pressure is increasing. Higher demands are being made of industry. A clean component produced efficiently and in an environmentally preferred manner (or at least in an environmentally acceptable manner) is a given in today's economy.

Performance, Reliability

Products are becoming smaller, with tighter tolerances and higher performance standards. Some products, such as implantable biomedical devices, are expected to perform for decades without a breakdown. Small amounts of soil and very tiny particles can irreparably damage the product.

To successfully remove the soils, you have to understand the various cleaning chemistries and cleaning equipment, and how they are combined and meshed with the overall build process.

Costs

Pressure to keep costs down increases constantly. The costs of the effective processes have tended to increase. Choosing the best option for the application can keep costs down.

Safety and Environmental Regulatory Requirements

The manufacturing community needs a wide selection of chemicals and processes to achieve better contamination control at lower costs. However, our understanding of health and the environment has led to restrictions on chemicals and processes. The manufacturer needs an understanding of atmospheric science and of the approaches used by regulatory agencies to foresee future trends.

Overview of This Book

Philosophy

In setting out to put together this comprehensive book on critical cleaning, I sought inputs from the experts in the field. Frequently these are people associated with vendors of cleaning equipment and/or cleaning agents. Naturally, each person's viewpoint is somewhat colored by their own portion of the market. However, on the whole, I was impressed with the scope and fairness of the material submitted. An attempt has been made to minimize use of brand names. In some cases, there are several contributors in a similar area. In general, my philosophy has been to include all but the most blatant material; by having a large number of contributors, a wide range of products and viewpoints are presented; the reader is expected to be intelligent enough to weigh the advantages and/or disadvantages of each approach for his or her own application.

Conclusions

While each application is very site specific, contamination control problems cut across industry lines. At the same time, each industry still tends to work in a separate little world. It is hoped that this book will provide a synthesis of cleaning approaches.

A diverse assortment of components and assemblies require critical or precision cleaning. Some examples include

Accelerometers
Automotive parts
Biomedical/surgical/dental devices (e.g., pacemakers)
Bearings
Computer hardware (metal, plastic, other composites—the insides of your computer and printer)
Consumer hardware (telephones)
Digital cameras
Disk drives

Electronics components
Flat panel displays
Gaskets
Gyroscopes
Motion picture film
Optics
Space exploration hardware
Wafers/semiconductors/microelectronics
Weapons, defense systems (missiles)

Acknowledgments

This book is the result of a phenomenal level of effort by those involved in the worlds of critical cleaning, surface preparation, and environmental issues. The information, expertise, and guidance provided by the contributing authors is invaluable. Dr. Ed Kanegsberg, business associate and spouse, provided support, encouragement, and invaluable participation in the editing process. He also provided the viewpoint and experiences of a physicist and practicing engineer. Bob Stern and the staff at CRC Press provided excellent guidance throughout the process.

I would also like to thank family members Deborah Kanegsberg, David Kanegsberg, Ruth Feinsilber, and Mimi and Murray Steigman for their patience and encouragement.

Finally, I would like to thank Dr. Shelley Ventura-Cohen, a wise colleague and adviser. She tells the story of her aunt, who, on observing Shelley staring blankly at a cookbook while an inert, raw chicken sat on the counter, exclaimed: "look at the chicken, not the book." Dear reader, critical cleaning, surface preparation, and contamination control are complex subjects, but they are also intensely practical subjects which relate to a product—your product. My suggestion, therefore, is to look at this book, and at the same time look at the chicken.

References

Carole LeBlanc, Toxics Use Reduction Institute, Lowell, MA, personal communication.

Dorothy Rosa, *A²C² Magazine*, personal communication.

Kanegsberg, B. Options in the high-precision cleaning industry: Overview of Contamination Control Working Group XIII, in *International CFC & Halon Alternatives Conference*, Washington, DC, October, 1993.

Petrulio, R. and Kanegsberg, B. Back to basics: The care and feeding of a vapor degreaser with new solvents, in *Presentation and Proceedings, Nepcon West '98*, Anaheim, CA, 1998.

Barbara Kanegsberg
BFK Solutions
Pacific Palisades, California

About the Second Edition

Philosophy

We want to help you clean critically, productively, and profitably; our goal was thus to make the second edition of the *Handbook for Critical Cleaning* even more comprehensive than the first edition. Contributors are experts in their field. We have included the viewpoints of manufacturers of parts/components of those who supply cleaning chemistries and cleaning systems, of people in regulatory agencies, and even of other consultants. We have minimized the use of brand names, but have included enough information to be unambiguous. Our philosophy is to include a range of viewpoints, some differing from our own. We urge you to make the optimal decision for your application.

Organization

Chapters in the Second Edition

The second edition of the *Handbook for Critical Cleaning* is substantially new. While we have reprinted a few classic chapters from the first edition, most chapters are new or have been substantially updated. We suggest that readers peruse not only the chapters related to their line of work and applications, but also look at what might at first glance appear to be unrelated applications. By providing a synthesis of cleaning approaches, we hope to help you make better decisions about your own cleaning processes.

We strive to achieve the impossible (or highly improbable)—a perfect balance of topics. After the publication of the first edition, we received comments that we did not devote enough space to aqueous processes, an approximately equal number of comments that we did not devote enough space to discussions on solvent processes, and assorted comments about a lack of attention to other advanced cleaning processes. We thank everyone for their comments; you are probably correct. Therefore, if you have a different viewpoint or unique cleaning application, let us know. This is how we keep learning and improving.

This series is divided into two books with five parts:

- Book 1: *Handbook for Critical Cleaning: Cleaning Agents and Systems*
  - Part I: Cleaning Agents
  - Part II: Cleaning Systems
- Book 2: *Handbook for Critical Cleaning: Applications, Processes, and Controls*
  - Part I: Process Implementation and Control
  - Part II: Applications
  - Part III: Safety and Regulations

Following is a capsule summary of each of the book chapters.

Book 1: *Handbook for Critical Cleaning: Cleaning Agents and Systems*

Part I: Cleaning Agents

An overview of cleaning agents is presented by the editor, Barbara Kanegsberg. In this expanded overview, Barbara attempts to capture the diversity of cleaning chemistry options and to put those options in perspective.

The part begins with a discussion on aqueous cleaning agents (see Chapter 1, Kanegsberg). Water is the most common cleaning agent. Michael Beeks and David Keller of Brulin & Company, a producer of aqueous cleaning equipment, expand and update their chapter from the first edition and give a comprehensive review of aqueous cleaning essentials (Chapter 2). Much of the information is also applicable to nonaqueous solvent cleaning.

Many of today's chemicals, both aqueous based and solvents, are blends. JoAnn Quitmeyer of Kyzen Corporation presents a new chapter that is a comprehensive review of cleaning agent chemistries, including single components and blends (Chapter 3).

John Burke of the Oakland Museum of Art, California, updates his particularly informative discourse on solubility and the techniques used to classify solvents (Chapter 4). It becomes clear from this chapter as to why certain solvents are applicable to removing certain types of soil.

John Owens of 3M updates his chapter on the hydrofluoroethers (HFEs), a class of solvents that have been introduced as replacements for the ozone-layer depleting chemicals (ODCs) (Chapter 5).

Joan Bartelt of DuPont updates the chapter by Abid Merchant (retired fom DuPont) that discusses the hydrofluorocarbons (HFCs), another class of ODC replacements (Chapter 6).

John Dingess and Richard Morford of EnviroTech International Inc. update the chapter by Ron Shubkin (retired from Albermarle Corporation and Poly Systems U.S.A. Inc.) on normal-propyl bromide (NPB), a substitute for the aggressive ODC solvent, 1-1-1-trichloroethane (Chapter 7).

Stephen P. Risotto, formerly of the Halogenated Solvents Industry Association (HSIA) and now with the American Chemistry Council, updates his contribution on the chlorinated solvents, a group of traditional solvents that are seeing a resurgence of use in certain applications (Chapter 8).

Ross Gustafson of Suncor Energy discusses critical cleaning applications of the bio-based D-limonene (Chapter 9).

Dan Skelly of Riverside Chemicals reviews benzotrifluorides, a group of VOC-exempt compounds (Chapter 10).

Part II: Cleaning Systems

This part reflects the wide range of process choices. The importance of drying is emphasized. Advanced and so-called nonchemical systems, such as CO_2 cleaning, steam cleaning, and plasma cleaning, are also covered. In these systems, the cleaning agent and the cleaning equipment are inseparable.

The part begins with an overview of cleaning systems contributed by the editors (Chapter 11). As with the overview for cleaning agents, this reviews processes that are treated in this book by other authors as well as those for which there are no additional chapters.

There are six chapters dealing with ultrasonics and the closely related megasonics technologies. The technology is widely used, and the diverse insights of the authors will be helpful to select equipment. John Fuchs, retired from Blackstone—Ney Ultrasonics, and Sami Awad of Ultrasonics Apps., LLC each give an overview of ultrasonics (Chapters 12 and 13). Sami Awad then teams up with K.R. Gopi, from Crest Ultrasonics Corp., to provide a new chapter on multiple frequency ultrasonics (Chapter 14). Mark Beck of Product Systems Inc. covers the basic technologies of megasonics (Chapter 15). The theory of cavitation has been absent from most discussions of critical cleaning geared to the manufacturing community. Further, the important yet elusive topic of ultrasonics metrics has seen much progress. Along these lines, we are pleased to present two new chapters. In the first chapter, Mark Hodnett of the National Physical Laboratories (U.K.) provides graphics covering theory and discusses a new technique for ultrasonics metrics along with case studies (Chapter 16). In the second chapter, Lawrence Azar of PPB Megasonics covers the principles and theoretical/mathematical basis of cavitation and discusses ultrasonics metrics (Chapter 17).

Edward Lamm of Branson Ultrasonics Corp. contributes a useful chapter, on optimizing the equipment design, covering solvent, aqueous, and semiaqueous cleaning equipment as well as rinsing, drying, automation, and other ancillary equipment (Chapter 18).

Ron Baldwin of Branson Ultrasonics Corp. contributes an important new chapter on equipment design for aqueous cleaning to help you during scale-up from laboratory cleaning to production cleaning (Chapter 19).

Dan Skelly of Riverside Chemicals contributes a chapter on equipment for cold cleaning, that is, where cleaning agents (notably solvents) are used below their boiling point (Chapter 20).

Richard Petrulio of B/E Aerospace provides a revised, expanded, and very readable chapter on the design of flushing systems (Chapter 21). This is one example where a company was able to design equipment for its own cleaning application. The chapter also provides very good guidance for the process of developing and testing a cleaning process.

Joe McChesney of Parts Cleaning Technologies updates and revises techniques for minimizing waste streams in solvent vapor degreasers, including methods for calculating the size or capacity of the required equipment (Chapter 22). Some recent case studies for the minimization of emissions have also been added.

Arthur Gillman of Unique Equipment Corporation contributes retrofitting vapor degreasers to allow the use of different cleaning chemicals or to meet newer emission control standards (Chapter 23). This option can obviate the need for new equipment.

John Durkee of precisioncleaning.com and Dr. Don Gray of the University of Rhode Island update their chapter on contained airless and airtight solvent systems, one approach for remaining in compliance with air regulations while using emissive chemicals (Chapter 24).

Wayne Mouser of Crest Ultrasonics Corp. contributes a new chapter on vapor phase organic solvent cleaning, a classic critical cleaning technique (Chapter 25).

In some cases, the cleaning agent and the cleaning equipment are inseparable. In particular, this is true for what are called "nonchemical" cleaning approaches. Several examples are provided in the next five chapters.

Ed Kanegsberg of BFK Solutions provides a new chapter, an overview to nonchemical cleaning, that addresses aspects covered in more detail by four of the authors and also reviews additional approaches, such as laser, UV/ozone, and fluidized dry bath cleaning (Chapter 26).

Jawn Swan of Crystal Mark, Inc. contributes a new chapter on micro-abrasive blasting, a technique with applications ranging from electronics and medical devices to architectural restoration (Chapter 27).

Robert Sherman of Applied Surface Technologies authors a new chapter on solid carbon dioxide cleaning, with applications for removing particles and small levels of soils from such critical surfaces as semiconductor wafers and precision optics (Chapter 28).

William Nelson of the U.S. EPA updates his chapter on supercritical and liquid CO_2 cleaning (Chapter 29).

William Moffat of Yield Environmental Systems (YES) teams with Kenneth Sautter, also of YES, to update the chapter on another approach to removing organics, plasma cleaning (Chapter 30).

Max Friedheim of PDQ Precision Inc. teams with his process engineer, Jose Gonzalez, to update and expand the chapter on the use of steam vapor cleaning for critical cleaning applications; additional case studies are included (Chapter 31).

John Russo of Separation Technologists has completely revised his chapter on selecting the best waste water treatment for aqueous operations (Chapter 32). This comprehensive chapter discusses pretreatment, posttreatment, and water recycling techniques.

Cleaning with liquids frequently means that drying is required. The final three chapters in Part II deal with this sometimes neglected process.

Barbara Kanegsberg of BFK Solutions provides an overview to drying (Chapter 33). Daniel VanderPyl of Sonic Air Systems updates his chapter on physical methods of drying (Chapter 34). Robert Polhamus of RLP Associates along with Phil Dale of Layton Technologies, Ltd. update their chapter on chemical displacement drying techniques (Chapter 35).

Book 2: *Handbook for Critical Cleaning: Applications, Processes, and Controls*

Part I: Process Implementation and Control

Part I integrates the topics of process selection and maintenance, contamination control, analytical techniques, and materials compatibility.

Barbara and Ed Kanegsberg lead off the part with a revised, expanded discussion of "How to Work with Vendors?" that applies to print, electronic, telephone, and face-to-face communication of information (Chapter 1). Barbara Kanegsberg continues with a new chapter, "The Balancing Act," discussing the technical, economic, political, and regulatory trade-offs and conflicts involved in developing and maintaining a process (Chapter 2).

Art Gillman of Unique Equipment Corporation contributes a new chapter, drawing on his many decades of experience as well as the experiences of his colleagues to present "Blunders, disasters, horror stories, and mistakes you can avoid," a compilation of cleaning lore that should be read and absorbed by all readers (Chapter 3).

Mike Callahan of Jacobs Engineering expands his chapter about optimizing and maintaining the process (Chapter 4). A number of topics such as fixturing, process monitoring, and process improvements are included.

Part I contains four new chapters related to clean room design, operation, and behavior. Controlling the cleaning environment improves the success of the cleaning process by minimizing product contamination. Scott Mackler of Cleanroom Consulting provides a comprehensive chapter on "Basis of design for life sciences cleanroom facilities" (Chapter 5). Kevina O'Donoghue of Specialised Sterile Environments brings a view from "across the pond" in Ireland with her chapter on "Validating and monitoring the cleanroom" (Chapter 6). Jan Eudy of Cintas Corporation provides a chapter on "Cleanroom management and gowning" (Chapter 7). Howard Siegerman of Siegerman & Assoc. and Karen Bonnell of Production Economics coauthor a chapter on the "Principles of wiping and cleaning validation" (Chapter 8). Ed Kanegsberg of BFK Solutions provides an overview of issues related to detection and measurement of contamination (Chapter 9). Finally, Ben Schiefelbein of RJ Lee Group provides a new, insightful chapter on the philosophy of and choices for analytical analysis with "Practical aspects of analyzing surfaces" (Chapter 10). Many of the techniques should be considered whether or not you operate in a clean room.

The next four chapters in this part address knowing when to clean, when the part is clean enough, and materials compatibility.

Mantosh Chawla of Photo Emission Tech. (PET), Inc. updates his chapter on the important topic of "How clean is clean? Measuring surface cleanliness and defining acceptable levels of cleanliness" (Chapter 11).

Two chapters are a must for those involved in process validation for implantable medical devices. Kierstan Andrascik of QVET Consulting brings her experience with analysis of medical devices in a new chapter on "Cleaning validations using extraction techniques" (Chapter 12). David Albert of NAMSA has expanded his chapter on "Testing methods for verifying medical device cleanliness" (Chapter 13).

Eric Eichinger of Boeing North America has expanded his chapter on the critical issue of materials compatibility both for metals and nonmetals (Chapter 14).

Part II: Applications

While each manufacturing situation may be thought of as unique, there are commonalities, and it can be helpful to explore common contamination problems in specific industrial sectors and to see how manufacturers in similar situations tackle cleaning problems. Therefore, the number of specific applications presented in this part has been expanded in the second edition.

Barbara Kanegsberg of BFK Solutions, with a contribution by Bev Christian of Research in Motion, provides "Clean critically: An overview of cleaning applications" (Chapter 15). Specific examples and case studies drawn from aerospace, electronics, and biomedicine are given.

A number of comprehensive new chapters are devoted to applications within the life sciences. John Broad of NAMSA and David Smith of Tissue Banks International provide a chapter on "Cleaning validation of reusable medical devices" (Chapter 16). Paul Lopolito of Steris Corporation provides insight into "Critical cleaning for pharmaceutical applications" (Chapter 17). Hein Timmerman of Diversey, Inc. in Belgium provides a chapter on "Cleaning in food processing" (Chapter 18).

Three new chapters reflect renewed importance of cleaning in electronic assembly applications. Mike Bixenman of Kyzen Corporation provides a comprehensive chapter on "Electronic assembly cleaning process considerations" (Chapter 19). Harald Wack of ZESTRON contributes a chapter on "Surfactant-free aqueous chemistries" (Chapter 20) and Helmut Schweigart, also of ZESTRON, writes about "Contamination-induced failure of electronic assemblies" (Chapter 21).

Ahmed Busnaina of Norteastern University treats the case of particle removal in his expanded chapter (Chapter 22). This is an effective treatment of surface physics presented in a readable and understandable manner.

Shawn Sahbari of Applied Chemical Laboratories, Mahmood Toofan of Semiconductor Analytical Services, and John Chu discuss the challenges faced in semiconducting wafer fabrication for aluminum interconnects (Chapter 23). In a new companion chapter, Shawn Sahbari and Mahmood Toofan discuss microelectronic cleaning with "Copper interconnect and particle cleaning" (Chapter 24).

The world of fine art is the subject of a new chapter, "The cleaning of paintings," by Chris Stavroudis, a paintings conservator in private practice, and Richard Wolbers of the University of Delaware (Chapter 25). Cleaning these valuable, critical surfaces involves careful formulation of cleaning chemistries as well as considerations involved in conserving surface qualities and attributes. If you are in manufacturing, peruse the chapter for ideas and approaches; there are more commonalities with the world of fine art than might be apparent on the surface.

Jason Marshall of Massachusetts Toxic Use Reduction Institute (TURI) provides a new "Road map for cleaning product selection for pollution prevention," outlining decisions that can be made to clean with less impact on workers and the environment (Chapter 26).

Bill Breault, Jay Soma, and Christine Fouts of Petroferm contribute a new focused case study on removing wax from aerospace build and assembly operations (Chapter 27). The approaches can be extrapolated to other operations.

Wayne Ziegler of the Army Research Laboratory and Tom Torres of the Naval Facilities Engineering Service Center (NFESC) team to provide a new chapter, "Implementation of environmentally preferable cleaning processes for military applications," that describes the efforts of the military Joint Services Solvent Substitution Working Group to implement processes across all military agencies with less use of hazardous or environmental degrading cleaning products while ensuring uncompromised performance (Chapter 28).

Part III: Safety and Regulations

Because cleaning almost always involves using materials or processes with environmental or safety "baggage," sucessfully navigating regulations and working with regulators has become part of the overall picture. This final part provides tools and approaches to achieving successful critical cleaning processes in a highly regulated world.

Barbara Kanegsberg begins with an expanded, frank overview of safety and environmental issues (Chapter 29).

Jim Unmack of Unmack Corporation contributes an updated and greatly expanded chapter that outlines health and safety aspects associated with cleaning processes (Chapter 30). Recognizing that manufacturing is a global issue, he provides insight on European as well as U.S. requirements.

Mohan Balagopalan of southern California's South Coast Air Quality Management District (SCAQMD) expands his thoughtful and frank discussion on working with regulators—from a regulator's viewpoint (Chapter 31).

Steve Andersen, recently retired from the U.S. EPA, teams with Margaret Sheppard of the U.S. EPA Significant New Alternatives Program (SNAP) to update and expand the discussion of how industry and government can work together, with a discussion on how lessons learned from the ODC phaseout can drive continuous environmental improvement (Chapter 32).

The book closes with an expansion of the chapter by Don Wuebbles of the University of Illinois-Urbana (Chapter 33). His chapter reviews "Screening techniques for environmental impact of cleaning agents."

Barbara Kanegsberg
Ed Kanegsberg
BFK Solutions
Pacific Palisades, California

Editors

Barbara Kanegsberg is the President of BFK Solutions LLC, Pacific Palisades, California, an independent consulting company established in 1994. BFK Solutions is now the industry leader in critical cleaning. As a recognized consultant in the areas of critical cleaning, contamination control, surface quality, and process validation, she helps companies optimize manufacturing cleaning processes, improve yield, resolve regulatory issues, and maintain trouble-free production. She has participated in projects that include aerospace/military equipment, electronics assembly, medical devices, engineered coatings, metals, pump repair, and optical and nanotechnological devices. She has also participated in a number of product development- and intellectual property-related projects for manufacturers of cleaning chemicals and industrial equipment. Prior to establishing BFK Solutions, she was involved in the substitution of ozone-depleting chemicals at Litton Industries. She also developed clinical diagnostic tests at BioScience Laboratories. She has a background in biology, biochemistry, and clinical chemistry.

Barbara is a recipient of the U.S. EPA Stratospheric Ozone Protection Award for her achievements in implementing effective, environment-friendly manufacturing processes. She has several publications to her credit in the areas of surface preparation, contamination control, critical cleaning, method validation, analytical techniques, and regulatory issues. She has organized and participated in numerous seminars, tutorials, and conferences, including programs at USC and UCLA. She regularly coauthors technical columns that appear in *Controlled Environments Magazine* and *Process Cleaning Magazine*. She also participates in standardization and guidance committees relating to aerospace, military, electronics, medical devices, and safety/regulatory issues. Examples include the JS3, (interagency military group), the ASTM Medical Device Cleanliness Testing Task Force, and the IPC group revising the cleaning/defluxing handbook.

She has a BA in Biology from Bryn Mawr College and MS in biochemistry from Rutgers University, New Jersey.

Barbara, "the cleaning lady," can be reached at 310-459-3614 or Barbara@bfksolutions.com.

Ed Kanegsberg is the Vice President of BFK Solutions, Pacific Palisades, California; he is also a chemical physicist and engineer who troubleshoots and solves manufacturing production problems. Ed is a recognized advisor and consultant in the areas of industrial cleaning process design and process performance. He uses his four decades of practical experience along with his background in physics and engineering to help companies solve production problems and optimize their cleaning and contamination control processes. As a member of the technical staff at Litton Guidance and Control Systems Division, he was responsible for precision instrument development and technology transfer from prototype to the production facility.

Ed is an educator and believes firmly that most of rocket science can be easily understood by non-rocket scientists. He has authored several technical articles and coauthors technical columns with Barbara. He has delivered numerous presentations and particularly enjoys discussing the physics of cleaning and successful automation. He has nine patents in high-reliability instrumentation and multiple "company private" inventions.

He has a BS and PhD from Massachusetts Institute of Technology and Rutgers University, New Jersey, respectively.

Ed, "the rocket scientist," can be reached at (310)459-3614 or Ed@bfksolutions.com.

Contributors' Bios

David E. Albert is a senior scientist at North American Science Associates (NAMSA). He was an adjunct professor at the University of Toledo and Lourdes College, where he taught pharmacology, pathophysiology, and biochemistry. He has been in the medical device industry for 30 years. He received his BS in pharmaceutical sciences from the University of Toledo, his MS in biochemistry from Bowling Green State University, Bowling Green, Ohio, and his doctorate in podiatric medicine from Ohio College of Podiatric Medicine, Cleveland, Ohio.

    Contact: 2261 Tracy Road, Northwood, Ohio 43619

    Phone: (419) 666-9455; e-mail: dalbert@namsa.com

Stephen O. Andersen retired in 2009 from the U.S. EPA Climate Protection Partnerships Division and is now director of research at the Institute for Governance & Sustainable Development (IGSD). At EPA, he helped organize ozone and climate protection partnerships, including the Industry Cooperative for Ozone Layer Protection and the Halon Alternatives Research Corporation, and negotiated voluntary agreement in solvent, foam, semiconductor, aerospace, and other sectors. He is the founding and continuing co-chair of the United Nations Montreal Protocol Technology and Economic Assessment Panel (TEAP). Dr. Andersen studied business administration as an undergraduate and received his MS and PhD in agricultural and natural resources economics from the University of California, Berkeley, California.

    Contact: PO Box 257, Barnard, Vermont 05031-0257

    Phone: (802) 234-5251; e-mail: soliverandersen@aol.com

Kierstan Andrascik founded QVET Consulting in 2009 to assist medical device manufacturers with their validation needs. She specializes in cleaning validations for both new and reprocessed medical devices. Previously, she worked at Nelson Labs, Salt Lake City, Utah where she served as study director covering a variety of activities involving testing, materials characterization, and new device cleaning validations. Kierstan has been actively serving on the ASTM Device Cleanliness Subcommittee since 2005. She received a certificate of achievement from ASTM in May 2007. She has a BS in chemistry and mathematics from Shepherd University, Shepherdstown, West Virginia.

    Contact: 1489 West 1200 North, Layton, Utah 84041

    Phone: (801) 682-8847; e-mail: kandrascik@comcast.net

Mohan Balagopalan is an Air Quality Analysis and Compliance supervisor with the South Coast Air Quality Management District (SCAQMD). He has 25 years of experience at SCAQMD, is involved in streamlining permitting activities, compliance enhancements and Administration. Mohan has a BS in mechanical engineering and an MBA. He is an instructor at the University of California, Riverside, California, for the Certificate Program for Air Quality and Clinical Instructor Department of Environmental and Occupational Health at Loma Linda University, California. He was the past president of the Southern California Society for Risk Analysis and the President-Elect for the Air and Management Association, West Coast Section.

Contact: 21865 Copley Dr. Diamond Bar, California 91765

Phone: (909) 396-2704; e-mail: mbalagopalan@aqmd.gov

Mike Bixenman, DBA, is the chief technology officer at Kyzen Corporation and has over 35 years of experience in the development of paint removal, paint and coatings, and precision cleaning agents. In 1988, Mike worked with Kyle Doyel to develop engineered cleaning agents for replacing ozone-depleting solvents, leading to the formation of Kyzen Corporation. Mike is the joint holder of several patents and has published nearly 100 technical articles on precision cleaning.

Contact: 430 Harding Industrial Drive, Nashville, Tennessee 37211

Phone: (615) 983-4530; e-mail: mikeb@kyzen.com

Karen F. Bonnell, Director of Application Research for Production Economics, is a consultant involved in designing, developing, and optimizing the performance of clean room consumable products. She has previously worked as a product manager for Illinois Tool Works, Texwipe Division, where she was involved in the development of innovative products, several of which are still to be patented. She has received one patent (5/2010). She received her BA in biology from Bryn Mawr College, Bryn Mawr, Pennsylvania and her MA in biochemistry from Temple University, Philadelphia, Pennsylvania.

Contact: 94 Woodland Road, Montvale, New Jersey 07645

Phone: (201)573-0926; e-mail: kfbonnell@gmail.com

Bill Breault is market manager for aerospace and defense at Petroferm Inc., Gurnee, Illinois. With over 13 years of experience in the precision cleaning industry, he has been involved in projects related to key account management, new business development, product line expansion and management, and OEM qualification testing and approvals. He received his BA in history from Hanover College, Hanover, Indiana.

Contact: 3938 Porett Drive, Gurnee, Illinois 60031

Phone: (859) 312-4602; e-mail: bbreault@petroferm.com

John J. Broad is a senior consultant at NAMSA's California laboratory, Irvine, California and a specialist microbiologist with the American Society for Microbiology (ASM), Washington, District of Columbia. John is affiliated with the American Society of Quality Control (ASQC) and is active in AAMI/ISO sterilization subcommittees. He has published articles for *Medical Device & Diagnostics, Controlled Environments, Medical Design Technology*, and *Biomedical Instrumentation and Technology*. He received his BS in microbiology from the California State Polytechnic University, Pomona, California and completed graduate work at the University of LaVerne, La Verne, California.

Contact: 9 Morgan, Irvine, California 92618

Phone: (949) 452-8015; e-mail: jbroad@namsa.com

Ahmed A. Busnaina, PhD, is William Lincoln Smith Professor and Director, NSF Nanoscale Science and Engineering Center for High-rate Nanomanufacturing and the NSF Center for Nano and Microcontamination Control at Northeastern University. He focuses his research on bringing valuable nanoscale discoveries from the lab into high-volume manufacturing. He has authored more than 200 papers in journals, proceedings, and conferences. He has also won a Fulbright Scholar Award.

Contact: 360 Huntington Avenue, 467 Egan Center, Norteastern University. Boston, Massachusetts 02115

Phone: (617) 373-2992; e-mail: a.busnaina@neu.edu

Mike Callahan is a principal chemical engineer with Jacobs Engineering, Pasadena, California. During his 30 years at Jacobs, he has been involved in a variety of projects, including the development of an EPA waste minimization manual. He is the coauthor of a book entitled *Hazardous Solvent Source Reduction*.

He has served on a California State Panel for Solvent Certification. He received a Bachelor of Science in Chemical Engineering from UCLA, Los Angeles, California and is a professional engineer registered in the State of California.

Contact: 1111 S. Arroyo Parkway, Pasadena, California 91105

Phone: (626) 568-7005; e-mail: mike.callahan@jacobs.com

Mantosh K. Chawla is president and cofounder of Photo Emission Tech., Inc. He has over 45 years of industrial experience including 26 years in surface cleanliness monitoring. He received his BS in mechanical engineering from Bradford University, England, and his MBA in finance from John Carroll University, University Heights, Ohio.

Contact: 760 Calle Plano, Camarillo, California 93012

Phone: (805) 482-5200; e-mail: mchawla@photoemission.com

Bev Christian has been at Research in Motion for the past 10 years where he is currently the director of materials interconnect research. He ran a materials lab at Nortel in Belleville, Ontario, and received several awards for his environmental work. His specific areas of interest include solderability, lead-free solders, environmental issues, and contamination. He is a member of ASM, SMTA, and IPC. Dr. Christian has been named a member of technical distinction of the SMTA.

Contact: 451 Phillips Street, Waterloo, Ontario N2L 3X2

Phone: (519) 888-7465x72468; e-mail: bchristian@rim.com

John Chu has over 20 years of experience in IC manufacturing at Intel, Samsung, and National Semi. As application manager at SVC/Shipley, he worked on polymer cleaning for Cu/Low-k technology and multilayer metals. He has won biodesign competition at Stanford entrepreuner week in 2008 on improving intragastric balloons. He is pursuing biomedical innovations through Stanford University seminars. He has a master's degree from CWRU in combining biomedical sensors into nanosized silicon ISFET.

Phone: (408) 923-2200; e-mail: john_chu@hotmail.com

Eric Eichinger works at Boeing in the Chemical Technology Group in Huntington Beach, California, where he is currently the manager of the Analytical Chemistry laboratory. He has over 20 years of experience in material substitution and source reduction for pollution prevention. He is the chairman of the Aerospace Chromate Elimination Team with experience in replacing ozone depleters and volatile organic compounds. He has a bachelor's degree from the University of California, Irvine, California.

Contact: 5301 Bolsa Avenue, Mail Code H021-F225, Huntington Beach, California 92647

Phone: (714) 529-6211; e-mail: cfc113@hotmail.com

Jan Eudy is corporate quality assurance manager for Cintas Corporation, Cincinnati, Ohio. A registered microbiologist, Jan oversees research and development, directs the quality system and ISO registration at all clean room locations, and supports validation and sterile services. She is a certified quality auditor. She is also an active member of several professional organizations and is a fellow and Institute of Environmental Science and Technology president emeritus. Jan has a degree in medical technology from the University of Wisconsin, Madison, Wisconsin with graduate studies in medical microbiology at Creighton University, Omaha, Nebraska.

Contact: 6800 Cintas Boulevard, Mason, Ohio 45040

Phone: (513) 573-4165; e-mail: EudyJ@cintas.com

Christine Fouts is the director of derivatives technology at Petroferm Inc., Gurnee, Illinois and has 20 years of experience in the electronics and precision cleaning industry. Over her tenure at Petroferm, she has been a research chemist, sales manager, product manager for precision cleaning products, and cleaning division business manager. In 2006, she returned to the lab in her current position. Christine received her PhD in chemistry from Emory University, Atlanta, Georgia.

Contact: 3938 Porett Drive, Gurnee, Illinois 60031

Phone: (847) 249-6758; e-mail: CFouts@Petroferm.com

Arthur Gillman is president of Unique Equipment Corporation and has over half a century of practical and theoretical experience in an array of critical cleaning applications. He has developed equipment for both aqueous- and solvent-based processes. He has been an advisor on several SCAQMD committees. He has also assisted numerous component and part manufacturers in areas ranging from benchtop applications to very large airless applications.

Contact: 2029 Verdugo Blvd., M/S 1005, Montrose, California 91020-1626

Phone: (818) 409-8900; e-mail: agillman@uniqueequip.com

Barbara Kanegsberg, president of BFK Solutions, is a recognized consultant in critical/industrial cleaning and contamination control. She helps companies optimize manufacturing cleaning processes, validate cleaning methods, improve yield, resolve regulatory issues, and maintain trouble-free production. She conducts dynamic workshops and training programs. Barbara received a U.S. EPA Stratospheric Ozone Protection Award. She received her BS in biology from Bryn Mawr College and her MS in biochemistry from Rutgers University.

Contact: 16924 Livorno Dr., Pacific Palisades, California 90272

Phone: (310)459-3614; e-mail: Barbara@Bfksolutions.com

Ed Kanegsberg, vice president of BFK Solutions, helps companies solve production problems and optimize their cleaning and contamination control processes. He has decades of experience in physics and engineering, including the transition of products from prototype to production. His writings emphasize the physics of cleaning and surface quality. He received his BS and PhD in physics from Massachusetts Institute of Technology and Rutgers University, respectively.

Contact: 16924 Livorno Dr., Pacific Palisades, California 90272

Phone: (310)459-3614; e-mail: Ed@bfksolutions.com

Paul Lopolito is a technical services specialist for STERIS Corporation's Life Sciences Division, Mentor, Ohio. He provides field support, site audits, training, and education for critical environment and process research cleaners. Paul has 13 years of experience supporting cGMP/ISO-regulated biopharmaceutical and biomedical device facilities as a technical services manager, manufacturing manager, and laboratory manager. He also has specific expertise in process cleaning and critical environment operations. Paul received his BA in biological sciences from Goucher College, Towson, Maryland.

Contact: 7405 Page Avenue, St. Louis, Missouri 63133-1032

Phone: (314) 290-4795; e-mail: paul_lopolito@steris.com

Scott E. Mackler is the founder and principal of Cleanroom Consulting, LLC, Pittsford, New York. He has 30 years of industry experience, including project planning and basis of design development, commercialization of new products, applications for process isolation and mini-environments, construction claims arbitration, vendor/contractor identification and qualification, RFP/RFQ preparation, project financial justification, and site selection services. Scott is a member of ASHRAE, SEMI, ISPE, PDA, and the IEST. He received his BSME from Rensselaer Polytechnic Institute, Troy, New York and his MBA from the University of Houston, Houston, Texas.

Contact: 31 Mitchell, Pittsford, New York 14534-2301

Phone: (585) 264-9430; e-mail: sem@cleanroom-consulting.com

Jason Marshall, laboratory director of the Massachusetts Toxic Use Reduction Institute (TURI), Lowell, Massachusetts helps companies evaluate the performance of cleaning chemistries and equipment. Projects include evaluation of bio-based products for janitorial applications in a hospital setting; and efforts to promote adoption of alternatives to trichloroethylene (TCE) in Massachusetts and Rhode Island. Dr. Marshall received his ScD in work environment occupational and environmental hygiene, his MS in environmental studies, and his BS in chemical engineering from the University of Massachusetts Lowell, Lowell, Massachusetts.

Contact: Univ. of Mass. Lowell, 1 University Avenue, Lowell, Massachusetts 01854

Phone: (978) 934-3133; e-mail: Jason_Marshall@uml.edu

Dr. Kevina O'Donoghue is a microbiologist at Specialised Sterile Environments, Galway, Ireland, and has worked with clean rooms for nearly 10 years. Beginning in diagnostics, working for Cambridge Diagnostics, and then moving to the medical devices, working for Abbott Vascular Galway, she was involved in an array of aspects of clean rooms such as monitoring, validation, auditing, and training. She developed and delivers an internationally accredited training programme on "contamination control practices within the clean room environment" and is one of the founders and facilitators of a clean room technology network group.

Contact: Unit 6D Mervue Industrial Estate, Mervue, Galway, Ireland
Phone: 00 353 876772526; e-mail: kevina@cleanrooms.ie

Shawn Sahbari is executive vice president at Applied Chemical Laboratories, Inc., Sunnyvale, California founded in 2006 by a team of Silicon Valley chemists and engineers with over 20 years of experience in research, development, and manufacturing of electronic chemicals and specialty materials for other industries.

Contact: 526 Almaor Avenue, Sunnyvale, California 94085
Phone: (408) 737-8880x208; e-mail: s.sahbari@appchem.com

Ben Schiefelbein, PhD, is regional sales manager at the Hayward, California, operations of RJ Lee Group. He has more than 30 years of industry and laboratory experience. His main field of specialization is the characterization of materials using surface analytical techniques and chemical instrumentation.

Contact: 3583 Investment Boulevard, Suite 7, Hayward, California 94545
Phone: (510) 567-0480; e-mail: bschiefelbein@RJLG.com

Dr. Helmut Schweigart is head of application and process technology at ZESTRON Europe, Ingolstadt, Germany, the world's largest precision cleaning services company. Schweigart has a diploma in mechanical engineering and did his theses concerning the reliability of electronic assemblies at the Technical University of Munich, Munich, Germany. He is chair of several working groups concerning corrosion and reliability of electronics as well as author of numerous publications.

Contact: Bunsenstr. 6, D-85053 Ingolstadt, Germany
Phone: +49(0) 841/635-29; e-mail: H.Schweigart@ZESTRON.com

Margaret Sheppard is an environmental scientist and team lead for the USEPA's Significant New Alternatives Policy Program (SNAP). She has been reviewing the health and environmental impacts of alternatives for ozone-depleting substances since 2000, with specialization in solvent cleaning. She has worked at EPA since 1990 on a number of issues, including protecting the ozone layer, reducing acid rain, improving air quality, and assessing risks from radioactive waste and from chemicals in drinking water. Margaret received her BA in physics from Wesleyan University and her MS in environmental planning and management from the Johns Hopkins University.

Contact: EPA, Stratospheric Protection Division; Mail Code 6205J; 1200 Pennsylvania Ave NW; Washington DC 20460
Phone: (202) 343-9163; e-mail: sheppard.margaret@epa.gov

Howard Siegerman, President of Siegerman & Associates LLC, is an independent consultant in contamination control in the electronic and life science markets. He was director of technology at ITW Texwipe, Kernersville, North Carolina and has held sales and marketing positions with Applied Materials, Materials Research Corporation, EG&G Princeton Applied Research, and Nicolet Scientific. Dr. Siegerman holds two patents and has written a book entitled *Wiping Surfaces Clean*, Vicon Publishing Inc., Amherst, New Hampshire, 2004. He received his BS, MA, and PhD in analytical chemistry from the University of Toronto, Toronto, Ontario.

Contact: 79 Alpine Terrace, Hillsdale, New Jersey 07642
Phone: (201) 666-2977; e-mail: hsiegerman@optimum.net

David A.B. Smith is the Director of Product Development with Tissue Banks International, San Rafael California and is well published in the engineering aspects of the surgical treatment of trauma, deformity, and pathology of the spine, as well as other areas of orthopedic surgery. He received his Bachelor's and Master's Degrees in Metallurgical and Materials Engineering from the Colorado School of Mines, Golden, Colorado. His other professional interests include novel biomaterials, spinal cord injury, and the application of metallurgical failure analysis to implants and surgical instruments.

Contact: 2597 Kerner Blvd, San Rafael, California 94901

Phone: (415) 455-9000; e-mail: David.Biomedengr@GMail.com

Jay Soma is the former R&D manager at Petroferm Inc., Gurnee, Illinois. He has 14 years of experience in research and development of products for use in electronics and aerospace cleaning; household, industrial, and institutional (HI&I) markets; personal care; firefighting foams; and water treatment and oil-field applications. Jay received his MS and PhD in chemical engineering from the University of Akron, Akron, Ohio and Tulane University, New Orleans, Louisiana, respectively.

Phone: (847) 557-2367; e-mail: jpsoma@gmail.com

Chris Stavroudis is a paintings conservator in private practice. He developed the modular cleaning program (MCP), a computer program that manages the chemistry of cleaning solutions used in conservation. He has presented a number of workshops on the theory of cleaning art surfaces and the MCP. He received his MS in conservation from the University of Delaware/Winterthur Program, Newark, Delaware in Art Conservation, and his BS and BA in chemistry and in art history, respectively, from the University of Arizona, Tucson, Arizona.

Contact: 1272 N. Flores Street, Los Angeles, California 90069

Phone: (323) 654-8748; e-mail: cstavrou@ix.netcom.com

Hein A. Timmerman is sector specialist and business development director for Europe, the Middle East, and Africa at Diversey Inc., Sturtevant, Wisconsin. He received his master's degree in food technology from the Technical University in Ghent and his MBA from the Economical Highschool in Brussels. He has 24 years of experience in the food processing industry, from engineering, project management, to sales and business development. Over the years, he has obtained an expert view on cleaning techniques in the food industry.

Contact: Koning Albertlaan 81, B-900 Gent, Belgium

Phone: +32-495-59-17-81; e-mail: hein.timmerman@diversey.com

Mahmood Toofan, PhD, is the technical director of Semiconductor Analytical Services (SAS), an analytical service laboratory offering consulting and cleaning services to the semiconductor industry. He has designed and developed a spray washing system and particle-free wafer washing and drying equipment. He has presented papers and holds patents in electrochemistry and microcontamination analytical processes. He received his BS from the National University of Tehran, Iran and his PhD from the University of California, Davis.

Contact: 1765 Landess Avenue # 20, Milpitas, California 95020

Phone: (408) 768-7773; e-mail: info@sas-page.com

Tom Torres is a chemical engineer at Naval Facilities Engineering Service Center (NFESC), Port Hueneme, California. He has been project leader and principal investigator of research, development, testing, and evaluation efforts for 25 years. He is cochair of the Joint Service Solvent Substitution Working Group responsible for coordinating solvent substitution within the Department of Defense. He performs waste characterization, feasibility studies, treatability studies, and cleaning specification development, and conducts test and evaluation of environmental treatment processes. He received his BS in chemical engineering from California State University, Northridge, California.

Contact: NAVFAC ESC, Code EV421/Tom Torres, 1100 23rd Avenue, Port Hueneme, California 93043

Phone: (805) 982-1658; e-mail: tom.torres@navy.mil

James L. Unmack, PE, CIH, CSP, is the president of Unmack Corporation, San Pedro, California and has over 40 years of health and safety, industrial hygiene, and environmental experience with government agencies and private industry. He has expertise in federal and state environmental and worker health and safety regulations, and has managed numerous programs to evaluate and ensure compliance. He is a graduate of the University of California at Berkeley, Berkeley, California in electrical engineering and Santa Clara University, Santa Clara, California in bioengineering.

Contact: 1379 Park Western Drive, PMB 282, San Pedro, California 90732-2217

Phone: (310) 377-2367; e-mail: jim@unmack.com

Dr. Harald Wack is the president of ZESTRON. He has been with the family-held business most of his professional life. He received his MS and PhD in organic chemistry from the Johns Hopkins University in Baltimore. He has 17 years of global experience in the electronics manufacturing industry. Dr. Wack holds numerous patents and has over 50 publications on topics related to precision cleaning. He is an active member in various industry organizations such as the IPC and SMTA.

Contact: 11285 Assett Loop, Manassas, Virginia 20109

Phone: (703) 393-9880; e-mail: h.wack@zestronusa.com

Richard C. Wolbers is associate professor and coordinator of science and adjunct paintings conservator at the University of Delaware, Newark, Delaware. His research includes work in developing cleaning systems for fine art materials. He has conducted workshops on his cleaning methods worldwide. In 2000, he published *Cleaning Paintings: Aqueous Methods* (Archetype Books, London). He received his BS in biochemistry from the University of California, San Diego, California, and his MS in art conservation from WUDPAC.

Contact: The Department of Art Conservation, 303 Old College, Newark, Delaware 19716

Phone: 302-888-4818; e-mail: wolbers@earthlink.net

Donald J. Wuebbles is the Harry E. Preble Endowed Professor, Department of Atmospheric Sciences and the Department of Electrical and Computer Engineering at the University of Illinois, Urbana, Illinois. He has written nearly 400 articles that cover interactions of atmospheric chemistry and physical processes affecting atmospheric composition, as well as effects of human activities on climate and natural phenomena. He received the U.S. EPA Stratospheric Ozone Protection Award and shares the 2007 Nobel Peace Prize for his work with the Intergovernmental Panel on Climate Change.

Contact: Department of Atmospheric Sciences, 105 S. Gregory St., Urbana, Illinois 61801

Phone: (217) 244-1568; e-mail: wuebbles@illinois.edu

Wayne Ziegler is a materials engineer at the Army Research Laboratory (ARL), Aberdeen Proving Ground, Maryland. He has over 25 years experience in solvent substitution, environmental technology evaluation, failure analysis, and materials information management. He is the Army lead for solvent substitution, cochair of the Joint Service Solvent Substitution Working Group (JS3), and project lead for the development of the Army Materials Selection & Analysis Tool (MSAT). He received his BS in materials engineering from Virginia Tech, Blacksburg, Virginia.

Contact: Attn: RDRL-WMM-D, Bldg 4600, APG, Maryland 21005

Phone: (410) 306-0746; e-mail: wayne.ziegler@us.army.mil

I

Process Implementation and Control

1 Evaluating, Choosing, and Implementing the Process: How to Get Vendors to Work with You *Barbara Kanegsberg* 3
Introduction • Before Contacting Cleaning Chemistry and Equipment Vendors • Record Your Impressions • Organize and Coordinate • Refine and Test • Make a Decision • References

2 Cleaning Agent Balancing Act *Barbara Kanegsberg* 15
The Quest for the Ideal Cleaning Agent • Physical/Chemical Properties • Regulatory Issues • Economic Constraints Impeding New Solvent Development • Future of Biobased Cleaning Agents • Green and Safe—Moving Targets • Moving Forward • References

3 Blunders, Disasters, Horror Stories, and Mistakes You Can Avoid
Arthur Gillman 29
Introduction • The Missing Expert: The Facilities Manager • We Thought of Everything, Just Build It! (*as Told to Me by a Dear Friend*) • We Know Exactly Where the Machine Goes, We Measured • The Right Chemistry • A Growth Spurt • The Case of the Mysterious Charred Solvent • Kanegsberg, Your Process Stopped Working • PSPS (Perfect Sample Part Syndrome) • Metallurgy? Chemistry? • But It Is Not on the Purchase Order! • The Perfect Cleaning Machine Specification • Saving the Strangest for Last: Two Examples of Physics 101 • Conclusion

4 Cleaning Practices and Pollution Prevention *Mike Callahan* 39
Overview • Cleaning Activities • Rinsate Treatment • Points to Remember • References

5 Basis of Design for Life Science Cleanroom Facilities *Scott E. Mackler* 61
Introduction • In the Beginning • Isolation Technology • Preconstruction Design Review • Primary Design Considerations • Contamination Control Criteria • Additional HVAC Considerations • Architectural Issues • Project Planning • Project Delivery • Basis of Design Preparation • Turnover Package • Commissioning • Conclusion • Further Readings

6   Validating and Monitoring the Cleanroom   *Kevina O'Donoghue*.................................89
Overview • Cleanroom Design • Principles of Cleanroom
Validation and Testing • Cleanroom Validation • Validation versus
Monitoring • Summary • References

**7   Cleanroom Management, Validation of a Cleanroom Garment System, and
Gowning Procedure**   *Jan Eudy*.................................103
Introduction • A Short Course in Particulation • Facility Design • Protecting
Cleanrooms from Human Contamination • Behavior • Cleanroom
Garments • Routine Monitoring of Sterilization Process • Gowning for the
Cleanroom • Housekeeping • Ongoing Assessments • Summary • References

8   Principles of Wiping and Cleaning Validation   *Karen F. Bonnell and
Howard Siegerman*.................................119
What Is Contamination? • How to Remove Contamination from the Cleanroom
or Mini-Environment • When Do You Clean? • Training • Cleaning
Validation • Conclusions • References

9   Overview to Analytical and Monitoring Techniques   *Ed Kanegsberg*.................................129
Why Should You Use Analytical Tests? • Visual • Standards • Standards
and Controls • In Situ versus Extractive Analysis • Airborne Molecular
Contamination • Conclusions • References

10   Practical Aspects of Analyzing Surfaces   *Ben Schiefelbein*.................................135
Introduction • Analytical Approach • Sample Handling and Packaging
Issues • Analytical Methods • Applicability of Different Analytical
Approaches • Special Topics • Summary • Bibliography

**11   How Clean Is Clean? Measuring Surface Cleanliness and Defining Acceptable
Levels of Cleanliness**   *Mantosh K. Chawla*.................................145
Introduction • Definitions • Types of Contamination • Why Monitor
Cleanliness? • Factors that Contribute to Inadequate Cleanliness • Factors to Be
Considered in the Selection of a Cleanliness Monitoring/Measurement Method • Types of
Cleanliness Measurement Methods • Most Common Verification/Measurement Methods
and Their Principles of Operation • Cost Impact of Cleanliness Levels • Methods for
Defining an Acceptable ("Optimum") Level of Cleanliness • In-Process or On-Line
Surface Cleanliness Monitoring • Conclusion • Bibliography

12   Cleaning Validations Using Extraction Techniques   *Kierstan Andrascik*.................................161
Introduction • Validating a Cleaning Process • Defining the Validation
Scope • Identifying the Contaminants • Choosing the Test Method • Choosing
the Extraction Solvent • Setting the Extraction Parameters • Validating the
Extraction • Establishing and Justifying Residue Limits • Conclusion • References

**13   Biomedical Applications: Testing Methods for Verifying Medical Device
Cleanliness**   *David E. Albert*.................................173
Introduction • Device Categories • Class I: Critical • Class II:
Semicritical • Class III: Noncritical • Cleaning and Sterilization
Residues • Analytical Testing • Chromatographic Analysis • Spectrophotometric
Analysis • Infrared • Ultraviolet/Visible Spectroscopy • Total Organic
Carbon • Biological Safety Assessment • Conclusion • References

14   Material Compatibility   *Eric Eichinger*.................................183
Overview • What Is Meant by Compatibility? • Small Investments in Compatibility Can
Eliminate Unwanted Problems • Who Should Consider Compatibility? • Compatibility
Data Are Most Useful When Considering a Change • Finding and Interpreting Good
Compatibility Data Can Be a Challenge • Compatibility Testing Is Easy • Example of a
Successful Compatibility Test • Take-Home Lessons • References

1

Evaluating, Choosing, and Implementing the Process: How to Get Vendors to Work with You*

Introduction 3
Before Contacting Cleaning Chemistry and Equipment Vendors..............3
Record Your Impressions...5
Cleaning Agents • Equipment • Drying
Organize and Coordinate 10
Refine and Test...12
Cleaning Evaluations by Suppliers
Make a Decision 13
References..13

Barbara Kanegsberg
BFK Solutions

Introduction

Each company has to select the best cleaning agent, cleaning equipment, and cleaning processes. The best of these is very application specific. Some essential steps to process implementation are summarized in Table 1.1. The most important steps are to understand the cleaning requirements, buy quality equipment and chemicals, test/evaluate/validate, get it in writing, and educate your workforce.

Every chemical and equipment vendor honestly sees their own product line as the optimal choice. You do not have to take their word for it. Cleaning and process equipment can no longer be considered commodity items; you have choices. For quite some time, both publicly and as part of projects, we have presented programs on how to work with chemical and equipment vendors (Kanegsberg 1997).

Before Contacting Cleaning Chemistry and Equipment Vendors

You may be addicted to conferences and trade shows, while others do web searches for new and exciting technologies. You may collect stack and stacks of product brochures, and others prefer meeting with sales reps directly, either in person or on the phone. We happen to like a combination of approaches and tend to favor phone and/or face-to-face meetings. This interactive communication saves time and helps avoid misunderstanding. In addition, it is easier to assure that the product being offered really suits your

---

* This chapter is based on the paper "Getting the Most Out of Precision Cleaning '97—How to Get Vendors to Work with You," presented at CleanTech '97, a conference sponsored by the Cleaning Technology Group, Witter Publishing Corp., April 1997, Cincinnati, OH.

TABLE 1.1   Steps to Process Implementation

Step	Comments
Understand your cleaning requirements	Do you need to clean?
	What are the critical steps?
Set up a cleaning team	Including management, technicians, and safety/ environmental people
Determine range of process choices	Avoid unnecessary restrictions
Contact cleaning agent and cleaning equipment vendors	Do this at the same time, not sequentially
Communicate your requirements to the vendors	Evaluate their responses
Evaluate the choices	Determine process costs, check references
Consider regulatory/performance requirements	Safety, environmental, customer, governmental
Test process efficacy	Use vendor test lab and/or lease equipment, consider corrosion and compatibility issues; spell out what should be tested, include controls, benchmarks
Evaluate process flexibility	Alternative cleaning chemistries, adaptability to newer product line
Determine full process costs	Include peripherals
Consider wash/rinse/dry steps	Budget for all necessary steps
Make a decision	The most important step
Get performance/design requirements in writing	"Don't worry" is not good enough
Educate/train employees	Document in writing
Buy quality equipment, chemicals	Insist on vendor support

needs if you get beyond the technical literature and/or advertisements on the Internet. Talking to a credible technical person also helps assure good product support.

Before you start, however, step back for a moment and take a look at the product line. Estimate throughput and product mix and flow (at present and in 3 years' time). Consider the major regulatory requirements for your area as well as critical company policy and customer requirements. Making a list of major process considerations prior to contacting sales reps or attending shows helps to clarify requirements both to the manufacturer's representatives and to yourself.

Hopefully, you do not have to consider all facets of process change on your own. Approaching process change as a team effort has been discussed on a number of occasions [1,2]. The team approach can save wasted time and effort, particularly where studies by chemical and equipment vendors are required [3,4]. If you are working with a team, you are fortunate, as it will be easier to optimize the process.

Based on team input and on your own considerations, a number of factors worth considering are likely to emerge, some of which are indicated in Tables 1.2 through 1.6. These tables are simply to help you focus on those factors that are most important for your particular application. I suggest you look over the tables, circling those things that are of most concern, and keep them in mind during your web searches, trade shows, and initial inquiries.

Become educated about a range of cleaning agents, cleaning equipment, sample handling equipment, and advanced processes such as CO_2, supercritical fluids, and plasma cleaning [5,6]. If management, the safety/environmental people or, most importantly, you box yourself in too much, too many cleaning options may be eliminated, to the detriment of company productivity.

In some cases, the company, particularly the safety and environmental group, may prohibit entire classes of cleaning agents. Often, the real problem is that additional record keeping and containment may be required. You may need to ask some pointed questions to determine the true regulatory requirements. It is important to choose the safest, most benign, most environmentally preferable process possible. However, because successful cleaning action inherently implies a degree of reactivity, virtually all processes have some safety and/or environmental baggage. With the proper containment and chemical handling systems, a range of products can be used responsibly.

TABLE 1.2 Preliminary Process Considerations: The Product Line (Suggestion: Circle the Factors Most Important for Your Process)

Factor	Considerations	Implications, Examples
Component, subassembly, part characteristics	Size (l × w × h); or continuous feed	Minimum tank size Optimal dimensions of cleaning tanks Type of equipment
	Materials of construction	Compatibility with cleaning, rinsing agents Maximum acceptable temperature Acceptability of ultrasonics
	Spacing of components (e.g., standoff of components from circuit board) Complex build, blind holes	Wettability Rinseability Force of cleaning action (ultrasonics, turbulation, spray) Type of drying equipment
Product mix	High throughput/high mix High throughput/low mix Intermittent changes in product line Major changes in product line expected	Cleaning agent Drying system Automation

TABLE 1.3 Preliminary Considerations: Throughput, Cleanliness Requirements (Suggestion: Circle the Factors Most Important for Your Process)

Factor	Considerations	Implications, Examples
Sample throughput	Approximate throughput Expected changes in 1–5 years	Maximum cycle time Drying requirements Soil loading Tank size
	Sample handling requirements	Conveyor belts Overhead robotics Fixturing Overall process flow (beyond cleaning steps)
Cleanliness requirements	Visual Performance Specific process requirements Surface soils Particulates (size range) Specific customer requirements	Specific testing needed Laboratory support from vendor Ongoing surface monitoring Recirculating cleaning agent with real-time particle monitor

Record Your Impressions

It is a good idea to take notes, particularly if you attend a trade show, an educational program, or meet with a sales representative. Taking notes does not have to be intrusive; even notes on the backs of business cards are okay, as long as you remember to keep them all in one place. You might ask: Why take notes when the rep will inundate you with cards or when swiping your card will fill your desk or e-mail inbox with brochures? There are several good reasons.

For one, taking notes helps you focus, prioritize, and note any strong positive or negative impressions about the product or vendor. By the time several weeks (or months) have passed, the superior technologies may no longer be obvious. In my experience, talking to a vendor with a notepad in hand tends to make vendors a bit more specific and careful in their claims. The notes help you document those claims. If nothing else, taking notes conveys a sense of purpose, allowing you to gather essential information.

TABLE 1.4   Preliminary Considerations: Budget, Workspace, Utility Constraints (Suggestion: Circle the Factors Most Important for Your Process)

Factor	Considerations	Implications, Examples
Budget	Chemicals	Initial cost
		Expected usage
	Equipment	Basic cost
		Cost for full system
	Consumables	Water filters
		Ion exchange resin
		Carbon filters
		Other consumables
	Water purification	Deionizers
	Solvent containment	Higher freeboard
		Fully contained system
	Waste stream	Evaporators
		Holding tanks
	Closed loop; recycling	Steam or vacuum distillation
Workspace limitations	Total footprint $(l \times w \times h)$ process equipment	Cleaning, rinsing, drying
		Conveyors
		Hoists, robotics
		Water recycling Evaporators
		Holding, separation tanks
		Solvent recyclers
Current workplace utilities	Electrical	May limit your equipment choices
	Water	
	Compressed air	
	Nitrogen lines	

TABLE 1.5   Safety, Regulatory, Company and Customer Requirements and Requests (Suggestion: Circle the Factors Most Important for Your Process)

Factor	Specifics
Environmental regulatory	Federal
	State
	Local
	Volatile organic compounds (VOCs)
	Hazardous air pollutants (HAPS)
	Record-keeping requirements
	Effluent requirements
	Cleaning agents
	Soils, trace metals
Safety regulatory	Right to know
	Chemical composition and handling
	Equipment
Corporate policy; customer policy or requirement	Ozone depleting chemicals
	HAPS
	VOCs
	Other specific requirements
Insurance	Containment of low-flash-point solvents
	Containment of toxics

TABLE 1.6 Additional Company Requirements

Factor	Specifics
Requests, requirements, strong prejudices	Cleaning agent odor
Management	Overwhelmingly positive or negative regarding
Safety/environmental	Aqueous
Production workers	Solvents
Customers	Ultrasonics
	Low flash point
	New technology
	Other considerations
Workforce sophistication	Chemical handling
	Adjustment to major process change
	Computer literacy

If you are at a trade show, you can multitask by simultaneously collecting technical information while engaging in the all-important collection of canvas bags, pens, chocolates, and travel drives.

Cleaning Agents

What used to be a narrow choice in cleaning agents and processes has exploded into an assortment of possibilities. The increasing trend toward blended solvents and the use of brand names that obscure the actual chemical composition leaves many components manufacturers in a state of semipermanent confusion. Unfortunately, engineers have been known to adopt a new cleaning agent and even report the process change at public forums, while having no idea if they were using an aqueous-based, hydrocarbon blend, nonlinear alcohol, or some other chemistry.

While it is not necessary to have an advanced degree in chemistry, an understanding of the basic nature of the cleaning agents being offered is essential to optimize any process and to achieve appropriate chemical handling. Sometimes cleaning agents are described by type:

- Aqueous (with cither an inorganic or an organic solvent base)
- Aqueous with significant amounts of organic solvents
- Solvents for liquid/vapor phase cleaning
- Higher boiling solvents
- Solvent blends (typically relatively high boiling), which can be rinsed with water and/or solvent, depending on the specific formulation
- Oxygenated solvents
- Biobased solvents

Peruse the cleaning agent section in *Handbook for Critical Cleaning: Cleaning Agents and Systems* for more detailed information.

The first thing to determine is what kind of cleaning agent is being offered. This is important, because all you may be told is that the product is advanced, carefully formulated, and removes all kinds of soil from all kinds of substrates really well, without any product damage, and with no safety or environmental issues. What the chemical manufacturer calls a degreasing agent may mean a blend of chlorinated solvents, a low-flash-point solvent, hydrocarbons, orange or pine terpenes, esters, nonlinear alcohols, hydrofluorocarbons, hydrofluoroethers, *n*-propyl bromide, parachlorobenzotrifluoride, aqueous/saponifiers, or some other chemistry. A checklist of some important questions to ask suppliers of cleaning chemicals is indicated in Table 1.7. The checklist is not meant to turn the discussion into a formal audit; let's instead call it an informal, friendly audit.

If a vendor offers a degreasing agent, it is important to determine exactly what that implies in terms of process change. If you plan to use current cleaning equipment, it must be adapted to the new cleaning

TABLE 1.7   Initial Evaluation Checklist for Cleaning Agent

Company (chemical vendor):

Date:

Contact: Phone:

What cleaning agents do you offer?

      Aqueous/saponifier

      Bicarbonate (liquid/solid)

      Semi-aqueous

      Orange terpene

      Pine terpene

      Hydrocarbon blend

      Ester blend

      Tetrahydrofurfuryl alcohol

      n-Methyl pyrollidone

      Alcohol (specify)

      Other solvent blend (specify)

      Microemulsion

      Solvent (specify)

      Chlorinated (specify)

      Brominated (specify)

      Combustible

      Flammable

      Parachlorobenzotrifluoride

      PFC

      HFC

      HFE

      HCFC 225

      VMS

      NMP

      other: _____

If it is proprietary, can we obtain the information after signing a confidentiality agreement?

Regulatory status:

      Ozone depletion potential (ODP)

      SNAP status: approved? pending?

      Volatile organic chemical (VOC)?

      Hazardous air pollutant (HAP)?

      Combustible?

      Flammable?

      Inhalation studies?   acute? 90 day?   inhalation level?

Please provide: MSDS

      Technical/applications sheet

Based on major soils and materials of construction, which products do you recommend?

Costs, performance:

      Cost per pound/per gallon

      Soil loading

      Cleaning/performance studies available?

      Compatibility information

What is the evaporation rate?

      Similar to alcohol? Similar to water?

TABLE 1.7 (continued)   Initial Evaluation Checklist for Cleaning Agent

Can you use liquid/vapor cleaning?

Can I use my current degreaser?

Do I need to rinse for my application?

Do I need a rust inhibitor in the rinse water?

Is the cleaning agent miscible in the rinse agent (i.e. will it dissolve, or will it form two layers?)

What process equipment do you recommend?

Is the equipment displayed at this show?

Can I retrofit my current equipment?

      Cleaning

      Ultrasonics

      Spray in air

      Spray under immersion

      Rinsing

      Drying

      Solvent containment

      Recycling/closed loop

      Other

Drying options:

      Air knives

      Forced hot air

      Vacuum drying

      Centrifugal drying

      Convection ovens

      Moisture absorbing materials

      Drying by solubilization (alcohols, etc.)

      Drying by displacement (surfactant-based solvents)

What kind of solvent containment will I need?

Vendor support:

Can you supply a sample for in-house testing?

Do you have an applications lab? Or can you perform testing at my site?

Name of chief chemist:

Phone:

Fax:

Can you do cleaning/soil loading studies for me?

Disposal of spent material:

      Waste stream treatment?

      On-site recycling?

      Take-back of spent solvent?

chemical. Ambitious adaptations, such as changing a vapor degreaser into a dip tank for a water-based cleaner or a high-boiling solvent are often not as straightforward as one might imagine. You have to consider the compatibility of the new cleaning agent with materials of construction of the cleaning equipment. Consider the process requirements and anticipated cleaning equipment at the same time as you evaluate the cleaning agent. Ask specific questions, and evaluate the answers critically.

High-boiling materials like aqueous and terpenes may not dry rapidly enough for your process requirements. They may leave significant cleaning agent residue—a single-tank system does not provide for rinsing. While manufacturers may claim that rinsing is not necessary, some users have found that rinsing is required for high-reliability applications. Further, aqueous or viscous blends may not penetrate spaces in low-standoff components, and it may be difficult to achieve cleaning and drying for complex assemblies with blind holes. Some cleaning agents have toxicity and/or regulatory issues. Many of the newer designer solvents require well-contained systems to be cost-effective.

You should be very suspicious of supposedly safe mystery mixtures. Insist on knowing what you are working with. Typically, you simply can't clean parts in chicken soup, and even chicken soup might have a substantial volatile organic compound (VOC) content. If the cleaning agent supplier will not conveniently and cheerfully supply a confidentiality agreement covering your company and pertinent consultants including lawyers, environmental advisors, and process development specialists, I suggest finding a different supplier.

Equipment

In evaluating any new piece of cleaning equipment, you need a ballpark idea of price, footprint, through-put including drying time, automation options, and suitable cleaning agents. Because cleaning equipment is so specialized and given the fear of sticker shock, the equipment manufacturer may be reluctant to provide pricing information, sometimes even claiming confidentiality. Rinsing and drying modules can double or triple both costs and space requirements.

A checklist of information to obtain from cleaning equipment manufacturers is indicated in Table 1.8. Again, it is neither necessary nor productive to ask all of these questions to all equipment manufacturers.

For large-scale processes, sample handling equipment can add more than intended height to the overall footprint. A number of components manufacturers have required last-minute workplace remodeling because the ceilings were too low. One needs to allow room for the overhead hoists and robotics to move.

In some cases, such as UV/ozone cleaning, plasma cleaning, and laser cleaning, the cleaning agent or cleaning action is effectively generated by the process. For liquid CO_2, supercritical CO_2, or CO_2 snow, the main concern is cleanliness of the CO_2 (or other gas or gas mixture under consideration). In such situations, controlling the process chemical is relatively simple.

More often, the optimal cleaning process is a combination of the cleaning agent and the cleaning equipment. To make an informed decision, it is important to learn as much about both aspects of the process at the same time. Because equipment manufacturers are often geared to the mechanical aspects of cleaning, they may downplay the importance of the chemical. Therefore, it is important to determine what cleaning agents can be used with the particular equipment in question.

Some cleaning equipment manufacturers either offer a line of cleaning agents or closely team with a small subset of cleaning agent manufacturers. In such instances, you may be strongly encouraged to purchase certain products. Ask why. There may be a legitimate stewardship or performance issues. On the other hand, be a bit cautious about situations where the choice of cleaning agents or brands of cleaning agents is restricted. With changing technology and regulations, you need options.

Drying

Given that many parts nest or have a complex structure, drying can be difficult. It is a factor that is often not considered in simple dip-tank conversions. In fact, inadequate drying can lead to recontaminated or damaged parts, and drying often becomes the limiting step in non-solvent production processes. As outlined in Table 1.8, a number of approaches to drying are available; all have their positive and negative features.

Organize and Coordinate

At some point, you need to organize the information obtained. If you are fortunate enough to be the main person in charge of upgrading the cleaning process, put the information into a somewhat digestible form. Note the key URLs. Summarize your notes. Do I hear grumbling and groaning? This does not

TABLE 1.8   Initial Evaluation Checklist for Cleaning Equipment

Company (Cleaning Equipment Vendor):

Date:

Contact: Phone:

Type of equipment:

Cleaning Agents/cleaning sequences which can be used:

Please give me some typical examples

Cycle time (minutes): range   average

Costs, availability, training:

 Base cost:

 Cost "fully loaded":

 Average system cost:

 Shipping, installation costs:

 Employee training availability, costs:

 Time to build from receipt of order (weeks):

Appearance/footprint, tank size

Overall impression

Materials of construction

Coved tank?

Welding

Footprint ($l \times w \times h$, in.)

Base equipment

Total equipment (including robotics, recycling, separation tanks)

Cleaning chamber capacity ($l \times w \times h$, in.)

Chamber capacity (gal)

Number of tanks or chambers needed

Fixturing: sizes, materials of construction

Vendor support and reliability:

 Can you supply a loaner model for in-house testing?

 Do you have an applications lab? Or can you perform testing at my site?

 Name of chief chemist or engineer:

  Phone:

  Fax:

 Can you do cleaning/drying/soil loading studies for me?

 Name, location of repair person

 Warranty

 UL/FM approval

 Have you worked with my local regulators?

 Can I get permitting in my area?

 What support can you provide?

Cleaning, rinsing, drying:

 Heat (adjustable temperature?)

 Spray (force of spray)

 Rotation

 Mechanical agitation

 Spray under immersion

 Centrifugal

 Immersion

(continued)

TABLE 1.8 (continued)   Initial Evaluation Checklist for Cleaning Equipment

Ultrasonics:
Standard or sweep
25, 40, 60 kHz, other
Over 100 kHz
Megasonics
Rinse action
Weir/sparger
Drying:
Air knives
Forced air
Vacuum
Other
Auxiliary equipment, pollution prevention:
   Do I need, do you recommend:
      Vapor containment
      Meets NeSHAP?
      Closed loop
      Hoist
      Batch robotics
      Parallel robotics
      Solvent/cleaning agent filtration
      Onboard recycling
      Vents
      Chillers
      Atmosphere inerting
      Fire suppression (nebulizers, CO_2)
      Separation tanks
      Waste stream treatment
      Deionization
      Reverse osmosis
      Other
Facility requirements for:
   Electrical
   Air
   Nitrogen
   Deionized water
   Water for chilling
   Venting
   Other requirements

have to be a major literary effort but will help you to focus. Put key product literature into an electronic or even a hard copy folder—whatever works for you. Then, work with your process change team, advisors, and even the insurance company. Check for impending regulatory changes. Ask the designers about impending product changes. Incorporate their ideas. If necessary, obtain clarification.

Refine and Test

A relatively short list of promising possibilities should emerge.

You must estimate how well the new process will perform for your application. This is no easy task; many engineers, chemists, and economists all struggle with this issue—there is no foolproof approach.

Further, if the equipment and chemical purchases are small, vendors may not be able to economically justify testing.

There are still several actions you should take, which are critical not only for small but also for larger equipment installations. Ask for an in-house demonstration. Ask for references, and then call those components manufacturers. Ask to visit some similar equipment installations. In my own experience, components manufacturers can provide a wealth of frank information about process performance, vendor response to crises, repairs, maintenance, and downtime. If you are still not ready to make a commitment, ask about leasing the equipment.

Cleaning Evaluations by Suppliers

One problem in setting up cleaning evaluations at the vendor facility is how to avoid a generic approach and a generic response. Often, the vendor will say: "Of course we can clean your component." You carefully pack up the samples and send them in for testing. The applications lab cleans the samples with the cleaning juice they most commonly use (never mind what you asked for) at the default settings (never mind that the temperature is too high for a critical component) and reports back that the parts were successfully cleaned. Even more frustrating, with the current pressure for process changeover, some components manufacturers have waited months for test results.

Specifying your requirements and expectations can help to avoid delays and meaningless testing. Contact the applications chemist; work directly with him or her. Tell the test lab, in writing, exactly how you want the tests to be run and how you want the parts to be handled. Be very specific about cleaning chemistries and process conditions. Provide representative soils and materials of construction. Note any known materials compatibility, temperature sensitivity, and related problems. Explain your cleaning standards, if possible. As a control or benchmark, provide the test lab with parts that you judge to be acceptably clean. Perhaps most important, asking for an unrealistic number of cleaning tests is not business-like and is counterproductive to your efforts. You want to achieve productive working relationships with your suppliers.

Make a Decision

You have the tools to make a decision; you have to move forward. This approach has proven successful (Kanegsberg et al. 1995) and, with economic and competitive constraints, it is even more important today.

Choose wisely. Purchase high-quality cleaning equipment and quality cleaning chemistries. This does not mean the most expensive chemical or the cleaning equipment with the most bells and whistles. It means selecting products and process features that best meet your requirements.

Above all, make a decision. It will not be a perfect decision, but make a decision. Getting stuck in a repetitive loop of evaluations is costly. Involve your management; justify the decision and move forward.

References

1. B. Kanegsberg, Getting the most out of precision cleaning '97—how to get vendors to work with you. Presented at *CleanTech '97*, Cincinnati, OH, April 1997.
2. B. Kanegsberg and E. Kanegsberg, Implement and Improve the Cleaning Process, presentation at Process Cleaning Expo, Louisville, KY, May 4, 2010.
3. B. Kanegsberg, H. Mallela, H. Dominguez, and W. G. Kenyon, Integrating precision de-oiling and defluxing processes in high volume manufacturing systems. *IPC Presentation and Proceedings*, San Diego, CA, May 1995.

4. See the section, Case Study: Cleaning in an Airless/Vacuum System Prior to Applying Engineered Coatings, in Chapter 15 of this book.
5. B. Kanegsberg, *Water Versus Solvents*, Part 1: A Complex Landscape, Controlled Environments Magazine, Sept. 2008.
6. B. Kanegsberg, *Water Versus Solvents*, Part 2: Process Trends, Selection, The Future, Controlled Environments Magazine, Oct. 2008.

2

Cleaning Agent Balancing Act

The Quest for the Ideal Cleaning Agent 15
Physical/Chemical Properties...17
Selectively Aggressive • Halogen-Free? • Flash Point and
Evaporation Rate
Regulatory Issues..19
Flash Point Solvent Choices and Flammability Characterization
Economic Constraints Impeding New Solvent Development...................19
Steps in Developing a New Solvent • Niche Market
Future of Biobased Cleaning Agents..20
Political Chemistry
Green and Safe—Moving Targets...22
Definitions • Green Evolution
Moving Forward 27
References..27

Barbara Kanegsberg
BFK Solutions

The Quest for the Ideal Cleaning Agent

For years, probably for decades, people have asked us about prospects for new, practical, innovative cleaning agents. Sometimes, they narrow the criteria to solvents (organic solvents), solvents that could be used undiluted (neat) or perhaps as the major component of a simple cleaning agent blend [1,2].

Some attributes of an ideal solvent in the second decade of the twenty-first century are summarized in Table 2.1. The attributes include technical/performance, economic, and regulatory factors. Shockingly (or not), we know of no cleaning agent or solvent that meets all of these requirements. Further, while we would like to be pleasantly surprised, it is unlikely that such a cleaning agent will be developed in the foreseeable future. Over the past decade, we have made a point of discussing this cleaning agent wish list with dozens of colleagues who are involved in cleaning agent and/or cleaning process development. So far, there are no signs of a magic answer, and with good reason.

Let us consider a few of the factors limiting the development of an ideal cleaning agent. By doing so, you will gain an appreciation of factors limiting cleaning agent development. Even more important, you will develop tools to make realistic selections of cleaning agents and cleaning processes. It is important to make informed choices about cleaning chemistries and to select attributes that are most important to the process. Table 2.1 highlights the commonly requested attributes and why the attributes are desirable.

TABLE 2.1   Future Cleaning Agent Wish List

Attribute	Why Attribute Is Wished for
Inexpensive	Low initial chemical cost Low in-use process cost Long process use/long bath life
Removes soil rapidly, effectively	Rapid, economical manufacturing High-quality product
Good wettability	Reaches component surfaces Enables effective soil removal, effective rinsing
Removes wide range of process soils	Maximizes process efficiency Adaptability to modifications in processing fluids (e.g., lubricants)
Single component, essentially used neat	Simplified acceptance testing Easier to monitor vendor/supplier performance
Good materials compatibility, wide range of materials	Adaptable to complex, multi-material components More cleaning process options (temperature, time, force)
Can be used with current cleaning process/equipment	Not capital intensive No extensive training
Evaporates rapidly	Rapid drying Thorough outgassing from porous substrates Rapid product assembly Avoids surface residue buildup
No residue	Avoids introducing contamination during the cleaning process
Nonflammable	Usable near live electronics assemblies Usable near other manufacturing processes [1] Minimizes worker safety issues Minimizes insurance costs
Nonhalogenated	Perceived to minimize regulatory scrutiny
Water based, low residue, noncorrosive, rapid drying	Desire for classic solvency combined with favorable regulatory/safety profile
Not a hazardous air pollutant	Minimizes interaction with regulatory agencies Avoids need to notify neighbors, sensitive receptors (e.g., schools) Less employee monitoring Fewer restrictions on the amount of material used Minimizes environmental liability
No/minimal enhancement of tropospheric ozone	Not classified as a volatile organic compound (VOC) or reactive organic compound (ROC) Minimizes use restrictions Minimizes interaction with regulatory agencies
No/minimal impact on stratospheric ozone	Minimize interaction with regulatory agencies Avoids loss of availability through use, sales, or restrictions, related to Montreal Protocol
Acceptable as substitute for ODC	Acceptable by U.S. EPA SNAP for sale, import, use
Not a greenhouse gas	Avoids potential loss of availability through use, sales, or production restrictions
Not toxic to workers	Minimizes needed process controls Minimizes worker safety controls (e.g., respirators) Minimizes employee monitoring, training Minimizes company liability
No residue harmful to living host	Important in implantable medical devices Facilitates interaction with pertinent regulatory agencies Assures reliable device performance Minimizes potential for lawsuits

TABLE 2.1 (continued)   Future Cleaning Agent Wish List

Attribute	Why Attribute Is Wished for
Biobased	Plant or animal derived product
	Sustainable production/renewable resource
	Perceived as environmentally preferred
	Legislative encouragement to use biobased
Local, national, and global environmental/worker safety acceptability; global availability	Efficient chemical control
	Efficient supply chain management
	Consistent process control
	Plant relocation results in minimal impact on cleaning process

Physical/Chemical Properties

Selectively Aggressive

We want an aggressive solvent or cleaning agent, one that will remove the soil in question rapidly and effectively.

"The soil" is often a process chemical used as part of manufacturing; defining the soil is becoming increasingly difficult. Most manufacturing soils are themselves complex, often proprietary mixtures. The properties of these process fluids are most often related to maximizing product performance or, sometimes, to complying with regulatory mandates.

For example, water-soluble lubricants have become important in manufacturing, in part for environmental considerations. These lubricants do indeed dissolve or disperse in water. However, the heat and forces involved in actual machining and manufacturing processes alter these lubricants; the best technical term for this soil is adherent, insoluble goo.

For many talented formulators, removal of the soil is not the foremost consideration. For example, a colleague and I coauthored a cleaning chapter for a lubricants handbook [3]. We ended up providing an overview of cleaning techniques. However, our initial intent was to base the cleaning techniques on suggestions by the formulators. To this end, we put together a series of questions and contacted them. This technique met with limited (make that zero) success. The concept of removing the lubricant simply did not compute. As I recall, the most common comment ran something along the lines of "We design these lubricants to meet the performance requirement of the product. Why would *anyone* want to remove this wonderful product?" Of course, we know that sometimes the most wonderful product becomes soil or matter out of place and you have to remove it.

To make matters even more amusing, in actual practice, soils consist of mixtures of these process chemicals in variable proportions. Process chemicals may change; complex supply chains take away even more control from the ultimate assembler or manufacturer.

So, we want an aggressive cleaning agent, one that removes an array of soils effectively. However, aggressive cleaning agents tend to exhibit increased potential for product damage, not to mention worker safety and environmental impacts. In fact, if the cleaning agents do not themselves show these problems, they may be used in aggressive cleaning processes, which may have increased the potential for product damage and worker damage.

More mild cleaning agents, more mild organic solvents tend to exhibit poor soil removal for most soils of interest. They often have a lower potential for product damage and show a more favorable worker safety profile. Some of the less-aggressive organic solvents may have environmental issues in the sense that they can be more stable, longer lasting in the environment, and may contribute to climate change.

Blends, whether water or solvent based, can exhibit synergy in terms of soil removal. Of course, with this comes increased potential for unexpected product damage. Further, there is the potential, a potential often not broached by regulatory agencies, for increased damage to workers and to the environment.

There are only so many ways that a molecule can be manipulated.

Many manufacturers expect that an aqueous cleaning agent will be self-rinsing and will dry rapidly. These are unrealistic expectations.

Halogen-Free?

Hydrocarbons (molecules that contain hydrogen and carbon) tend to be flammable and have a limited solvency range. Chlorine, fluorine, bromine, and, to a limited extent, iodine are elements that are important in organic (carbon based) solvents. Chlorine is also important in some aqueous cleaning and disinfecting chemistries.

Molecules in which chlorine and bromine replace hydrogen tend to be more aggressive solvents; they dissolve more soils than do hydrocarbons. Fully fluorinated solvents, like the perfluorinated carbons (PFCs), and partially fluorinated solvents like the hydrofluorocarbons (HFCs) and hydrofluoroethers (HFEs) tend to be very mild solvents for most industrial soils of interest. (By the way, "per" means that there are no hydrogens, just fluorine and carbon in the molecule. Similarly, in perchloroethylene, we have all chlorines.)

Some highly halogenated organic compounds do not support flame propagation, like HFE 7200, n-propyl bromide, and perchloroethylene. When the carbon halogen bond is attacked, halogen-free radicals are released. These free radicals are considered relatively stable and may stop the chain reaction of flame propagation.

There are drawbacks. As would be expected from aggressive solvents, some chlorinated solvents are under regulatory scrutiny because of worker safety and neighborhood safety concerns. While we know of no outright ban on perchloroethylene (PCE, tetrachloroethylene) for precision cleaning applications in the United States, there are restrictions. In addition, one might expect that upcoming bans on use in commercial dry cleaning may eventually result in reduced availability and higher costs.

In terms of flammability, you cannot assume that all halogenated compounds will not have a flash point; you have to consider the properties of the individual halogenated compound. For example, trans-1,2-dichloroethylene (trans) has a flash point in the range of 2.2°C (36°F). However, it forms azeotropes with HFCs and HFEs that do not exhibit a flash point. Because it is a fairly aggressive solvent with a kauri-butanol value of 117, it has found utility in cleaning applications.

Some halogenated solvents are referred to as ozone depleting compounds (ODCs) because they contribute to the destruction of the stratospheric ozone layer and have been phased out via the international Montreal Protocol. In the United States, depending on the compound in question, there have been production phaseouts, import restrictions, as well as sales and usage restrictions. The regulations are complex and sometimes confusing and usage specific; manufacturers should use ozone depleters in a mindful manner.

Flash Point and Evaporation Rate

If you exclude halogens, it is difficult to achieve rapid evaporation with a high flash point. High-boiling cleaning agents can be an option, but the manufacturers then have to consider residue and wettability.

There could be other options. Ionic liquids are sometimes invoked as alternatives. They are liquid salts consisting of ions and ion pairs. They may contain organic and/or inorganic salts, some of which may be halogenated [4]. The term liquid is relative. While some are liquid at room temperature, others are liquid at over 100°C or even 200°C, depending on who is defining the term. They are inherently slow to evaporate and materials compatibility may be an issue. Similarly, some phosphorus compounds have been used as flame retardants, but they are not volatile. With all proposed cleaning agent compounds, the toxicological properties would have to be studied and the results successfully defended to the regulatory community.

Regulatory Issues

Regulatory issues, including worker safety, community exposure, regional air/water concerns, and global environmental safety, drive the way we conduct manufacturing processes. Regulatory constraints may also limit the quality of our products.

Flash Point Solvent Choices and Flammability Characterization

On the surface, determining a flash point would seem to be very straightforward and cut and dried. In the United States, both the Occupational Safety and Health Administration (OSHA) and the Department of Transportation (DOT) define "flash point" as the minimum temperature at which a liquid gives off a vapor in sufficient concentration to ignite when tested by specific methods as set forth in the regulations [5].

Determination of the presence or absence of a flash point is very method dependent. As a chemist, I can assert that over the years, I have caused many liquids to catch fire both inadvertently and in attempts to determine flash points. Manufacturers need to be aware of the potential for flammability, even with chemicals that are deemed to have no flash point and no flammability.

There are standardized methods for flash point determination [5]. In the United States, American Society for Testing and Materials (ASTM) methods are commonly used, for example, ASTM D56 for low viscosity fluids, ASTM D93 for high-viscosity fluids, and ASTM D3278. While they are clearly written, many of the standard methods contain a number of decision points in terms of exactly how to run the test as well as descriptions of what a positive flash point looks like can be misunderstood, unless an experienced analyst runs the test. In some instances, an enhanced pilot light flame may be observed. However, a true flash point is taken to mean that the vapors sustain combustion on their own.

In addition, methods deemed acceptable differ markedly depending on local and national practices, policies, and regulations. In some instances, the preferred methods for testing have changed over the years. Flash point alone may not tell the full story. For example, the Limits of Flammability Test (ASTM E681) is appropriate, and chemical producers may also use the Minimum Ignition Energy Test (ASTM E528).

When using any organic solvent or any blend that releases volatile organic vapors, it is important to consider issues relating to flammability. This includes the flash point and the flammable range. It also should include an analysis of process specifics and other processes performed in the vicinity.

Economic Constraints Impeding New Solvent Development

There is a tendency to expect, or at least to hope, that chemical companies will somehow develop a magic formula, a new molecule, that will solve all critical cleaning problems.

Steps in Developing a New Solvent

One way to understand the costs involved in introducing a new cleaning solvent is to look at an overview of the steps. There is an initial screening survey that includes economic feasibility, toxicity screening, and synthetic pathway. Many candidates are weeded out by computer prediction. After the more promising candidates are selected, performance testing is required, including laboratory and beta site testing. Market studies are conducted; and these studies include alternative uses of the feedstock required for the proposed product. Toxicity testing can be exceedingly complex and costly. There are also global, national, and, in some cases, local regulatory submissions, marketing, advertising. Last but not least comes lobbying and advocacy; more about lobbying in the Political Chemistry section.

Developing new solvents represents a tremendous allocation of resources. Estimates vary from a low of $5–$10 million to estimates in the billions. Animal testing is costly. One small study was estimated at over $600,000. One recent submission to the EPA, covering a category of related chemicals, was estimated to cost over $40 million. It is difficult to estimate costs, because they are often split among business units for blowing agents, solvents, and aerosols. Costs may even be split among companies. Of necessity, the research and development money, the money for toxicity testing, and the resources to work with regulatory agencies all go to the areas of heavy solvent consumption. Such areas include refrigeration, foam blowing, dry cleaning, and assorted synthetic processes.

The bottom line is that if the cleaning niche is small and the chance of recovering costs is low, then the product probably will not be developed.

Niche Market

Those concerned with solvent cleaning are only a small blip on the radar screen of most chemical companies. This small blip can be large in absolute terms of investment; however, the amount and the emphasis is, of necessity, relatively small.

Further, cleaning is a niche market. The industrial cleaning market, including general cleaning, and precision or critical cleaning, might best be described as a series of sub-blips. "Old-style" cleaning, used through the mid-1980s was about 50% solvent based and 50% aqueous based (water alone, or water with chemical additives). For general solvent cleaning, chlorinated solvents, hydrocarbons, and assorted flammable solvents, including hazardous air pollutants were used. For precision cleaning, the standard protocol included cleaning sequences of CFC 113, trichloroethane, and isopropyl alcohol along with acetone as a final clean/rinse/dry. Over the past two decades, the number of potential cleaning agents has grown significantly. However, these include not only single solvents but, to an increasing extent, solvent blends and blends containing or soluble in water.

Even without regulatory constraints involved in the introduction of new chemicals, manufacturing requirements are themselves so diverse that it may be difficult for chemical manufacturers to see a market trend favoring the development of what we might find to be a useful cleaning agent. We use increasingly diverse and creative materials of construction, so no one cleaning agent is likely to be compatible with all of the substrates in use by all manufacturers. We have varying requirements as to the meaning of "how clean is clean enough."

Chemicals marketed as cleaning solvents are often actually expanded uses of chemicals developed for other purposes. HCFC 123 and HCFC 365 were developed as refrigerants. Parachlorobenzotrifluoride is a synthetic by-product of the production of other chemicals. N-propyl bromide is used in the production of ibuprofen. HCFC 225 is used as a feedstock in the production of other chemicals. Most biobased products were developed as food. The use of the products as cleaning agents and lubricants is a relatively recent introduction.

Future of Biobased Cleaning Agents

Biobased agents are derived from plants and are sustainable resources. There has been growing (excuse the pun) interest in biobased products for critical cleaning. Biobased chemicals have proven useful in cleaning applications. As discussed in Chapter 1 (Kanegsberg) of *Handbook for Critical Cleaning: Cleaning Agents and Systems*, methyl soyate has received a considerable amount of attention both neat and in blends, and ethyl lactate (from corn feedstock) has proven to have good penetration properties to assist the effectiveness of removal of paints and coatings. Chapter 7 (Dingess, Morford, and Shubkin) of *Handbook for Critical Cleaning: Cleaning Agents and Systems* covers attributes of a cleaning agent that has found utility in general metal cleaning, optics, and aerospace applications. Beyond D-limonene, methyl soyate, and ethyl lactate, there are other possible biobased candidates for critical cleaning applications [6]. For example, beyond soy derived products, there are other fatty acid methyl esters. The length of the carbon chain and any modifications to the carbon chain of the fatty acid portion of the ester (often designated "R" to indicate some unspecified carbon-containing material) will influence the solvency properties of the molecule. Short chain "R" groups could be derived from coconut or linseed oil. Long chains include castor, tallow, canola, and palm oil. If low-cost palm oil at different levels of saturation could be obtained, this would influence the physical and solvency properties of the molecule. Castor oil could be a particularly interesting candidate for a methyl ester, because the "R" chain has a hydroxyl group. This hydroxyl group makes the molecule more polar, more like water. There could be a potential for removing water-based lubricants, for example.

The extent of development of such cleaning agents, as with other cleaning agents, depends on a number of factors such as development costs, production costs, regulatory factors, performance, adaptability to cleaning equipment/processes, and market demand. In terms of regulatory drivers, achieving VOC exemption and/or having a low relative reactivity level are essential considerations.

Political Chemistry

Many cleaning agents have come under regulatory scrutiny. This means that chemical companies may spend untold millions of dollars testing and developing a new cleaning agent only to find that the product is subject to so many restrictions that it is not profitable to produce the product. This scrutiny may be due in part to the fact that solvents used neat or as blends of two to three components are very visible to the regulatory community. Certainly, safety and environmental groups may become involved.

One additional factor is political chemistry; it is not unknown for chemical companies to extol the favorable safety and environmental attributes of their cleaning agents to regulatory agencies and also for them to inform regulatory agencies about what they see as problems with chemicals produced by the competition. Using a political chemistry approach, chemicals produced by the competition will have unacceptable safety and environmental impacts; one's own chemicals will be favored. Those working in regulatory agencies may recognize this lamentable practice for what it is, but they can understandably take such assertions under consideration.

There is the perception among some regulatory agencies that restricting the availability, allowable exposure, or allowable emissions of a particular chemical will make the world, neighborhoods, or individual workers safer. We know, of course, that not just the chemical per se but also the method of handling the chemical determines the impact on workers and on the environment. However, in developing regulations, agencies tend to look, in part, at the level at which a chemical is being used. Therefore, the success of a cleaning agent will generate visibility to regulatory agencies, either directly or because the competition tells the regulatory agency about the "new" or "emerging" danger posed by the cleaning agent to worker safety or to the environment.

Given an atmosphere of political chemistry, what chemical company will be brave enough to introduce even a moderately aggressive solvent? Very, very mild solvents will be favored. In addition, blends are apt to be favored, because each ingredient may be used only in small amounts. In complex blends, if one chemical becomes politically incorrect, another similar chemical can be substituted. This could adversely impact the process, and it might not be better for workers or for the environment. However, unless the level of a particular chemical attracts the attention of regulators, the product may be relatively safe from extinction.

It should be added that this same aura of political chemistry has invaded the arena of process equipment. There have been attempts over the years to show that one particular technology is favorable over another. This is good, except that there are too often attempts to show that competing technologies ought not to expect to have favorable regulatory status or even that they should be considered to be unacceptable.

Perhaps regulatory standards and metrics of the future will improve the situation. Perhaps political chemistry, particularly negative political chemistry will become obsolete, particularly as chemical suppliers and providers of cleaning equipment options realize that such efforts are harmful not only to other providers of cleaning agents and cleaning equipment, not only to those engaged in cleaning products but, ultimately, also to their own efforts. Perhaps, this hope is almost as realistic as is the quest for a perfect solvent.

Green and Safe—Moving Targets

Definitions

We use terms like "green," "environmentally preferred," "sustainable," "biobased," and "safe." Sometimes, the terms are used interchangeably. However, their meaning and interpretation really depend on one's viewpoint and sometimes on the regulatory constraints and/or the company stewardship program to be met. The concept of "green" may be met with unbridled enthusiasm, with skepticism, or even with contempt. The following thoughts on green cleaning reflect my viewpoint, as an individual who has been involved in precision and industrial cleaning activities for decades [7].*

Green Evolution

My view of green cleaning has evolved; and will no doubt continue to change. About a generation ago, in the late 1980s, I became involved in ("was coerced into" might be a more apt description) industrial and precision cleaning as a consequence of the move away from ozone depleting chemicals, specifically CFC-113 (chlorofluorocarbon 113, popularly known under the trade names Freon or Genesolv-113) and TCA (1,1,1-trichloroethane). Both were considered relatively benign to workers; they were inexpensive, plentiful, and widely used in industry. Unfortunately, given their molecular stability, they reach the stratosphere, releasing chlorine free radicals that destroy the protective ozone layer.

Both were phased out of production under the Montreal Protocol; the phaseout presented an acute problem for industry. Typical precision or critical cleaning processes were straightforward; processes depended on repeated spraying, ultrasonic cleaning, and vapor phase degreasing with CFC-113 and/or TCA. Perhaps a bit of isopropyl alcohol might be used; a final rinse/drying step in acetone was popular. Aqueous cleaning was often consigned to industrial, rather than critical, cleaning applications.

As the availability of CFC-113 and TCA decreased and the costs increased, a wide variety of aqueous and solvent-based cleaning chemistries became available. So-called nonchemical cleaning such as CO_2 snow and plasma cleaning were considered. It gradually dawned on many in manufacturing that cleaning is not a chemical, it is a process. In electronics, water-washable and so-called no-clean fluxes were introduced. Cleaning processes were tested and validated; crucial customer requirements were met; irrelevant customer requirements were negotiated away. The manufacturing world did not wither and die; it thrived.

It Is Not Easy Being Green

We helped to protect the ozone layer. In fact, the U.S. EPA "Stratospheric Ozone Protection" award is proudly displayed in my office. Protecting the ozone layer, while necessary, was not sufficient. The concept of green still eludes us. The stakes are higher; the pathway to success is more complex.

Many of the chemicals that were initially instituted as replacements for ozone depleters have themselves come under fire by safety and/or environmental regulatory agencies. We protected the stratospheric ozone layer but often adopted substitutes that impact tropospheric (smog producing) ozone. Aqueous processes were adopted, but the impact of waste streams were either not understood or not adequately dealt with. As more studies were preformed, issues of worker safety and neighborhood safety

---

* This section is based on material published in *Controlled Environments Magazine®*. Reprinted with permission.

increased. With increasing regulatory scrutiny and given the costs to develop new cleaning chemistries, finding appropriate cleaning agents has become a challenge.

Environmental and worker safety regulations are becoming increasingly stringent. The "regulatory distress" of a given cleaning agent depends on a complex blend of local, regional, and national regulations. A given chemical may be either favored or essentially banned, depending on where you alight on this planet. Effective cleaning agents that can be readily adopted are decreasing and may be site specific.

At the same time, manufacturers are faced with increasingly tough performance requirements. Miniature and micro-components as well as nano-based products all involve surfaces with exacting qualities and properties. Critical or precision cleaning activities have increased.

Perhaps, in response to this complexity, there is a growing trend in some regulatory and industrial groups toward a more holistic view. The concept of cleaning operations that utilize principles of pollution prevention has been supplanted by the concept of sustainable cleaning processes. The concept of "cradle to cradle" is supplanting that of "cradle to grave" to describe manufacturing operations.

Meeting the Regulations

The reality is that the "command and control" regulatory approach is still the norm. Given limited technical and economic resources, many companies struggle to understand, interpret, and meet a complex and perhaps conflicting set of safety and environmental regulations.

Certainly, meeting or exceeding the regulations is necessary, but it is not sufficient.

The regulations are complex and ever changing. A company may adopt what appears to be a sustainable, environmentally preferred cleaning process, only to discover within a few years that their efforts are insufficient or even counterproductive.

VOCs

For many manufacturers, green cleaning has become using a specific chemical or group of cleaning agents, or avoiding a specific chemical or group of chemicals. For example, some manufacturers have adopted all aqueous processes, or they use acetone or some other VOC-exempt compound extensively.

Acetone?

Volatile organic compounds (VOCs) produce smog, but not all VOCs are created equal. While acetone has been declared "VOC exempt" at the Federal level [8], the exemption does not mean that acetone does not contribute to smog. It does, but at a rate below a threshold set by the EPA.

As a result of the exemption, acetone has been adopted extensively in areas of poor air quality. In terms of air pollution, in some locales it is treated essentially like water, so people use a great deal of acetone. Aside from issues of materials compatibility and flammability, this means that a great deal of acetone is emitted to the air. It is, in a way, like a reduced, but not a zero, calorie cookie. If you eat enough of them, you can still gain weight. Emit enough acetone, and you still produce smog. Given the lack of availability of other VOC-exempt options that are cost-effective, reasonably aggressive, and volatile, my colleagues and I have observed significant increases in acetone use in cleaning.

Other Approaches to VOC Reduction

Other scenarios might reduce the current dependence on acetone for cleaning. All have pros and cons. Using smaller amounts of nonexempt VOCs may result in more effective cleaning, with less solvent usage, and less smog production. However, many local regulatory agencies take a dim view of this approach, preferring to show decreases in the amount of VOCs in their area. That is, a reduction in the inventory of VOCs is used as a measure of how well the agencies themselves are doing.

Another possibility is to foster aqueous cleaning agents with the VOC content "as applied" below a certain limit. This approach has been used in Southern California. In some areas and applications, the vapor pressure of the cleaning agent is considered. Not all agencies find this to be appropriate, preferring to effectively expunge nonexempt organic compounds from the list of options for industry.

Alternatively, there is the concept of relative reactivity [9], where the inherent smog-producing potential of all organic compounds is considered, whether or not these chemicals are aqueous-based or solvent-based process materials or cleaning agents. There are some moves toward adopting relative reactivity, particularly in California. At the same time, there is some concern, particularly among formulators, that using relative reactivity might result in a record-keeping quagmire. With the proliferation of spreadsheets and related programs, others see those concerns as perhaps less relevant than, say 10 years ago. Additionally, there could be an option for suppliers to use either the VOC/exempt approach (an either/or, "line in the sand" approach), or to use the more detailed relative reactivity approach.

The bottom line is that perhaps inventive approaches to how the government looks at VOCs could result in greener, more sustainable, and more productive cleaning processes.

Toxicity

Cleaning chemicals have a specific job to do—to break chemical bonds that cause soils to adhere. Since many or most soils are organic, chemicals that are effective for cleaning can interact negatively with biological entities like people. It is, therefore, no surprise that many of the most effective cleaning chemistries have toxicity.

One solution, adopted or favored by many regulatory agencies, is to ban or severely restrict the use of many toxic chemicals. This can have, however, an analogous effect as the use of acetone to reduce VOCs. That is, manufacturers might be driven to use more of a less toxic material when the smaller amount of the toxic, appropriately contained, would be both cost-effective and minimize the environmental footprint. The key phrase is "appropriately contained". There are many effective containment processes, with controls, to allow usage with low risk to nearby workers or the community.

Lowering the Environmental Footprint of the Cleaning Process

So how do you do green cleaning? What follows are a few of my own suggestions that may lower the environmental footprint of your cleaning process and move the operation toward green, sustainable cleaning.

Do Less Cleaning

By this, I do not mean to promote leaving undesirable residue on the component. Quite the contrary. Time invested in planning the product and planning the assembly process can yield benefits in decreased need for cleaning.

Product Design

In designing new products, it is typical to consider such factors as performance, cost, miniaturization, and the nature of materials of construction. Historically, the ability to assemble the product and to avoid surface contamination has not been high on the list of requirements. I find it encouraging that designers are now asking to be educated in precision cleaning, contamination control, and surface quality. Designers are collaborating with those who will actually fabricate the product. They are learning about the physical and chemical properties and limitations of aqueous and solvent cleaning agents.

Companies with the goal of using water-based cleaning agents exclusively would do well to consider that the surface tension of aqueous cleaning agents limits the spacing of components, the population density of electronics assemblies. Aqueous cleaning agents have to be rinsed with water, so the surface tension of water has to be factored in. Coordinate the product design with the anticipated cleaning process. A process may be environmentally preferred. However, an inefficient, ineffective cleaning process is not good for the environment; it is not good for product quality; it is not good for the bottom line.

What if the product has a configuration that makes aqueous cleaning impractical? I think it is prudent to understand this at the design stage and to factor in the cleaning equipment and environmental controls as part of the initial design review.

Greener Soils and Cleaning Agents

A processing agent that is essential at a given point in the fabrication process eventually has to be treated as soil, or matter out of place. It needs to be removed to achieve a low-residue surface. It stands to reason that soils that are more readily removed tend to be greener, in the sense that cleaning steps may be reduced or eliminated. The replacement of rosin flux with "No-clean" or low-residue fluxes is a historical example of substituting a soil that decreased the need for defluxing (i.e., cleaning) of electronics assemblies. Certain solder fluxes leave a tolerable level of residue for many applications. For some applications, water or water with a low level of additives is used to decrease the level of even the "no-clean" flux residue. Further, with increased miniaturization and higher performance expectations, critical cleaning of electronics assemblies is again becoming a fact of life. In such cases, green cleaning means designing and controlling processes in an effective manner.

Moving to water-soluble metalworking fluids decreases the amount of cleaning—or not. While water-soluble lubricants are inherently more readily removed by water, issues may arise as a result of machining in that the heat and forces involved in the machining process can change the metalworking fluid into something that is not readily soluble in water.

Biobased cleaning agents are being promoted as a green approach to cleaning chemicals. However, it should be noted that the word "biobased" is not a synonym for "effective." Some biobased agents are effective in a number of cleaning applications. However, they are primarily large molecules that can leave residues unless thoroughly rinsed. In addition, many biobased processes are VOCs. Further, the fact that a chemical is derived from a natural source and that it may have been used at limited concentration for decades without apparent harm to the worker or the environment does not mean we are "out of the woods" in terms of toxicity. Toxicity studies are important even for biobased materials. When plant-derived chemicals are concentrated and then used in cleaning processes along with heat and force, we have to consider issues such as worker exposure, community exposure, and waste stream management as well as the impact on the local and global environment.

Green and Document the Supply Chain Cleaning Processes

Assure that your suppliers clean the component or subassembly promptly. Prompt and effective cleaning by your suppliers minimizes adherent residue [10]. With proper testing and documentation, you may be able to eliminate cleaning steps.

Requiring that your suppliers document the cleaning processes and demonstrate effectiveness of the processes can help green your own facility, and, based on experience, it will help you to produce a better quality product, decrease failures, and ultimately will save your company money.

There is a proviso. It is tempting for manufacturers of critical products for applications like medical devices or aerospace components to specify a level of cleanliness achievable only by using very aggressive and heavily regulated chemicals. While such chemicals can be used in a nonemissive manner with minimal impact on employees, some smaller operations may not be in a position to invest in the appropriate cleaning systems. In my opinion, you are not being green if you require your suppliers to be irresponsible to their workers, their neighbors, or the environment. Instead, require that your suppliers and sub-vendors document that cleaning processes meet or exceed environmental and safety regulations.

In-House Cleaning Processes

Worker Safety, Local Environmental Concerns, and Global Environmental Concerns

Green is not the same as safe. Initiatives involving worker safety and "green" chemistry may actually be at odds with each other. A chemical may be relatively benign in terms of the impact on the individual worker, but may have a long atmospheric lifetime, and therefore endanger the planet. As recent history illustrates, chemicals that do not impact stratospheric ozone (upper ozone, good ozone) may increase tropospheric ozone or smog, and vice versa.

You have to obey all applicable safety and environmental regulations, and this can be a challenge because regulations may impel you in conflicting directions. In addition, by corporate policy, you may be restricted to certain cleaning agents.

Consider the Cleaning Process, Not Just the Cleaning Agent

Based on my experience, you cannot be green and you cannot assure the safety of the worker based on selecting a supposedly safe and/or green cleaning agent.

You have to consider the process, i.e., the way the cleaning agent is used. Concentration, temperature, time, mechanical force, all contribute to the overall safety and environmental preferability of the process. Consider rinse steps, water usage, product rework rate. A supposedly green cleaning process that puts a high proportion of your product in the landfill may not be green for your purposes.

Put Your Cleaning Process on a Diet

To me, being truly green means minimizing the impact of your process on workers and on the environment. This does not mean reusing cleaning agents in a way that could potentially contaminate the product. However, consider such approaches as

- Water conservation, recycling, and closed-loop processes
- Solvent conservation, in-process recycling, on-site recycling
- Well-contained systems
- Energy use reduction
- Design processes for containment
- Consider efficacy of cleaning

We have published or presented case studies illustrating the environmental, economic, quality, and worker safety benefits associated with containing the cleaning process [11]. Reduction in energy usage, as it applies to cleaning and manufacturing processes, has great potential, particularly in the design of cleaning process equipment [12].

Lean and Green

Lean cleaning ought to include green cleaning. Particularly in the current competitive and intense economic climate, the reality is that an inefficient, ineffective green cleaning process will not be widely adopted. It probably should not be adopted, not in any economic climate. Ideally, green and lean ought to be interchangeable, and at least some regulatory agencies are highlighting the connection. The EPA, in fact, offers "The Lean and Environment Toolkit" [13]. In addition, the EPA Partnership Programs [14] provide ways for firms, organizations, and individuals to team with the EPA to foster green or sustainable efforts. Most people are familiar with the "Energy Star" program due to home appliances with the label "Energy Star." There are many more partnership programs, including "Labs21," "Green Engineering," and "Design for the Environment."

Whether or not your company chooses to partner with a regulatory agency, the concept of lean and green cleaning should not be ignored.

Green Reflections

The concept of green cleaning will continue to evolve. Regulators explain what you have to do, and perhaps they provide guidelines as well. Because regulations will evolve, sometimes in a conflicting manner, the most reasonable approach to insulating against changing regulations is to adopt flexible processes, processes typically not dependent on a single cleaning agent. Characterizing and validating the cleaning processes and documenting acceptable residue levels will also help in coping with the need for the seemingly inevitable periodic process change, the change in response to the new regulations.

I have focused on only a few aspects of what "green cleaning" is or perhaps ought to be. Some of the many additional aspects of green cleaning include

- Minimizing the compounds with long atmospheric lifetimes
- Replacing current ozone depleting compounds
- Optimizing the energy efficiency of the cleaning process
- Minimizing water usage
- Assessing environmental persistence in soil, water

In addition, those in manufacturing, those who develop and more importantly utilize cleaning technologies have every business, every responsibility, to provide input and guidance to help green cleaning evolve.

In my view, green cleaning is not about a green chemical. Any effective chemical, including a bio-based chemical, is likely to have some sort of environmental or safety baggage. While many might argue the point, there are no ultimate green chemicals, any more than there are any "ultimate foods," foods that alone contribute to health and happiness. Ok, perhaps dark chocolate is an exception. Most cleaning chemicals can be used safely and with respect for the environment, under appropriate conditions.

Perhaps green cleaning will become more about the cleaning process, more about incorporating industrial cleaning, precision cleaning, critical cleaning—whatever we want to call it—into the overall manufacturing process. For the manufacturer, in determining green cleaning, it is critical to factor in worker safety, efficiency, process costs, and product quality. Green cleaning is a goal we will probably reach asymptotically, and I suspect that green cleaning may ultimately merge with lean manufacturing.

Moving Forward

Cleaning is not a chemical, it is a process. Although I would like to be wrong, I have concluded that an ideal cleaning agent is unlikely to be developed. We are in an era of specialization, of niche markets, of ever-changing technical and regulatory requirements. The cleaning process that is correct for your competition may not be the optimal one for you. Therefore, as you make your way through the Handbook, consider all the options, including options that may not currently be used for your application. Consider the option not to clean, but be aware that the choice to clean or not at any given step is an important technical decision.

References

1. Kanegsberg, B. Prospects for practical, innovative solvents. Joint Service Solvent Substitution Working Group (JS3) NAVAIR Washington Liaison Office. Crystal City, VA, April 27, 2004.
2. Kanegsberg, B. Perfect cleaning agents. *CleanTech05*, Chicago, IL, 2005.
3. Shubkin, R. and B. Kanegsberg. Critical cleaning of advanced lubricants from surfaces. In *Synthetics, Mineral Oils and Bio-Based Lubricants Chemistry and Technology*, ed. L.R. Rudnick, pp. 663–682. Boca Raton, FL: CRC Taylor & Francis Group, 2006.
4. Johnson, K. E. What's an ionic liquid? *Interface, The Electrochemical Society*, Spring 2007: 38–41.
5. Shubkin, R. and B. Kanegsberg. Solvent flammability basics. *CleanTech Magazine (Witter)* 3(11), November/December 2003: 17.
6. Chang, J. Director innovations and formulations, Florida Chemical Company. March 11, 2010.
7. Kanegsberg, B. Green cleaning. *Controlled Environments Magazine*, May 2009: 8–15.
8. EPA. *More Information on HCFC's*. December 8, 2009. http://www.epa.gov/Ozone/title6/phaseout/hcfcuses.html (accessed February 2010).

9. Carter, W. P. L. Development of ozone reactivity scales for volatile organic compounds. *J. Air Waste Manage. Assoc.* 44, 1994: 881–899.

10. National Manufacturing Week. Lean cleaning with your global supply chain, Half Day Workshop, Rosemont, IL, September 24, 2007.

11. Case study: Cleaning process prior to PVD of critical metal substrates. Bob Dowell, Plasma Technology; Steve Norris, Plasma Technology; Jim Unmack, Unmack Corporation; and Barbara Kanegsberg, BFK Solutions, Presentation and Proceedings, *CleanTech03*, Chicago, IL, March, 2003.

12. Costs of cleaning. Prepared for University of Massachusetts Lowell, Toxics Use Reduction Institute, Presentation, *CleanTech 2001*, Rosemont, IL, May 2001.

13. EPA. The lean and environment toolkit. http://www.epa.gov/lean/toolkit/index.htm

14. The EPA gateway for partnerships programs. http://www.epa.gov/partners/

3

Blunders, Disasters, Horror Stories, and Mistakes You Can Avoid

Introduction 29
The Missing Expert: The Facilities Manager 30
Dropped Ball • "Not to Worry" • Assumptions
We Thought of Everything, Just Build It! (*as Told to Me by a Dear Friend*)..31
The Fun Begins
We Know Exactly Where the Machine Goes, We Measured.....................32
Transportation • The Door
The Right Chemistry...32
Cleaning Solvent versus Degreasing Solvent versus Vapor Degreasing
Solvent, Whatever! • The Wrong Chemistry? The Right Technician
A Growth Spurt 33
The Case of the Mysterious Charred Solvent 34
Kanegsberg, Your Process Stopped Working 34
PSPS (Perfect Sample Part Syndrome) 35
Metallurgy? Chemistry? 35
But It Is Not on the Purchase Order! 36
A Cleaning System Costs *How* Much? We Will Build One Ourselves!
The Perfect Cleaning Machine Specification 36
Saving the Strangest for Last: Two Examples of Physics 101 37
Air Pressure • Remedial Physics?
Conclusion 38

Arthur Gillman
Unique Equipment
Corporation

Introduction

In my over 50 years of experience in parts cleaning applications, you might easily expect some stories of installations and decisions gone wrong. You would be correct. The purpose of this collection of cautionary tales is to alert the user and potential user of cleaning processes to examine areas of decision making, evaluation, installation, and operation that might, at first, seem obvious. It is not my intention to embarrass the user or anyone in the supply line. Mistakes happen. Perhaps, one or more of the listed events will alert someone on the decision team to review what could become a critical issue.

I have solicited comments and stories from trusted sources. Most were pleased to share stories, but were hesitant to be identified for fear of upsetting a relationship. Therefore, all related stories will be identified as anonymous.

The truth is that good companies with good, smart people can become too narrowly focused, make too many assumptions, rely on a trusted friend who knows almost enough, or just plain forget an important detail and wham! A disaster in the making.

The Missing Expert: The Facilities Manager

To begin, let us discuss who should be at decision-making meeting(s). Ok, you have various engineering disciplines purchasing, production, and perhaps a vice president of operations. All excellent choices but who is often forgotten that just might save the day? The facilities manager! Perhaps you call them maintenance personnel. I do not care what they are called, the folks that see to it there is enough and the right kind of electricity, water, venting, draining, floor support, door openings, and many more details must be involved, and early on!

Here are a couple of episodes that could have been avoided if only a competent facilities manager had been included.

Dropped Ball

A manufacture of lighting components needed a large capacity conveyor-type aqueous cleaning unit to produce spot-free parts. Spot free was the focus. Because of the fear of rework (spots), the company insisted on an extra heated D.I. rinse and filtered drying. Not an extraordinary request considering the importance of their focus and the cost of rework (hand wiping). However, after the order was placed, the company became even more nervous and added two more heated D.I. rinses and a longer filtered dryer. Drawings and new specifications were sent and approved. The unit was built, shipped, and a crew was dispatched for installation.

The machine arrived in three pieces and was spotted one section at a time. By the time the third section was in place, it was obvious that there simply was not enough room on the plant floor. Electricity, exhaust, water, and drains were already installed and were now in the wrong places. Why? Input from the facilities department was never included in the Engineering Change Notice (ECN). Facilities did not have the original drawings, so they could not prepare the area. It gets worse! Since part of the ECN included two added rinses and a larger heated drier, the power requirement for the added changes had never been estimated. But wait, there is more! The new electric requirement could not be altered because of wire size and contact requirements.

During that era, companies had to beg power companies to install more power, and this company had already had all the power they thought they would need. Now, there was inadequate power at the pole!

The machine was installed 6 months later, by which time several engineers were no longer employed. If only a competent facilities engineer had been consulted during the entire process, a huge mistake and tremendous cost overrun could have been avoided. As a sad postscript, the company transferred their production overseas within 6 months after installation.

"Not to Worry"

Since we are on the subject of facilities, here is another tale of horror! This tale comes from a "friendly" competitor whom I have known and respected for over 30 years. We received an order and installed an aqueous conveyor unit for defluxing printed circuit boards. The company had no water drains in their cement floors and decided to pump the drain water overhead out of the building to the available drain. This seemed to be cheaper than digging up the cement floor and adding a drain line. The height of the overhead drain line was 50 ft and the length was about 200 ft. Probably doable, wouldn't you think?

When the installation crew arrived, everything was complete and ready for equipment startup. The voltage and motor phase were checked; the water source was moved the final 5 ft to the machine and connected. The drain line was also connected, but there was concern that it might not be large enough. "Not to worry" the customer said, "our specifications said the drain water waste was 3–6 gpm, and their line could handle 10 gpm."

The machine was turned on in sections. The wash tank heated, the drier worked.

Disaster! A few seconds after the rinse water was activated and reached the drain pump, the overhead drain plumbing began to buck and whip creating a terrible racket. In seconds, the plastic (yes, plastic) pipe ruptured, and it began to rain water from 50 ft above and over a 200 ft area. The rain fell onto the electronic assembly area. There was water and mist everywhere. People were screaming and rushing for the exit. By the time the drain pump was turned off (it was mounted on a wall some distance away), it was too late. The entire plant had to be shut down.

This goes back to the issue of "competent facilities personnel." I still know companies who do not treat facilities personnel with the respect they deserve.

Assumptions

A medical device manufacturer ordered a medium size vapor degreaser with an added solvent still and automation. The voltage/phase specified was 480/3. The unit was built and delivered.

There was just one problem; the entire building was wired for 208 V. The engineer in charge was new and just could not imagine a 50,000 ft^2 building with no 480 V. Yes, the problem was corrected with a transformer, but what a wasted expense! The value of competent facility engineering cannot be emphasized strongly enough.

We Thought of Everything, Just Build It!
(as Told to Me by a Dear Friend)

Some time ago, a major military subcontractor wanted to deflux some very expensive printed circuit board assemblies. The engineers were secretive about the nature of the boards, and the vendor was advised to quote and build the cleaning equipment to specification and not ask too many questions.

Order taken, equipment built and delivered.

The somewhat unusual aspect of the equipment was that it was designed to use isopropyl alcohol in a conveyor unit that would be steam heated. The vendor gently suggested dealing with a number of issues but was told "we know what we are doing, build it to spec…" So they did.

The installation went well from the equipment point of view. The vendor left with the unit running and the company waiting for sign off approval by the fire department and company insurance carriers.

The Fun Begins

A second site visit by the equipment supplier was agreed on to provide operator and maintenance training. After a couple of weeks when the equipment supplier did not receive a call to arrange for training, the customer was called. "Seems to be a slight problem or two," they were told. The fire department would not sign off unless and until the outside wall was replaced with a breakaway wall in case of explosion. Further, on the other side of this wall was a parking lot, for executives!

Of course, it does not stop there! The customer was also informed by one of their insurance carriers that their choice of steam generator was sized such that they would be required to have a full-time "certified" operator.

Several months later, we received a call to come back for the training session. Results of the initial run were not satisfactory. A brown, ugly, and perhaps corrosive residue appeared. It quickly covered the printed circuit board assemblies and the machine. What had happened? They had changed flux between the time of the original test cleaning and production. There was so much secrecy that the team failed to act as a team. Somehow, embarrassing just does not cover the situation.

The bottom line in this horror story was that the company abandoned the entire process of cleaning with alcohol. The manufacturer instructed the equipment vendor, at great expense, to modify the new defluxing machine so that it would work with an established, nonflammable flux remover. There were significant, avoidable costs including the initial capital investment in a system for low-flashpoint solvents, engineering costs, equipment modification costs, and the costs of lost productivity. At the end of the entire exercise, they could actually clean boards and get back valued executive parking spaces. All of this happened because the electronics manufacturer insisted they knew everything; and they refused to work with the equipment vendor. Oh well.

We Know Exactly Where the Machine Goes, We Measured

Transportation

This one is a fairly simple example of a situation that happens far too often.

A critical parts manufacturer ordered a cleaning system that was configured to exact dimensions (L × W × H) in order to fit into the very limited space of the lab. Facilities (electrical, water, etc.) were installed according to the supplied drawings. All went well until the delivery truck pulled up at the dock.

You see, the facilities group at this company was different from the transportation group. It was the responsibility of the latter group to convey the machine from the dock to the lab. There was just one little problem. The lab was located in the middle of the building and the machine (which could not be taken apart without major problems) was too big to fit down any of the aisles! Of course, the transportation group was not brought in during the order process; their job does not begin until the crate is delivered. After clearing out the entire lab, cutting and welding, and reassembling, the job was finally completed with a 6 month delay. My understanding is that the transportation group is now under the direction of facilities.

The Door

A variation on the same theme was communicated to me by a longtime leader in the cleaning equipment field. He reminds all of us that part of due diligence to achieve a successful installation ought to include a walk through from the dock to the installation site. Take note of aisle space, door dimensions, and to look for any other possible obstructions. This is great advice as you will see from the following story.

An equipment supplier tried to deliver a cleaning machine only to find that the rear door was too small. He volunteered to cut a larger hole to allow access and then to install a new door once the installation was complete. What a guy! What relief! Who could ask for anything more?

Unfortunately, structural design was not one of his strong suits. After cutting a larger hole around the door, the roof fell in! Fortunately, no one was hurt. However, the business was forced to close until repairs could be completed. On balance, a walk through the plant before the cleaning equipment is delivered looks like a great idea.

The Right Chemistry

Cleaning Solvent versus Degreasing Solvent versus Vapor Degreasing Solvent, Whatever!

We received a call from a customer who claimed to need either replacement heaters or a heater controller for his vapor degreaser. They were not sure which they needed. When asked for the symptoms, we were told that the solvent was heating but not boiling. We checked the purchase records. Based on the design of the unit and the solvent being used, there should have been no heating problem. We asked if they were still using the same solvent. They replied yes, but it was a different brand. We asked what brand and formula number they had switched to. We are familiar with most solvent companies and products;

none of the information provided by the client was familiar. We asked why they had switched and were told they saved a "ton" on the cost of solvent. Now we were suspicious. We first asked them to read what was on the drum label. The label stated brand name, formula number, and the phrase "Degreasing Solvent." We were pretty certain we had the answer but asked for the MSDS. An examination of the MSDS revealed that this solvent had a boiling point of 300+ F degrees and was therefore only suitable for "cold" cleaning.

It turns out the decision was made by purchasing based on a local chemical company's assurance that the solvent strength was the same as what they were using (true) and that it was suitable for "degreasing." It was also true that the solvent strength of the two solvents was equivalent, and the new solvent could remove oils and greases. Unfortunately, the concept of "vapor degreasing" was never brought up.

The inexpensive solvent was not the right one for the cleaning system. Correcting the mistake was time-consuming and costly, and involved draining the new, supposedly economical solvent, cleaning the vapor degreaser, pointedly lecturing the customer about vapor degreasing, and purchasing the original solvent.

The Wrong Chemistry? The Right Technician

This customer had a successfully operating vapor degreaser system that they used for removing rosin flux from printed circuit board assemblies. The assemblies were used in computers and, therefore, reliability was essential. For a variety of reasons, including politics (and perhaps political correctness), they decided to switch to aqueous cleaning. They batch tested the chemistry and selected the make and model of a conveyor-style cleaning machine. During startup, the assemblies appeared to be clean and dry.

Everything looked great until an inspector ran up and announced that the wax-covered capacitors were being stripped of the wax. Not acceptable! Engineering was embarrassed because the test assemblies they had used for the evaluation did not include the wax covered component.

While the engineers convened to discuss an emergency change of either the cleaning chemistry or the component, one of our technicians looked at the board, and did another test run, this time with the heated drier off. Eureka! The wax remained intact on the capacitor. It turned out that the temperature of the drier was high enough to melt the wax off. A temperature adjustment solved the problem. Our whole installation crew was treated to lunch!

A Growth Spurt

A customer in Canada seemed to be successfully using a vapor degreaser for defluxing. We received a call from our rep requesting we do a site visit with the goal of upgrading to a large, inline solvent unit. We agreed, pleased that the customer had increased his business enough to warrant a unit of that size and expense. Our rep said, "Oh no, production has not increased at all, they just want an inline cleaning system to save solvent cost." Since they were our customer, we knew that they had a moderate-sized vapor degreaser. Moving up to a fully automated conveyor type unit to save on solvent expense did not make sense. Our rep said, "Just get up here and you will see." We suspected a massive solvent leak but made the trip anyway.

As we entered the building lobby, we could smell the solvent. Now we were really worried. Back in the cleaning area, we were genuinely shocked.

Mounted at the unload area of the vapor degreaser, and propped up at an angle so that it would drain back into the sump, was a 10 ft stainless steel ribbed counter top of the type used in commercial kitchens. A very large man wearing a mask was dunking a full basket of printed circuit board assemblies first into the boil tank, and then into the rinse tank. As the final step, he would sling the contents of the basket up onto the large, angled counter. The boards stuck to the counter. The excess solvent (a very expensive, very volatile fluorocarbon) ran back down the counter and into the vapor degreaser with a whoosh and a cloud of fumes. Well, some of it did.

After viewing this catastrophe, we adjourned to the conference room. We asked about current solvent usage. At one to two 55 gal drums per week, the solvent costs justified the purchase of an inline unit. In addition, their current level of production was beyond the capacity of their existing vapor degreaser. The new inline system was ordered and installed without incident. The customer was thrilled with the achieved level of cleanliness and with the solvent savings. Even the receptionist thanked us.

What had gone wrong? Nothing really. The capacity of the original cleaning system was well matched to the original throughput. However, over time, production crept up until a crisis point was reached. When our rep became involved, no one at the company considered production to have increased dramatically, but it had.

The Case of the Mysterious Charred Solvent

We received a local service call with the complaint that the solvent in the vapor degreaser had "charred." Not that the solvent level had gotten low and burned the bottom, the entire batch of solvents in both tanks was "charred." When we asked for a description of "charred," the customer described a solvent that was black and that contained tiny bits of carbon. The customer wanted to know what we were going to do about it.

You probably know that solvent simply does not char. But, it was a curious event, and the customer was local. When we looked at the cleaning system, it was exactly as described—black solvent with specs of carbon everywhere!

Nothing that the customer was cleaning could account for the mess. Our service technician asked when the condition was first noticed. He was told it must have happened during the night because it was first noticed at the start of the day shift. However, there was no night shift, and the cleaning system was padlocked. Hmmm.

Our service technician asked if anyone was in the plant at night. Just the janitor, he was told. Does the janitor have a key to the vapor degreaser? No.

Our technician suggested that the customer visit the plant at night. Guess what? It turns out that the janitor was removing the cover hinges from the padlocked degreasing system and using the vapor degreaser to clean motor cycle engine parts. After he was done, he would replace the cover. Case closed.

Kanegsberg, Your Process Stopped Working

(This example, also involving auto parts, was contributed by one of the editors of this book (BK). In this case, it was not a nefarious, hidden use, but a continuance of what had been accepted practice.)

In an effort to move a client from vapor degreasing to an in-line aqueous process, i.e., the defluxing process, I had compared perhaps half a dozen aqueous and semi-aqueous options. The client wanted choices. A semi-aqueous process gave the best performance. The equipment was qualified and installed with minimal indigestion, and the vendor provided the training/education sessions for the operators. This system included post-process filtration to assure that the spent rinse solutions met Publicly Owned Treatment Works (POTW) requirements.

All was serene for about 3 weeks. Then, at the beginning of the new month, early on a Friday morning, the irate messages began. "Kanegsberg, the process doesn't work worth (expletive)." "The cleaning agent is all gone, and all the post-process filters are ruined." Decorum does not allow me to convey all of the messages. Of course, I had to resolve the problem.

BK: Had anyone on the third shift evaded the training sessions?
No.
BK: Did anything unusual happen the night before?
Just the usual preparation for the first weekend of the month drag races.
BK: And just how do you prepare for these drag races?

We all bring in parts of our race cars for cleaning. The old system worked great for that and we figured since the new system worked so well with the electronics assemblies, we could continue to do a great job on our race cars.

Resolution: We explained that while vapor degreasing is very forgiving, cleaning automotive parts in an in-line semi-aqueous cleaning system is not a good idea.

P.S. Several managers also quietly assured me that they would no longer attempt to get stains off of their neckties using process equipment.

PSPS (Perfect Sample Part Syndrome)

This is an annoyingly common problem. In evaluating a new cleaning system, due diligence sometimes requires that sample parts be cleaned. However, making a decision based on sample part cleaning without factoring other requirements can lead to disaster.

A machine parts company needed a cleaning system to meet their rapidly growing production requirements. The equipment vendor performed the evaluation. The sample parts were acceptably clean, and the vendor provided a quotation. The equipment vendor was informed that the prospective customer had found a cleaning unit from a second supplier, and that the cleaning system was smaller and cost one-third what he proposed. This less-costly equipment was new; sample parts had been tested and found to be acceptably clean.

What happened was that a small batch washer had been used to do the test cleaning. Since the results were excellent, the purchase was made. The installation, which was not hassle-free, included a floor drain cut from 40 feet of cement trenching. In addition, it became apparent quite quickly that the wash sump was woefully undersized and the rinse sump even worse. After 17 min of production, the wash sump required changing, and no more than one basket of parts could be rinsed effectively before changeout. In short order, defeat was admitted and new equipment was ordered. The production manager and the project engineer were no longer at the facility.

Metallurgy? Chemistry?

A major military facility ordered two solvent vapor degreasers. They were shipped and installed successfully. Approximately 4 months later, the customer reported a serious problem. Both units were corroded and leaking. As might be expected, the customer requested a conference call. The vendor team included the engineering manager, the production manager, the quality control manager, and the plant manager. The customer team included their engineering representative along with a chemist and their plant manager.

The customer was furious and demanded to know exactly what tests were run to certify that the metal exposed to the solvent was in fact series 300 stainless steel. The vendor replied that all they ever used was series 300 stainless steel. "Yes," the customer said, "but what proof do you have?" The vendor replied that they could easily obtain the certification.

Next, the vendor asked a simple, incisive question: Exactly what, how, and when did this massive corrosion take place? The answer was that the cleaning systems ran perfectly without symptoms for approximately three and a half months, and then almost overnight began to corrode. Soon after that the leaks started. The vendor asked if the solvent had been changed. "No", the customer screamed, "you simply did not use the proper stainless steel!" The vendor asked if they had changed the brand of solvent. After some discussion, the response was, "Well, yes. The buyer found a cheaper version." The vendor suggested that while he obtained certification of the stainless steel, the customer find out if the solvent being used was stabilized and certified for use in a vapor degreaser.

Twenty-four hours later, the customer called to apologize. The replacement solvent was not stabilized and had gone acid. The case closed, but with a horrible result and expense.

But It Is Not on the Purchase Order!

The key to this story is somewhere between the obvious (but to who?) and having a sixth sense.

A very large company ordered a custom cleaning system. Several visits to the customer were made to insure that everything was perfect. When the equipment was ready, the customer, including a large contingent of engineers and executives, flew to the factory for approval. When they arrived and introductions were made, the customer was taken to the factory floor where their equipment was ready for inspection. The big boss said, "Where is our equipment?" Puzzled, the vendor said, "It is right here in front of us." The boss said, "Can't possibly be our equipment." Now scared, the vendor asked, "Why not?" The boss said, "Because it is not blue." The vendor said, "But, no color was specified on the purchase order!" The boss said, "Your team has been to our factory several times. Have you not noticed that 'everything' in this company is blue?" Now, we are at the heart of the issue. There is notice and there is "notice." The bottom line was that the vendor, at their expense, disassembled the equipment, painted all the parts customer blue, reassembled it, and a new inspection date was set. All went well after that except the profit margin. Who was at fault? Who cares? This is one of those details that are not likely to happen more than once with any given vendor.

Editor's note: One of the editors (BK) recalls a somewhat similar incident where a company evaluated five competitive pieces of cleaning equipment. She submitted a thick report, with one clear winner. However, the equipment that was purchased was the second choice in terms of expected performance, and it performed well, but not optimally. Why was it chosen? The color matched the company's decor!

Take home lesson: If you need a certain color, ask for paint chips. Ask for racing stripes, if necessary. If you are a vendor, determine if there are any esthetics involved with the down select.

A Cleaning System Costs *How* Much? We Will Build One Ourselves!

A vapor degreaser is indeed a simple apparatus. However, execution requires knowledge and experience. A local customer was overwhelmed by the quoted price. The problem was that the customer could not understand why he could not hook up a heater to a 55 gal drum, run a cooling coil around the top, and viola! A vapor degreaser that would save the company a ton of money. No amount of explanation or caution would change his mind.

About 2 weeks later, the local paper reported a fire that demolished an entire building on the same block as the customer. I could not resist, I drove out. Sure enough, it was indeed the same customer. When I arrived there were fire department inspectors on the scene. I asked if they knew the cause of the fire. They replied that it was apparently a malfunction of a cleaning machine made from a drum.

The Perfect Cleaning Machine Specification

We received an RFQ for a cleaning machine to wash, rinse, and dry 2000 lb of steel machined parts with a size range of 2″ diameter to 12″ diameter and a height ranging from 3″ to 15″.
The following requirements were specified:

- No exhaust will be allowed
- No waste drain will be allowed
- Electricity available is 208 V 1 Ph, at 30 A
- No rust will be tolerated
- No chemistry that is either acid or alkaline will be allowed
- No solvents will be allowed
- Floor space allowed would be a maximum of 36″ × 84″ × 72″
- Noise level must be <50 dB

When we tried to reach the specifying engineer, we were informed that he was out of the country for 30 days on a family emergency, and that the quote had to be submitted and approved before he returned.

We were aware that other companies had received the same specification, and we were certain that no one could respond to the quote as written. We explained to the customer that to the best of our knowledge, no existing cleaning process would satisfy all of their requirements. They replied that they could not, and would not, change anything.

With a sense of humor, we sent the company a certified letter explaining that we were working on a process that would "scan" their oily parts with an energy source causing the oil to drop off as a fine grain, odorless material onto a filter cloth below the parts. We claimed the process would take 15 s, including loading and unloading. We stated that because of the proprietary nature of the process, we would require a nondisclosure agreement to be signed by an officer of the company.

Imagine our shock when we received a letter of agreement via overnight delivery. We were forced to tell the customer that the process was simply not far enough along in the development process to quote.

P.S. The "miracle" process is still not sufficiently developed. I never did learn what they ordered.

Saving the Strangest for Last: Two Examples of Physics 101

Air Pressure

A customer in New England ordered a conveyor-style vapor spray degreaser. What was a bit strange was that they insisted that the refrigerated cooling system be air-cooled. We explained that meant more horsepower and a lot more BTUs in the room. They insisted that is what they wanted. The unit was built and installed during the winter, with no problems.

About 6 months later, we received a very unusual call. It seems that the day before, the customer had installed a sheet metal hood that attached from the output of the cooling system condenser up to the roof. The idea had been to use the heat output of the condenser to heat the room during the winter and then vent the air from the condenser out of the room during the summer. There was just one problem; they had not added any air into the room to replace what was being vented.

Because the doors to the room opened outward, the room sealed itself as soon as the machine was turned on. The doors were literally "vacuum sealed" shut. In order for personnel to come and go, the machine had to be turned off and then back on. The call to us was to find out if there was anything they could do other than remove the vent to the roof. Sorry, no.

Remedial Physics?

And finally, a lab at a major university ordered a small vapor degreaser. They planned to use a solvent with a boiling point of 165°F. After a few weeks, we received a call from the chairman of the physics department. He explained that he was using the vapor degreaser to remove fat from specimen bones. This was a very unusual but technically sound application. The problem was that the bones were overheating and the good doctor wanted to reduce the boiling point of the solvent. He explained that when he tried to accomplish this by adjusting the temperature, the boiling action slowed but the temperature remained the same.

His question: How could he lower the boiling point of the solvent? I was at a loss for words. I did not want to insult the head of the physics department by explaining that the boiling point is constant at a given pressure. I gently led him down a path to comprehension. I asked him if it would be possible to seal the room. He said that it would be but wondered what would be accomplished. I explained that it would be necessary to lower the pressure in the room in order to lower the boiling point. After a long silence, he responded, "You know, I have a PhD in chemistry and in physics and I forgot physics 101. Thank you for your help, I will search for a solvent with a lower boiling point."

There are many basics involved in parts cleaning, and the basics are easily forgotten.

Conclusion

As outrageous as these stories sound and as utterly avoidable as most of the blunders are, every one of them actually happened. The problems happened at different companies using different equipment at different times for different reasons; there were different excuses and all were avoidable.

I humbly suggest that any process that involves consideration of a capital expenditure be treated with more caution. Get more input from all departments involved, including and especially facilities. Then keep asking what has or is changing. Be prepared for the unexpected. When all is said and done, you really can only do the best you can.

If you, dear reader, would like to share a story of your own, please contact me.

4

Cleaning Practices and Pollution Prevention

Overview 39
Cleaning Activities 40
Good Operating Practices • Process Improvements
Rinsate Treatment...51
Good Operating Practices • Preliminary Treatment • Treatment
Technologies • Tracking Incidental Pollutants
Points to Remember...59
References...60

Mike Callahan
Jacobs Engineering

Overview

The Pollution Prevention Act of 1990 clearly established pollution prevention (P2) as the Nation's preferred approach to environmental protection and waste management. While prior protection and management regulations focused on hazardous waste (i.e., waste minimization), the act applies to all forms of pollution. As stated

> The Congress hereby declares it to be the national policy of the United States that pollution should be prevented or reduced at the source whenever feasible; pollution that cannot be prevented should be recycled in an environmentally safe manner whenever feasible; pollution that cannot be prevented or recycled should be treated in an environmentally safe manner whenever feasible; and disposal or other release into the environment should be employed only as a last resort and should be conducted in an environmentally safe manner.

The adoption of pollution prevention practices in the field of cleaning should follow a prescribed sequence or hierarchy of investigation. This hierarchy starts with the most preferable approach of prevention, followed by recycling and/or reuse, and ending with treatment and disposal. Specific P2 steps that follow this hierarchy are

- Eliminate the need to clean by eliminating upstream contamination or by relaxing an overly critical cleanliness requirement. Establish how clean the parts need to be.
- Modify the part or contaminant so that a less-hazardous cleaning material or method may be used. The removal of cutting oil often requires the use of a solvent. By switching from oil- to water-based lubricants, cloth wiping or water rinsing may be viable.
- Select the least hazardous cleaning materials and/or methods that achieve the required level of cleanliness. In terms of potential impact, a general sequence may be water and steam, wet or dry abrasives, aqueous cleaners, naturally derived solvents, petroleum distillates and organic solvents, and then halogenated solvents. But keep in mind that actual impacts are highly site and application specific.

- Optimize and maintain the process so that the greatest amount of soil is removed for a given amount of cleaning material consumed. Section "Cleaning Activities" discusses ways to optimize and maintain the cleaning process.
- Keep all resulting wastes segregated to promote recycling, and do not allow halogenated solvents to mix with nonhalogenated solvents and waste oils. One source of halogen contamination is the use of automotive aerosol cleaners over cold cleaning tanks.
- Recycle and/or reuse spent cleaners to the extent practical. Most solvents may be recycled via distillation—aqueous cleaners via ultrafiltration. Maintenance operations may be able to reuse the spent cleaner from critical cleaning activities, but reuse will depend on the nature of the contaminants present in the cleaner.
- Optimize and maintain the effluent handling systems. All air and wastewater discharges from the facility must comply with strict discharge prohibitions and limitations imposed by Federal, state, and local rules and regulations. Properly designed and well-maintained effluent handling systems insure compliance and achieve P2 through reduced energy and material consumption.

The following sections discuss P2 measures that apply to the cleaning process and to the treatment of rinse water from the cleaning process. Each section looks at the optimization and maintenance of the process by means of good operating practice and system improvement. Good operating practice typically involves a change in procedure, while system improvement often involves equipment modification. A discussion of ways to recycle spent solvents via distillation and aqueous cleaners via ultrafiltration is presented elsewhere in this book.

Cleaning Activities

Good Operating Practices

While much has been written about good operating practices, they are still the most cost-effective and underused means to improve overall process efficiency. The enforcement of these practices can result in reduced chemical use, reduced emissions, improved cleaning efficiency, and a safer working environment. Common practices include the clear definition of worker responsibilities; providing adequate training; following proper procedures for start-up, shutdown, maintenance, and repair; and segregating wastes to promote recycling and avoid treatment problems.

Improve Part Drainage

Drag-out, the solution that remains on parts as they are pulled out of the bath, is one of the major reasons for material loss associated with cleaning. High levels of drag-out represent an expensive loss of cleaner. This drag-out may also interfere with subsequent processing operations, thereby requiring the parts to be rinsed with clean water after the cleaning bath. Rinsing then leads to the generation of dirty rinse water and the need to treat this water so as not to discharge excessive dirt and cleaning solution to sewer.

Unfortunately, the complete elimination of drag-out is not possible. The physical shape of some parts will catch and trap fluids that can only be removed by rinsing with another fluid. This is common practice when attempting to clean bent tubing or components with narrow passages. Specific cures may be possible, but what works for one part may not work for others. Hence, the search for ways to reduce drag-out must be viewed as an attempt to achieve the optimum answer as opposed to the absolute best answer.

To find ways to reduce drag-out, it is often helpful to understand how drag-out occurs. Drag-out may be viewed as two sequential processes involving liquid removal followed by drainage. The thickness of a liquid film that clings to a flat vertical surface as it is removed from a process bath is mainly a function of the liquids inertia or resistance to flow (i.e., its viscosity) and the speed of removal. Parts removed quickly from the bath will drag out much more liquid than parts that are removed slowly.

Equations have been proposed for the modeling of withdrawal losses due to drag-out on flat metal panels, and the one developed by Kushner[1] is presented below. Kushner developed his model by means of dimensional analysis using drag-out data presented by Soderberg[2] and others. While the Kushner model may not be as robust as other more complex models, it can be used to explain many observations made in the field. The equation relates the average film thickness to the speed of withdrawal and the viscosity and density of the solution. The initial equation is

$$f = K \times \left(\frac{(U \times \mu)}{(\rho \times g)} \right)^{m}$$ (4.1)

where
   f is the average film thickness, cm
   U is the speed of withdrawal, cm/s
   μ is the viscosity of the solution, poise or gm/cm/s
   ρ is the density of the solution, gm/cm^3
   g is the gravity, 980 cm/s^2
   K, m are the constants, empirically derived

And the final equation, with the empirically derived K and m values inserted, and the terms rearranged to calculate volume rather than film thickness is

$$V = 0.02 \times A \times \left(\frac{(l \times \mu)}{(t \times \rho)} \right)^{1/2}$$ (4.2)

where
   V is the volume of liquid, cm^3
   A is the area of wetted surface, cm^2
   l is the vertical length of the panel, cm
   t is the time of withdrawal, s

The major point to note in Equation 4.2 is that for a given fluid held at constant temperature and concentration, withdrawal losses are a function of wetted area, vertical length, and the time of withdrawal. The slower a part is withdrawn from a bath (i.e., as t increases), the less the liquid will cling. For flat parts with no recesses or crevices, a slow rate of withdrawal will minimize drag-out and reduce the need for prolonged draining. The issue of withdrawal time versus drainage time will be discussed shortly.

To illustrate the impact withdrawal rate plays in the determination of initial drag-out, data presented by Soderberg may be employed. Soderberg reported a drag-out value of 10.8 mL for a zinc cyanide plating solution clinging to a 4 in. wide by 12 in. long steel panel. The panel was removed from the bath at a rate of 0.5 s/ft (or 2 ft/s) and allowed to drain briefly for 0.5 s. The 10.8 mL value equates to a drag-out of 4.28 gal of solution per 1000 ft^2 of wetted surface.

Knowing the volume of solution removed and the time of withdrawal, and holding all other variables constant (i.e., wetted area, vertical length, viscosity, and density), Equation 4.2 may be rearranged as shown and the system constant C determined:

$$V = C \times \left(\frac{1}{t} \right)^{1/2}$$ (4.3)

$$C = V \times (t)^{1/2}$$ (4.4)

For a drag-out value of 4.28 gal per 1000 ft^2 and a withdrawal time of 0.5 s, a system constant (C) of 3.0 is derived. Equation 4.4 may now be used to estimate drag-out losses as a function of withdrawal time. For a withdrawal time of 10 s (or a speed of 0.1 ft/s for the 1 ft long panel), drag-out is estimated to be 0.96 gal per 1000 ft^2. This represents a 78% reduction in drag-out compared to the base case.

Increasing withdrawal time and reducing withdrawal speed may be achieved by reprogramming the conveyor system or by changing gears in the hoist. Manual operations are most difficult to control since few operators are capable of slowly lifting a heavy part out of a bath. Not only is this act uncomfortable, it is unsafe. The weight places a tremendous strain on the back of the operator increasing the potential for injury, and the added time increases worker exposure to heat and fumes. To avoid such risks, the operator should be provided with an overhead hoist to raise and lower the parts into and out of the bath.

Following withdrawal, drainage is the next part of the process that determines the overall liquid loss due to drag-out. All of the previously mentioned physical factors apply, but now the surface tension of the liquid also has an effect. Surface tension may cause the liquid to be retained within crevices and small holes and openings. To overcome the effect of surface tension, wetting agents may be added to the cleaner and rinse water.

To illustrate the effect increased drainage time plays in the determination of overall losses, the drag-out data developed by Soderberg are again employed. Table 4.1 lists three sets of data representing 0.5, 2, and 10 s of drainage following various withdrawal rates. The 2 and 10 s drainage data are from Soderberg while the 0.5 s data are derived from Equation 4.4. (Note that the original data were reported in units of milliliter but have been converted to units of gallons per 1000 ft^2.)

The data shows that increased drainage time can have a pronounced effect dependent on the initial amount of drag-out present (i.e., dependent on the time of withdrawal). Drainage time is most critical when parts are removed quickly. When parts are removed slowly, the amount of liquid initially removed is minimized and drainage time is of less importance. The data shows that similar amounts of drag-out occur for a withdrawal time of 0.5 and 10 s of drainage (10.5 s overall), as it does for a withdrawal time of 5 and 0.5 s of drainage (5.5 s overall). The benefit of long drainage times is negligible when flat panels are removed at a rate slower than 4–6 s/ft.

TABLE 4.1 Effect of Time of Withdrawal and Drainage Time on Drag-Out

Time of Withdrawal (s/ft)	Drag-Out (gal/1000 ft^2) at Drainage Time (s)		
	0.5 s[a]	2 s[b]	10 s[b]
0.5	4.28	2.32	1.35
2	2.14	2.12	1.31
4	1.51	1.68	1.26
6	1.24	1.21	1.17
8	1.07	1.10	1.10
10	0.96	0.86	0.86
15	0.78	0.69	0.69
20	0.68	0.60	0.60

Source: Callahan, M.S. and Green, B., *Hazardous Solvent Source Reduction*, McGraw Hill Publishing Co, New York, 1995.

[a] Predicted drag-out values derived from the Kushner equation. See text for explanation.

[b] Drag-out values derived from Soderberg for a zinc cyanide plating solution draining from a 4-in. × 12-in. panel.

TABLE 4.2 Effect of Part Configuration and Drainage on Drag-Out

Part Configuration	Drag-Out Loss (gal/1000 ft²)
Vertical, well drained	0.4
Vertical, poorly drained	2.0
Vertical, very poorly drained	4.0
Horizontal, well drained	0.8
Horizontal, very poorly drained	10.0
Cup-shaped, very poorly drained	8–24+

Source: Soderberg, G., Drag-out, *Proceedings of the Twenty-Fourth Annual Convention of the American Electroplaters Society,* Cleveland, OH, 1936, vol. 24, pp. 233–249.

However, few parts are truly flat and draining free. The real-world importance of drainage time may be seriously understated in the work conducted with flat panels. Liquid may pool on a flat horizontal surface and be carried out of the bath. Hollow recesses may fill with liquid that becomes trapped. Such parts may require rotation after withdrawal to allow the trapped liquid to drain. Soderberg investigated the effect part configuration plays in the determination of overall losses due to drag-out. His often quoted estimates of drag-out for various part configurations are presented in Table 4.2.

By following proper racking practices, drag-out can be reduced by 50% or more. Proper racking may involve the orientation of parts inside a basket so as not to collect and retain liquid. Drinking glasses placed inside a dish washer should always be placed upside-down so as to promote drainage. Dishes are best cleaned by stacking them vertically. The cleaning of industrial parts is no different. For complex shapes, the determination of the optimum racking position will likely require experimentation.

Modification of the part to promote drainage (e.g., by adding several drainage holes) may be a possibility, but tends to be a limited solution in practice. Bigge and Graham[3] present an example where 0.25-in. holes were added to the ends of a "wrap-around" automobile bumper. The hole in the lower bend allowed cupped solution to drain while the upper hole allowed trapped air to vent from the underside of the bend. Another example was the replacement of a hollow channel fabricated from rolled steel with a die cast channel. The die cast channel, being solid, eliminated the problem of solution entrapment.

In addition to good operating practice, process improvements may be employed to reduce the amount of drag-out leaving the tank. Drainage boards should be installed between tanks so that the drag-out that drips from the parts is captured and returned. Drag-out can be knocked off parts by means of burp bars, air knives, or wipers. Air knives are useful in cooling very hot parts that must be subsequently handled, but they may also create problems due to the drying of solution prior to rinsing, increased air emissions, and soiling if the air source is not oil free. For parts that undergo rinsing, a low volume water spray placed over the cleaning bath can be used to capture and return solution to the bath. While each part may represent a unique challenge to minimizing drag-out, the guidelines presented in Table 4.3 have been found to be effective.

Monitor Cleaning Performance

The rigorous monitoring of cleaning performance is a given practice in critical cleaning. Many techniques and devices have been developed to ensure critically clean parts. Some of the surface analysis techniques capable of detecting contamination at the microgram per square centimeter level include Auger electron spectroscopy (AES), electron spectroscopy for chemical analysis (ESCA), Fourier transform infrared spectroscopy (FTIR), secondary ion mass spectroscopy (SIMS), and optically stimulated electron emission (OSEE).[4] While suitable for a well-equipped and staffed lab, these techniques are too expensive and too complex for the majority of cleaning operations.

TABLE 4.3   General Guidelines to Minimize Drag-Out

Favor slower withdrawal speed over longer drainage time. For a fixed time cycle, spend two-thirds of the time withdrawing the part from the bath and one-third of the time draining over the bath

Drain parts over the tank or use drainage boards to extend the effective length of the tank. Drainage boards may be installed inside manual tanks to serve as a scrubbing table. Make sure drainage boards slope back toward the tank so that drained liquid is returned

Do not rack parts directly over one another. Drippage from the top parts may dirty the parts below and cause spotting or staining

Orient part surfaces as close to vertical as possible. Rack parts with their lower edge tilted from the horizontal. Run-off should be from a corner rather than an entire edge. Burp bars may help to knock solution off the part after removal from the tank

Ensure that all cavities and recessed pockets are oriented downward. Rotation of parts during withdrawal and drainage may help

Use the correct size and type of basket so as to minimize the wetted surface area of the basket. A wire mesh basket can retain much liquid if the mesh is too tight. Never use porous materials such as ropes or cloth bags to hold the parts

Source: Callahan, M. S. and Green, B., *Hazardous Solvent Source Reduction*, McGraw Hill Publishing Co, New York, 1995.

Instead, maintenance and industrial cleaning operations often rely upon a variety of methods that infer cleanliness rather than measure the level of contamination directly. What these methods lack as to sensitivity is compensated for by their low cost and ease of implementation. One of the simplest means of monitoring cleaning performance is to run test coupons through the bath and note any change in cleaning efficiency. Coupon tests are often used to rule out the cleaning bath as a source of trouble following a sudden increase in rejects.

Direct visual inspection and paper tissue wiping are two widely used methods to denote cleaning effectiveness. The tests are limited to visible soils (i.e., grease, soot, and particulate matter), and the use of a microscope by a trained individual increases the validity of the test. The paper wipe method is sensitive to the pressure applied while wiping, and the method is most sensitive when performed on a wet surface.

Water spray atomization (ASTM F-21-65 (2007), the Standard Test Method for Hydrophobic Surface Films by the Atomizer Test) and water break (ASTM F-22–02 (2007), the Standard Test Method for Hydrophobic Surface Films by the Water-Break Test) are two quick and effective ways to determine cleanliness. The methods are based on the observation that water will not bead on a clean surface, it will form a continuous film. Cleaned parts may be dipped into water and the flow behavior of the water as it drains from the part, noted. In the spray method, the formation of beaded areas denotes the presence of hydrophobic (water-fearing or oil-based) soils. Both of these methods are most effective when performed on flat parts and by trained observers.

The paint adhesion method (ASTM D3359-09, Standard Test Methods for Measuring Adhesion by Tape Test) is based on the observation that an improperly cleaned part will not provide a good surface for adhesion. Clean, dry parts are painted and allowed to cure for a specified length of time. A baking cycle may be employed to speed curing. The paint is then crosscut with a razor blade and a piece of adhesive tape placed over the area. The cuts should be of equal width, forming a checkerboard pattern. The depth of cut should be down to the substrate. The tape is lifted off at a steady rate of pull and the number of squares remaining on the part serves as an indication of cleanliness.

Gravimetric measurement involves the soiling of test panels with a known amount of contaminant and then weighing each panel before and after cleaning. Cleaning effectiveness is reported as the weight or percentage of contaminant removed. The direct measure of removal has good sensitivity, but it does not truly indicate surface cleanliness. A residual film of cleaning solution, the presence of metal and cleaner reaction by-products, or metal removal/etching by the solution can all affect weighing and skew the results.

For determining the removal efficiency of water-soluble ionic contaminants, the cleaned parts may be rinsed with deionized water and the water monitored for a decrease in resistivity. This method is commonly used to test soldered printed circuit boards for the removal of ionic flux. Deionized water rinse and resistivity measurement may also be performed after the water break test to check for ionic surfactant contamination due to residual cleaner.

Monitor Bath Quality

In addition to monitoring soil removal efficiency and cleaning effectiveness, there are a number of measurements that may be used to directly monitor bath quality. Such measurements include soil loading and acid acceptance for solvents and alkalinity for aqueous cleaners. Since bath quality is monitored by taking a bath sample directly, as opposed to performance that is based on the condition of the parts after cleaning, these tests may show up potential problems before an increase in rejects occurs.

Solvent Monitoring

Field test kits have been developed by the military for monitoring soil loading in maintenance cold cleaning baths.[5] The military found that leaving the decision to replace the solvent up to the operator either resulted in excessive solvent use or questionable cleaning performance. Some operators would tolerate a very high level of contamination while others would change-out the solvent as soon as they noted a slight decrease in the rate of cleaning. The intent of developing a simple field kit was to place the decision for solvent replacement on a more quantifiable and consistent basis. The most reliable combination of monitoring tests included light transmittance, electrical conductivity, and specific gravity.

For solvents used in vapor degreasing, soil loading may be determined by a change in specific gravity, by gravimetric analysis, and by monitoring the boiling sump temperature. High soil loading often requires an increase in temperature to maintain adequate vapor generation. For industrial cleaning, vapor degreasing solvents may be replaced when they become loaded with 25% or more oil.[6] This loading typically equates to a 6°F (3.3°C) increase in boiling point. For critical cleaning applications, a solvent distillation unit may be employed to prevent oil from building up in the boiling sump.

Solvents used in vapor degreasers need to be routinely checked for acid acceptance. Both 1,1,1 trichloroethane and methylene chloride are highly susceptible to breakdown in the presence of catalytic contaminants such as aluminum or zinc fines. Other causes include the degreasing of wet parts and failure to routinely remove lost metal parts, oils, fines, and sludge from the boiling sump. When solvent break-down occurs, hydrochloric acid is formed that may etch metal parts and quickly cause extensive equipment damage. Acid formation is also a potential problem with vapor degreasing solvents that employ normal propyl bromide or nPB.[7]

To prevent solvent breakdown and acid formation, stabilizers are added to the virgin solvent by the producer. A number of causes may result in stabilizer loss, and the most direct way to determine stability is to perform an acid acceptance test (ASTM D2106-07e1 Standard Test Methods for Determination of Amine Acid Acceptance (Alkalinity) of Halogenated Organic Solvents). This test should be performed monthly or quarterly as conditions warrant. Users of perchloroethylene and trichloroethylene should also perform acid acceptance tests if the solvent is reclaimed or used for an extensive length of time.

Acid acceptance is typically reported as the milligrams of potassium hydroxide needed to neutralize the acid in 1 g of solvent. The titration is often performed on a water extract of the solvent (i.e., the solvent is mixed with an equal volume of water and the separated water phase is then titrated to determine acidity). A water extraction test can only be performed on a water-immiscible solvent such as a halogenated solvent or petroleum distillate.

While much less of a problem, acid formation may also occur in a nonhalogenated solvent. The exposure of the solvent to air and failure to routinely remove sludge from the bath may lead to the formation of organic acids. Many of the soils typically encountered in an industrial setting are acidic, and these too can lead to acid buildup. One way to determine the acid level of a solvent is to mix a solvent sample with water, separate the water extract, and then measure the pH of the extract. A standardized procedure for this test is available (ASTM D2110-00 (2006), the Standard Test Method for pH of Water Extractions of Halogenated Organic Solvents and their Admixtures).

To monitor the effectiveness of particulate removal, the solvent used for cleaning may be passed through a membrane filter and a particulate count performed on the retained matter. This method can also be used for testing the particulate removal efficiency of aqueous cleaners. Standardized procedures for these tests are available (ASTM F311-08, Standardized Practice for Processing Aerospace Liquid Samples for Particulate Contamination Analysis Using Membrane Filters and ASTM F312-08, Standard Test Methods for Microscopical Sizing and Counting Particles from Aerospace Fluids on Membrane Filters).

Aqueous Cleaner Monitoring

The quality of an aqueous cleaner is commonly monitored by measuring alkalinity or acidity by titration. Since many alkaline cleaners have a reserve capacity for producing hydroxide ions, titrations often involve determining both the free and total alkalinity. Titration is performed by first adding a few drops of indicator to a fixed volume of sample, and then adding a standard acid solution until a color change takes place. The amount of acid added is a measure of free alkalinity. A second indicator is then used and the acid addition repeated to determine total alkalinity.

By comparing free alkalinity to total alkalinity over time, changes in the quality of the bath can be monitored. Changes in the ratio of free to total alkalinity may be due to loading of the bath with acidic soil, excessive drag-out, the use of hard water for makeup, and failure to make proper additions. Liquid-based cleaners are often favored over dry or powder-based cleaners because they are easier to handle and they ensure better mixing when added to the bath.

The performance of an aqueous bath is a complex function of four key variables: concentration, temperature, time, and agitation. The failure to maintain and control all four variables can result in poor cleaning performance. By increasing any one variable, cleaning performance can often be improved and a problem corrected, but there are limits. Overall performance may actually decline if certain variables are pushed beyond the limit.

Increasing the concentration of the cleaning agent can improve the rate of cleaning, increase soil loading, and reduce the need for frequent additions. On the negative side, a high concentration of cleaner can result in tarnished or etched parts and increased drag-out due to an increase in bath viscosity, and if the cleaner is difficult to rinse off, rejects may increase as the cleaner now becomes a soil in subsequent processing operations. In some critical cleaning operations, where the amount of soil to be removed is slight, using a very low concentration of cleaner offers the best overall performance.

As with increased concentration, the cleaning rate may be improved by increasing the bath temperature. Solid soils, such as fats and waxes, must be heated above their melting point before they can be removed. A higher operating temperature can often be used to improve a poorly operating bath. However, at too high a temperature, the parts may warp or soils may set, making them harder to remove. A higher operating temperature also results in increased energy use and evaporative loss. This may lead to a quicker buildup of solids in the bath if hard water is used for makeup.

Finally, close monitoring of the cleaning bath can often help identify potential problems before they show up. One can determine if unknown dilution is occurring by monitoring bath strength. This is a common problem with poorly maintained automatic makeup controls. A slow leak of water into the tank results in the loss of cleaner to drain and in poor cleaning performance. By routinely checking bath strength both before and after a period of inactivity, such as a weekend or holiday, problems with makeup water valve leakage can often be identified.

Housekeeping and Loss Prevention

The inadvertent spillage of chemicals represents a direct loss in process efficiency and often results in the generation of hazardous waste. While efforts should be made to recover as much of the spilled material as possible so that it may be used for its intended purpose, this is not always possible. The recovered material may be too contaminated for use or it may now be in a form too difficult to recover. Spills and leaks represent a decrease in process efficiency since more material must be purchased for a given

TABLE 4.4 General Housekeeping and Loss Prevention Practices

Conduct periodic inventories and use materials on a first-in first-out basis. The use of off-spec material can result in poor product quality and reduced bath life

Conduct periodic sampling of supplies held in bulk storage to confirm that they still meet specification. Potential contaminants include water due to atmospheric moisture or rain and iron due to storage tank/bin corrosion

Improve storage facilities by converting fixed roof tanks to a floating roof (less product loss due to evaporation and less potential for contamination by rain or moisture) and by replacing rusted storage units with plastic or lined containers

Maintain the physical integrity of storage tanks over time. Perform monthly visual inspections of all tanks and vessels, especially the weld seams. Take corrective action at the first sign of weeping or corrosion. Routinely test level sensors and overflow alarms

Install sufficient secondary containment to hold the stored materials in the event of equipment failure or accident

Store materials inside the shop or provide outdoor covered storage. Area should be well lit and secure. Provide dikes and berms to contain material in the event of a spill. A concrete pad is better than asphalt and both are better than bare soil. Hold drums on spill pallets

In the event of a spill, avoid the urge to grab a hose and flush it away (unless there are safety reasons to do so). After containment, try to recover as much of the material for reuse and then use dry absorbents

amount of cleaning. In addition to the costs associated with raw material procurement, cleanup, and waste disposal, serious leaks and spills may result in fines due to regulatory enforcement actions.

While the potential for leaks and spills can never be completely eliminated, there are numerous ways in which this potential can be reduced and mitigated. One way is to implement and support a strong maintenance program, whether preventive or predictive. Good maintenance programs can be instrumental in cutting production costs stemming from expensive equipment repairs, excessive waste generation and disposal due to upsets, and business interruptions. Table 4.4 lists some ways to avoid waste generation through good housekeeping and loss prevention practices.

Material Substitution

Material substitution seeks to eliminate the use of hazardous cleaners by switching to a product that is less hazardous but equally effective. A substitute cleaner, regardless of how "safe" or less hazardous it may be, is not viable if it does not achieve the desired level of clean or if the amount of cleaner consumed is excessive. Both complaints are often heard when a shop tries to make a simple "drop-in" replacement. Few, if any, substitutions are truly "drop-in," and the new product is bound to fail in a system optimized for the old product. Adequate planning, training, and good product support are key elements to successful material substitution.

Table 4.5 presents a listing of some material substitutions that have been successfully made in the metal finishing industry. Concerns over worker health and safety, wastewater treatment costs, regulatory compliance costs, and potential liability are prompting many sites to investigate these practices. Major limitations include the need for tighter process control, worker retraining, and issues of product quality. However, with management commitment, many limitations of the past have been successfully overcome and the development of new practices is expected to continue.

Process Improvements

Process improvements often involve the physical modification of the existing equipment or the installation of additional equipment. Many of these methods can provide a substantial increase in process efficiency, but the condition of the existing equipment and/or design of the facility may prohibit their use. In these cases, replacement of the system with a new one may be a more viable and cost-effective approach than trying to retrofit the system. Some examples of cleaning process improvements that have been shown to be successful in reducing rejects and the amount of waste generated are listed in Table 4.6. These examples serve as a brief overview of the discussions that follow in this section.

TABLE 4.5 Material Substitution Practices in Metal Finishing

Non-cyanide-based copper plating solutions are available for steel, brass, lead-tin alloy, zinc die cast, and zincated aluminum. Cyanide-free metal stripping solutions are also available. By eliminating cyanide, the need for two-stage chlorination during rinsate treatment is avoided

Cadmium can be replaced by electroplating with zinc alloy or by the ion vapor deposition of aluminum. Since cadmium solutions contain cyanide, conversion eliminates two undesirable pollutants

Hexavalent chromium is used in hard plating and substitution is difficult due to physical performance requirements. Some decorative plating has switched to the use of trivalent chromium and new chrome spray painting systems have entered the market

Hexavalent chromium conversion coatings are widely applied to magnesium, aluminum, and zinc. Sulfuric acid anodizing has been tried as a substitute, but the coating is thicker and more brittle, which may be an issue depending on the application

Many of the aqueous cleaners and degreasers once used by aerospace and the military contained hexavalent chromium as a corrosion inhibitor. Due to health and safety concerns, non-chromate versions of these products have been developed and military specifications (MIL-Specs) have been revised to allow for their use

Halogenated solvents were once widely used for vapor degreasing. Many shops have switched to alternative nonhalogenated solvents or to aqueous cleaners while some have installed zero-emission systems. In the practice of cold cleaning, some local air quality management districts have forced total conversion to aqueous cleaners

TABLE 4.6 Examples of Cleaning Bath Process Improvements

Use demineralized water to replenish evaporative losses. Hardness, salts, sulfates, and chlorides contained in tap water can increase over time and lead to rejects. For a heated tank subject to tight control, hardness limits may quickly go out of bounds due to the use of tap water

Improve the heating system to avoid localized hot spots. Overheating can lead to the formation of degradation products and an increase in heater coil fouling. Once fouled, the tank must be taken out of service because the coil cannot provide sufficient heat to keep the tank at temperature

Periodically check the steam heating system for leaks. Process chemicals may enter the condensate lines during shutdown, resulting in contamination of the discharge. Steam may leak into the tank diluting the cleaner

Routinely test bath to ensure that all constituents are within their proper operating range. Very low concentration levels may indicate excessive drag-out, a stuck open makeup water valve, or a steam heating coil leak. For facilities without a lab, the solution supplier may offer testing

To prevent overfilling a bath with makeup water, automatic control valves and high/low level alarms should be installed. Solid-state sensors are preferable to float valves because they are less prone to salt buildup and fouling

Agitation is used to keep soils suspended and prevent their deposition onto the parts. Air agitation is widely used but has limitations. Compressed air seldom provides adequate agitation, it increases the rate of heat loss, increases evaporative loss, and may contain oil. Mechanical agitation via in-tank pumps is favored over air agitation

Use Demineralized Water for Makeup

Tap water is often added to a cleaning bath to compensate for losses due to evaporation and drag-out. Since tap water contains dissolved minerals that remain behind when the water evaporates, these minerals may precipitate and form solids if their concentration gets too high. One way to avoid this problem is to accept a high amount of drag-out, so as to purge the bath of collected soil and solids, but this practice is wasteful.

Another approach is to use an aqueous cleaner that is formulated for hard water. These cleaners contain chelators that tie up and prevent the minerals from forming solids. The use of chelators to prevent precipitation is a good approach as long as the chelators remain active and do not create a problem elsewhere. Chelators may complex with heavy metals such as lead and make their subsequent treatment more difficult.

In addition to chelators, sodium hydroxide (caustic soda) is a common component of aqueous cleaners, and it will react with minerals in the water to form solids. It has been reported that as much as 10%–25% of a caustic cleaner may be consumed by this softening effect.[8] This leads to more frequent bath replacement and higher operating costs. To avoid the consumption of cleaner due to reactions with minerals, and to avoid the need to use a chelated cleaner, the tap water should be treated to remove dissolved minerals prior to use. Viable ways to remove the dissolved minerals include reverse osmosis (RO) and ion exchange (IX).

The use of demineralized water for bath makeup and for rinsing may also benefit the wastewater treatment system. Much of the hazardous sludge produced by a treatment system consists of nonhazardous minerals such as calcium and magnesium hydroxides. These minerals enter the system by way of the raw water supply. Chemical analysis of wastewater treatment sludge from shops using tap water often shows the dry sludge to consist of 80% or more calcium and magnesium minerals. By removing these minerals from the water before it is used, the volume of hazardous sludge generated during rinsate treatment is reduced.

Increase Tank Agitation

Mechanical agitation, either in the form of a liquid jet pump, spraying, basket rotation, basket raising and lowering, ultrasonic cavitation, or manual brushing is an effective way to increase cleaning efficiency. Many solvents used in cold cleaning are replaced when the speed at which they clean is no longer acceptable to the operator. This may correspond to a contaminant level of 2%–3% oil by volume while by providing mechanical agitation, soil loading levels as high as 10% can be tolerated.[9] Disadvantages of mechanical agitation include higher electrical costs, more equipment to maintain, and increased worker exposure depending on the method employed.

One development in the field of tank agitation is the use of small, seal-less, in-tank pumps. By placing the pump inside the tank, energy losses due to piping pressure drops and the potential for solution spillage due to pipe or seal leakage is eliminated. Reusable filter units may be attached to the pump inlet so as to provide solution filtration and agitation at the same time.

To achieve an even greater level of agitation, ultrasonic agitation may be considered. This technique uses high frequency sound waves to create a cavitation effect in the liquid. Small vapor bubbles are formed in the liquid by the rarefaction (low pressure) sound waves and the subsequent pressure waves result in bubble collapse. Very high pressures and temperatures are created within the bubbles as they collapse and implode. This cavitation effect blasts the soil away from all surfaces of the part in contact with the liquid.

Ultrasonic agitation is more difficult to implement on an existing tank than in-tank agitation. Ultrasonic transducers must be tightly coupled to the tank to minimize mechanical stress and tank wall erosion. Holding fixtures are often prone to attack due to repeated exposure. For a new system, erosion and deterioration due to ultrasonic agitation is directly dependent on weld and fabrication quality. System life should be at least 5 years for properly built equipment. The cost for an ultrasonic system varies widely depending on tank size and level of control. Small tabletop- or desk-sized systems may be purchased for less than $1000 (no pumps, filters, etc.).

Parts fabricated from materials that are poor conductors of sound may not be effectively cleaned in an ultrasonically agitated system. Instead, they may absorb the sonic energy and deaden the effect. The same is true for the materials used to fabricate baskets and part holders. Parts held in a glass beaker are effectively cleaned since the glass walls of the beaker will transmit the sonic waves. With a polyethylene beaker, the intensity of the sound will be attenuated or reduced by 25%–50%. Similar effects may be noted with metal mesh baskets. Fine mesh (300 mesh) and coarse mesh baskets allow the sound to pass while medium mesh (40–60 mesh) baskets tend to absorb the most energy.

The reader should note that the use of air sparging to agitate the bath was deliberately excluded from the discussion. While sparging is relatively easy to implement, there are not many benefits to support its use. Sparging is an inefficient mixing method and it does not provide the level of agitation needed to insure complete mixing. The injection of air beneath the parts can lift them off the rack. Entrapped air can prevent the cleaning solution from reaching recessed areas, hence increasing rejects. And if the air is not oil-free, the parts can be soiled, which defeats the purpose of cleaning.

Air agitation also results in very high operating costs for energy and water. Cleaning chemical usage and spent bath treatment costs can also increase. The air leaving the surface of the bath is saturated with water vapor that carries away a large amount of heat. Extra energy must be used to replace this heat and to heat the incoming makeup water. As more water is added, the mineral impurities increase, consuming the cleaner. Air sparging is a common practice best left in the past.

Remove Soils from Bath

Common ways to remove solid soils from cleaning baths include gravity separation, decantation, centrifugation, and filtration. These techniques are employed in both solvent- and aqueous-based systems to remove solids and extend bath life. The process of filtration removes the insoluble particulate matter from a fluid by means of entrapment in a porous medium. It is often used to extend the life of a cold cleaning bath or to continuously remove metal fines and sludge from a vapor degreaser sump. Common styles include a bag and disposable cartridge.

The expanding use of ultrafiltration and microfiltration allows finely emulsified oils to be removed from aqueous cleaning solutions. With solvents, no degree of filtering will remove dissolved soils from the bath. To recover a clean solvent, the spent bath may often be recycled via distillation. The condensed solvent vapor, now free of oil, may undergo further filtering to remove any particulates carried over in the vapor.

While standard filtration does not remove soluble contaminants such as dissolved oils, it can be used to remove insoluble dirt and semisolid grease. Passing the dirty solvent through a fine metal screen may remove the grease before it has a chance to dissolve and load the bath. Routine screening is an effective way to extend the life of a cold cleaning bath.

A microfiltration system can remove soil to a much finer degree than standard filtration. Most often, a vapor degreaser is equipped with a standard 5 or $10\,\mu m$ filter. Large particles are trapped in the filter but smaller particles pass through and collect in the sump where they will eventually contaminate the solvent vapor. A microfiltration system can remove particles down to less than $0.1\,\mu m$ in size that minimizes the potential for carryover. Water and organic acids may also be removed by the system.

Moving beyond microfiltration, membrane and ultrafiltration can remove emulsified oil and grease from an aqueous cleaning solution.[10] Organic molecules with a molecular weight of 500 or more, and solid particles as fine as $0.01–0.003\,\mu m$, can be readily removed. Membrane filtration is sometimes so effective that it will remove surfactants and other additives from the cleaner. Thus, cleaner selection for a given system often requires prior design knowledge and/or pilot testing.

Employ Two-Stage Cleaning

Two-stage countercurrent cleaning can reduce the amount of solvent or cleaner used per unit of production. The process involves soaking the parts in a tank of dirty solvent followed by rinsing and cleaning the parts in clean solvent. The dirty solvent removes most of the oil and grease and the clean solvent ensures that the parts are clean. When the clean solvent reaches some level that triggers change-out, the dirty solvent is removed and replaced with the formerly clean solvent. The now empty clean-side tank is refilled with fresh solvent. The example below shows the level of reduction possible by converting a single-stage cold cleaning tank to a two-stage one.

Assume that a 100 gal tank of solvent is disposed of weekly with 2% oil. At this level, the parts will drag out enough oil to be considered too dirty for subsequent use. The parts are small so that the existing tank can be divided into two 50 gal tanks by welding in a divider. With the new system, the parts are soaked in the dirty tank where 80% of the oil is removed. This is followed by a rinse and spray flush with the clean solvent.

Given an oil loading rate of 2 gal/week and 80% removal in the dirty stage, it will take 2.5 weeks for the clean solvent to reach 2% oil. After disposing of the dirty solvent and transferring the clean solvent into the dirty stage, the clean stage will then be filled with fresh solvent. To provide a margin of safety, the facility elects to change-out the dirty stage every 2 weeks. This means that 50 gal of dirty solvent will be generated every 2 weeks compared to the previous 100 gal/week, a reduction of 75%.

TABLE 4.7 Water Conservation Measures

Select the minimum-size tank in which the parts can be rinsed and use the same size tank for the entire process line, where practical

Locate the clean water inlet below the surface and at the opposite side of the rinse tank, away from the outlet to avoid short-circuiting

Use solid-state level sensors to control makeup water rather than a float actuated switch since they are less prone to fouling. Routinely check bath strength to make sure that the supply valve is not leaking

Use multiple rinse tanks in a counter-flow configuration. Compared to a single stage rinse, a counter-flow rinse can substantially reduce water use. Look into reusing the spent rinse water as process bath makeup. Ways to estimate the savings achievable via different rinse tank configurations are available[16]

Install flow-restrictors on water supply lines to limit the discharge. Stop water flow during idle time by use of production-activated controls, conductivity probes, and/or timers

Investigate the use of ion exchange (IX) systems to remove soluble metals from rinse streams so that the water can be reused. Offsite regeneration of the IX resin may also allow the site to reduce or eliminate the need for rinsate treatment

For maintenance cleaning activities, install shutoff spray nozzles on the ends of all hoses. Workers who have to walk back and forth from the end of a long hose to the shutoff valve are likely to let the hose run. In a case where the resulting runoff must be treated, the savings in avoided treatment costs will be substantial

Adopt Water Conservation Measures

Water conservation measures reduce the volumetric load placed on the wastewater treatment system, increase the potential for reuse of recovered drag-out, and make it more practical to use demineralized water for all process baths and rinses. Some locations may be required to adopt water conservation measures as an operating permit condition. No longer can a facility use large quantities of fresh water as a means of insuring properly rinsed parts. Table 4.7 presents a brief listing of some measures that can be adopted to conserve water.

Rinsate Treatment*

Pollution prevention, as defined by the Pollution Prevention Act of 1990, encompasses not just the initial activities that create waste, but all activities associated with management and disposal of the waste. This "cradle to grave" philosophy means that the designers of waste generating activities, such as critical cleaning, need to be sensitive and aware as to how their wastes may impact the downstream treatment process. Similar to the application of P2 practices to critical cleaning activities, P2 practices may be applied to treatment activities so as to ensure compliance and to reduce operating costs.

For sites experiencing treatment compliance problems, one should seek out help. Treatment system providers and chemical suppliers can often provide help in getting one's system back into compliance. Another excellent resource for information regarding treatment system operations is the Office of Water Programs at California State University Sacramento.[11] The U.S. Department of Defense has adopted the use of the Sacramento manuals and has issued a companion handbook to address specific issues that affect military facilities.[12] In seeking out help, do not overlook the local permitting agency. Much of the information presented in this section was developed by the author as part of an industry outreach program sponsored by the City of Los Angeles.[13]

Good Operating Practices

Once a treatment system has been installed, good operating practices should be followed to avoid compliance problems and to maintain the efficiency of the system. Many of these practices are procedural

---

* In critical cleaning, the term "pretreatment" is used to denote the removal of impurities from the raw water supply prior to use. This same term is also used in wastewater discharge regulations to denote the on-site removal of hazardous components from aqueous streams prior to discharge. To avoid confusion, this section uses the term "treatment" to denote the on-site handling of discharged streams.

in nature and can be implemented with no outlay of capital. Keeping informed as to one's duties and responsibilities is the first line of defense in avoiding a compliance issue.

Knowledge of Prohibitions and Limits

The Federal Water Pollution Control Act of 1972, as amended by the Clean Water Act of 1977, and the Water Quality Act of 1987, gives the United States Environmental Protection Agency (USEPA) the authority to regulate the discharge of pollutants to waters of the United States. The National Pollutant Discharge Elimination System (NPDES) permit program (40 CFR 122) was established under this Act. An NPDES permit must be obtained for all direct discharges to surface waters but is not required for indirect discharges to a Publicly Owned Treatment Works (POTW). Most facilities engaged in critical, industrial, and maintenance cleaning are indirect dischargers permitted by the local authority.

All facilities that discharge wastewater to the environment, either directly or indirectly, must comply with general discharge prohibitions and limits. The USEPA General Pretreatment Regulations (40 CFR 403) establishes two sets of standards; Prohibited Discharge and Industrial Categorical Discharge. The Prohibited Discharge Standards specify conditions and pollutants that can interfere with the performance of the POTW. Industrial Categorical Discharge Standards, for more than 50 industrial categories, regulate the discharge of conventional pollutants such as pH, oil and grease, biochemical oxygen demand (BOD), suspended solids, and priority toxic pollutants.

Take a Proactive Approach

The best way to minimize or avoid the impacts associated with noncompliance is to take a proactive approach. Facilities that actively seek out and correct the potential causes of upset seldom find themselves operating in "panic" mode. Such reactive maintenance and repair can be quite costly when one considers the overall cost. Worker overtime, the need to rent emergency response equipment, production downtime, and possible legal fines can far outweigh the costs associated with taking a proactive approach. Table 4.8 provides some examples.

In the Event of Upset

Upsets are bound to occur and when they do, the severity of penalty or sanction that may be imposed will greatly depend on your response and the effect the upset has on the POTW. Most industrial users will have several opportunities to resolve an issue of noncompliance before any escalated administrative enforcement action is taken.

The first level of enforcement action regarding a noncompliance problem is the issuance of a Notice of Violation (NOV). If the proper corrective action to achieve compliance is taken, no further action will occur. If continued noncompliance persists beyond the NOV stage, then the enforcement action may escalate to the issuance of an administrative order (Cease and Desist or Compliance Order). If noncompliance continues to persist, the local agency in charge may escalate the enforcement action and suspend or revoke one's industrial wastewater discharge permit, levy administrative fines and penalties, and apply cost recovery.

TABLE 4.8   Ways to Be Proactive

Routinely evaluate P2 and recycling opportunities to reduce or eliminate the load placed on the treatment system. Once an option is implemented, monitor and track its performance. If you do not get the results expected, investigate why and take additional corrective action
Recognize that eventually all equipment becomes worn-out or obsolete and must be replaced. Some factors to consider include efficiency, physical integrity, safety, and replacement part cost/availability. The replacement of old equipment with new is often justified if one considers the total life cycle cost
Adopt a reliability-centered maintenance approach. This approach seeks out the root cause of equipment failure so that corrective measures can be enacted to prevent reoccurrence. Equipment is monitored so that the level of maintenance provided is based on actual need rather than routine

In severe cases of willful noncompliance and repeat offense, civil penalties and criminal prosecution may result. Such penalties and prosecution may also result from failure to respond to an upset in a responsible way.

Set Standards for Rinsate Reuse

In an effort to conserve water, facilities should investigate the option of treating the rinsate for reuse in the cleaning process. Sometimes the dirty rinsate from a critical cleaning operation will be relatively clean and the potential contaminants known. For a shop that uses deionized or demineralized water for rinsing, it may be easier to treat the dirty rinsate for reuse than it is to initially treat the raw water supply. Reverse osmosis (RO) and ion exchange (IX) systems are often used to recover highly purified water.

One potential obstacle to the reuse of water is the lack of a standard or specification regarding water quality. This may often be the case in maintenance cleaning where the only qualifier is that the fresh water must be potable. Depending on shop location, the quality of potable water can vary widely and it may be a poor choice for cleaning equipment that is subject to chloride attack. The specifications for potable water are broad, and have been developed to address health and aesthetic concerns, not issues related to equipment reliability and safety.

Table 4.9 is based on water quality criteria identified from available literature. The column titled "plating" is a general water quality guideline that should be acceptable for most plating baths (note: one should always check with their solution supplier to determine if there are any specific contaminants of concern). The column titled "aircraft" is a USAF guideline reported for aircraft cleaning and corrosion control.[14] In discussions with corrosion control specialists from the U.S. Army Corrosion Control Program Office, they indicate that the limits set for chlorides and TDS are too high and that future limits will be lower. The last column titled "future" reflects the expected limits as assumed by the author.

In procuring a recycling system, it is important to establish some type of water quality parameters for judging performance. While many vendors of wash rack recycling systems claim that their systems produce water of sufficient quality for reuse, few will provide a guarantee of meeting a specific standard. Most systems are capable of removing oil, grease, and suspended solids, but they do not remove dissolved solids, chlorides, or surfactants. The removal of surfactant is very difficult to achieve and just a few parts per million of nonionic surfactant in the rinsate can cause foaming. To control the level of dissolved solids and chlorides, a portion of the treated rinsate must be bled to sewer. Thus, setting too high a goal for the recycling system such as "zero" discharge is prone to failure.

TABLE 4.9 Water Quality Criteria for Reuse

Analysis	Units	Plating	Aircraft	Future
Chlorides	mg/L	25	400 max	200 max
pH	—	8.23	6.5–8.5	6.5–8.5
Total dissolved solids	mg/L	270	500 max	300 max
Total suspended solids	mg/L	<10	5 max	5 max
Hardness (CaCO$_3$)	mg/L	100	75–150	125 max
Biological oxygen demand (BOD)	mg/L	—	5 max	5 max
Total petroleum hydrocarbon (TPH)	mg/L	—	10 max	none
Langelier saturation index (LSI)	—	0.26	+0.0	+0.0

Note: The LSI value should be slightly positive. The recycled water should also be adequately disinfected to control the growth of microorganisms.

Preliminary Treatment

Proper wastewater treatment should always begin ahead of, or upstream of, the treatment process. Most treatment processes are highly sensitive to flow and load variations and the presence of solids. Efforts to eliminate or reduce variations can mean the difference between compliance and noncompliance. Preliminary treatment methods include stream segregation, flow equalization, and screening.

Stream Segregation

Stream segregation is important because most treatment systems operate best when handling a limited range of pollutants. The mixing of streams, such as those containing heavy metals with those containing chelating compounds, can make treatment difficult. Such mixtures require additional treatment steps that add complexity and cost to the overall system. Points to remember regarding stream segregation are presented in Table 4.10.

In addition to easing the complexity of the treatment process, segregation allows streams to be treated individually. The treatment of individual streams often results in better compliance because less pollutant is present in the final effluent. This effect of better pollutant removal via stream segregation is illustrated by this example.

A 5 gallon per minute (gpm) stream with 100 parts per million (ppm) copper is discharged to a pretreatment system handling 500 gpm. Assuming no other sources of copper, the 500 gpm of influent will contain 1 ppm copper. Pretreatment via pH adjustment and precipitation will reduce the 1 ppm of copper to 0.4 ppm (the solubility limit at a pH of 8–11).

Now assume that a small system is installed for the copper-bearing stream. The resulting effluent of 5 gpm will contain 0.4 ppm copper (the solubility limit), but this level will drop to 0.004 ppm (4 ppb) in the final combined effluent. Thus, stream segregation increases the copper removal efficiency from 60% to 99.6%.

Screening of Solids

When large solids are present in the influent, they should be screened out prior to treatment. Failure to remove these solids ahead of treatment can lead to poor mixing and reaction, increased chemical usage, and higher maintenance costs. The removal of these solids from the effluent may also be more difficult due to physical changes in their size and handling properties. Fine screens may be used to remove solids such as sand, silt, and fibers that exceed 75 μm in size. Screening is not a common treatment activity for aqueous wastes derived from critical cleaning.

Flow Equalization

After segregation and screening (if needed), some means of flow equalization should be provided. Biological processes are very sensitive to changes in flow and pollutant load because they are slow to react. A sudden increase in the influent flow or load may pass through the system untreated. Particularly for physical–chemical treatment, variations in flow can result in the under or over use of chemical additives. This occurs because the addition of the chemical is most often based on flow while the actual need is based on pollutant load.

TABLE 4.10   Points to Remember about Stream Segregation

Concentrated metal-bearing solutions should be batch treated and the effluent bled into the existing wastewater treatment system. Large and sudden bath dumps are sure to result in upsets and noncompliance
Acidic and alkaline streams should not be discharged to the treatment system in the same piping. The mixing of these streams can result in the formation of solids that can foul pumps and plug lines. The exothermic reaction that occurs can also damage the piping and in extreme cases, cause an explosion
Waste streams containing cyanide must be piped separately to the treatment system and not be allowed to mix with other streams. The mixing of cyanide with low pH (i.e., acidic) streams can result in the release of toxic hydrogen cyanide gas

TABLE 4.11 Will Flow Equalization Benefit Your System?

Follow these steps to determine if flow equalization will help in your efforts to maintain wastewater treatment compliance:

1. Monitor the flow of wastewater into your treatment system for one or more cycles of operation. One cycle of operation may be an 8-h shift or a 24-h day depending on your operation
2. From the monitoring data, calculate the average flow rate and the standard deviation of flow. Divide the standard deviation by the average flow rate and multiply by 100. This term is your coefficient of variability in percent
3. If your coefficient of variability is greater than 25%, then you may want to consider installing in-line or sideline flow equalization. Do not overlook good operating practices as a way to lower your coefficient of variability
4. Repeat steps 1 through 3 using monitoring data for a critical pollutant or a representative parameter (e.g., pH, conductivity, TDS). If the coefficient of variability exceeds 25%, consider in-line flow equalization

TABLE 4.12 Factors That Make Flow Equalization Successful

Solids should be removed from the wastewater via screening prior to the flow equalization tank. Solids inside the tank may settle out and reduce the liquid capacity of the tank. Abrasive solids may cause excessive wear

Proper and complete mixing is required for successful flow equalization. Tanks should be provided with internal baffles to promote mixing. Large tanks may require mixers and/or pump operated jets

Tank inlets should be located near the mixers to disperse the flow as soon as possible. The flow outlet should be configured to prevent short-circuiting of the incoming wastewater

Streams containing volatile solvents should be segregated from the main wastewater flow. Solvent vapors may create an explosive atmosphere inside the equalization tank or trigger the need for emission controls

For biodegradable wastes, mixing and aeration may be required to prevent septic conditions. Septic conditions can lead to the conversion of benign sulfates into harmful sulfides. Hydrogen sulfide is a highly corrosive and toxic gas

The equalization of incoming wastewater flow is an economical way to improve the performance of any treatment system. Flow equalization reduces the variation in hydraulic and pollutant loads that can result in upset and pass-through. With flow equalization, downstream equipment operates more efficiently and with less stress. This can result in power savings and reduced maintenance costs. For a planned system, the inclusion of equalization can help reduce the peak design load. This, in turn, reduces the need for oversizing the system and saves capital expense.

There are two common methods used to provide flow equalization: in-line and sideline. In-line flow equalization takes all the incoming flow and passes it through a large holding tank. This tank provides sufficient surge capacity so that the wastewater can be discharged to the treatment system at a constant rate. In sideline equalization, flows that exceed the average rate or pollutant load are diverted to a holding tank for eventual return to the system.

While both methods are effective in equalizing the hydraulic load, the in-line method is best for equalizing the pollutant load. Table 4.11 provides a method for determining if a system will benefit from flow equalization. Table 4.12 lists factors to make flow equalization successful.

Treatment Technologies

Depending on the nature of the hazardous constituents present in the final rinsate, one or more treatment technologies may be used to render the wastewater acceptable for discharge. Common technologies include pH adjustment, heavy metal precipitation, suspended solids removal, oil and grease separation, biochemical oxygen demand (BOD) reduction, and toxic organic treatment. These technologies may also be practiced upstream, either directly on the process bath to extend its life or on the rinse system so as to reduce overall water demand. Thus, the distinction between recycling and treatment is most often based on the disposition of the treated effluent rather than the actual technology involved.

Other treatment technologies found in manufacturing facilities that are engaged in electroplating, metal finishing, and industrial cleaning include emulsion breaking, chromium reduction, and cyanide destruction. These technologies are most often used to batch treat a spent cleaning or process bath with the resulting effluent then slowly fed to the rinsate treatment system. Since the treatment of these concentrated baths requires specialized attention and equipment so as not to upset the rinsate treatment system, many small quantity generators elect to send their spent baths off-site for treatment. These technologies are therefore excluded from further discussion.

pH Adjustment

The most common method of adjusting pH is by adding an acid or base. To raise pH, basic compounds such as sodium hydroxide (caustic soda), calcium oxide (lime), calcium hydroxide (hydrated lime), or sodium carbonate (soda ash) are used. To lower pH, sulfuric or hydrochloric acid may be used. For ease of ordering, handling, and safety, some facilities use their existing process chemicals for pH adjustment.

The pH value of neutral water is seven, with a value less than 2 representing a very acidic condition and a value greater than 12 being very basic. The pH value is derived from the negative logarithm of the hydrogen ion concentration. To measure pH, a specially designed meter employing a glass bulb sensor is used. The sensor is placed in the sample and the meter compares the signal from the sensor to the signal from an internal reference cell.

Common sources of error include improper calibration, failure to adjust for temperature effects, bulb damage or fouling, the presence of interfering ions, and lack of a representative sample. A slime layer of 1 mm over the sensor can slow response time from a few seconds to several minutes or more. Trapped solution can cause the readings to freeze. Litmus paper may be used for spot checks, but reliable process control requires continuous monitoring.

Heavy Metal Precipitation

Heavy metals tend to be most soluble at acidic or low pH conditions. By raising pH via the addition of caustic or lime, solubility is reduced and the metals precipitate out of solution as insoluble hydroxides. Under most conditions, hydroxide precipitation is consistently capable of meeting the parts per million discharge limits set by local ordinance.

The overall process consists of adjusting the pH to an optimum value to precipitate the metals of concern followed by multistage coagulation and flocculation, and then clarification. Additional operations include effluent polishing and sludge dewatering. If the water contains hexavalent chromium or cyanide, specialized treatment is required prior to precipitation.

After precipitation, the charge that is present on the formed particles must be neutralized so that coagulation and flocculation can occur. Charge neutralization may require the use of aluminum sulfate, ferric chloride, or calcium hydroxide. Rapid mixing is important to ensure complete mixing and intimate contact between particles. This period of rapid mixing must be followed by a period of gentle mixing to allow time for flocculation. An organic polymer may be added during this stage to speed the process. As the insoluble particles flocculate and grow larger, their settling velocity increases and they are more likely to settle.

Solids removal is typically performed by gravity separation in a clarifier. The solids settle out and they are removed via the clarifier underflow as sludge. To minimize the weight of sludge sent off-site for disposal or recovery, the sludge is dewatered and the filtrate is returned to the head of the treatment process. Overflow from the clarifier may be polished by passing the effluent through a screen or filter prior to discharge.

Multistage treatment may be required if trying to remove metals with different precipitation points. Each treatment stage should be equipped with an individual pH control and a settling unit. The determination of optimum pH should be established by jar tests. The jar tests should use chemicals from the line and not lab grade chemicals because results may differ. Routine jar tests can help confirm the proper rate of polymer use. Solids carryover from the clarifier may occur if the detention time is less than the settling rate. Reducing flow or adding a second unit in parallel will increase detention.

Suspended Solids Removal

Specific gravity and particle size are two key factors that determine the settling rate of solids. Large and heavy solids can be readily removed via gravity sedimentation while the removal of small and light (i.e., suspended) solids requires screening or filtration. Screening is often used to remove suspended solids in the 20–50 μm size range prior to discharge. To remove suspended solids in the 5–10 μm size range, the wastewater may be filtered.

To remove suspended solids that are smaller than 2–4 μm in size, various membrane filtration systems may be employed. Microfiltration (down to 0.1 μm), ultrafiltration (down to 0.005 μm), and reverse osmosis (salts and dissolved minerals) are often used to treat various effluents so they can be reused or recycled back to the originating bath. Their major disadvantages compared to conventional filtering include higher cost, susceptibility to chemical attack, and fouling potential.

Physical Oil and Grease Separation

A variety of methods are used to remove oil and grease from wastewater streams. Grease traps and gravity separators are often used to remove biodegradable oil and grease of animal or plant origin. These units remove the free-floating oil and grease with the oily water and then discharge it to sewer for biological treatment at the POTW.

Removal efficiency for a gravity separator is a function of residence time, specific gravity of the oil, oil droplet size, fluid salinity, and temperature. The wastewater velocity through the unit needs to be kept low, typically less than 3 ft/min so as to prevent turbulent mixing. A well-designed separator can remove oil particles with a diameter greater than 150 μm and reduce the oil content of the effluent to 100 mg/L or less.

To increase separation efficiency, a gravity separator may be equipped with a series of parallel plates. The plates increase the surface settling area inside the unit and enhance contact between the oil particles. Oleophilic (oil-attracting) materials such as polyethylene, fiberglass, or nylon are used to construct the plates. The plates are held in a stainless steel frame or pack so they can be easily removed for cleaning. Parallel plate separators can generally remove droplets of free-floating oil greater than 60 μm in diameter and achieve an effluent limit as low as 50 mg/L.

Mechanical impingement devices induce coalescence of dispersed oil droplets in the form of a filter or cartridge-type emulsion breaker. These units may be used as the final step in an oily water separation system consisting of solids filtration and free oil removal. Cartridge units typically contain a medium having numerous small (25 μm) irregular passages through which the wastewater flows. The emulsion is broken by impingement of the oil droplets on the surface of the medium. The cartridge can be backwashed to control fouling or is replaced.

Pressure vessels containing a solid-phase adsorbent may be used to remove low levels of chemically or mechanically emulsified oil, usually as a final polishing step. The first vessel may contain a blend of gravel, sand, and anthracite for the removal of suspended solids followed by a second pressure vessel containing activated carbon. The carbon can remove solvents, diesel fuel, and surfactants. Some systems may use activated clay in place of or ahead of the carbon bed.

Biochemical Oxygen Demand Reduction

Biochemical treatment systems are widely used to treat dirty wastewater with high biological oxygen demand (BOD). High levels of BOD may be due to the presence of emulsified oil and grease or other biodegradable organic matter. Heavy industrial and maintenance cleaning are two activities that generate a high BOD effluent. Some electronic facilities also produce a high BOD effluent due to the discharge of solvents and organic materials in their wastewater streams.

Most BOD reduction systems operate under aerobic conditions with air providing oxygen to the microorganisms. Common system designs used at many small- to intermediate-sized facilities include a trickling filter and a rotating biological contactor (RBC). A trickling filter may be used alone to remove low levels of BOD or it may be used as the first stage of a combined system if BOD levels are high. Small to medium sized plants that generate wastewater high in BOD often employ a high rate or roughing filter that is followed by several stages of RBC.

A trickling filter consists of a large open tank filled with rock, plastic, redwood, or other support media on which the biomass grows. The wastewater is sprayed at a controlled rate down upon the media and BOD is reduced as the water "trickles" through the filter. Much of the water that collects in the sump is returned to the top of the filter so as to maintain a constant hydraulic load and prevent drying of the biomass. Excess water is discharged from the sump to a clarifier for solids (i.e., biomass) removal.

In an RBC, the rigid media consists of large diameter plastic disks mounted on a long horizontal shaft. The disks slowly rotate at a speed of one to two rev/min with their bottom edge dipping into the wastewater. Biomass grows on the disks and the constant rotation brings the biomass into contact with the wastewater and air. Multiple stages of RBCs are often used in series and in parallel.

Many wash rack recycling systems that rely upon biological treatment have entered the market in recent years. Some systems use bacteria as the primary means of degrading and breaking down hydrocarbons while others use bacteria as a final treatment step and to control odor. Odor control is achieved by the growth of aerobic bacteria that suppress the growth of anaerobic bacteria that are responsible for creating odors. Periodic dosing of the system with fresh bacteria and nutrient is often required.

Toxic Organic Pretreatment

Seven federally regulated industrial categories have treatment standards established for total toxic organics (TTO).[15] The term TTO is defined as the sum of specific toxic organic compounds present in the discharge at a concentration of 0.01 mg/L (10 ppb) or more. The seven categories include electroplating (40 CFR 413), metal finishing (40 CFR 433), electrical and electronic components (40 CFR 469), metal molding and casting (40 CFR 464), coil coating (40 CFR 465), aluminum forming (40 CFR 467), and copper forming (40 CFR 468).

In the field of critical cleaning, the need for TO treatment must be made on a case-by-case basis. Some shops may decide to switch away from cleaning solvents that contain a TO such as benzene or halogenated solvent. Other shops may decide to continue to use these solvents but to improve housekeeping so that they do not enter the wastewater stream. Such housekeeping measures may not be possible when the parts require rinsing as part of the cleaning process.

Common treatment methods for the removal of TO include air stripping, steam stripping, and aqueous phase carbon. Air stripping or carbon adsorption is often used to remove TO prior to sewer release while steam stripping is used prior to biological treatment. All three methods physically remove the TO from the aqueous stream and additional operations are required to manage the removed TO. Such operations include off-gas incineration, the recycling or incineration of recovered liquids, or carbon reactivation.

Tracking Incidental Pollutants

Sometimes, the source of a noncompliance problem is due to the discharge of an incidental pollutant. One such example might be the discharge of copper from a facility that is not engaged in metal plating or finishing. In this situation, potential sources could be the incoming raw water supply, the leaching of copper from water supply pipes, from acidic storm water washing over copper flashing, or from the use of a copper-containing biocide to control fouling in the cooling tower. While the list of suggestions

TABLE 4.13 Tracking Incidental Pollutants

Copper and lead are commonly present in water supplies due to corrosion of the piping distribution system. Zinc compounds used to control corrosion can also be found in the water supply. Shallow groundwater supplies may draw in pollutants such as heavy metals, solvents, or petroleum products

Boiler blowdown and cooling tower discharge may contain high levels of copper, nickel, and zinc due to uncontrolled corrosion. Biocides used to control biological growth in the cooling water system may contain copper or chlorinated organics

Chemical supplies may contain impurities that can upset the production process and the treatment system. Sulfuric acid may contain high levels of iron or heavy metals due to metal corrosion during storage, or it may be due to prior use for steel pickling

Storm water falling on the roof of a process building may pick up pollutants during the first flush. Pollutants in sewer lines may be flushed out during the first rain

As part of maintenance, careful consideration should be given to selecting the proper materials of construction. Brass valves should not be used in alkaline service because they will leach zinc and lead. Cheap plastic piping may contain high levels of plasticizer or lead that will leach out over time

Janitorial chemicals often contain chlorinated and other hazardous compounds that may end up in the wastewater stream through improper use or sewer cross-connection. The use of safer substitutes is not without potential problems because they may contain complexing agents such as ammonia or EDTA that make heavy metal removal more difficult

presented in Table 4.13 is far from complete, they should offer some help to facilities faced with tracking down an incidental pollutant.

Points to Remember

Optimizing and maintaining the critical cleaning and rinsate treatment process is an ever-constant challenge. New processes may be introduced as products change, soils change, or regulations change and either prohibit the use of proven chemicals or trigger the need for tighter effluent control. But with all this change, there is still a constant approach to insuring compliance and that you are running the best operation possible. Key points to remember include

- Eliminate the need to clean by eliminating upstream contamination or by relaxing an overly critical cleanliness requirement. Modify the part or contaminant when possible so that a less-hazardous cleaning material or method may be used.
- Select the least hazardous cleaning materials and/or methods that achieve the required level of cleanliness. Once selected, optimize and maintain the process so that the greatest amount of soil is removed for a given amount of cleaning material consumed. For water-based cleaning systems, use demineralized water to extend bath life, promote recycling, and reduce sludge generation.
- Keep all wastes segregated to promote recycling. Recycle and/or reuse spent cleaners to the extent practical. Consider distillation for solvents and ultrafiltration for aqueous cleaners. Adopt water conservation measures and investigate ways to reuse rinsate.
- Optimize and maintain the effluent handling system. The same approach taken with cleaning applies to treatment. Follow good operating practices to maintain compliance, conserve water to minimize treatment volume, and provide flow equalization to eliminate large swings in flow and load.
- Periodically evaluate all your systems for the opportunity to upgrade or replace. The need for system replacement may be triggered by a major change in production throughput (either up or down), an inability of the old system to meet a new specification, the inability to obtain vendor support and/or repair parts, concerns over operator safety due to physical condition of the system, or a change in environmental regulations that prohibits the continued use of the system.

References

1. Kushner, J. B. A three part series of articles on rinsing. In *Metal Finishing*. Part I, vol. 49(11), 1951; part II, vol. 49(12), 1951; and part III, vol. 50(1), Elsevier Science, New York, 1952.
2. Soderberg, G. Drag-out. *Proceedings of the Twenty-Fourth Annual Convention of the American Electroplaters Society*, Cleveland, OH, 1936, vol. 24, pp. 233–249.
3. Bigge, D. M. and Graham, A. K. In A.K. Graham (ed.). *Design for Plating in Electroplating Engineering Handbook*, 3rd edn., Chap. 2. Van Nostrand Reinhold Company, New York, 1971.
4. Alconox, Inc. *The Aqueous Cleaning Handbook: A Guide to Critical-Cleaning Procedures, Techniques, and Validation*, 4th Printing, 2005. Available on-line at www.alconox.com
5. Donahue, B. A., Tarrer, A. R., Dharmavaram, S., and Joshi, S. B. *Used Solvent Testing and Reclamation, Volume I: Cold-Cleaning Solvents*. USA-CERL Technical Report N-89/03, vol. 1 and AFESC Report ESL-TR-88-03. AD-A204 731, NTIS, Spring Field, VA. December 1988.
6. DOW. *Economical and Efficient Degreasing with Chlorinated Solvents from Dow*. Form No. 100-06096-1199 AMS, November 1999.
7. Micro Care Corporation. *BromoTest™ Solvent Test Kit for Bromothane Cleaners*. Product Spec PS-104, M2S-BA, issued, December, 2003.
8. Spring, S. *Metal Cleaning*. Reinhold Publishing Corp., New York, 1963.
9. USEPA. *Waste Minimization in Metal Parts Cleaning*. USEPA Office of Solid Waste and Emergency Response. EPA/530-SW-89-049, Washington DC, September 1989.
10. Osmonics. *Alkaline Cleaner Recycling Handbook*, 2002. *Membrane Filtration Handbook*, 2nd edn., Minnetonka, MN, November 2001.
11. California State University Sacramento, Office of Water Programs, Sacramento, CA. A series of training manuals that include the following titles: (a) *Operation of Wastewater Treatment Plants*, vols. 1 and 2, 7th edn., 2008; (b) *Operation and Maintenance of Wastewater Collection Systems*, vols. 1 and 2, 6th edn., 2003; (c) *Industrial Waste Treatment*, vols. 1 and 2, 3rd edn., 2003. (d) *Advanced Waste Treatment*, 5th edn., 2006; (e) *Treatment of Metal Wastestreams*, 3rd edn., 2003; (f) *Pretreatment Facility Inspection*, 3rd edn., 1996.
12. USDOD. *Wastewater Treatment System Operations and Maintenance Augmenting Handbook*. MIL-HDBK-1138. U.S. Department of Defense, Tyndall AFB, FL, October 31, 1997.
13. IWMD. *Pretreatment Program Compliance Guide*. City of Los Angeles Industrial Waste Management Division, Los Angeles, CA, 2001 edn.
14. USDOD. *Cleaning and Corrosion Control*, vol. II, Aircraft. NAVAIR 01-1A-509-2, TM 1-1500-344-23-2. Naval Air Systems Command, Technical Data and Engineering Service, North Island, CA, March 1, 2005.
15. IWMD. *Guide to Preparing a Streamlined Toxic Organic Management Plan*. City of Los Angeles Industrial Waste Management Division, Los Angeles, CA, 2001 edn.
16. Kushner, J. B. and Arthur, S. *Water and Waste Control for the Plating Shop*, 3rd edn. Hanser Gardner Publications Inc., Cincinnati, OH, 1994.

5

Basis of Design for Life Science Cleanroom Facilities

Introduction 61
In the Beginning 61
Isolation Technology...63
Preconstruction Design Review 64
Typical Facility File Table of Contents
Primary Design Considerations 66
Contamination Control Criteria 66
ISO Standards and Cleanroom Classifications • ISO 14644 • ISO 14698
Additional HVAC Considerations 68
Architectural Issues..69
Project Planning 70
Project Delivery 71
Basis of Design Preparation..72
Product Flows • Component Flows • Operator
Intrusions • Equipment • General
Turnover Package 75
Responsibilities
Commissioning...85
Direct Impact Systems
Conclusion 86
Further Readings 88

Scott E. Mackler
Cleanroom Consulting, LLC

Introduction

Without the implementation of proper front-end assessment and planning, no project will be successful. Life science cleanroom facilities delivery requires a systematic methodology from concept through commissioning. For fast track projects with tight deadlines, the issues are compounded and assurance that a consistent, validatable standard has been applied to design and construction must be developed a priori—or it will not occur.

In the Beginning

The successful design and realization of a validatable facility does not happen by chance. Early coordination and communication among all parties involved is essential, from process design and scale-up through cGMP (Current Good Manufacturing Practices) layout, preconstruction design review,

The "Basis of Design" document

☐ Process description and process flow diagrams
☐ cGMP floor plan and general equipment arrangement
☐ Sized major process equipment list, w/utilities requirements/consumption
☐ Sized process support services utilities list
☐ Functionality flow diagrams (process, people, product, material, components, air, waste)
☐ HVAC zoning and room classification, incl. microbial limits
☐ Budget quality (±20%) cost screening estimate
☐ Preliminary "scope of work" descripition (design/specify/furnish/install/inspect/test/balance/certify/guaranty/challenge/quality)
☐ Complete list of regulatory authorities that will perform inspections

FIGURE 5.1 Important elements: Basis of design plan. Detailed documents outlining requirements and expectations should be developed prior to cleanroom construction.

commissioning, validation, and on-site inspection by regulatory authorities. A complete "Basis of Design" document is necessary to optimize the utilization of scarce resources. Time spent in design optimization at the front end, well in advance of the "bricks and mortar" stage, will be saved many times over. It is important to remember that regulatory guidelines specify performance, not methods.

The two most important keys to project success are a well-defined scope and early, extensive planning. It is in the conceptual stage where we can best establish design criteria and a solid basis of design plan (Figure 5.1) so that the project will benefit from accurate budgets and a firm execution plan developed together with the process owner as an integrated team. This diagnostic approach minimizes project risk and prevents costly and unforeseen "adders" once construction begins. It is in the conceptual stage when we are still just dealing with "lines on paper" that the most money can be saved at the least actual "out-of-pocket" expenditure through thorough engineering planning and design.

The project validation master plan should be established immediately after the completion of the conceptual design phase. Commissioning and validation protocols should be written during the design phase of the project.

Early planning should include a substantial amount of "value engineering," a term often misconstrued as elimination of scope or the substitution of lower quality materials. The owner and designer have an obligation, early on, to agree as to what is meant by value engineering.

Definition: Value engineering is the systematic analysis of design alternatives that provide maximum value at minimum cost.

For example, the use of existing warehouse space can save time and money over the construction of a new building. Warehouse buildings provide good access, open floor space, and are likely designed for future expansion.

On the other hand, warehouse space is typically not air-conditioned. The roof and floor slab may not be designed to handle the weight of cleanroom ceilings, ductwork, HVAC (heating, ventilating, and air-conditioning) equipment and heavy process equipment, and utilities service to a warehouse will almost always have to be upgraded to support energy-intensive cleanroom applications.

A balance sheet analysis, however, may prove that by providing independent structural support and HVAC for the cleanroom facility, off-the-balance-sheet financing in the form of a true operating lease, or accelerated depreciation, versus real property, are both viable and attractive options for the owner.

Operating companies having experience in facilities delivery will know to specify a usable interstitial space of around 10–12 ft from BOS (bottom of steel). While at first blush this might seem excessive, the benefits in terms of initial constructability and ongoing maintenance cannot be overstated. HVAC

HVAC capital and operating cost comparison			
	Chilled water	Air cooled DX	Water cooled DX
Chillers	134k	NA	NA
MUA AHU	53k	178k	140k
Recirc units	69.5k	102k	102k
Piping	88k	5k	69k
Total capital	$344k	$285k	$311k
Total energy (300 tons at 0.08/kWh)	$168k	$252k	$189k
First year, excluding maintenance	$512k	$537k	$500k

FIGURE 5.2 Example, value engineering analysis, central chilled glycol system versus individual air cooled direct expansion units. The central glycol system was chosen because, while it has the highest capital, it has the lowest energy. DX = direct expansion. To be valid, the analysis must be performed for each specific application.

duct routing with respect to structural and architectural elements, along with the requirement to place all cleanroom mechanical equipment requiring periodic service indoors (such as bag-in/bag-out filters on 100% once thru exhaust rooms), often leads designers to underestimate the optimal height for interstitial space. Walkable ceilings eliminate the need for maintenance catwalks over cleanrooms, and can help to lower the required height of a new building as well. In an example of value engineering trade-offs one should analyze early on, placement of individual air handling units (AHUs) directly over the clean space will limit ductwork runs but increase structural steel requirements. For a commercial-scale, therapeutic protein–manufacturing facility, a remote equipment loft design was utilized. While longer duct runs and larger fan capacity were required, structural loading for the large clean containment area was reduced, and no building columns intruded into the clean space. Building columns within the controlled area were clearly prohibited by the operating company as they would serve to limit rearrangement flexibility in the event that re-configuration might be required to serve new clients down the road.

In another project, one for a large robotized lyophilization facility, a capital and operating cost analysis favored the use of a central chilled glycol system as opposed to individual air cooled direct expansion units (Figure 5.2). This was a brand new facility utilizing greenfield construction. Quite the opposite conclusion was reached for a nearby sterile fill suite retrofit project (located on the same site).

Isolation Technology

Isolation technology is the placement of a barrier between a process and its operators. The purpose of the barrier may be to protect the process and its materials from the effects of the operators, or it may be to protect the operators from the effects of the process; in some cases it may seek to do both. The barrier may be total, so that the process is always behind either a physical wall or at least under HEPA filters, or it may be partial, so that the process may be separated only, for instance, by engineered airflow.

The application of this technology tends to focus on one or more specific product offerings and claims abound ranging from "fill aseptically to SAL (sterility assurance level) 10^{-6} in your living room" to "isolators promote poor aseptic technique and increased hazards, and are suitable only for sterility testing."

Obviously, somewhere in between these two extremes lies the truth; and it is a temporal situation requiring constant revisiting. The key elements, once again, are a careful analysis of process and facilities requirements, and the implementation of a diagnostic approach. The integration of modern barrier technology into cGMP facilities can yield measurable benefits when a thoughtful, unbiased assessment is performed. Experience and careful analysis leads to the correct balance between operational procedures and hardware features.

In one recent project for the production of clinical trials quantities of an immunodiagnostic, barrier technology facilitated a "single corridor" style layout since personnel did not need to come in contact with contaminated materials. The facility was required to comply fully with cGMPs and the fermentation suite was designed as BL1-LS. (*Note:* Biosafety Level 1, Large Scale Production. Levels are based on guidelines from the National Institutes of Health. Level 1 spaces are for working with organisms that pose at most minimal safety to workers or to the environment.) An analysis to assess the feasibility of segregating operations by barrier techniques, as well as time and protocol was performed. The reduced capital cost for the facility outweighed the added cost of the isolators. In addition, cell banking, seed preparation, cell culture, and cell separation were carried out in the same room, each within an isolated unit.

As a general rule for aseptic designs, we find that isolator-based fill machines show no capital disadvantage for high-speed line applications in multiline facilities (and there are clear operating cost advantages), and offer excellent control of cross contamination for multiproduct facilities.

Unfortunately, when developing economics for single-line facilities, there does not seem to be any generic-type sterile challenge data available that would allow us to take credit for higher yields, reliability, or safety with the isolator-based design. The isolator-based fill line is somewhat less flexible than a conventional line, should reconfiguration be required. Thus it is not usually favored in process development applications unless one is dealing with highly toxic or potent compounds. Isolator validation will cost more than validation for a conventional fill line, and take longer.

Process isolation should be evaluated and incorporated whenever there is a need to protect the product, such as in aseptic filling, or the personnel and environment—such as in potent drug or live biologicals production. Frequently, this is NOT a financial decision as much as it is a decision to incorporate "best available technology" and "build quality into the product." Any time you can remove a contamination source from the production process, it is an obvious plus to do so!

Because we still need to prove controlled conditions in the event of a breach, and have cGMP functional flows in an validatable facility, most isolator-based life science facilities today are being designed to at least an EU Class "C" or "D" per the EU GMPS, Annex 1, Manufacture of Sterile Medicinal Products under "Operational Conditions."

Preconstruction Design Review

Let us review participation in a facility design review meeting in which the plans for a new commercial-scale cell culture cleanroom, including HVAC and process support services, are to be vetted by the FDA's CBER (Center for Biologics Evaluation and Research). The sponsor company intends to use the new "greenfield" cleanroom facility for commercial production of a purified bulk API (active pharmaceutical ingredient) that is presently in Phase III clinical trials as an anticancer therapeutic. The new cleanroom consists of a unidirectional flow layout with each suite benefiting from dedicated airlock/gown-up ingress and egress and completely separate and isolated HVAC systems that provide independent control (and eliminate any danger of cross-contamination) for pre and post-viral inactivation, as well as for media and buffer prep, wash areas, raw material receipt, quarantine and release, and bulk filling.

In this case, the facility is designed to test and certify to ISO-14644 Grade 7, "operational" for both the upstream and downstream production areas. Bulk API filling is performed in a small suite located just off the downstream (post viral inactivation) processing area, and final fill and finish is a small volume fill performed under an ISO Grade 5 (Class 100) laminar flow hood.

The FDA review meeting followed an agenda that was based on our completed Facility File document, which included the following areas of concern from a facility design and regulatory standpoint:

Typical Facility File Table of Contents

1. Introduction
General overview of project and process. What the facility is intended to accomplish and what the final goals are for the project, including throughput and any future expansion requirements.

2. Regulatory design requirements
A brief discussion of the regulatory standards and guidelines that the facility, utilities, and process will be designed to meet including local, state, federal, and overseas organizations (e.g., EMEA [European Medicines Agency], HPB [Health Protection Branch, Canada], MCA [Medicines Control Agency], ICH [International Conference on Harmonization], JIS [Japanese Institute of Standards] as well as general industry groups such as ISPE [International Society for Pharmaceutical Engineering]).

3. Products and processes
A general description of the product(s) to be manufactured, an outline of the type of processes to be performed including a description of such areas as warehousing, quality control, quality assurance, mechanical spaces, raw material receipt, quarantine, and storage.

4. Containment and segregation
A general description of containment and segregation requirements based on process and/or product needs, including requirements for flexibility. Any special design requirements, e.g., hazardous materials, are listed here.

5. Facility and major utilities (conceptual design)
A general description of the facility including GMP and non-GMP areas, and appropriate architectural finishes. Also a description of critical utilities, e.g., HVAC (heating ventilation air-conditioning), WFI (water for injection), USP (United States Pharmacopoeia) water, and CDA (clean dry air), is important, as well as support services such as chilled water and glycol. Information is presented in drawing and matrix format.

5.1   Process description and process flow diagrams
5.2   Floor plan and general equipment arrangement
     5.2.1   Individual drawings
5.3   Major process equipment and utilities requirements
     5.3.1   Process equipment matrix
5.4   Major critical and support utilities
     5.4.1   Utilities matrix
5.5   Functionality flows (personnel, product, materials, waste, components, supplies)
     5.5.1   Crossing matrix
     5.5.2   Individual layouts for each functional flow
     5.5.3   Facility HVAC zoning, room pressurizations, temperature, RH, cleanliness classifications, biosafety and microbial levels
     5.5.4   Individual layout drawings
5.6   Architectural and mechanical/electrical requirements
     5.6.1   Room criteria matrix
     5.6.2   Room data sheets
     5.6.2   Reflected ceiling plan
     5.6.3   Architectural details
     5.6.4   HVAC flow diagrams, heat load calculation and equipment selections, mechanical details
     5.6.5   Electrical power one line diagram (simplified electrical diagram)

Primary Design Considerations

The envelope and the environmental controls directly affect product quality and reliability. Safety and effectiveness must be designed and built into the room, not tested in or inspected into the product or facility as an afterthought. Often designers forget that extra, expensive real estate is required for unidirectional people/product/materials/ process/waste flows. Separation through layout is more reliable than separation through procedure. This is especially true whenever we talk of clean containment (or negative pressure cleanrooms) for toxic or potent compounds, for live biologicals production or for cell and gene therapy applications. Many times in containment we are also faced with having to achieve aseptic processing conditions—a very interesting problem when dealing with negative pressure!

All these issues lead us to have to consider the following design criteria:

- Material and equipment access and cleaning
- Operator access, gown-up, and procedures
- Utility penetrations
- Spill containment
- Doors, windows, pass-throughs
- Architectural finishes
- Temperature, humidity, pressure
- Air flow and direction, cleanliness and microbial counts
- Future expansion or introduction of new equipment

As well as a hierarchy for containment technologies:

- Robotics
- Barrier
- Directionalized laminar flow
- Local exhaust
- General exhaust
- Open operations

Only absolutely necessary equipment and supplies should be brought into containment areas. Product will normally follow a logical flow in the direction of increasing cleanliness, and waste will normally leave through a decontamination autoclave. Consideration must also be given to cleaning and decontamination procedures in case of spills or accidents that may require decontamination of the ductwork. Provisions should be made for disposal and sterilization of all waste as well as the sterilization of materials, production equipment and the work area itself.

Keeping all this in mind, we also need to consider where we are in the drug development phase in planning our facility. For early clinical production, the relatively small volume demands and low numbers of operating personnel may allow GMP requirements to be met using protocols and procedures rather than capital-intensive equipment and facilities. We need to be diagnostic in our approach to facilities solutions!

Contamination Control Criteria

Terminal ceiling HEPA (high-efficiency particulate air) units and low-wall returns, standard in today's cleanrooms, eliminate the need for welded stainless steel ductwork. Note that the nominal CFM per filter can range lower than the typical manufacturers' suggested ratings. This derating is an example of "building quality into the room." See IEST RP-012 for additional discussions on recommended average air velocities and air change rates. By adding more filters than might be expected, we gain

- Longer HEPA filter life
- Lower HEPA filter pressure drop
- Better air distribution and elimination of drafts and temperature or humidity gradients.

In a Class 10,000 Category 3 containment facility, the number of air changes may be reduced to half the conventional value, due to energy savings when forced to utilize 100% once-through fresh air. But in this case, do not forget to utilize 95% ASHRAE rated pre-filtration! Remember, "air changes per hour" is a calculated, dependent variable, and it has not, to our knowledge, been canonized in any sacred texts.

As another case in point, one should consider which regulatory standards are pertinent to the facility validation? And do you appreciate the cost impact?

ISO Standards and Cleanroom Classifications

On November 29, 2001, the U.S. General Services Administration (GSA) officially announced that Federal Standard 209E, airborne particulate cleanliness classes in cleanrooms and clean zones, had been canceled and is now superseded by the ISO Standards for Cleanrooms and Associated Controlled Environments, ISO 14644-1, "Classification of Air Cleanliness," and ISO 14644-2 "Specifications for Testing and Monitoring to prove continued compliance with ISO 14644-1."

The GSA action was the result of a recommendation made by IEST Working Group CC-100 to "sunset" Federal Standard 209E in favor of the ISO documents. 209E was a very useful standard that defined the minimum acceptable criteria for U.S. government contracts across virtually all industries. It had become widely accepted for use in private contracting as well and served as the de facto key reference to consult when quantifying the particulate cleanliness of a clean space. The ISO documents are equally useful and further serve to promote global industrial harmonization for cleanroom cleanliness classification.

FS 209E is NOT a document controlled by the FDA and thus the FDA was not compelled to issue a change notice (209E and the ISO standards are NOT cGMP guidance documents). The USP had referenced 209E in <1116>.

Cleanliness class designations and quantity have changed from FS209E in 14644-1. Along with the obvious change to a metric measure of air volume, ISO 14644-1 adds three additional classes—two cleaner than Class 10 and one dirtier than Class 100,000.

ISO also forces the contractual partners to specify (1) the particle size of interest and (2) the state of cleanroom occupancy for certification, i.e., "as-built" (a completed room with all services connected and functional, but without production equipment or personnel within the facility), "at-rest" (a condition where all the services are connected, all the equipment is installed and operating to an agreed manner, but no personnel are present), or "operational" (all equipment is installed and is functioning to an agreed format, and a specified number of personnel are present working to an agreed procedure). Particle Count Tests for classifications less than or equal to ISO 5 will be required every 6 months and for classifications greater than ISO 5, testing will be required every 12 months.

Air pressure difference for all classes will be required every 12 months as well as airflow recertification. And further, installed filter leakage for all classes is recommended every 24 months as is containment leakage testing, recovery testing, and airflow visualization. Intervals between retests can be extended provided that frequent, periodic monitoring of the working environment is performed and demonstrates conclusively that effective controls exist.

As with 209E, the ISO documents should not be misunderstood as conferring cGMP conformance on aseptically processed products. There is no change in the FDA regulations and guidance. Inherent to the ISO documents is the recognition that minimum requirements do not constitute a universally applicable "one size fits all" solution. Neither 209E nor 14644 provide all of the information needed when one is developing a protocol to qualify an HVAC system. For aseptic processing, clean area classification should not only be based on the nominal test grid locations prescribed by ISO 14644-2 (which typically calls for less initial testing points than did FS209E!), but one must include consideration of:

1. Various locations that have been carefully chosen on the basis of the risk that they pose to the operation and the product
2. Microbiological monitoring data obtained from these and other appropriate locations in a given critical area

ISO/TC 209 has developed multipart documents encompassing the standardization of equipment, facilities, and operational methods. The cornerstones of the standard are ISO 14644: Cleanrooms and controlled environments, Part 1: Classification of air cleanliness and Part 2: Specifications for testing and monitoring to prove continued compliance with ISO 14644-1.

ISO 14644

ISO 14644 currently consists of the following parts, under the general title Cleanrooms and associated controlled environments:

- Part 1: Classification of air cleanliness
- Part 2: Specifications for testing and monitoring to prove continued compliance with ISO 14644-1
- Part 3: Test methods
- Part 4: Design, construction, and start-up
- Part 5: Operations
- Part 6: Vocabulary
- Part 7: Separative devices (clean air hoods, glove boxes, isolators, and minienvironments)
- Part 8: Classification of airborne molecular contamination
- Part 9: Classification of surface particle cleanliness

ISO 14698

In addition to ISO 14644, we must also consider the recently released ISO 14698. ISO 14698 consists of the following parts, under the general title Cleanrooms and associated controlled environments:

- Part 1: Biocontamination control—General principles
- Part 2: Biocontamination control—Evaluation and interpretation of biocontamination data

Additional HVAC Considerations

The need for flexibility, prevention of cross contamination and the requirement for individual room pressure, temperature, and humidity control often results in a more costly HVAC system for biotechnology facilities versus the classic large volume systems commonly applied in the typical chemical-based pharmaceutical facility. Most biotechnology facilities consist of several suites, each made up of multiple small rooms.

Generally, the closer that the HVAC equipment can be located to the processing areas, and the smaller the volume of 100% once-through air, the lower the installed cost. Pre-engineered, prefabricated, pretested packaged double-wall HVAC equipment—having a relatively small footprint—often provides an advantage in this application. Most packaged HVAC units today utilize direct drive variable frequency control plug fans, lending flexibility to the initial placement orientation and making post-installation balancing easier and faster.

Also, 100% once-through outside air containment facility applications require more extensive MUA (makeup air) pretreatment than that needed for recirculating-type systems. Outside air intakes should be located on the side of the building exposed to prevailing winds and all ductwork should be leak-tested in situ. Individual room pressure control is typically provided by variable frequency direct drive-controlled exhaust fans, with all of the room exhaust receiving HEPA filtration before discharge into the atmosphere.

When multiple small HVAC units are used, a central station PC-based operator interface for control and monitoring must be provided. This interface should support full graphics, system start/stop and status, and access to individual zone devices and environmental conditions. The computer must also provide PID (proportional integral derivative) settings for all control loops and mechanical devices including the ability to change set points, view actual outputs, force outputs for OQ (operational qualification) protocol challenge tests and provide alarm management, data trending, report generation, remote monitoring, networking, and data transfer.

There are NO industry standards (e.g., ISO) that dictate air change rates, although the 1987 cGMP guidelines did call out a minimum of 20 ACPH for controlled areas (typically Class 100K, ISO 8) and EC Regs require a minimum of 20 ACPH in "every room for which particulate levels must be controlled." The current FDA guidelines on "Sterile Drug Products Produced by Aseptic Processing" states "an adequate air change rate should be established for a cleanroom. For Class 100K supporting rooms, air flow sufficient to achieve at least 20 ACPH is typically acceptable..."

ACPH (air changes per hour) is often treated inappropriately as an independent variable as a result of these regulatory influences, and of course ACPH is actually a dependent variable, i.e., ACPH = (Avg. room velocity × 60)/(room height) for a unidirectional flow room and therefore similarly, for a mixed or turbulent flow room, Room CFM = (ACPH × room volume)/60. (Note: average room velocity is a very appropriate descriptor for unidirectional flow situations; however, for mixed or turbulent flow rooms, there is no truly representative average room velocity, so CFM and ACPH become the metrics of interest).

ACPH clearly lead us to conclude how the room will perform in the event of a contamination excursion (i.e., "recovery time"). However, to be able to place some meaningful performance criteria on the initial room (and in the absence of accurate data on particulate generation, which could allow us to perform a mass balance on particles) we must rely on experience and consider the flow rate through the filter, which is a function of filter face velocity rather than average room velocity.

To achieve a high-quality cGMP cleanroom, and one that will consistently perform below desired operational alert and action levels for aseptic processing it necessary to de-rate the HEPA filters (say down to ~350 cfm/filter versus the manufacturer's rating of typically around 750 cfm) and to place MORE filters in the ceiling than one would expect for a non-cGMP cleanroom of similar classification—eliminating the effects of poor distribution due to influences such as drafts, deadspots, etc.

Equally important is the placement of returns, to be able to achieve suitable airflow sufficient to sweep particulates away from the product. Dynamic flow modeling will provide a useful check on the validity of the proposed system configuration.

Architectural Issues

In the design of controlled environments, the use of rounded corners and smooth surfaces to prevent dust shelving and facilitate cleaning, as well as maintaining the integrity of the atmospheric seal, are major concerns. Transitions from walls to floor, walls to ceiling, as well as glazings and corner junctions must be rounded, completely airtight, and smooth. All surfaces must be impervious to moisture, bacterially inert, chemically resistant, and non-particle shedding. These are the pivotal criteria for the elimination of areas that may harbor microbes.

Prefabricated cGMP wall panels are far more appropriate for cGMP facilities than the conventional gypsum board style construction. Gypsum board on stud walls is not impermeable to moisture, and, over time, the "gypboard" inevitably shifts on the stud and cracking occurs. With daily cleaning, solutions will permeate the gypboard and create a non-validatable condition. And unlike concrete masonry unit construction, technical access for process support services is typically provided within the modular wall systems junction, permitting the chasing of small-diameter piping and process and convenience electrical, computer cabling, etc., throughout the entire cleanroom envelope—both walls and ceilings—at each and every juncture.

No exposed piping or conduit is permitted in a GMP facility and prefabricated panels—with integral utility chases—eliminate the mystery of trying to find services that have been "buried" in gypboard or CMU construction by the contractor during initial facility construction.

Details such as fully cleanable double-wall return chases and technical access panels for areas with high densities of process support services are standard details that do not have to be redesigned from scratch each time. When utilized in a design/build facility delivery package, this prefabricated envelope system eliminates the finger pointing that so often occurs on so-called "plan and spec" projects between the architect/engineering firm that prepared the design, and the general contractor responsible for execution.

One should always confirm that prefabricated panels have been subjected to rigorous strength and durability testing including smoke and fire testing, immersion in various hydrocarbon solvents, bodily fluids and typical cleaning, disinfecting, and sterilizing solutions including cationic surfactants, hydrogen peroxide, peroxyacetic acid, sodium hypochlorite, dilute Clorox solutions and sporocidin, all without noticeable effect.

Project Planning

For a 60,000 ft² multiproducts facility built inside a steel building, pre-engineered construction reduced the time required versus that of conventional construction by better than 6 months. After completion of the design phase, wall and ceiling panels, and complete packaged HVAC systems were factory fabricated while site activities such as foundation preparation, shell erection, and roughing-in of process support services was taking place.

Once in the field, the execution phase utilizing pre-engineered or "modular" technology is inherently much faster as modular facilities are built from the ceiling down. Immediately after the roughing in of process support services and major duct trunks, a walkable ceiling system can be hung to provide a working platform to separate activities above the ceiling from those going on below. This permits the constructor to practice parallel trades, e.g., the mechanical work above can be completed while the architectural installation below proceeds. The net result is that field construction time is markedly shortened.

By the way, walkable ceilings are worth every penny! The primary benefit of the walkable ceiling is that there will never be a need to send a maintenance person into the cGMP environment to change a light bulb or calibrate a sensor. As a result there is no need to shut down, clean, and resanitize when performing routine maintenance. The loss of production following routine maintenance can be carefully designed out of aseptic and clean containment facilities. Walkable steel panel ceilings are far better suited than gyp board or tee grid and lay-in tile ceilings for cleanrooms and clean containment. Lay-in tiles and/or membrane ceilings are difficult to clean and cannot provide a crack-free, cleanable surface. Tiles can lift and break sterility, resulting in a total loss of product. According to the FDA, "When in doubt, throw it out!" and we have seen numerous examples of cGMP violations related to lay-in tile ceilings.

Cleanroom replaceable filter media housings are well applied in cGMP facilities. During the routine 6 month DOP integrity testing required to maintain validation, a single technician can, from roomside, inject DOP upstream of the media pack, measure upstream concentration, and scan the filter face for leaks. Should media replacement be required, it can be done with the filter housing left intact in the ceiling, thus never having to open up the room or the ductwork to ambient conditions after HEPA housing installation and ductwork pressure testing have been completed.

A well-documented preventive maintenance program is critical for maintaining systems reliability, which leads to the three inviolate rules for preventive maintenance:

- If it ain't written, it don't exist.
- If it is still, calibrate it.
- If it moves, train it!

The prefabricated cleanroom is also "built clean." Conventional construction, with its spackling, taping, sanding, painting, noxious compounds, and waste disposal issues risks building sources of contamination into the room. Think of it as a form of facilities "Russian Roulette." For example, epoxy or polyester coatings that cover gypboard outgas when new; and because they dry out over time, such coatings crack and shed with rather predictable frequency. Particles sealed into the epoxy during construction will likewise shed when the wall is cleaned or simply leaned against.

With pre-engineered technology, validation proceeds quite a bit faster as well. Since standard, repeatable solutions are utilized, a natural by-product of the pre-engineered design/build approach to facilities delivery is that most of the data required to satisfy installation qualification and operational qualification (IQ/OQ) protocols becomes available upon substantial completion of the facility. In traditional construction, the data needed to satisfy validation protocols is usually left to the end of the project when each supplier submits portions of the documentation to a contractor who then must assemble and write specific maintenance schedules and operating instructions.

Nonprogressive, demountable wall panels provide for flexibility in moving equipment in and out of rooms, particularly where such equipment may be too large to pass through standard-size doors. With no vertical or horizontal members or studs anywhere in the controlled area envelope, maximum flexibility is clearly ensured and bulkhead mounting of any shape or size of process equipment is feasible.

Project Delivery

For another project, the owner decided to deliver the project as a design/build, utilizing a team, including vendors, brought together immediately following the completion of the schematic phase. The team, whose members were selected based on technology and prior favorable experience, worked together to develop a conceptual design and a very high-quality budget estimate. Upon project authorization, the team carried this fast track project to completion on time and on budget, accomplishing what many said could not be done.

The project structure depicted in Figure 5.3 might, at first glance, appear similar to a construction management type of arrangement, but in fact what we have created here is a "virtual" design–build company working together as a team. A strong owner is required to manage such a team.

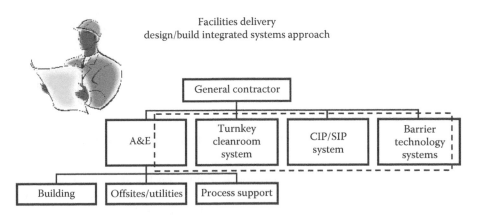

FIGURE 5.3 Example: A virtual design–build company, cleanroom for a pharmaceutical application. (A&E, architecture & engineering; SIP, sterilization in place; CIP, clean in place.)

Bringing the team together early for conceptual design eliminated the most common objection inherent to design/build facilities delivery, i.e., not enough owner control over content.

In the design/build approach, there is a SINGLE CONTRACT ONLY, between the owner and the design/builder. There are no separate, sequential contracts for architects, engineers, and construction managers or general contractors. The design/builder, and NOT the owner, is at risk for quality, schedule, and cost. Because a substantial amount of engineering is required up-front to get accurate cost, schedule, and scope prior to letting a design/build contract, this is often done via a "bridging" or conceptual design contract, in which the design/builder will prepare a complete conceptual design including an accurate (±10%–15%) estimate, a firm scope of work, and a realistic Gantt chart schedule.

Every site has to be evaluated on its own merits. While "stickbuilt" is generally lower first cost (excluding architectural and engineering fees, which can be substantial), so-called "modular" or pre-engineered cleanroom solutions may often qualify for accelerated depreciation versus built-in-place construction. Therefore, on an overall evaluated basis, the modular solution is less expensive. However, it takes a thoughtful analysis to ensure that the correct decision has been made. Modular construction will always be faster than stick-built, and we do not know of any industries today where "time to market" is *not* money!

Basis of Design Preparation

Often the most challenging issue in facility design and construction is the lack of firm process knowledge. Often the procedures are new, and will only be truly tested after the facility is in use. Therefore, all layouts are predictions; and most predictions are inaccurate. In 1986, Louis Sullivan, the Chicago high-rise designer, wrote, "Form ever follows function." I think that Winston Churchill really hit the nail on the head (no pun intended) with his quote "We shape our buildings, and afterwards our buildings shape us."

The FDA cGMPs provide only broad guidance and not very many specific requirements. An example of a GMP expectation that applies to buildings and facilities:

The facility is designed, constructed, and maintained to prevent mix-ups and contamination of the drug product, and cross-contamination between drug products.

GMP requirements specific to this GMP expectation are indicated in the following excerpts from the Code of Federal Regulations:

- Any building used in the manufacture, processing, packing, or holding of a drug product shall be of a suitable size, construction and location to facilitate cleaning, maintenance and proper operation. [CFR Title 21 Part 211.42(a)]
- Operations shall be performed within specifically defined areas of adequate size. There shall be separate or defined areas for the firm's operations to prevent contamination or mix-ups... [211.42(c)]
- Any building used in the manufacture...of a drug product shall be maintained in a good state of repair. [211.58]
- Prevention of mix-ups and cross-contamination by physical or spatial separation from operations on other drug products. [211.130(a)]

So, how does one get specific? Well, pharmaceutical facilities must be designed around the process, and the superordinate goal is a facility that "meets or exceeds the specification." It is in the conceptual stage of the project where the greatest impact on cost can be made. As cleanroom designers and builders, it is in the conceptual stage where we normally begin working with an owner, and/or the owner's representative, to develop an optimized facility. In this preliminary planning stage, modular construction offers an advantage in the preparation of accurate budget estimates. Input for the owners cost model should not vary significantly from programming through commissioning, given relatively consistent scope.

The ISPE (International Society for Pharmaceutical Engineering) *Baseline Guide for Sterile Manufacturing Facilities* asks these questions in its Appendix 3, Section 4:

Product Flows

- At what point does the product become sterile?
- How does it enter the aseptic manufacturing area?
- At what point is the product exposed to the environment?
- How is the product placed into its final enclosure?
- Does the product have to be transferred in its final enclosure before it is sealed?
- How is the product protected until it is sealed?
- At what point is the product considered sealed into its final enclosure?
- How does the product leave the aseptic manufacturing area?

Component Flows

- Do the components need washing?
- Do the components need sterilization?
- How do the components enter the aseptic manufacturing area?
- Do the components need cooling in the aseptic area?
- How are the components fed into the filling machine?
- How is the sterile stopper bowl protected, where is it located?
- How are the components handled after filling and sealing?

Operator Intrusions

- At what points in the process do the operators intervene with the product?
- At what points in the process do the operators intervene with the product's contact components?
- How are the components and product transferred and handled within the aseptic manufacturing area?
- How many operators are required in the preparation area?
- How many operators are required in the aseptic manufacturing area?
- Where will operators stand in the aseptic area under normal operation?

Equipment

- What type of washing equipment is used before sterilization of components?
- What type of sterilization equipment is used to transfer components into the aseptic area?
- Is any accumulation of sterilized product final enclosure required?
- Do any parts of the equipment produce large amounts of particulate loads (will this be considered "background," what are the particulates, are there any OSHA regulations that must be considered)?
- Do the equipment items which have exposed sterilized components or product need regular operator intervention?
- How is equipment maintained, is it from within the aseptic area or from outside the area?

General

- What other items need to enter the aseptic manufacturing area?
- How do other items enter the aseptic area?
- Is there any storage requirements of product contact parts (machine parts, filters, etc.) within the aseptic area?
- Cleaning sterilization regimen?
- Required hours of operation of the facility?

It is amazing how many times we are asked to design and build a facility, before the questions above have been answered. In some cases, it is not even known that these questions need to be asked! The good news is that with the ability to re-arrange, expand, provide "external to the envelope" maintenance access, and even relocate, the owner of a modular facility will have the versatility to adapt to changing process and product requirements quickly, systematically, and cost effectively.

As stated in the chapter introduction, a basis of design document (BOD) consists of following listed minimum facility conceptual design elements:

1. Process description and process flow diagrams
2. cGMP floor plan and general equipment arrangement
3. Sized major process equipment list with utilities requirements/consumption
4. Sized process support services utilities list (e.g., WFI)
5. Functionality flow diagrams (process, people, product, material, components, waste, and directionality of air flows)
6. HVAC zoning and room classifications, including MICROBIAL limits
7. Budget quality cost screening estimate
8. Scope of work matrix (this matrix includes the following elements: design/specify/furnish/install/inspect/test/balance/certify/guaranty/challenge/qualify. It identifies explicitly who is responsible for every element of the HVAC, envelope, and process.)
9. Realistic project schedule from kick-off through validation
10. List of the appropriate/applicable regulatory authorities and jurisdictional venues for which the facility will have to be validated

FDA Reviews

This BOD information will provide the sponsor with the information required to support FDA reviews. There are defined hold points during a project when a meeting with the FDA will be enlightening:

- Design review—concept sketches, flow diagrams, tentative floor plans (should not be conducted prior to completion of conceptual design).
- Preconstruction review—study plans, elevations, HVAC diagrams, all equipment, layouts, process support services (detail design review).
- Construction/equipment installation and qualification review—FDA will review portions of the facility on-site while under construction.
- Preproduction review—typical FDA field inspection.

From the FDA Center for Biologics Evaluation and Research (CBER) Division of Manufacturing and Product Quality (DMPQ), one of our clients received the following list intended as a general overview of information that FDA has found useful to discuss during a pre-facility meeting. This list should not be considered all-inclusive:

1. Purpose of the meeting, with particular questions to be addressed by CBER (if necessary)
2. A summary of your organization, size, location(s)
3. A description of the product(s) and a brief description of the manufacturing process

4. For multiuse facilities
   a. Products to be manufactured
   b. Description of equipment areas that are dedicated to one product or shared with other products
   c. Product changeover procedures
   d. Segregation of product and personnel (where appropriate)
   e. Cleaning validation and product residual testing plans
5. A product process flow chart
6. Floor plans with the following flows described: manufacturing, personnel, waste
7. An overview or the HVAC system, including
   a. A floor plan showing AHUs (zones covered by each unit)
   b. A floor plan showing pressure differentials for containment/protection
   c. A diagram showing air quality classifications of rooms and areas
   d. A description of the operation of the system (air changes, makeup air, single pass, etc.)
8. A description of the support systems (WFI, clean steam, gases) and monitoring of these support systems (for instance, frequency and type of monitoring for the WFI system)
9. Procedures, equipment, or facilities that minimize contamination or cross-contamination (method of transfers such as steam in place or aseptic connections, gowning procedures, etc.)
10. For drug product manufacturing (final filled vials), a description of the process, including preparation of components, should be described. An overview of the validation of aseptic processes, including media fills, should also be discussed
11. Environmental monitoring program, including
    a. Description of the locations and frequency of the monitoring
    b. Action/alert limits
    c. Viable/nonviable particulate monitoring
    d. Surface monitoring
12. A brief description of validation procedures (IQ, OQ, PQ—installation qualification, operation qualification, performance qualification or verification) including the validation master plan, if appropriate
13. Timeline and type of scheduled submissions
14. Description of licensing issues (shared or cooperative manufacturing, contract locations)
15. Other issues such as viral inactivation procedures, scale-up or anticipated modifications, production schedules, etc.

Turnover Package

After completion of a BOD, and assuming that we do not have to go back and do significant "value-engineering" (although at least now we have a rational, analytical basis for value engineering), we can proceed to the generation of the elements of system detail design. System Detail Design is often best presented as the documentation deliverables required at project closeout.

The following defines a typical Turnover Package (TOP). This is the type of documentation that is to be provided under good engineering and construction practices. The goal is to identify documentation that will readily allow operating companies to demonstrate that the vendor/contractor provided goods and services that comply with good manufacturing and good laboratory practices.

The TOP provides the information that will allow the owner to validate/qualify, operate and maintain the equipment and facilities. TOPs are useful in that they (Table 5.1) provide the information necessary to write and execute the IQ and OQ protocols.

Responsibilities

Proper documentation is as important as the final equipment or facility.

TABLE 5.1 Systems That Typically Should Have TOPs

Architectural finishes	Labelers
Blenders	Lyophilizers/freeze dryers
Bioreactors	Mixing tanks
Capping machines	Modular cleanrooms
Chromatography/separation systems	Neutralization systems
Chilled water/circulating glycol	Plant steam and condensate
Coldrooms	Process tanks/pressure vessels
Clean steam	Refrigerators
Clean-in-place (CIP)	Separation/purification equipment
Compressed air	Specialty gas systems
RO/DI water systems	Steam generators/pure steam systems
Dry Heat sterilizers/ovens	Steam sterilizers
Fermentation equipment/bioreactors	Stopper washers
Freezers	Solvent recovery systems
Filling equipment	Tower water systems
Filtration/purification	Waste inactivation systems/"kill"
HVAC: environmental controls	Systems
Incubators/environmentally controlled chambers	Vial inspection systems
Water for injection (WFI) Systems/pure water systems	Vial washers
	Washers (glass, tank, carboys, etc.)

The design/build team must ensure the assembly of detailed testing, inspection, and procurement packages that reflect as-built conditions. This includes construction drawings, specifications, and electronic files that are compatible with the owner's CAD and word processing systems.

The design/build team will be responsible for TOP documents for all contracted services and equipment. This includes all work provided by the subcontractors.

Formal acceptance by the owner serves to document that all TOP requirements have been realized.

Changes from original design documents must be reflected in the final TOP documentation. This is normally provided in the form of "as built" and "record" drawings and specifications.

Architectural

The architectural turnover package includes documentation prepared during design and construction of the facilities. These are the documents that should be the result of the SOPs (standard operating procedure) of the design architect, owner, general contractor, and subcontractors. These documents record the design and installation checkout of the various finishes (Tables 5.2 and 5.3), furnishings, and specialty (Table 5.4) items for facilities that will be subject to validation testing and regulatory inspections.

The following are guidelines to inclusions in a comprehensive package of drawings and specifications:

1. Matrix index—lists the required data contained in each section
2. Summary of work—written description defining the project scope
3. Architectural specifications
4. Record drawings (as-built)
5. Mechanical completion letter with "Certificate of Occupancy" (upon completion of punch list items)
6. Room data sheets—defines equipment, utility requirements, cleanliness class, room finishes, air balance reports, room static pressures

TABLE 5.2 Architectural Finishes

Element	Comments
Cleaning agents and procedures	MSDS sheets and manufacturer's instructions
Glazing components	Types, cleaning procedures
Chemical resistances, wash down capabilities	Walls, floors, and ceilings
Hardware	Types, cleaning procedures
Wall protection system	E.g., impact panels, bumper guards, door panels
Installation inspection	
Room finish schedule	Matrix format
Floor loading capacities	Defined by structural determinations

TABLE 5.3 Furnishings

Laboratory casework drawings; plans and elevations
Lighting levels: design and as-found conditions
Floor mats
Fire extinguishers, cabinets and accessories
Storage cabinets—flammable and hazardous materials storage
Countertops—chemical resistance
Specialty stainless steel construction
Installation inspections
Emergency eye wash stations and showers

TABLE 5.4 Specialties

Sealants and caulking	Clean room curtains
Pass-throughs	Biosafety cabinets: testing and certification documents
Cold boxes and environmentally controlled rooms	White boards/bulletin boards
Scales	Special signage
Fume hoods: testing and certification documents	Moisture proofing of slab; use of diaphragm materials
Moisture proofing of roof; use of diaphragm materials	Special equipment or features

7. Code and regulatory analysis—type of construction, occupancy classifications, occupant load data, seismic criteria, ADA (Americans with Disabilities Act) requirements, biosafety level, hazardous materials inventory
8. Warranties and certifications
9. Operation and maintenance manuals
   a. Walls, floors, and ceiling finishes: cleaning and repair procedures
   b. Wall protection systems: installation instructions, impact resistance features and cleaning procedures
   c. Cold boxes and environmentally controlled rooms: system operation and capacity, monitoring devices, lights, materials, access, and methods of construction
10. Purchase orders, field orders, and change orders in chronological order

TABLE 5.5 Basic Specifications of Packaged Equipment

Specification of operation	Operation and maintenance manuals
	Equipment specifications
	Description of operation
	Trouble shooting guide
	Preventative maintenance schedule
Safety	Engineering data and reference
	Spare parts list
	Vendor data, if applicable

Equipment

Packaged equipment is process equipment consisting of multiple components that could be made to operate independently of any other equipment (Tables 5.5, 5.6, and 5.11).

Table 5.7 describes documentation required for the design, fabrication, and installation of vendor equipment, free-standing pieces of equipment purchased independently of a vendor package or system. Examples of equipment for which this guide would apply are pumps, heat exchangers, and vessels. This guide would not necessarily apply to a vendor skid or package that consists of multiple pieces of equipment, instrumentation, piping, and/or controls from various vendors or "original equipment manufacturers" (OEM).

HVAC

This section covers the documentation required for the design, fabrication, and installation of HVAC systems. Deliverables required for the HVAC portion of the job are indicated in Table 5.8.

Piping

This section covers the turnover documentation prepared during the design, fabrication, and installation of sanitary and non-sanitary piping systems. The documentation should be a sequential accumulation of all inspections, testing, and daily activity as defined in the project specifications and contractors' standard operating procedure (SOP).

Sanitary Piping Systems

Sanitary piping systems refer to piping systems that directly contact raw material streams, intermediate product streams, finished product streams, pure water streams, and pure steam used for sterilization–sanitization purposes.

Documentation   The following documentation should be contained within separate sections in a three (3) ring binder formulated by the contractor or third party inspection firm.

Material inspection record of all materials prior to use
Weld procedure for type of welds to be joined qualified to section IX of ASME Code
Welder performance qualification record for each welder to section IX of ASME Code
Purge gas quality with certificate of analysis for each container used on project
Record of weld and inspection for each weld within the system
Disposition of any rejected weld
Record of weld coupon samples
Record Isometric drawings with location of welds and weld numbers
Record of slope verification

TABLE 5.6 Testing of Packaged Equipment

Testing/ Documentation	Test/Guidelines/Drawings	Details
Controls testing	Factory acceptance test	Check off sheets Package check sign off input/output (I/O) verification results Alarm set-points testing and results
	Site acceptance test	Check off sheet Package check sign off Input/output (I/O) verification results Alarm set-points testing and results
Installation documentation	Installation guidelines	Proper unloading and uncrating techniques Recommendations for handling and setting in place Utility connection details Grounding requirements Mechanical isolation details
	Pre-start-up checklist	Control setup Field adjustments
	On-site operational test	If applicable Performed and documented by supplier
	Piping and instrumentation diagram	Recommended utility Requirements (layout, flow rates) Instrument list
	Equipment arrangement/layouts control panel layouts P&ID symbols and definitions Control system wiring diagrams Electrical wiring diagrams, schematics	
	As-built drawings	Signed/dated by manufacturer/fabricator
Additional documentation	Material specifications/certifications	All materials used (tubing, fittings)
	Sanitary pipe documents Welding procedures specified and used Weld log books/documentation Welder qualifications Traceable instrument Calibration (NIST) Passivation reports Cleaning procedures Leak test reports	
	Code certifications	E.g., ASME, Seismic, NEC

Record of pressure test
Certificate of compliance for components and devices
Record of equipment calibration
Record of passivation

The above items should be included as a minimum but not limited to this listing.

Documentation Format Each document/form should contain an area to identify who performed the task, initials of the inspector, and if the inspection is acceptable or rejected. Project specifications and contractor SOPs should identify the requirements for each report. Third party inspection agencies generally provide their own forms and reports.

TABLE 5.7   Required Vendor Equipment Documentation

Category	Documents/data
Procurement documents	Purchase order
	Equipment data sheets
	Equipment specifications
	P&ID (process and instrumentation diagram) issued for construction, if applicable
Vendor data	Certified fabrication drawings
	ASME (American Society of Mechanical Engineers) code Certifications (U-1 form)
	Written verification of conformance to Seismic Zone requirements
	Complete list of materials, including insulation
	Certified material mill test reports
	Completed data sheets
	Cut sheets
	Welding procedures
	Welder certifications
	Polishing procedures
Inspection and testing	Factory acceptance test procedures and documentation of results
	Cleaning/passivation procedures and documentation
	Documentation of surface finish inspection
	Site acceptance procedures and documentation of results
Operating and maintenance (O&M) manuals	Installation instructions
	Operating procedures
	Preventive maintenance procedures/schedules
	Spare parts list
Other	A. As-built P&IDs, if applicable
	B. As-built specifications

TABLE 5.8   HVAC Documentation

Category	Documentation
Specifications and design drawings	Specifications and drawings in as-built state
	Instrumentation list and data sheets
	Air handler fan curves and data sheets
	Coil data
	Exhaust fan curves and data sheets
	Blower fan curves and data sheets
	Thermostat specifications
Ductwork documentation	Cleaning
	Pressure testing
	Layout drawings in as-built state
Controls and software system description	Sequence of operation
	Functional description/schematics
	Control logic
	Start-up, operating and maintenance manuals
Startup/commissioning documentation	Air balance (TAB) report with room static pressures
	Wet side balancing: hydraulics/hydronics
	HEPA filter testing and certification (DOP challenges and velocity/air supply rates)
	Room air change rate calculations
	Calibration certification/factory testing

Non-Sanitary Piping Systems

Non-sanitary systems are defined as utility piping in facilities that manufacture FDA-regulated products. Utility piping systems may or may not require documentation. Review the project specifications for specific documentation requirements.

Documentation Recommendation It is recommended that welders qualified to section IX of the ASME Code be used for welding. The following documents are recommended to be within the turnover package:

> Material test reports for piping and associated pipe components
> Letter of compliance for valves and specialty items
> Welder performance qualification record
> Weld inspection records
> Flushing of system(s)
> Record of pressure test
> Verification of Slope check
>> Hanger/supports
>> Commodity labels with arrows

The above items are recommended but not limited to the listing.

Electrical

An electrical turnover package includes documentation related to electrical distribution systems and components. The electrical distribution system is defined as power wiring beginning from the local utility supplier and that which distributes power to points of use within a designated area, facility or equipment skid. The electrical distribution system includes components that condition, terminate, and disconnect electrical power and are typically supplied by the Electrical Contractor and Equipment Manufacturers. All testing is performed according to NETA, ASTM, ANSI, IEEE, and local permitting guides. Documentation can be checklists and/or signed and dated marked-up drawings. This level of documentation is required on the systems indicated in Table 5.9.

Specifications and as-built drawings in the package include electrical specifications, single-line diagrams/schematics, process motor wiring diagrams, shop drawings, point-to-point wiring diagrams, and local control panel cross-references for point-of-use locations.

The following summarizes the pertinent certifications, testing, reports, and checklists.

Electrical equipment inspection report:

A. Equipment grounding verification
B. Wiring tags and termination tags

TABLE 5.9 Systems Requiring Electrical Turnover Documentation Package

Emergency power
Uninterrupted power supply (UPS)
Grounding
Lighting systems
Fire alarm system and voice line transmission capability
Intrusion alarm system and voice line transmission capability
Specialty low voltage systems
Specialty voice/data transmission and networking runs

Cable termination inspection report:

A. Verification of physical, electrical, and mechanical condition
B. Statements of proper anchorage, required clearances, physical damage, and proper alignment
C. Documentation of cleaning of switch gear
D. Documentation of inspection of insulators for evidence of physical damage or contaminated surfaces
E. 12 kV and greater: documentation of fire taping, suspension correctness, correctness of termination
F. 12 kV cabling factory test reports
G. Megger test reports

Switch gear test report:

A. Document switch gear data (identification, manufacturer, serial numbers).
B. Document inspection for physical, electrical, and mechanical condition.
C. Document checks for proper anchorage, required clearances, physical damage, and proper alignment.
D. Document cleaning of switch gear.
E. Document inspection of insulators for evidence of physical damage or contaminated surfaces.
F. Document performance of insulation resistance testing on each BUSS section, phase-to-phase testing, and phase-to-ground testing.
G. Perform control wiring performance tests to verify satisfactory performance of each control feature and provide documentation.

Transformer test report:

A. Note transformer data (identification, manufacturer, serial number).
B. Document physical, electrical, and mechanical condition.
C. Document checks for proper anchorage, required clearance, physical damage, and proper alignment.
D. Document turn ratio.
E. Document that mounting is correct.
F. Document correctness of grounding.

Motor checklist:

A. Confirm motors were sized to specifications; could be in tabular form.
B. Verify protection procedures have been followed.
C. Document rotation test results.

Breaker testing report:

A. Note breaker data (identification, manufacturer, serial numbers).
B. Document physical, electrical, and mechanical condition.
C. Document checks for proper anchorage, required clearances, physical damage, and proper alignment.
D. Verify cleaning of breakers.
E. Document inspection of insulators for evidence of physical damage or contaminated surfaces.
F. Document trip testing of breakers.

Motor control center (MCC) test report:

A. Note switch gear data (identification, manufacturer, serial numbers).
B. Document inspections for physical, electrical, and mechanical condition.

C. Document checks for proper anchorage, required clearances, physical damage, and proper alignment.
D. Document cleaning of switch gear.
E. Document inspection of insulators for evidence of physical damage or contaminated surfaces.
F. Perform insulation resistance testing on each BUSS section, phase-to-phase, phase-to-ground testing, and document results.
G. Perform control wiring performance tests to verify satisfactory performance of each control feature and document results.
H. Test over load relay and document results.

Panel board checklists:

A. Document inspections for physical, electrical, and mechanical condition.
B. Check for proper anchorage, required clearances, physical damage, and proper alignment, and document results.
C. Document cleaning of breakers.
D. Document inspection of insulators for evidence of physical damage or contaminated surfaces.
E. Supply detailed listings of breaker labels.

Circuit and outlet assignment checklists:

A. Verify outlets are labeled and referenced to a circuit panel.
B. Verify that emergency circuits are designated.
C. Verify that dedicated circuits are labeled.
D. Document that special grounding requirements have been satisfied.

Emergency power load test report:

A. Document testing under connected load conditions.
B. Document testing with an external load at 150% of connected load.

UPS load test:

A. Document testing under connected load conditions.
B. Document testing with an external load at 150% of connected load.

Ground fault (GFI) test report:

A. Document inspection for physical damage and compliance with plans and specifications.
B. Document inspections of neutral main bonding connections for zero sequence sensing system is grounded on line side of sensor, ground connection is made ahead of the neutral disconnect link, ground conductor (neutral) is solidly grounded.
C. Document inspection of neutral main bonding connection to ensure adequate capacity for system.
D. Document inspection of zero sequence systems for symmetrical alignment of core balance transformers about all current carrying conductors.
E. Document testing of the relay pickup current by primary injection at the sensor and operate the circuit interrupting device.
F. Test the relay timing by injecting 150%, 200%, and 300% of pickup current and report findings.

Emergency generator test report:

A. Megger armature windings and report results.
B. Calibrate over current relay and provide documentation.
C. Test for vibration acceptability and document results.

D. Check voltage and amperage output and document results.

E. Document that testing for surge test is within specified limits.

Wire type and size verifications:

A. Check wire type and size; compare them to as-built specifications and document.

Transformer load test reports:

A. Note transformer data (identification, manufacturer, serial numbers).

B. Document inspection for physical, electrical, and mechanical condition.

C. Document testing for rate load correctness.

D. Test primary and secondary phase A, B, C, and neutral; test voltage, amperage, RMS, peak and frequency, and document all results.

Panel load test reports:

A. Document panel data (identification, manufacturer, serial numbers)

B. Document panel rating test results.

C. Document panel load test data.

Calibration certifications (on power mains) documentation:

A. Voltage meters: ±0.5%

B. Amperage meters: ±0.5%

C. Kilowatt hour meters: ±0.5%

Support documentation:

A. Document panel and equipment identification matrix

B. Reference specific cut sheets.

C. Document utility confirmation-transformer, line size, and bill rate.

D. Installation/operation/maintenance manuals.

E. Reference purchase orders.

F. Specify recommended spare parts lists.

Controls and Software

This section covers the turnover of critical documentation prepared during the design and fabrication of the instrumentation and control systems. These documents record the design and checkout of the instrumentation and control systems that have been specified and installed.

Controls and Instruments

Deliverables required for the controls portion of job consist of the following:

Equipment specifications

As-built data sheets on all the instruments (normally to the ISA format).

Instrument list for all instruments on the project: instrument numbers, manufacturers, serial numbers, panel locations, etc.

Panel drawings—electrical and assembly
Cable schedules (if applicable)
Loop drawings
Manufacturers manuals and cut sheets
Operation and maintenance manuals
Completed bench check forms

Completed loop checkout forms
Completed field calibration forms (if applicable)
Control system architecture

Software

Deliverables required for the software portion of job consist of the following:

Software Design Description This document is the first one developed and is the controlling document. It is typically developed from the P&IDs and the process description. It specifically details the control system operation as well as describing the operator interface. It includes software listings, critical parameters, security-access levels, system recovery capabilities, and screen hierarchy.

Software Test Plans This document tests the integrity, functionality and security of the code as it relates to the software design description. In addition to the normal operation, the software's response to abnormal conditions is tested. This tests the software and control system as completely as possible using simulated inputs and viewing outputs.

Operational Tests This document tests the functionality of the software controls and equipment. This will functionally test the entire system prior to turnover to the owner.

Programming Documentation This is the normal documentation printout of the logic, controller configuration, operator interface, and the distributed control system (DCS) printouts. These should have comments explaining each section of code in plain English so a nonprogrammer can determine the function of each section of code. In ladder logic, this is handled by Rung Comments, in C code, basic, etc. This is handled by "Comments." In some systems that do not support these types of commenting, a separate document may need to be provided explaining the sections of code. Systems such as DCS configuration do not require additional commenting.

Commissioning

Commissioning is the process of ensuring that all building and process systems are designed, installed, functionally tested, and capable of operation in conformance with the design intent. Commissioning process steps generally include system documentation, equipment start-up, control system calibration, testing and balancing, performance testing, and release of the systems to the owner for validation. Some important elements of a commissioning plan are summarized in Table 5.10.

Direct Impact Systems

"Direct impact" systems are those that directly affect product quality, and these MUST be validated. All other systems (i.e., "indirect impact" and "no impact") need be commissioned only. Indirect impact systems support direct impact systems.

For systems that need to be validated, this occurs by executing protocols that serve to qualify those particular systems. If the commissioning activities for these direct impact systems are executed properly, much of the work done to commission these systems can also be used to qualify the systems. For indirect and no impact systems that do not require formal protocols, commissioning can utilize standard SMACNA (Sheet Metal and Air Conditioning Contractors' National Association), IEEE (Institute of Electrical and Electronic Engineers, Inc), etc., checklists and forms. Vendors of packaged equipment can often satisfy large portions of their commissioning requirements through FAT (Factory Acceptance Test) and SAT (Site Acceptance Test) execution.

Overall, the project and its various stages, with respect to recommended validation activities, can be envisioned easily with a simple Gantt chart, Table 5.11.

TABLE 5.10　Elements of a Commissioning Plan

Hydronic balance testing	Verification of the proper operation of coils, air handlers, fans and filters, including test sequences, shut-down and start-up
Sound measurement testing	
Vibration testing	
Alarms and interlocks testing	P&ID walkdown, resulting in an "as-built" P&ID
Air flow rate testing in ductwork	Utilities check
Air volume supply and return, testing and balancing	Instrument calibration
Fan RPM and amperage	Electrical power tests
Temperature, humidity (coil duties) and static pressure testing (duct leakage)	Motor run tests
	Lubrication checks
Differential pressure testing and Balancing	Isometric drawing checks
Loop checks	Safety checks
HEPA filter integrity testing	

TABLE 5.11　Example of Validation Activities for Life Science Cleanroom Construction

Validation Time Line					
Conceptual Design	Preliminary Engineering	Detailed Engineering	Construction	Validation and Startup	Operation
Validation input					
	Master plan				
		FDA meeting			
	Engineering design review				
	Protocol development				
			Construction/ installation review		
			Commissioning and Installation qualification		
				Operational qualification	
				Process qualification	
				Summary reports	
				FDA review	
			Change control		

Conclusion

The manufacture of pharmaceutical and biotechnology products requires that the appropriate level of quality be designed and constructed into the facility and into the systems that support the production process.

The goal is to avoid receiving a warning letter or "Notice of Inspectional Observations" (often referred to as a Form 483 or simply a 483) from the FDA. A review of recent 483 observations and warning letters indicates that the FDA's current compliance focus is on "inadequate facility design" and "environmental and personnel monitoring." One can only conclude that the skill level, training, and attitude of the personnel involved were inadequate in the cases cited with respect to the obvious requirement to minimize particulate, microbial, and pyrogen contamination.

And, why worry about HVAC? Does not everyone know how to design and build these ubiquitous systems? At the 1997 PDA (Parenteral Drug Association) Spring Conference in San Diego ("Directions for Parenterals"), Jeffery Yuen of the FDA presented a session entitled "Inspectional Findings Today." He stated that

> HVAC issues have received much attention…Inadequate IQ/OQ/P&IDs have led to poorly defined HVAC systems which (in turn) led to poor validation of these (same) systems. This is especially the case with many older, redesigned facilities.
>
> HEPA leak testing criteria, for example, should specify what is considered to be a major leak; smoke studies performed under static and dynamic conditions should demonstrate unilaminar airflow at the critical height of the fill; specifications on HEPA to HEPA variations in air velocity should be established since widely divergent velocities have the potential for turbulent air flows; and companies must be prepared to explain why and how they deal with the situation.

At the same meeting, Cristina Kerry of CBER presented some common 483 inspectional observations. Kerry began with personnel issues, noting observations related to improper control of area access, improper gowning, inadequate process, and GMP training. Regarding building and facilities, she said that

> …older manufacturing sites are not being maintained, and facilities designed for single products are being used for multiple product production. There have been documented failures in provisions for separating or defining areas to prevent contamination, cross-contamination, or mix-ups. Inspections have also revealed poorly designed or inappropriate air handling systems, a lack of temperature/humidity monitoring, and poorly designed water systems.

How do these situations develop? While from time to time the European Union may provide some specific recommendations to meet cGMPs, the FDA normally does not dictate how a specific outcome is to be achieved. A classic example comes to mind when recalling a number of unidirectional flow aseptic fill cleanrooms that, once built, proved unable to meet validation requirements. In most of these situations, the aseptic fill application was treated as if it were simply a Class 100 particle count requirement without regard for the critically important airflow patterns required to ensure that exposed products and components are protected from contamination.

This control aspect, as demonstrated by a suitable and appropriate flow of air that "washes" the surfaces within the critical zone and exhibits unidirectional flow, is essential for proper performance and is more important than any superimposed velocity value.

When observing an operation, the FDA assesses whether the design creates potential contamination routes, e.g.,

- Does the design adequately incorporate appropriate separation and control measures for the differing levels of air quality required by a particular operation?
- Are material choices (i.e., composition of material and surface quality) consistent with the need for cleaning, sanitization, and sterilization?
- Is a maintenance program in place that appropriately addresses the gradual breakdowns in facility infrastructure?

Keeping all this in mind, we also need to consider where we are in the drug development phase in planning our facility. For early clinical production, the relatively small volume demands and low numbers of operating personnel may allow GMP requirements to be met using protocols and procedures rather than capital-intensive equipment and facilities. We need to be diagnostic in our approach to facilities solutions!

The successful design and realization of a cleanroom or containment project will not happen by chance. The time spent in the preparation of as complete a "Basis of Design" as is practicable will be saved many times over during the actual implementation.

Facilities are always asked to provide greater throughput and are utilized longer than originally anticipated. It is increasingly difficult to predict whether or not today's plant will meet tomorrow's facility requirements. The design goal today should be to provide the owner with future planned-engineered contingency. There is a universal rule—"All Buildings Grow"—and today, probably more money is being spent on changing existing facilities than on building new ones.

Further Readings

EC guide to good manufacturing practice. Revision to Annex 1.

Federal Standards, FED-STD-209E. Airborne particle cleanliness classes in clean rooms and clean zones, September 1992.

Food and Drug Administration. http://www.fda.gov/foi/warning.htm.

Food and Drug Administration 21 CFR (Code of Federal Regulations) Parts 210 and 211. Current Good Manufacturing Practice.

Food and Drug Administration. Sterile drug products produced by aseptic processing.

R. L. Friedman and S. C. Mahoney *Risk Factors in Aseptic Processing*, Food & Drug Administration, Center for Drug Evaluation and Research, American Pharmaceutical Review, Spring 2003.

ISO-14644-1. Cleanrooms and associated controlled environments, Part 1: Classification of air cleanliness.

ISO-14644-2. Cleanrooms and associated controlled environments, Part 2: Testing and monitoring to prove compliance with ISO 14644-1.

ISPE Baseline Guide, Vol. #3, Sterile manufacturing facilities.

ISPE Baseline Guide, Vol #6, Biopharmaceutical manufacturing facilities.

ISPE Baseline Guide, Vol #5, Commissioning and qualification.

PDA Draft Technical Report. Points to consider for aseptic processing.

PDA Technical Report No.13. Fundamentals of an environmental monitoring program.

Product Quality Research Institute. Aseptic Processing Work Group. Final Report, March 12, 2003.

USP 26 <1116>. *Microbiological Evaluation of Cleanrooms and other Controlled Environments*. USP, Rockville, MD.

6

Validating and Monitoring the Cleanroom

Overview 89
Cleanroom Design 89
Principles of Cleanroom Validation and Testing 90
Cleanroom Validation...90
So How Does One Determine What Tests Need to Be Performed
during a Cleanroom Validation? • Microbial Validation Testing •
Validation Report
Validation versus Monitoring 95
Routine Particulate Monitoring • Establishing an Environmental
Monitoring Program
Summary 100
References..101

Kevina O'Donoghue
*Specialised Sterile
Environments*

Overview

Many have fallen into the trap of spending huge amounts of money building a cleanroom with little thought and understanding of what is required to get it up and running and the subsequent maintenance of it. It is vital, even at the start, to see that the design of the cleanroom, with regard to the contamination controls required specifically for the product that is to be manufactured within it, is exactly right. Too often not enough time and thought are put into this initial phase with serious consequences with regard to cost of modifying or redesigning the facility and having protocols that cannot be effectively implemented. Validation of the cleanroom takes time, patience, and understanding to complete but is vital to get right the first time. This will ensure that the cleanroom is controlled from the start, that the process within the room is not posing a risk to the environment, that personnel working within the cleanroom are following protocol, and that products manufactured or processes performed within it will be consistently and reproducibly safe and of the highest quality. The two main parts to consider when performing a cleanroom validation are certification including classification and measuring microbial contamination levels. Once validation has been completed, a comprehensive plan should be established for the routine monitoring and maintenance of the cleanroom. This chapter discusses designing, validating, and monitoring of cleanrooms.

Cleanroom Design

A large number of new clients contact the author when they have their cleanrooms built to find out that they may not need a cleanroom of such a high class, that the support rooms may be too small for the number of people going through, that they have no place for waste or finished products to leave the cleanroom except through the gowning room, that they have no material transfer room included, that they are unsure of what kind of services will be required in the cleanroom that will lead to further drilling and construction at a future date, and so on. In other words, they have a cleanroom built that will now not service the requirements of their current product. Making changes at this stage can become very costly and inconvenient in comparison to building and designing it correctly in the first place.

The process itself should be the main driver of the design of the cleanroom. The process should dictate the level of contamination control required, and hence the class of the cleanroom, not the industry (Hansz 2008). Unlike many facilities, a cleanroom must be designed from the inside out. All efforts to control contamination within a cleanroom are directed at people; therefore the flow and movement of people within the cleanroom must be taken into consideration during the design. In the author's opinion, this is one of the most neglected aspects of cleanroom design. The process within the cleanroom should have a layout that minimizes the movement of personnel as much as possible.

Supporting rooms are vital for a cleanroom to work effectively; i.e., a spacious gowning room is required for cleanroom personnel to enter and leave concurrently and gown without difficulty; the gowning room must allow enough space to store all required apparel. A material transfer room or hatch that incorporates a wipe down area is fundamental. All materials, paperwork, and equipment should enter the cleanroom through the material transfer room or hatch after receiving a thorough clean down. A separate area or transfer hatch that allows the transport of finished products out of the cleanroom may also be incorporated into the design. One of the most forgotten areas during the design stage is how waste is removed from the cleanroom. If no facility has been put in place during the design to allow waste to leave the cleanroom area, more often than not, it is removed through the gowning room. This undesirable practice in turn will affect the process flow and can have serious implications on contamination levels within the gowning area.

The greater the considerations as to the exact contamination controls that are built into the cleanroom design, the more appropriate the specific cleanroom facility design will be for the manufacturing process and operations in question. The appropriate level of quality should be designed and constructed into the facility and systems that support the production process. The development of the cleanroom design, based on the contamination controls required for the process, should necessitate the development of the operational protocols, the first one being the cleanroom validation protocol.

Principles of Cleanroom Validation and Testing

In order to show that the cleanroom environment is in control, it is necessary to demonstrate that the air supplied to the cleanroom is of sufficient quantity to dilute or remove the contamination generated within the cleanroom and that the air supplied to the cleanroom is of a quality that will not add significantly to the contamination levels within the cleanroom. It is important that the air moves in the correct direction from clean to less clean areas and that the air movement within the cleanroom demonstrates that there are no areas within the room with high concentrations of contamination (Whyte 2010). It is also important to demonstrate, from a microbial perspective, that operations and manufacturing can be performed within an environment that meets its microbial contamination criteria.

Cleanroom Validation

Within highly regulated environments such as medical device and pharmaceutical industries, there is a requirement to provide an appropriate amount of assurance that critical processes can be performed within controlled conditions in order to produce a final product that is of eminent quality, reliable, and safe for the end user.

Despite popular assumptions, cleanroom validation is, unfortunately, more than just counting particles. It encompasses many different tests that have to be carried out in various cleanroom states in order to show that the cleanroom is fit for its intended use and that the cleanroom meets the required classification.

TABLE 6.1   Different Phases and Occupancy States of Testing Performed during a Cleanroom Validation

Testing Phase Terms	Occupancy State	Test Results
As built testing—phase 1, installation qualification, post-build approval	Testing is performed where the installation is complete with all services connected and functioning but with no production equipment, materials, or personnel present	Proves that the environment was correctly installed and meets its intended design specification
At rest testing—phase 2, operational qualification	Testing is performed when the equipment is installed and operating in a manner agreed upon by the customer and supplier, but with no personnel present	Proves that all parts of the installation operate together to achieve the required conditions. This information can be used as baseline data indicating normal conditions of the cleanroom environment before production commences
Operational testing—phase 3, performance qualification	Testing is performed when the installation is complete and functioning in the specified manner, with the specified number of personnel present and working in the manner agreed upon	Proves that the completed installation achieves the required operational performance with the specified process and a maximum number of personnel working within the environment

Source: Data taken from IS EN ISO 14644-1 classification of air cleanliness.

There are an assortment of standards and guidance documents available to help with this, such as IEST CC006.3 Testing Cleanrooms, IEST CC001.4 HEPA and ULPA Filters, IEST CC012.1, *Considerations in Cleanroom Design*, and the *European Union Guide to Good Manufacturing Practices* (EU GGMP).

However, IS EN ISO 14644 is the prime standard adhered to for validation of cleanroom environments. This standard does not consider specific processes within the cleanroom. Rather, it provides constructional guidance for the start-up and qualification of the cleanroom environment detailing the basic elements needed to ensure continued satisfactory operation and maintenance.

According to the IS EN ISO 14644-2 standard, every time a cleanroom is put into operation initially or changes its intended use, a validation must be performed. The initial setup of a cleanroom requires a validation to be performed over a specified period of time to ensure that the cleanroom is functioning as required over the given period of time. Over this period of time, historical data are collected to ensure that the cleanroom is performing effectively. Changes made to an existing cleanroom require an assessment of how the changes will affect the cleanroom as this in turn will decide how extensive the revalidation will be. Either way, this testing certifies that the cleanroom environment meets the stated standards, protocols, and design criteria.

As mentioned, validation testing is performed when the cleanroom is in different phases or occupancy states. The various terms used for the phases or occupancy states are detailed in Table 6.1.

Therefore, when a cleanroom is certified to a specific class, the room performs to a standard that meets or exceeds the performance of that class under a specific occupancy status.

So How Does One Determine What Tests Need to Be Performed during a Cleanroom Validation?

The majority of companies get an independent testing and certification body to outline an appropriate testing program and perform the required testing. IS EN ISO 14644-1 determines the type and frequency of testing required to conform to the standard. Some tests are mandatory and some are optional. A typical testing program will include the following tests:

Airflow Volume and Velocity Tests

The more clean air supplied to the cleanroom, the cleaner the room will be. A cleanroom must have sufficient clean air supplied to dilute or remove any airborne contamination that may be present. This air supply to the cleanroom is often reported as air changes per hour. Air change rates within a cleanroom will usually be equal to and above 20/h; however, this measurement should be based on the level of contamination control required within the cleanroom, the number of people present or the level of activity within the room, the size of the room, and the process itself. In unidirectional cleanrooms, air supply velocity is measured. Measuring of the airflow is applicable in all three of the designated occupancy states.

The FDA recommends a velocity of 0.45 m/s ±20% and the EU GGMP suggests a range of 0.36–0.54 m/s. These are guidance values. The air supply volumes and velocities should be decided at the design stage.

HEPA/ULPA Filter Installation Leak Testing

High-efficiency air filters that have been installed within the cleanroom need to be checked to ensure that they are efficiently removing particles from the supply air and will allow an air supply of high quality to enter the cleanroom. Filter systems, therefore, must be tested to ensure there are no leaks or faults in the filters themselves or in their housings.

Air Movement Visualization

Visual airflow patterns are a function of airflow velocity, airflow direction, room design, HVAC layout, air change rates, and equipment layout (Cleanzone Technology). Demonstrating the air movement within the critical areas of the cleanroom, in particular with unidirectional airflow, is important to ensure that airborne contamination has been swept away from personnel and the product, thus minimizing the risk to the finished product.

The FDA Guidance for Industry suggests that

> Proper design and control prevents turbulence and stagnant air in the critical area. Once relevant parameters are established, it is crucial that airflow patterns be evaluated for turbulence or eddy currents that can act as a channel or reservoir for air contaminants. In situ air pattern analysis should be conducted at the critical area to demonstrate unidirectional airflow and sweeping action over and away from the product under dynamic conditions. The studies should be well documented with written conclusions and include evaluation of the impact of aseptic manipulations and equipment design. Videotape or other recording mechanisms have been found to be useful aides in assessing airflow initially as well as facilitating evaluation of subsequent equipment configuration changes.

Airflow patterns can also be observed in turbulently ventilated cleanrooms demonstrating that the air is well mixed within the cleanroom.

Room Recovery

Room recovery tests are generally performed in the "as built" state but can be performed in the "at rest" state also. This test demonstrates the rate at which an area within the cleanroom will recover if it becomes contaminated. It should be noted, however, that if contamination does occur within areas of the cleanroom, production/manufacturing should not recommence after the recovery rate time has lapsed without double checking that particle counts and pressures in the area are within the specification and that the area is clean.

Room Pressurization

Cleanrooms are generally positively pressurized; i.e., there is a higher volume of air entering the cleanroom than the volume extracted. This leads to a buildup of pressure within the cleanroom. The highest pressure should be in the cleanest required area or the room that contains the most critical process. When an opening occurs within the room, there is an outward flow of air from the cleanest area to a less clean area and so on until it reaches the external environment. A pressure differential then exists between the cleanest area and the less clean area minimizing the risk of contaminated air or air from a lower-class area flowing back into the cleanroom.

The IS EN ISO 14644-4 details pressure differentials between adjacent clean zones of different cleanliness levels to lie within the range of 5–20 Pa. The FDA guideline suggests that a positive pressure differential of at least 10–15 Pa should be maintained between adjacent rooms of differing classification.

This differential air pressure test demonstrates that the pressure differences between areas are acceptable and that the air is flowing in the correct direction. This test is applicable in each of the three occupancy states.

Once these tests have been completed with acceptable results, then the concentration of particles and levels of bacterial contamination should be measured to ensure they meet the required standards and protocols.

Airborne Particle Count Test

The airborne particle count test verifies that the cleanroom, personnel, equipment, and process is performing to the intended classification or the clean level. The classification level of the cleanroom is based on the number of particulates present per cubic meter. This particulate level will change dramatically from when the room is first built compared to when the room is in full production due to increased activity from equipment and personnel which are all dispersing particles.

Before testing for the level of particles, a few points to consider are as follows:

Particle Size

The particle size to be measured within the cleanroom should be considered.

Sampling Locations

The minimum number of sample locations taken within the cleanroom is calculated by taking the square root of the area of the cleanroom in square meters. This number should be rounded up to the nearest whole number (IS EN ISO 14644-1).

Air Sampling Volume

A calculation is documented in IS EN ISO 14644-1 for determining the minimum sampling volume to collect at each sampling location.

Statistical Analysis

When the number of locations sampled is more than 1 and less than 10, the mean, standard deviation, and 95% upper confidence limit (UCL) from the average particle concentrations for all locations within the cleanroom must be calculated for each particle size. Calculating the 95% UCL is often found to be problematic. The calculation is avoided by sampling more than 9 locations in the cleanroom or using a particle counter that will automatically display the calculation. It is important to note that samples should be taken within the same time period, i.e., on a given day as opposed to over a week or a month.

If there are ≥10 sampling locations or < 2 sampling locations, no statistical analysis is required. In this case the average of the sampling locations for each particle size is calculated. If only one sampling location is required, a minimum of three single sample volumes at that location is required and the average of the sampling locations for each particle size calculated.

Acceptance Criteria

According to IS EN ISO 14644-1, the cleanroom has met its classification if the average particle concentration at each location or the 95% UCL is below the specified limit. The acceptable particle limits for each classification as detailed in IS EN ISO 14644-1 are shown in Table 6.2.

Cleanroom Classification within Pharmaceutical Facilities

The prime standard adhered to by pharmaceutical industries is the EU GGMP, even though IS EN ISO 14644 still applies for the cleanroom certification. Annexure 1 of the EU GGMP was updated in 2008 to

TABLE 6.2   Selected Airborne Particulate Cleanliness Classes for Cleanrooms and Clean Zones

ISO Classification Number (N)	Maximum Concentration Limits (Particles/m³ of Air) for Particles Equal to and Larger Than the Considered Sizes Shown Below					
	0.1 (µm)	0.2 (µm)	0.3 (µm)	0.5 (µm)	1 (µm)	5 (µm)
ISO Class 1	10	2				
ISO Class 2	100	24	10	4		
ISO Class 3	1,000	237	102	35	8	
ISO Class 4	10,000	2,370	1,020	352	83	
ISO Class 5	100,000	23,700	10,200	3,520	832	29
ISO Class 6	1,000,000	237,000	102,000	35,200	8,320	293
ISO Class 7				352,000	83,200	2,930
ISO Class 8				3,520,000	832,000	29,300
ISO Class 9				35,200,000	8,320,000	293,000

Source: IS EN ISO 14644-1: Classification of air cleanliness.

Note: Uncertainties related to the measurement process require that concentration data with no more than three significant figures be used in determining the classification level.

TABLE 6.3   Particle Classifications per Grade

	Maximum Permitted Number of Particles/m³ Equal to or Greater Than the Size Tabulated			
	At Rest		In Operation	
Grade	0.5 µm	5.0 µm	0.5 µm	5.0 µm
A	3,520	20	3,520	20
B	3,520	29	352,000	2,900
C	352,000	2,900	3,520,000	29,000
D	3,520,000	29,000	Not defined	Not defined

Source: EU guidelines to good manufacturing practices: medicinal products for human and veterinary use. Annex 1 (2008). European Commission's Health and Consumers Directorate-General, © European Communities, 2003.

include a modified IS EN ISO 14644 standard that addresses sterile medicinal products. A table of cleanroom certification values that roughly translate into the IS EN ISO 14644 standard was defined (Table 6.3).

The sample volumes/location should not be less than 1 m³ for Grade A and B areas and is preferable in Grade C areas also. It is advised that a minimum length of sample tubing be used due to the potential loss of large-sized particles (5.0 µm) in the sample tubing.

It should be noted that classification results for sterile medicinal products are very different from those used for operational environmental monitoring and these should be clearly differentiated. A comprehensive report detailing the certification testing and classification of the cleanroom should be documented and supplied by the testing company. A continuous operational environmental monitoring plan should be defined based on the class of the cleanroom and the nature and risks of the product manufactured. Records should be kept of all continuous monitoring results separate to the certification report.

Additional Tests

Additional tests such as measuring temperature, relative humidity, sound, lighting, and vibration levels may be required depending on the process within the cleanroom.

Microbial concentrations within the cleanroom will be ascertained when all the above tests are satisfactory.

Microbial Validation Testing

Even when the cleanroom has been classified and certified that it is working within the required acceptance criteria, a microbial validation, particularly for medical, pharmaceutical, and related applications, should still be performed.

As part of the cleanroom performance validation, the level of microbial contamination should also be measured at the "in operation" state in order to demonstrate compliance and, in some cases, gain historical data to establish alert and action limits. Microbial testing should be performed when there is maximum activity within the cleanroom and when the maximum number of personnel is present. A sampling map detailing types and locations of sampling should be prepared. Depending on the class cleanroom, microbial acceptance criteria may be dictated by the standards or by a process risk assessment. Either way, microbial contamination should be monitored initially to gain an insight into the levels of microbial contamination present within the cleanroom during operations and to demonstrate compliance. The establishment of a comprehensive environmental monitoring program will be discussed further in this chapter.

Validation Report

A complete cleanroom validation report should comprise two parts.

Part 1 is the certification and classification report that is often obtained from an external contractor. This report should detail the occupancy states at which the tests were performed, the acceptance criteria, and the results of tests performed such as air changes/h or air velocity, filter leak tests, airflow visualization, and particle counts. All calculations performed and raw data should be detailed where possible.

Part 2 should take into consideration the microbiological contamination levels within the cleanroom under both static (unmanned) and dynamic (manned) conditions. Alongside the viable monitoring, nonviable particulate levels and pressure differentials should also be measured within the same time frame as these critical parameters can influence the contamination levels of the cleanroom.

The testing performed within each area of the cleanroom, the testing conditions, i.e., static or dynamic conditions, test methods for each test type, sampling map/plan clearly showing the sampling points within the cleanroom, acceptance criteria to be met, and calibration certificates for all equipment required should be attached as part of the final validation report. The cleanroom is now fully validated and will be governed by a routine sampling program to check that it is performing to the standard required with regard to both nonviable and viable contamination levels.

Validation versus Monitoring

Cleanroom certification and validation as discussed previously, involves checking the room for various criteria to ensure that it is built to a specific set of requirements. However, after the room has been certified and validated, it must be monitored periodically, relative to risk, to prove that a clean manufacturing environment can be maintained throughout its life. Monitoring of the cleanroom is important to show that the cleanroom is performing satisfactorily under dynamic conditions, i.e., that all aspects of the construction and supporting equipment are fully operational and performing at the same level as when the room was certified, that the process within the room is not posing a risk to the environment, and that personnel working within the cleanroom are following protocol.

As discussed in the section "Microbial Validation Testing", a comprehensive environmental monitoring program should be in place for the routine monitoring of the cleanroom environment and supporting areas. This program should include the monitoring of airborne viable and nonviable particulates, pressure differentials, as well as surface microbial contaminants on equipment, product contact surfaces, walls and floors, and personnel.

TABLE 6.4   Required and Optional Testing Requirements

	Schedule of Tests to Demonstrate Continuing Compliance		
Test Parameter	Class	Maximum Time Interval (months)	Reference Test Procedure
Particle counts Test	≤ISO Class 5	6	IS EN ISO 14644-1
	>ISO Class 5	12	
Pressure differentials	All classes	12	IS EN ISO 14644-3 Annex B.5
Airflow volume or airflow velocity	All classes	12	IS EN ISO 14644-3 Annex B.4
	Schedule of Additional and Optional Tests		
Installed filter leak test	All classes	24	IS EN ISO 14644-3 Annex B.6
Containment leakage	All classes	24	IS EN ISO 14644-3 Annex B.14
Recovery	All classes	24	IS EN ISO 14644-3 Annex B.13
Airflow visualization	All classes	24	IS EN ISO 14644-3 Annex B.7

Source: IS EN ISO 14644-2: Specifications for testing and monitoring to prove continued compliance with ISO 14644-1.

Routine Particulate Monitoring

Table 6.4 illustrates the required and optional tests and the maximum time interval for monitoring to demonstrate continuing compliance. Particle concentration testing is the most important analysis that must be carried out to demonstrate that the cleanroom continues to comply with IS EN ISO 14644-1.

In the author's opinion, it is advisable to perform more frequent demonstrations of compliance than those outlined in Table 6.4. If a situation occurs where the cleanroom does not meet its specification or is not in compliance, then the quality of all products, materials manufactured or processes performed within the cleanroom since the last demonstration of compliance are questionable. The more often demonstration of compliance is displayed, the smaller the loss of downtime, product and materials if out-of-specification results are obtained.

All grade cleanroom environments should be routinely monitored while they are in operation. The locations chosen for particulate monitoring should be selected based upon a risk assessment of the area with the certification and classification results in mind.

For Grade A cleanrooms, particle monitoring should be performed for the full duration of critical processing including equipment assembly and setup stages. Exceptions are those justified by contaminants in the process that would damage the particle counter or present a hazard. Examples of exceptions are live organisms or radiological hazards. Grade A cleanrooms should be monitored continuously so that all interventions, transient events, and any system deterioration would be captured and alarms triggered if alert limits are exceeded. It is recommended by the EU GGMP that a similar system be used for Grade B cleanrooms also.

Appropriate alert and action limits should be set relative to the results of particulate and microbiological monitoring. If these limits are exceeded, operating procedures should prescribe corrective action. Monitoring systems must rapidly detect and record any changes that might compromise the cleanroom environment or product/process and alert the relevant personnel of such changes immediately.

Establishing an Environmental Monitoring Program

Before microbial testing is performed, an environmental monitoring plan should be derived. There are many different guidance documents that deal with environmental monitoring, such as the *EU Guide to Good Manufacturing Practice*, Annexure 1, USP 1116, the *FDA Aseptic Processing Guide*, IS EN ISO 14698, but these can be ambiguous and leave a lot of scope for interpretation. There are no commonly accepted levels of environmental testing probably due to the large and varied number of industries, processes, and subprocesses to which these regulations must apply. The key points of these guidelines should be taken into account, but a commonsense approach is sometimes required for the application to your specific process and facility. In the author's opinion, a good environmental monitoring plan should include the following:

Types of Sampling Performed

The types of sampling that will be performed should be decided. The main sampling types are surface sampling, air sampling, and personnel sampling.

Contact plates are used to sample flat surfaces such as work benches, equipment, floors, and walls. These agar plates have a dome-shaped layer of agar that is pressed gently onto the surface. Organisms on the surface will be transferred onto the agar. This method is considered quantitative as it measures the number of viable organisms on a defined surface. Surfaces that are not flat, smooth, or easy to access can be sampled using a moistened swab. The swabs can be used to inoculate a plate immediately or can be immersed in a diluent before inoculation.

There are two methods for air sampling to choose from: active and passive. Active air sampling is considered to be quantitative as it measures the number of viable organisms in a defined volume of air, i.e., per cubic meter. Several types of active air samplers are available. Passive air sampling is not considered to be quantitative. It is performed by placing an agar plate exposed on the test area for a period of time (no longer than 4 h due to desiccation of the medium that may reduce recovery rates). Viable air sampling gives a snapshot in time of the microbiological status of the cleanroom under dynamic conditions.

Personnel monitoring is important as part of aseptic and GMP training. This type of monitoring is performed by placing some or all pads of the fingers (not the tips of the nails) onto the agar surface. Gloves or finger cots should be changed after this testing and hands washed before recommencing work in the cleanroom. Monitoring of both the garments and hands indicates the level of contamination on these areas before, during, or after working within the cleanroom.

Personnel are one of the greatest sources of contamination within a cleanroom. Poor handwashing can pose a high risk to product or processes within the cleanroom, as hands are used to touch many different materials and surfaces within the cleanroom that could be a potential source of contamination. Therefore, personnel monitoring of hands, gloves or finger cots, and other cleanroom apparel is an obvious tool in assessing contamination risks.

Sampling Locations

The sampling locations within the cleanroom and supporting areas for each sampling type should be decided. A risk assessment of the process in operation within the cleanroom is vital as it will establish the areas within the cleanroom that pose the greatest risk with regard to contamination.

Sampling Map

A map of the cleanroom and supporting rooms should be drawn up detailing the exact sampling locations.

Sampling Frequency

The sampling frequency for each sampling type within the different areas should be decided. In high-class cleanrooms, the frequency of monitoring may be continuous, i.e., during each batch or critical operations. In lower-class cleanrooms, monitoring may take place monthly or quarterly depending on the associated risk.

Interpretation of Results

In some cases, companies do not have their own in-house microbiologist for the environmental monitoring of the cleanrooms. In such instances, personnel who are responsible for this monitoring may not fully understand the importance of performing appropriate microbial testing, the meaning of the results, and the impact of the results on the final product or process. Personnel looking after microbial testing must be trained, fully understand the testing performed, be able to interpret the results, and know the impact the results may have on the finished product.

Organisms that can sometimes be detected within cleanrooms are *Staphylococcus aureus* as it is a commensal of the human skin and therefore can enter the cleanroom via people. Bacillus species are spore-forming bacteria that can be very difficult to eradicate from the cleanroom once they enter. This organism can enter cleanrooms on materials/items transferred into the cleanroom that have not been cleaned effectively or by torn shoe covers. Depending on the quantity of spore-forming bacteria present in the cleanroom, this could seriously compromise the sterility assurance level of the end product or process and could shut down production within facilities. Other organisms such as gram negative bacteria, for example, *Pseudomonas* spp., *Escherichia coli* spp., and *Salmonella* spp., must under no circumstances enter the cleanroom environment as they produce endotoxins that cannot be removed by sterilization techniques. Such organisms could very seriously compromise the health of a patient leading to even fatality. The biggest source of these organisms is water. Figure 6.1 shows *Pseudomonas aeruginosa* bacteria growing on a surface.

There seems to be an attitude present in some lower-class cleanrooms that products going for sterilization will ultimately be free from all contamination and that the sterilization process will kill "everything." Be assured that this is not the case at all. Sterilization, as discussed, will not inactivate or eradicate endotoxins that can be fatal to the end user.

Also, if a product is not clean before it goes to the sterilization process, then it can still have a negative impact on the host. The host is not sterile, so any contaminants such as oils or residues on the product can provide "food" for the host bacteria leading to toxicity or infection of the host. Contaminants can interact with the host immune system and can interfere with device interaction within the host. It is, therefore, important to physically and effectively remove undesirable surface material from these products.

With sterilization or disinfection, the issue is rendering the contaminant nonviable. With cleaning, the issue is physically removing contaminants from the surface without changing that surface in a way that interferes with product performance (Broad and Kanegsberg 2007).

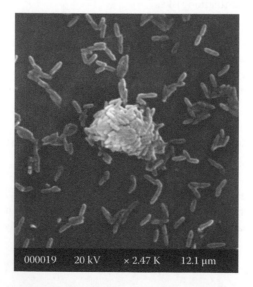

FIGURE 6.1 *Pseudomonas aeruginosa* biofilm forming on a surface. (From O'Donoghue, K., Physical and biological characterisation of biofilms and bioaerosols, PhD thesis, National University of Ireland, Galway, 2004.)

Establishing Alert and Action Limits

A method for establishing alert and action limits should be detailed. Cleanrooms that are in compliance with the *EU Guide to Good Manufacturing Practice*, Annexure 1, have recommended limits stated for microbial contamination within a Grade A to Grade D cleanroom (reference Table 6.5). However, cleanrooms that do not adhere to Annexure 1 have very little guidance as to the microbial levels that should be met.

In this case, microbial acceptance criteria must be obtained based on the requirements and contamination risk factors associated with the process. From the author's experience working with different cleanroom industries, there are two ways this information can be obtained. The first method is by monitoring microbial contamination levels over a period of consecutive days in the dynamic/occupied state. The second method is by monitoring microbial contamination levels routinely over a period of months. Data gathered from both these methods can be used to establish temporary alert and action limits until further historical data have been obtained to set more permanent limits. Where possible, limits should be as strict as those in the standard.

Some cleanrooms will monitor the microbial contamination levels in the static state first in order to establish a baseline status showing the contamination levels in the room without people or the process in operation.

When validating the microbial contamination levels, whether to observe if they are in compliance with the standard or to establish historical data, other parameters should be monitored at the same time. For example, particulate levels and pressures should also be taken into consideration at the time of microbial monitoring as these may have a direct impact on the level of contamination within the room. For example, particulate levels that are over the action limit and a decrease in positive pressure could be an underlying reason for an increase in contamination levels.

Actions to Be Taken for Out-of-Spec Results

Comprehensive procedures should be in place detailing actions to be taken if results are over the alert or action limits. Typical microflora should be identified in order to determine the risk posed by the out-of-spec result and help identify the root cause. In the author's opinion, if results are above alert and action levels, colony morphology and colony identification should be performed, respectively. The data obtained from both counts and organism identification lead to a greater overall understanding of the microbial presence within the cleanroom environment and can be used to identify high-risk areas as well as to eliminate potential routes and sources of contamination.

Trending and Documenting of Results

The aim of an environmental monitoring program is to detect any changes in the numbers and types of organisms present within the cleanroom outside of the "norm." Results should be trended on a continuous basis highlighting any results that are over alert or action limits. A library of organisms that

TABLE 6.5   Recommended Limits for Microbial Contamination

Grade	Air Sample (cfu/m³)	Settle Plates Diameter 90 mm (cfu/4 h)	Contact Plates Diameter 55 mm (cfu/Plate)	Glove Print Five Fingers (cfu/Glove)
A	<1	<1	<1	<1
B	10	5	5	5
C	100	50	25	—
D	200	100	50	—

Source: EU guidelines to good manufacturing practices: medicinal products for human and veterinary use. Annex 1 (2008). European Commission's Health and Consumers Directorate-General, © European Communities, 2003.

Note: These are average values and settle plates may be exposed for less than 4 h.

have been identified within the cleanroom should be established and maintained. This library is a good source of information with regard to the contamination found within the cleanroom. Such a library increases awareness of the types of organisms entering the cleanroom, the preventive action that can be implemented to reduce these organisms from entering, or decreasing the quantities of these types of organisms present, and understanding the impacts of this type of contamination in the environment and on the end product.

It is worth spending the time defining a good environmental monitoring program. Microbiological results and trends provide vital information on the quality of the cleanroom environment, minimize the risk of release of a potentially contaminated product or process, and prevent future contamination by detecting adverse trends. It also serves as a reminder to cleanroom personnel of the importance of their actions, behavior, and good manufacturing practices within the cleanroom environment.

Often, microbiology testing is observed as a technically specific task that is not understood by personnel trained in other departments. Therefore, it is often not seen as a priority. This testing is paramount for all products entering the human body as well as for assessing how the cleanroom environment itself is performing.

It is important that the seriousness of microbial testing is understood and that the results of iterative communication with experts in this area is received by pertinent technicians and management. It is critical to interpret the results obtained correctly and understand the impacts if the results do not meet those expected.

Summary

Cleanrooms are costly to build and maintain. Therefore, they must be customized to produce or manufacture a specific product/service. Cleanrooms are used primarily for the production of products that are subject to special requirements that have been established to minimize risks of particulate, microbial, or pyrogen contamination. Quality assurance is imperative and this type of manufacture must strictly follow carefully established and validated methods of preparation and procedure within an environment that has itself met the appropriate validation criteria. Sole reliance on terminal sterility should not be taken for granted.

When building or extending a cleanroom, it is paramount to spend time on the design to get it right from the beginning and ensure all aspects of the cleanroom are addressed. A poorly designed cleanroom will lead to increased costs as many changes will be required to customize the cleanroom to the process or bring the cleanroom up to specification. The most common mistake personnel make is starting the cleanroom validation process without the appropriate knowledge and expectations. This can result in further associated costs, an increase in timelines, a cleanroom that may not be working to its required classification, and a higher degree of required maintenance.

Thorough procedures and policies are required for the routine monitoring of contamination levels within the cleanroom. These procedures require strict adherence. In order to achieve a high-standard cleanroom with regard to controlling contamination on a day-to-day basis, all personnel, regardless of frequency of entry or job classification, need to be given a basic understanding and fundamental awareness of contamination control. Personnel must understand how contamination can enter the cleanroom through the specific processes in place within the facility, how in-house processes can be improved, and how contamination can be reduced to a minimal level within the facility. Personnel must gain an understanding of why there are such strict controls in place within this working environment as opposed to any other working environment. This understanding and awareness can sometimes be difficult to achieve because within cleanrooms we are dealing with particles and organisms that we cannot see. Very often, the saying "out of sight, out of mind" comes into play. From the author's experience in this field, the key to controlling a large amount of contamination from poor personnel activities and practices is to develop knowledgeable, aware, and empowered employees.

Consistent, quality education to achieve this understanding and awareness, along with support and commitment from management down through all levels of staff can yield many benefits including a substantial decrease in product rejects, a corresponding increase in profitability, and an output of high-quality products/services.

References

Broad, J. and Kanegsberg, B. Minimising viable and non viable contamination: Standards and guidelines for medical device manufacturers. *Controlled Environments Magazine*, Sept 2007. Available at www.cemag.us

Cleanzone Technology Ltd., Testing & Certification Company, Dungarvan, Co. Waterford, Ireland. Available at www.cleanzone.ie

O'Donoghue, K. Physical and biological characterisation of biofilms and bioaerosols, National University of Ireland, Galway, 2004.

EU guidelines to good manufacturing practices: medicinal products for human and veterinary use. Annex 1, 2008. European Commission's Health and Consumers Directorate-General, © European Communities, 2003.

FDA Guidance for Industry. Sterile drug products produced by aseptic processing—Current good manufacturing practices, Sept 2004.

IEST CC006.3. Testing cleanrooms.

IEST CC001.4. HEPA and ULPA filters.

IEST CC012.1. Considerations in cleanroom design.

IS EN ISO 14644-1. Classification of air cleanliness.

IS EN ISO 14644-2. Specifications for testing and monitoring to prove continued compliance with ISO 14644-1.

IS EN ISO 14644-3. Test methods.

IS EN ISO 14644-4. Design, construction and start-up.

Thomas, H.E. No 'absolutes' for cleanliness in ISO 5 design. *Controlled Environments Magazine*, Feb 2008. Available at www.cemag.us

Whyte, W. *Cleanroom Technology: Fundamentals of Design, Testing and Operation*. Chichester, U.K.: John Wiley & Sons, Second Edition, 2010.

7

Cleanroom Management, Validation of a Cleanroom Garment System, and Gowning Procedure

Introduction 103
A Short Course in Particulation 104
Facility Design 104
Protecting Cleanrooms from Human Contamination 106
Behavior 107
Cleanroom Garments 108
Evaluation of Fabrics Using ASTM and AATCC Test Methods at 1×,
50×, and 100× • Evaluation of Seams and Components via RP-3
Recommendations • Evaluation of Ability of Garment System to Entrain
Particles Using Body Box Testing at 1×, 50×, and 100× • Evaluation of
the Cleaning of the Garment System • Validating a Cleanroom Garment
System Supplier • Validation of Gamma Radiation Sterilization of
Cleanroom Apparel • Gamma Subcontractor Qualification
Routine Monitoring of Sterilization Process 111
Certificate of Sterility
Gowning for the Cleanroom 112
Sterile Gowning Procedure
Housekeeping 114
Ongoing Assessments 115
Summary 116
References 116
Standards and Recommended Practices • Journal Articles

Jan Eudy
Cintas Corporation

Introduction

The cleanroom environment exists for the protection of the product being manufactured by the organization. An effective cleanroom environment is only as good as the cleanroom management program used to ensure that the environment is performing as expected. The management program is a comprehensive system that keeps the cleanroom environment free of any viable or nonviable airborne particle that could contaminate the product being manufactured.

The implementation of a cleanroom management program compliant with current good manufacturing practices (cGMP) includes the following:

- Validation of the design of the cleanroom
- Designation of linear flow of product and personnel
- Written procedures and document control
- Change control
- Training and auditing for compliance

A balanced cleanroom management program is a comprehensive set of standard operating procedures (SOPs) that outline the requirements necessary to maintain the cleanroom environment successfully. The program is defined as "balanced" because if one component of the cleanroom management program is not working properly, the entire program is disrupted, thereby allowing for possible contamination of the cleanroom. A balanced cleanroom management program takes into account the following requirements:

- Reducing potential contamination by designing a facility and cleanroom with a low-risk floor plan
- Creating linear flow of both product and personnel
- Protecting from human contamination by choosing the right uniform program, requiring adherence to gowning procedures, and setting guidelines for personnel behavior
- Utilizing the correct types of cleanroom supplies for each application
- Creating housekeeping procedures that maintain a consistent level of cleanliness
- Requiring continuous improvement through ongoing assessments of the cleanroom environment and auditing and trending of test results

A Short Course in Particulation

The size of particles that are a contamination hazard to the cleanroom environment is measured in micrometers. A micrometer is one millionth of a meter. Some common cleanroom contaminants and their micrometer sizes are

Human hair: 70–100 μm
Human skin flakes: 0.4–10 μm
Pollen: 5–100 μm
Mold: 2–20 μm
Smoke: 0.01–1 μm
Household dust: 0.05–100 μm
Bacteria: 0.25–10 μm

Facility Design

There are many sources of contamination to the cleanroom environment. Equipment generates particles through friction, heat, and exhaust. Additionally, equipment generates static electricity, which causes unwanted particle migration due to electro-inductive forces. Cleanroom processes can cause contamination by generating by-products (outgassing) that are exposed to the cleanroom surfaces and air. Incoming components used in production may also contain contaminants that will compromise the integrity of the cleanroom environment.

Even though cleanroom structures and furnishings are designed to be low particle generators, particles may still be released into the cleanroom environment by a rubbing action over the surfaces.

The cleanroom is a controlled environment in which all incoming air, water, and chemicals are filtered to meet high standards of purity. Temperature, humidity, air changes, and air pressure are also controlled. For the comfort of the personnel working inside the cleanroom, temperatures are usually maintained between 60°F and 70°F, and the relative humidity is usually between 30% and 50%. The recommended number of air changes for ISO Class 5 cleanrooms is >500 air changes per hour. The EU mandates that the pressure differential between lower classified cleanrooms should be ≥0.05 in. of water. Smoke studies show control of airflow direction. These parameters must be monitored and measured regularly to develop rational cleanroom operating parameters.

Cleanrooms are tested and certified to a level of cleanliness at three different levels:

1. As built
2. At rest
3. Operational

Tests performed may be all or some of the following tests listed:

- Airflow velocity and uniformity test
- Filter installation leak test
- Airborne particle count test
- Enclosure pressurization test
- Temperature uniformity and humidity test
- Enclosure induction leak test
- Lighting level and uniformity test
- Sound level test
- Airflow pattern testing
- High-efficiency particulate air/ultralow particulate air (HEPA/ULPA) filter leak test
- Calculations of number of air changes per hour
- Recovery test

The recommended practice IEST-RP-CC006, "Testing Cleanrooms," provides the testing procedures to perform the testing mentioned in the preceding text.

All cleanrooms must be certified by a qualified certification company as per ISO 14644–1, "Cleanrooms and associated controlled environments—Part 1: Classification of air cleanliness."

The recommended practice IEST-RP-CC012, "Considerations in Cleanroom Design," provides the formula for calculating the number of air changes per hour (Table 7.1), as well as typical airflow velocities and air changes per hour for different cleanroom classifications. To calculate air changes per hour

Air changes per hour = Average airflow velocity × room area × 60 min/h/room volume

Monitoring and trending these data will enable cleanroom operators to design a prevention response, recovery, and reaction plan.

TABLE 7.1 Recommended Target Air Changes per Hour as per IEST

ISO Classification	Average Airflow Velocity	Air Changes per Hour
1&2	0.305–0.508 m/s (60–100 fpm)	360–600
3	0.305–0.457 m/s (60–90 fpm)	360–540
4	0.254–0.457 m/s (50–90 fpm)	300–540
5	0.203–0.406 m/s (40–80 fpm)	240–480
6	0.127–0.203 m/s (25–40 fpm)	150–240
7	0.051–0.076 m/s (10–15 fpm)	60–90
8	0.005–0.041 m/s (1–8 fpm)	5–48

Additionally, if the ISO Class 5 (or better) cleanroom is not using a continuous monitoring system, it must be recertified every 6 months as per ISO 14644–2, "Cleanrooms and associated controlled environments—Part 2: Specifications for testing and monitoring to prove continued compliance with ISO 14644–1." The three parameters that must be measured are

1. Airborne particle count test
2. Air pressure differential
3. Airflow

Even the best facility design and the most stringent airflow system do not guarantee that the product will be protected from contamination. Of all of the components that comprise cleanroom operations and processes, humans are the easiest to control, yet contribute the most contamination. We have a choice to be part of the problem or part of the solution.

Protecting Cleanrooms from Human Contamination

The cleanroom industry is acutely aware of the many possible sources of contamination that threaten cleanroom operations. The most significant threat is also the threat that is easiest to control—the humans working in the cleanroom.

As cleanroom operators work inside the cleanroom, they generate millions of particles with every movement. Schlerin photographs show particles emitted from the human body. Particles migrate up through the cleanroom apparel toward the head and fall down the legs during cleanroom activities.

Examples of generation of particles of size 0.3 μm and greater per minute are

Action	Particles Generated at 0.3 μm/min
Motionless, sitting, or standing	100,000
Head, arm, neck, leg motion	500,000
All of the above with foot motion	1,000,000
Standing to sitting position and vice versa	2,500,000
Walking at 2.0 mph	5,000,000
Walking at 3.5 mph	7,500,000
Walking at 5.0 mph	10,000,000

The above number of particles includes both viable and nonviable particles generated by cleanroom operators working within the cleanroom environment. Bacteria, molds, and yeast are viable particles that are chemically active. By-products of bacterial, mold, and yeast growth and replication cause a variety of contamination scenarios, each of which is based on the nature of the chemicals released during generation.

Additionally, humans release elemental chemicals that can cause contamination. Examples of elemental chemical contamination are

- Spittle: potassium, chloride, phosphorus, magnesium, and sodium
- Dandruff: calcium, chloride, carbon, and nitrogen
- Perspiration: sodium, potassium, chloride, sulfur, aluminum, carbon, and nitrogen
- Fingerprints: sodium, potassium, chloride, and phosphorus

One of the most significant methods for reducing human contamination in the cleanroom is through a complete cleanroom uniform program. Cleanroom apparel is designed to capture and entrain particles and not allow contaminants to be dispersed into the cleanroom environment. Apparel protects from numerous contaminants that are generated from the human body, including

- Viable particles such as bacteria and yeasts
- Nonviable particles such as hair, dead skin cells, and dandruff
- Elements such as sodium, potassium, chloride, and magnesium

The Institute for Environmental Sciences and Technology (IEST) publishes recommended practices that contain information for developing a comprehensive cleanroom management program. IEST-RP-CC027, "Personnel and Practices and Procedures in Cleanrooms and Controlled Environments," specifically addresses aspects to control human contamination in the cleanroom. This recommended practice addresses hiring and training of cleanroom personnel, health and hygiene expectations of cleanroom personnel, behavior expectations of cleanroom personnel, and monitoring and auditing the efficacy of the cleanroom management program.

Behavior

Meticulous hiring practices for cleanroom personnel include screening potential operators for adaptability to the cleanroom. Applicants tend to be screened out if they are smokers; have facial hair; are sensitive to heat, cold, or humidity; and have any seasonal allergies (including skin allergies). In addition, during the interview process it is critical that a realistic job preview is provided to the applicant, so the applicant and hiring manager can test for comfort in the cleanroom uniform and the cleanroom environment. This is a good method for identifying problematic characteristics such as claustrophobia.

Just as all operators are thoroughly trained in occupational skills and knowledge of the product being produced, cleanroom operators must be trained in the unique practices and behavior required for working in the environment. Training is mandatory for any and all personnel who frequently or infrequently work in the cleanroom. The training program should include

- Basic contamination control
- Behavior in the cleanroom
- Personal hygiene
- Gowning procedures
- Cleanroom housekeeping procedures
- Cleanroom testing procedures

Testing and certification of cleanroom operators and subsequent observation of operators after training ensures that the training has been implemented and is effective.

One area of training is setting the expectations for personal health and hygiene. Cleanroom operators are instructed that they must engage in daily bathing or showering, shaving, brushing of teeth and hair, and application of non-silicone-containing skin moisturizers to reduce skin flakes. Hair gels, hair sprays, aromatic after-shave lotions or body lotions are not cleanroom compatible. Cosmetics are prohibited because in addition to the gross particle generation, cosmetics release elemental chemicals such as iron, aluminum, silicone, carbon, titanium, magnesium, potassium, sulfur, and calcium. At work, all employees must wash their hands with soap and water after eating or using the toilet. Cleanroom-compatible hand cream may be applied prior to gowning if using donning gloves during the gowning process.

As previously mentioned, any activity by the cleanroom operator generates millions of viable and nonviable particles. Therefore, it is imperative to limit talking and actions in the cleanroom to only those required for the manufacturing of the product. Running, horseplay, and other nonprofessional activities are not permitted. Other behavioral requirements include, but are not limited to, the following pointers:

- Smoking is not allowed inside the manufacturing facility, including all cleanroom areas. Note: Smokers release particles for at least one-half hour after smoking one cigarette. It is recommended that all smokers drink an 8 oz glass of water prior to returning to the cleanroom after smoking a cigarette.
- Nothing is allowed inside the cleanroom complex that is not required in the cleanroom manufacturing process. This includes personal items such as jewelry or keys, cosmetics, tobacco, and food or drink in any form. Hair may not be combed in the cleanroom gowning area.

- Only cleanroom-compatible, indelible ballpoint ink pens are allowed inside the cleanroom for recording data on cleanroom-compatible paper and clipboards.
- While working in the cleanroom, mannerisms such as scratching head or rubbing hands are to be avoided. Cleanroom personnel may not access the inside of the cleanroom uniform.
- The use of facial tissues is prohibited in the cleanroom. If one must use a cleanroom-compatible non-linting tissue, it must be used only outside the gowning area and disposed of appropriately in a waste receptacle.
- All doors must remain closed when not entering or exiting. Emergency doors may be fitted with a visual and audible alarm to enforce compliance.

Cleanroom Garments

IEST-RP-CC003, "Garment systems considerations for cleanrooms and other controlled environments," addresses gowning requirements for all cleanroom classifications as well as additional requirements for aseptic gowning.

It is important to note that because the human body produces so many contaminants in such large quantities, the cleanroom apparel may be overwhelmed. Therefore, change frequencies and garment system configurations must be evaluated for the room cleanliness that you want to achieve.

IEST also publishes the recommended practice for garments, IEST-RP-CC003, "Garment considerations for cleanrooms and other controlled environments." This document is a tremendous resource for determining the fabric, garment system, and garment configuration for cleanroom applications. This recommended practice provides guidance for the selection of fabric, garment construction, cleaning and maintenance of cleanroom garments, and testing of cleanroom apparel and components for use in aseptic and non-aseptic cleanroom environments. During the validation of the cleanroom garment system, the information gathered regarding the fabric and the construction of the cleanroom garment system is evaluated during the installation qualification phase of validation. The documentation of the cleaning, maintenance, and testing of the cleanroom garment system comprises the operation and performance qualification phases of the validation.

Evaluation of Fabrics Using ASTM and AATCC Test Methods at 1×, 50×, and 100×

Over the years, the contamination control industry has evolved unique, innovative fabrics and apparel to encapsulate humans working in cleanrooms, thereby protecting the product and the processes from possible deleterious contamination. There are several test methods devised by the American Society for Testing and Materials (ASTM) and by the American Association of Textile Colorists and Chemists (AATCC) to evaluate these new fabrics.

- The weight of the fabric determines its strength and durability; however, a lighter fabric contributes to operator comfort. Additionally, the grab tensile and tongue tear tests give an indication of the strength and durability of the fabric.
- The pore size is an indicator of barrier efficiency. The smaller the pore size, the more particles that will be entrained. Therefore, the evaluation of the pore size is important to the evaluation of the fabric used in construction of the cleanroom garment.
- The moisture vapor transmission rate (MVTR) evaluates the ability to move moisture through the fabric and translates to more comfort to the operator. Moisture buildup causes the operator to feel hot due to the increase in humidity between the fabric and the body.
- Air permeability is the ability of a fabric to allow air to pass though it, which is quantified by the volume-to-time ratio per area. Airflow in heating and cooling processes, such as the cooling process of the body, contains contaminants that can be transferred to the product. The lower the

permeability or transfer of air from within the garment to the outside, the lower the contamination to the product.
- There are several tests to determine the splash resistance of the fabric or its ability to resist absorption of liquids. These characteristics allow the operator to be better protected from spills in the cleanroom environment.
- Static decay testing and surface resistivity testing are performed to document that the fabric is static dissipative. Fabrics outside of the static dissipative range of 10^5–10^{11} Ω^{-2} (this unit of measurement refers to the ESD thread gridlines in the fabric) may cause an electrical discharge and subsequent product failure.

All testing of fabrics should be performed over time and with exposure to gamma radiation. The results over time should not be significantly different from the original results, thereby demonstrating the durability of the fabric characteristics over time.

These same tests may be used in the evaluation of the garment system (fabric and components of garments) to withstand chemicals used in the cleaning of the cleanrooms, the cleaning of the garments, the application of gamma radiation, and even autoclaving in some cases.

Evaluation of Seams and Components via RP-3 Recommendations

Currently, all reusable cleanroom garments are constructed of 99% polyester and 1% durable carbon threads and cleanroom-compatible snaps, zippers, and binding. These garment systems are lightweight, non-linting, and economical and control both nonviable and viable particle contamination. The IEST document details recommended seam construction and components for cleanroom garments.

Evaluation of Ability of Garment System to Entrain Particles Using Body Box Testing at 1×, 50×, and 100×

All cleanroom garment systems will deteriorate over time due to wear, wash/dry/wear cycles, and sterilization. The ability of the garment system as a barrier to contamination and its filtration efficacy is evaluated in a "body box" test. The "body box" is a mini-cleanroom. The particle cleanliness of the area is determined by typical room particle measurement with a particle counter and probe. This is the background. The operator inside the body box wearing the garment system performs a series of prescribed movements to the prescribed cadence of a metronome. The particle measurement during the prescribed movements determines the efficacy of the garment system.

Evaluation of the Cleaning of the Garment System

The IEST-RP-CC003 details recommended parameters for the cleaning of cleanroom garments, and revised the performance of the Helmke tumble test for particle cleanliness (Table 7.2). This revised version has established test parameters that, when followed precisely, produce test results that are more robust, repeatable, and reproducible over various test laboratory settings. The Helmke tumble test is specifically designed to test the particle shedding of a garment over time. This test evaluates the integrity of the garment, as well as the overall ability of the cleanroom garment laundry to render the garment item "particulately clean." The Helmke tumble test evaluates particle shed of the entire cleanroom coverall, frock, or hood at 0.3 μm and larger. The ASTM F51 test evaluates the particle and fiber characteristics at a larger micrometer particle (≥5 μm). This test is less reproducible due to technician variability over various laboratory settings.

Additionally, extraction testing can be performed to determine if residual elements and/or compounds are present in the cleanroom garments after cleanroom laundering.

TABLE 7.2   Helmke Garment Cleanliness Classification: IEST-RP-CC003.3 Cleanliness Classification Chart

Category	Garment Type	Particle Emission Rate, G Particles/min	
		0.3 μm and Larger	0.5 μm and Larger
I	1 Frock	<1,700	<1,000
I	1 Coverall	<2,000	<1,200
I	3 Hoods	<780	<450
II	1 Frock	1,700–17,000	1,000–10,000
II	1 Coverall	2,000–20,000	1,200–12,000
II	3 Hoods	780–7,800	450–4,500
III	1 Frock	17,000–170,000	10,000–100,000
III	1 Coverall	20,000–200,000	12,000–120,000
III	3 Hoods	7,800–78,000	4,500–45,000

Note: The particle emission rates shown for each of the garment types are proportional to the respective areas of fabric involved. The areas of the garments considered in the preparation of the table are as follows:

Garment Type	Average Area, m² (Both Sides)	Average Area, ft² (Both Sides)
Frock[a]	4.63	49.8
Coverall[a]	5.99	64.4
Hood[a]	1.03	11.0

[a] Medium-sized garments.

Validating a Cleanroom Garment System Supplier

There are numerous steps involved in validating a cleanroom garment system supplier. These steps include the following:

- Complete an installation qualification that audits the garment system supplier and evaluates their qualifying tests and testing results.
- Perform an operation qualification that includes a trial at the customer site and evaluation of the customer-qualifying tests and results.
- Conduct a performance qualification that includes evaluation of the performance of the fabric and garment system over time within the customer's cleanroom.

All of these steps are necessary to ensure that your garment system meets your expectations and can fully service your apparel needs.

In conjunction with the garment configurations recommended by IEST-RP-CC003, the appropriate cleanroom fabric, findings (i.e., snaps, zippers), and garment style must be determined by the cleanroom manufacturer. Cleanroom fabric should be evaluated for small pore size to entrain particles, comfort to the wearer, durability, and presence of cleanroom-compatible, gamma-compatible carbon thread to impart static dissipative qualities, thereby minimizing the effects of triboelectric charging and preventing unwanted particle migration due to electro-inductive forces in the cleanroom.

Validation of Gamma Radiation Sterilization of Cleanroom Apparel

Gamma radiation sterilization is the most common method of sterilization of cleanroom apparel. Gamma radiation sterilizes the garments using Cobalt 60 ionizing energy. This method has a high penetration of materials and has lethal effects on microbial populations at very low levels of radiation. Gamma radiation sterilization is a reproducible and reliable process that is based on the exposure time of the product to the gamma source.

The ANSI/AAMI/ISO 11137-2006 standard establishes the requirements for quarterly dose audits of the bioburden and sterility testing. The established dose is based on the bioburden and the associated sterility assurance level (SAL). The recommended SAL per ANSI/AAMI ST 67:2003, "Sterilization of healthcare products—Requirements for products labeled 'STERILE'," for cleanroom apparel is 10^{-6} SAL or one positive in one million garment items.

One must also validate the packaging of the cleanroom apparel as per IEST-RP-CC032, "Flexible packaging materials for use in cleanrooms and other controlled environments," determine the maximum tolerated dose of gamma radiation, and test for strength, discoloration; and durability. Additional shelf-life testing is performed to ensure sterility for an extended period of time by performing an aerosol challenge, bubble testing, and burst testing.

Gamma Subcontractor Qualification

The gamma subcontractor must be audited and approved by the cleanroom apparel provider. This is usually an on-site audit that verifies the equipment validation and dose mapping performed by the gamma irradiator to determine sites of minimum and maximum dose ranges. The dose is variable with respect to the density of the packaging, and it is, therefore, imperative that the density remains uniform in the mapped product.

Routine Monitoring of Sterilization Process

Colorimetric indicators may be used as a visual aid to show that the garments have been exposed to irradiation. Indicators for gamma radiation change from yellow to red and are typically placed on inner and outer packaging. If an indicator is exposed to isopropyl alcohol (IPA) such as in wiping the external packaging prior to placing the garment item in the sterile gowning area, the IPA can cause the red indicator (one that has been exposed to irradiation) to turn back to yellow. Additionally, exposure to ultraviolet lights such as used in warehouses as insecticutors will turn the colorimetric indicators from yellow to red, prior to gamma radiation. Therefore, the recommended practice is to only use the dosimetric report and Certificate of Gamma Radiation Processing provided by the contract sterilizer and the Certificate of Sterilization Verification provided by the cleanroom garment laundry as final proof of sterilization of the batch lot of cleanroom garments. The use of biological indicators inside the product to be gamma radiated, such as *Bacillus pumulis* spores, is not recommended as final proof of sterility.

Certificate of Sterility

Each lot of garments gamma processed will be issued a certificate of sterility based on the evaluation of the dosimetry data results of the irradiation run. A typical certificate of sterility will contain the following information:

- Product description or catalog number
- Lot number
- Irradiation run number
- Date of irradiation
- Statement of how the product was sterilized and how the process was validated
- Minimum and maximum specified dose
- Minimum and maximum delivered dose
- Signature, date, and title of the individual performing the dosimetry testing
- Signature, date, and title of the individual performing the review of test data

Gowning for the Cleanroom

It is imperative that all operators be trained in proper donning and doffing techniques specific to the cleanroom classification and cleanroom manufacturing operations. Only approved cleanroom apparel is donned before entering the cleanroom. This apparel must be worn correctly to be effective in encapsulating contamination emanating from the cleanroom operator. The cleanroom management system should include a gowning procedure for the cleanroom. Best practices in cleanroom management indicate that hanging gowning posters in the gowning room will help ensure that all cleanroom operators are following the proper steps and gowning according to protocol. A sample poster (Figure 7.1) illustrates the many steps required for gowning (non-sterile cleanroom). An example of an employee after donning cleanroom apparel is illustrated in Figure 7.2.

Gowning procedures should also be documented in the cleanroom management program. The following is a sample process for gowning in a sterile cleanroom. This information is provided to show the recommended documentation method.

Sterile Gowning Procedure

Donning

1. Wash and dry hands.
2. Enter non-sterile gowning area.
3. Put on bouffant. Ensure all hair is entrained under bouffant.
4. Walk over tacky mat to remove excess soil from shoes.
5. Put on disposable shoe covers.
6. Select appropriately-sized sterile coverall, hood, and boots.
7. Put on first set of sterile gloves. Sanitize gloves with sterile IPA.
8. Walk to sterile gowning area.
9. Slide hood from sterile bag. Touch only inside of the hood. Don sterile hood. Completely cover bouffant. Fit snugly with vertical and horizontal snaps in back of hood. Don sterile facemask. Sanitize gloves with sterile IPA.
10. To don coverall, first grasp inside neck of coverall and slide coverall from sterile bag. Allow coverall to unfold. Unzip zipper to the full length. Begin rolling coverall backward into a tube, gathering the sleeves inside the roll. Gather up one leg. Place foot in one leg and pull up to thigh. Do the same with the other leg. Pull coverall up to waist. Begin unrolling top of coverall. Slide in one arm, and roll coverall over back and shoulders. Slide in the other arm. Zip up coverall. Snap at collar and ankles. Sanitize gloves with sterile IPA.
    Note: Do not let the coverall touch the floor or walls.
11. To don boots, remove one boot from sterile bag by placing hand on inside of boot top. Put on boot and step to sterile side of gown area. Place hand inside of other boot top. Remove other boot from sterile bag. Put on boot and step to sterile side of gown area. Pull up boot tops. Snap around calf area and snap to back of coverall. Sanitize gloves with sterile IPA.
12. Don sterile goggles or shield.
13. Apply sterile IPA to donning gloves. Put on second pair of sterile gloves over first pair. Roll gloves over cuffs of coverall.

Doffing

Reverse the gowning procedure. Place disposable items in trash. Place reusable coverall, hood, and boot in proper receptacle. Place goggle in separate receptacle. Exit de-gowning area.

Gowning procedures for the
non-sterile cleanroom

IEBT-RP-CC053.3 "Garment system considerations for cleanrooms and other controlled environments" should be consulted for guidance to the
specification, testing, selection, and maintenance of apparel and accessories appropriate for use in cleanrooms and controlled environments.

FIGURE 7.1 Sample poster: steps in appropriate gowning.

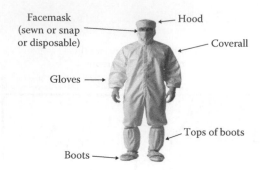

FIGURE 7.2   Cleanroom apparel.

Housekeeping

An integral piece of the cleanroom management program is cleaning of the cleanroom. IEST-RP-CC018, "Cleanroom housekeeping," details all aspects of proper cleaning of the cleanroom, including selection of cleaning materials, equipment, and cleaning agents, and auditing the cleaning procedures. Figure 7.3 is an example of a checklist for auditing cleanroom housekeeping. A documented and controlled cleanroom cleaning program should be established and maintained to ensure the integrity of the cleanroom environment. Selection of cleanroom cleaning materials, equipment, and cleaning agents should be appropriate to the type of cleaning required (i.e., aseptic versus non-aseptic). All cleaning agents and subsequent cleaning protocols should be validated to ensure efficacy of cleaning.

The selection and validation of the cleaning and disinfecting chemicals should include

- Product compatibility
- Process compatibility
- Cleanroom compatibility
- Environmental compatibility

IEST-RP-CC023, "Microorganisms in Cleanrooms," provides guidance on selection of cleaning agents, disinfectants, and sanitizers in cleanrooms. In addition to the disinfectant guidelines, there is information regarding the parameters and effectiveness of the different cleaning chemicals and applications and important characteristics of the different cleaning chemicals.

If the area surrounding the cleanroom is kept clean, the ability to keep the cleanroom clean is enhanced. Sticky, disposable entrance mats will contain gross dirt from shoes from the areas surrounding the cleanroom. The sheets should be changed frequently, as soiling occurs. The mat should never be "yanked." The soiled sheet should be removed by gently peeling all four corners toward the middle and lifting to form a pocket. This lifting action should be accomplished in a deliberate and slow manner to minimize static charge generation and unwanted particle contamination into the air.

When cleaning any surface in the cleanroom, a deliberate top-to-bottom motion is used. On vertical surfaces, start at the top and move downward with overlapping strokes. On horizontal surfaces, start at the rear of the surface and wipe forward with overlapping strokes. These cleaning motions ensure that the entire surface has been cleaned and that any particles (viable or nonviable) on the surface are captured and removed. These motions apply to cleaning any surface, including tables, walls, floors, and equipment. Disinfectant solutions are applied in the same manner; however, for disinfection, the solution must remain on the surface for the appropriate dwell time. It is very important to remember that wipers should be selected specifically for the cleaning requirements and cleaning chemicals. Rules for wiping a surface are as follows:

Gown room

- [ ] Baseboards clean
- [ ] Benches clean
- [ ] Corners clean
- [ ] Doors and doorframes clean
- [ ] Floors mopped
- [ ] Floors vacuumed
- [ ] Glass clean
- [ ] Horizontal surfaces clean
- [ ] Supplies stocked in shelves
- [ ] Rack and Legs cleaned
- [ ] Tacky mats cleaned
- [ ] Tacky mats placed properly
- [ ] Trash cans cleaned and liners replaced
- [ ] Return air vents clean
- [ ] Walls clean

Packaging area

- [ ] Baseboards clean
- [ ] Corners clean
- [ ] Horizontal surfaces clean
- [ ] Light fixtures clean
- [ ] Glass clean

Exterior corridor

- [ ] Baseboards clean
- [ ] Corners clean
- [ ] Doors and doorframes clean
- [ ] Floor swept/mopped
- [ ] Horizontal surfaces clean
- [ ] Light fixtures clean
- [ ] Overhead piping cleaned and labeled
- [ ] Tacky mats clean
- [ ] Tacky mats placed properly
- [ ] Trash cans cleaned and liners replaced
- [ ] Windows clean

Cleanroom area 1

- [ ] Baseboards clean
- [ ] Work Tables clean
- [ ] Corners clean
- [ ] Doors and doorframes clean
- [ ] Pass-through clean
- [ ] Floor free of debris
- [ ] Floors mopped
- [ ] Floors vaccumed
- [ ] Garments on hangers
- [ ] Glass clean
- [ ] Gown racks clean
- [ ] Garment carts clean
- [ ] Horizontal surfaces clean
- [ ] Return air vents clean
- [ ] Supplies stocked
- [ ] Trash cans clean; liners replaced
- [ ] Tables and Legs clean
- [ ] Walls clean

Protocol adherance

- [ ] Bouffant cover all hair
- [ ] Employees mop properly
- [ ] Facial in proper place
- [ ] Gowned/gloved properly
- [ ] No candy, gum, drinks, etc.
- [ ] No make-up
- [ ] No paper products in cleanroom
- [ ] No spraymakers in cleanroom
- [ ] No tobacco products in cleanroom
- [ ] Only CR approved paper
- [ ] Only CR approved pens

Cleanroom area 2

- [ ] Baseboards clean
- [ ] Carts cleaned
- [ ] Corners cleaned
- [ ] Doors and doorframes clean
- [ ] Floor mopped
- [ ] Glass clean
- [ ] Walls clean
- [ ] Return Air Vents clean

Cleanroom area 3

- [ ] Baseboards clean
- [ ] Corners clean
- [ ] Floor mopped
- [ ] Floor vacummed
- [ ] Supplies stocked
- [ ] Table surface and legs clean
- [ ] Return Air Vents clean

Housekeeping supplies

- [ ] All containers labeled
- [ ] Cleanroom chemicals approved
- [ ] Equipment clean
- [ ] Equipment in good condition
- [ ] Equipment stored properly
- [ ] Used mops segregated

FIGURE 7.3 Example of a Cleanroom Housekeeping Checklist.

1. Only use the surface of a wiper one time.
2. Never reuse a wiper.
3. Only use a cleanroom wiper validated and approved by the quality department of your organization.
4. Follow the required SOPs for cleaning, as established by the validation process.

Ongoing Assessments

Monitoring and auditing the cleanroom management program ensures that the documented procedures and protocols are understood, implemented, and effective at all levels within the cleanroom manufacuring process. The environmental monitoring sites should include all contamination sources:

- People
- Cleanroom air and surfaces

- Deionized (DI) water
- Equipment surfaces and entrapment areas
- Chemicals
- Incoming components

The audit of the environmental monitoring program should provide documented and impartial evidence that the cleanroom management program is robust and reproducible. Tracking and trending this data will allow establishment of alert and action limits and show shifts in the program, which can be addressed prior to compromising the entire cleanroom management program.

Summary

The contamination control industry standards and recommended practices enable the manufacturer to create a comprehensive cleanroom management program. The advantages of controlling human contamination within a balanced cleanroom management program are

- Knowledgeable and empowered employees
- Controlled cleanroom environment
- Quality-assured processes
- Decrease in product rejects
- Increase in profitability

Just as humans are the greatest potential contamination risk, they are the greatest resource for contamination control. A thorough and comprehensive training program detailing all aspects of cleanroom management will empower the cleanroom operators to control the degree of contamination during the production process.

References

Standards and Recommended Practices

1. ISO 14644-1-1999, Cleanrooms and associated controlled environments—Part 1: Classification of air cleanliness.
2. ISO 14644-2-2000, Cleanrooms and associated controlled environments—Part 2: Specifications for testing and monitoring to prove continued compliance with ISO 14644-1.
3. Institute of Environmental Sciences and Technology, Garment considerations for cleanrooms and other controlled environments, IEST-RP-CC003.
4. Institute of Environmental Sciences and Technology, Evaluating wiping materials used in cleanrooms and other controlled environments, IEST-RP-CC004.
5. Institute of Environmental Sciences and Technology, Cleanroom gloves and finger cots, IEST-RP-CC005.
6. Institute of Environmental Sciences and Technology, Testing cleanrooms, IEST-RP-CC006.
7. Institute of Environmental Sciences and Technology, Considerations in cleanroom design, IEST-RP-CC012.
8. Institute of Environmental Sciences and Technology, Cleanroom housekeeping, IEST-RP-CC018.
9. Institute of Environmental Sciences and Technology, Electrostatic charge in cleanrooms and other controlled environments, IEST-RP-CC022.
10. Institute of Environmental Sciences and Technology, Microorganisms in cleanrooms, IEST-RP-CC023.
11. Institute of Environmental Sciences and Technology, Cleanroom operations, IEST-RP-CC026.

12. Institute of Environmental Sciences and Technology, Personnel practices and procedures in cleanrooms and controlled environments, IEST-RP-CC027.
13. ANSI/AAMI/ISO 11137-1:2006 Sterilization of healthcare products—Radiation—Part 1: Requirements for development, validation and routine control of a sterilization process for medical devices.
14. ANSI/AAMI/ISO 11137-2:2006 Sterilization of healthcare products—Radiation—Part 2: Establishing the dose.
15. ANSI/AAMI ST 67:2003 Sterilization of healthcare products—Requirements for products labeled "Sterile".
16. ASTM F-51-00 (reapproved 2007) standard test method for sizing and counting particulate contaminant in and on clean room garments.

Journal Articles

1. Eudy, J., Human contamination. *A2C2*; 9, April 2003.
2. Eudy, J., Apparel considerations for cleanrooms. *A2C2*; March 2004.
3. Eudy, J., Dress up your cleanroom, contamination control; Winter 2008.
4. Eudy, J., Ask Jan column, *Controlled Environments Magazine*; Quarterly 2003–2009.

8

Principles of Wiping and Cleaning Validation

What Is Contamination? 119
How to Remove Contamination from the Cleanroom or
Mini-Environment 120
When Do You Clean? 123
Training...124
Cleaning Validation...124
Cleaning Validation for Semiconductors • Cleaning Validation for
Pharmaceutical Manufacturing • Cleaning Validation for Aerospace
Manufacturing
Conclusions..126
References...126

Karen F. Bonnell
Production Economics

Howard Siegerman
Siegerman &
Associates LLC

In a cleanroom or mini-environment, there is a constant battle against contamination. This is a battle that must be fought again and again: every day, every shift, between production campaigns, and after spills.

These environments must be cleaned on a regular basis, following the standard operating procedures (SOPs) that have been set up for the cleaning process. The cost of not fighting the battle successfully could be disastrous: bad or damaged products or products that could potentially damage or harm other products or people. In the semiconductor industry, where the products are chips with measurements in nanometers, small amounts of particles can ruin entire runs. Contamination control for the medical device industry (with products such as replacement knees and hips, catheters, and stents) is critical to avoid microbial contamination. Wiping down products prior to sterilization will help remove particles, microbial contamination, and cleaning residues.

What Is Contamination?

What, exactly, is contamination? In the broadest sense of the word, contamination is anything that should not be present (similar to weeds being plants that are in the wrong place). To be more specific, contamination includes particles, fibers, and residues (typically chemical—sometimes from the cleaning process itself).

Contamination comes from a number of sources: people, processes, and objects. People are the greatest generators of particles and fibers: flakes of skin, hair, bits and pieces of garments, and airborne molecular contaminants. Personnel in a cleanroom need to wear protective garments to protect the environment and the products: gowns, booties, bouffants, beard covers, hoods, masks, and gloves. Every movement in a cleanroom generates particles—even sitting generates particles [1]. It is very important to limit movement; when movement is required, it should be slow and deliberate.

Processes create contamination. In microelectronics, processes such as chemical vapor deposition, ion implantation, and plasma processing can all result in contamination settling on semiconductor wafers or hard-drive discs. In a pharmaceutical plant, the process of creating a tablet from a powder can result in airborne contamination. In addition, any movement, either by people or by equipment parts within a cleanroom or mini-environment, can generate particles.

Objects can also be a source of contamination: any textured unsealed surface (such as an upholstered chair) can be a generator of particles and fibers. Objects brought into a cleanroom may effectively contaminate the environment [2].

The goal of the cleaning battle is to remove existing contamination without adding new contamination. Part of this goal is to eliminate cross-contamination, by not dragging contamination from one area to another.

How to Remove Contamination from the Cleanroom or Mini-Environment

Particles (contamination) adhere to surfaces via capillary adhesion forces. Removing surface contamination requires effort and energy. The act of wiping provides the mechanical energy necessary to remove particles from a surface. The fibers of a wiper act as scrapers on a surface, loosening particles and other contamination. Solvents and cleaning solutions—disinfectant, isopropyl alcohol (IPA), deionized water (DIW) with a surfactant, or deionized water alone—help to solubilize chemical contaminants and reduce the energy necessary for particle removal [3]. Proper wiping techniques will remove contaminants and contain them within the wiper itself, without redepositing them elsewhere.

The use of disinfectants is advised for surfaces that are contaminated with microorganisms. When used according to recommended SOPs (applied using wipers, mops, or spray bottles), disinfectants kill the microorganisms but do not remove the resulting dead organisms. After waiting the recommended period for the disinfectant to be most effective (dwell time), wiping down the disinfected surfaces with a wiper dampened with IPA or DIW is crucial. The wiping action removes the residual disinfectant along with the small corpses.

The choice of cleaning weapons in the contamination war depends upon what must be cleaned and where it is located.

All surfaces in a cleanroom environment (or even a controlled environment) need to be cleaned to maintain the desired level of contamination control. "All surfaces" includes floors, walls, ceilings, windows, pass-throughs, process equipment, and instruments. Any surface that is exposed during the activity within the cleanroom (such as the interior of the pass-throughs) must also be cleaned.

Surfaces that are within arm's reach can be cleaned using wipers and solutions appropriate to the environment. These surfaces will include counter tops, exteriors of cabinets and process equipment, most surfaces of laminar flow hoods, biosafety cabinets and isolators, pass-throughs, etc. (Figures 8.1 and 8.2).

The weapons of choice are cleanroom wipers appropriate to the ISO class of the area to be cleaned and a cleaning agent such as a disinfectant solution or an isopropyl alcohol solution. Wipers should be "low linting"—that is, wipers should not leave particles and fibers behind when they are used to clean surfaces. In a minimally controlled environment (e.g., ISO Class 5-8), a nonwoven polyester/cellulose wiper can be used. However, in areas where a more stringent environment is necessary (ISO Class 3-5), a knit polyester wiper is indicated. Polyester wipers can have edges that are cut, sealed, or border sealed, depending upon the cleanliness requirements.

Wipers either can be purchased pre-wetted or can be wetted at the time of use with the solution of choice. Be certain that the wipers are not wetted to the point of dripping—an overly wet wiper can re-disperse contaminants and can increase the chances of the solution itself becoming a contamination source.

Wipers should be folded so that the entire surface of the wiper can be used. Typically, wipers are folded into quarters and then refolded after every wipe of the surface to contain and remove the contamination.

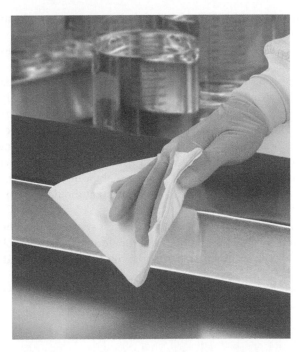

FIGURE 8.1 Readily accessible surfaces can be cleaned with appropriate wipers and cleaning agents. (Courtesy of ITW Texwipe, Kernersville, NC.)

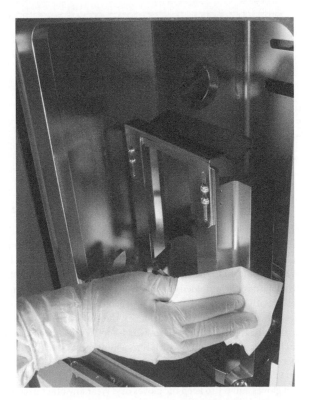

FIGURE 8.2 Vertical surfaces should also be wiped to remove contaminants. (Courtesy of ITW Texwipe, Kernersville, NC.)

In this manner, there are eight different surfaces of the wiper exposed and capable of being used for wiping. Wiping should be done in overlapping linear strokes, from clean to dirty or dry to wet [4]. After all eight surfaces have been used, the wiper (and its captured contamination) should be discarded in an appropriate container (Figure 8.3).

That's all very well and good for the surfaces that can be reached. But what about those surfaces beyond arm's length? Mops are effective extension devices, as they are essentially wipers on long sticks.

Mops are traditionally used for the floor. String mops of the edgeless variety with tubular knitted yarn that have captured ends (to minimize release of particles and fibers) are considered ideal. String mops should be rinsed frequently using two to three buckets to accomplish rinsing. The mop should first be dunked into the "dirty" bucket, agitated, and then wrung out. Then the process should be repeated with the second (and third, if present) buckets to ensure that the dirt the mop has just picked up is not redeposited. The bucket water/cleaning solution or disinfectant should also be changed frequently to avoid reintroducing contamination.

Floors can also be cleaned using flat mops with a bucketless system. When a bucketless mopping system is used, the mop covers are either pre-wetted with the cleaning or rinsing solutions, or the operator sprays the mop covers with a cleaning solution after the covers have been placed on the mop head. The

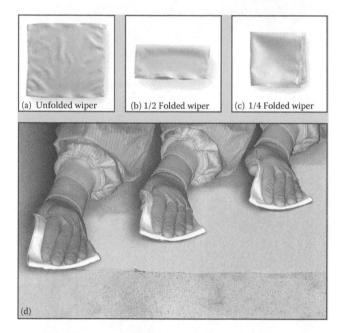

FIGURE 8.3 Appropriate techniques for wiping contaminants and spills are essential. Cleanroom wiping guide: 1. Follow relevant site protocol (procedures for safety, contamination, etc.) and wear cleanroom gloves. 2. Fold wiper in mid-air into quarter folds (a through c). This will produce several clean surface areas and allow better contact with the surface to be wiped. 3. When wiping, hold the wiper so that the folded edge is toward the area to be wiped. Hold the unfolded edges in your hand. Group the unfolded edges between thumb and forefinger. 4. Use either a pre-wetted wiper or a dry wiper moistened with an appropriate cleaning agent. 5. Wipe in one direction, overlapping wiped area by 10%–25%. 6. Wipe from cleanest to least clean regions of the surface being wiped. Wipe systematically, for example, from top to bottom, far to near (d). 7. Keep track of which surfaces have been cleaned and which wiper areas are unused. 8. Always use the cleanest surfaces of the wiper. If re-wiping use a clean portion of the wiper, not the used wiper area. 9. Dispose of wipers according to site procedures. Wiping wet spills: 1. Identify the spilled liquid. Follow the material safety data sheet (MSDS). 2. Choose wiper and gloves that will not be degraded by the liquid. 3. For hazardous spills, wear two pairs of gloves and try to keep the gloves dry. Wear any other necessary protective gear. 4. Use dry wipers to wipe spills up immediately. Then clean the affected surfaces by following steps 1–9 in the "Cleanroom wiping guide" above. 5. Dispose of wipers according to site procedures. (Courtesy of ITW Texwipe, Kernersville, NC.)

FIGURE 8.4 Cleaning difficult-to-reach places is expedited by using an isolator cleaning tool. (Courtesy of ITW Texwipe, Kernersville, NC.)

mopping should be done in linear overlapping strokes, from clean to dirty, dry to wet—the same procedure that is recommended for wiping. In a cleanroom, the corners furthest from the entry doors are usually the cleanest areas since they have the least foot traffic. Aisles and entry doors sustain the greatest traffic, and are usually the dirtiest areas. So mopping is done from far corners to aisles and entry doors (cleanest to dirtiest).

Using a bucketless mopping system to clean the walls and ceilings, as well as the surfaces of large tools, eliminates dripping or pooling of wetting solutions as the mops can be optimally wetted. Large areas can be cleaned efficiently with a bucketless system as the mop covers are light and easy to change. Bucketless systems also eliminate the need to frequently change large volumes of water and/or solutions, which are necessary steps when using string mops.

When very small spaces need cleaning and are difficult to reach with wipers (such as back walls and far corners of the deck in mini-environments), an isolator cleaning tool (a small mop) with replaceable mop covers is employed (Figure 8.4).

Swabs wetted with a cleaning solution or IPA or DIW can be used to address very tight areas. Some possible candidates for swab cleaning are vents in the back wall or front surface, the junction of the walls with the floor and ceiling of a cabinet, surface dimples surrounding rivets or screws, and connectors to hoses or filters, among others (Figure 8.5).

When Do You Clean?

Cleaning (decontaminating) should be done according to protocols and SOPs that have been established and agreed to, such as ASTM E2042 [5] and IEST RP-18.2 [6]. ASTM E2042 includes a table ("Examples of Cleaning Frequencies") with recommended frequency of cleaning corresponding to the ISO Class of the environment. At a minimum, cleaning should be done at the beginning of each work shift and at the

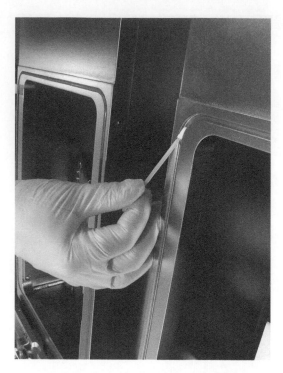

FIGURE 8.5   Swabs can be used to clean very tight areas. (Courtesy of ITW Texwipe, Kernersville, NC.)

change of a production campaign (if applicable). Cleaning should also be done when visible residues are present or when there is a spill.

Training

Training, based on written protocols, is essential for proper cleaning of a cleanroom [7]. Cleanroom personnel should have a good understanding of how and what they are to clean, and why. They will perform much better if they have knowledge of what they are expected to do.

Written protocols define the SOPs describing cleaning methodology. SOPs tell personnel what needs to be cleaned, what tools (wipers, mops, swabs, cleaning agents) are to be used, how often to clean, and how to do the cleaning.

Using these SOPs, personnel need to be trained and then tested to ensure proficiency in cleaning and maintenance. In addition, personnel should be retrained on a regular schedule. The written SOPs should be available at all times. Visual aids, such as posters or graphs, can be used to reinforce the training [8].

Cleaning Validation

After environmental or process equipment surfaces have been cleaned, the remaining task is to verify that no detectable contamination remains on these surfaces—hence the name "cleaning validation." The term cleaning validation in the context of this discussion should be distinguished from other uses of the term. For example, cleaning validation is a term that is also used in determining residue levels on critical products, notably medical devices. In that context, specific stringent requirements such as risk analyses, analytical testing after exhaustive extraction, and sign-off protocols may be in order.

This task can be simple or complex. The simplest approach to cleaning verification is to examine the cleaned surface visually (bright light illumination helps here) to confirm that all surface residue has been removed. Illuminating the cleaned surface with ultraviolet light (guard against eye damage!) can help in the visualization of certain soils, since they will appear as bright spots on a purple background. Wiping cleaned surfaces with black wipers can also help to visualize light colored residues that might otherwise remain undetected.

Visual examination can verify that the surface has less than approximately $1-4\,\mu g/cm^2$ of surface residues. Obviously, surfaces on which residues are visually detected must be recleaned. "Cleaning is continued up to the absence of visible residues" should be a standard operating procedure.

Visual examination of cleaned surfaces is necessary but may not be entirely sufficient for some industries. Some manufacturing processes or products are so critical that verification of the minimization of specific contaminants is required post-cleaning. Examples include the following:

- Minimization of surface particles for semiconductor manufacturing
- Minimization of product or cleaning agent residues in pharmaceutical manufacturing
- Minimization of surface nonvolatile residue for aerospace manufacturing

Cleaning Validation for Semiconductors

In the manufacture of semiconductors, particles can cause "killer" defects in the device and must be controlled to extremely low levels. Often, the particles that must be detected are smaller than the $50\,\mu m$ resolution of the human eye. Jarmin [9] describes the use of a Pentagon QIII instrument that measures surface particles by drawing them up off the surface through a vacuum port, then counting and analyzing the particles according to size. Particle sizes in the range of $0.3-10\,\mu m$ can be detected. This device is promoted by its manufacturer [10] as a means of ensuring that semiconductor process tools have been cleaned to the requisite level prior to being put back into production. Armentrout [11,12] has shown improvements in semiconductor product yield and equipment uptime with the use of this instrument.

Cleaning Validation for Pharmaceutical Manufacturing

In pharmaceutical manufacturing, the cleaning of process equipment and environmental surfaces occurs on a regularly scheduled basis, usually at the completion of a production lot consisting of thousands of manufactured units (vials, bottles, filled syringes, etc.). Once the cleaning of the requisite surfaces is completed, it is necessary to perform a cleaning validation to ensure that detectable residues of either the manufactured product or cleaning agents are below acceptable levels and will not contaminate the subsequently manufactured product.

This involves sampling the surfaces (usually with ultraclean swabs) in a reproducible manner to remove any surface residues, extracting those residues into a suitable solvent and then analyzing the solvent to determine the identity and quantity of any contaminants. Details of the sampling procedure, including choice of swab fabric, degree of swab wetting, and swabbing mechanics, are critical for success in cleaning validation [13].

Selection of the extracting liquid is determined by the chemistry of the suspected contaminant and by the analytical procedure to be employed. As an example, surfactant solutions often employed as cleaning agents can leave behind residues on the cleaned surface. Those residues are picked up by the sampling swabs and then extracted from the swabs into high-grade deionized water, or into a dilute acid or base (e.g., 0.01 M HCl or 0.1 M NaOH). The liquid is then analyzed by total organic carbon (TOC) instrumentation. The higher the TOC content, the higher the surfactant content of the residue. The lack of specific functional groups on surfactant molecules makes them difficult to analyze by traditional chromatographic or spectroscopic techniques, so TOC is used instead. Product residues (i.e., residues from the manufactured pharmaceutical) on swabs can often be dissolved into organic solvents (e.g., toluene) prior to analysis by chromatography or spectroscopy.

Cleaning Validation for Aerospace Manufacturing

Strictly speaking, the residues discussed in the pharmaceutical section are nonvolatile residues (NVRs), since these are the substances that remain on surfaces after liquids have evaporated. In that section, the emphasis is on identifying and quantitating the residues found, in order to verify that the cleaning process is sufficiently thorough that subsequent production lots would not be contaminated.

In this section, the emphasis is focused on quantitating NVR levels on specific surfaces. Identification of the residue type is not as critical here, unless there is concern about contamination from certain types of airborne molecular contaminants (AMCs), such as phthalate esters emanating from plastics [14].

The NVRs treated in this section typically refer to substances that will interfere with the optical properties of sensors, lenses, etc.—the type of equipment found in aerospace payloads. Obviously, it is impractical, if not impossible, to reclean payloads in space, so the NVR sampling methods are critical for success. Indeed, the American Society for Testing Materials (ASTM) has developed a number of standards [15,16] describing the procedures to be followed.

Satellites, launch vehicles, and associated payload hardware are scrupulously cleaned prior to launch, to remove surface residues. Then, the surfaces are sampled with ultraclean wipers (multiply extracted with solvents to minimize NVR originating from the wipers) dampened with extraordinarily pure solvents. After surface sampling, the wipers are then extracted with these same high-grade solvents to determine the residue quantity [17,18].

Conclusions

Cleanrooms or mini-environments must be kept free of contamination in order to maintain functionality. Once surface contamination has been defined, several wiping methods for contamination removal can be used; the wiping method chosen depends upon the placement of the contamination:

1. Wipers
   a. Dry
   b. Dry with a solubilizing agent (disinfectant, IPA, DIW with surfactant, DIW)
   c. Pre-wetted with a solubilizing agent
2. Mops
   a. String mops and buckets
   b. Flat mops with wetted mop covers (bucketless systems)
   c. Isolator cleaning tools (small mops)
3. Swabs wetted with cleaning solution, IPA, or DIW

After cleaning, cleaning validation should be done to verify that there is no detectable contamination remaining on the cleaned surfaces. Surface sampling can be done by

1. The Pentagon QIII instrument (semiconductor industry)
2. Ultraclean swabs (pharmaceutical industry)
3. Ultraclean wipers for NVR residues (aerospace industry)

These sampling agents are then analyzed for contamination residues.

References

1. J. Eudy. Human contamination. *Controlled Environments Magazine*, April 2003.
2. B. Kanegsberg and Ed. Kanegsberg. Contamination control in and out of the cleanroom: Spots before your eyes. *Controlled Environments Magazine*, May 2009.
3. D. Cooper. Cleaning aseptic fill areas. *Pharmaceutical Technology*, February 1996.

4. D. Cooper. Wipe first, clean later. *A2C2 Magazine* [now *Controlled Environments Magazine*], June 1998.
5. Standard practice for cleaning and maintaining controlled areas and clean rooms, ASTM E2042-04. ASTM International, West Conshohocken, PA, 2004.
6. Cleanroom housekeeping—Operating and monitoring procedures, Document IEST-RP-CC018.2. Institute of Environmental Science and Technology, Rolling Meadows, IL.
7. I. M. Wallis. Decontaminating the cleanroom requires good science. *Cleanrooms*, July 1997.
8. H. Siegerman. Is this a convenient time to clean? *Controlled Environments*, January 2007.
9. G. Jarmin. Particle testing for cleanroom forms and labels. *Controlled Environments*, January 2002.
10. http://www.pen-tec.com/products/q3-surface-particle-detector
11. L. Armentrout. Analysis of tool contamination. *Future Fab International*, 7, 95.
12. L. Armentrout. A surface particle study of wafers in a boxless environment. *Future Fab International*, 8, 2000.
13. H. Siegerman et al. How to succeed in the search for nothing: Effective swabbing techniques for cleaning validation, *SterilTechnik*, 1, 2006; *Controlled Environments*, February 2007.
14. The rate of deposition of nonvolatile residue in cleanrooms, IEST-RP-CC016. Institute for Environmental Sciences and Technology, Mount Prospect, IL.
15. Handling, transporting, and installing nonvolatile residue (NVR) sample plates used in environmentally controlled areas for spacecraft, E1234. American Society for Testing Materials, West Conshohocken, PA.
16. Test method for gravimetric determination of nonvolatile residue (NVR) in environmentally controlled areas for spacecraft, E-1235. American Society for Testing Materials, West Conshohocken, PA.
17. Standard test method for gravimetric determination of nonvolatile residue from cleanroom wipers, E1560-06. American Society for Testing Materials, West Conshohocken, PA.
18. Standard test method for nonvolatile residue of solvent extract from aerospace components (using flash evaporator), *F331-05*. American Society for Testing Materials, West Conshohocken, PA.

9

Overview to Analytical and Monitoring Techniques

Why Should You Use Analytical Tests?..129
Visual 129
Definitive Analysis
Standards 130
Standards and Controls 131
In Situ versus Extractive Analysis...131
Airborne Molecular Contamination..132
Conclusions...133
References..133

Ed Kanegsberg
BFK Solutions

Why Should You Use Analytical Tests?

Troubleshooting, including the use of analytical tests, becomes a necessity when process parameters go awry. However, even when the process is working or appears to be optimum, process parameters should be monitored and quantified periodically.

When a process is being developed or modified, it is crucial to validate or qualify the process parameters. This acts as a benchmark to keep a process on track or serves as a point of reference for process revision. Such testing can demonstrate and document process quality and consistency. It can also act as an indicator of its safety and environmental acceptability and help assure customers (internal and external), regulatory agencies, and, especially, you yourself that the process is in control.

Analytical techniques are aptly summarized by Dr. Ben Schiefelbein, one of the contributing authors, who said that analytical tests involve "somehow perturbing the area to be analyzed and observing what results" (Kanegsberg and Kanegsberg, 2007).

Visual

Perturbing the area may be as simple as looking at it. The first task in determining whether a cleaning process has done its job is to look at the result. For many processes, the standard of "clean" is that it "is visually clean," that is, to the experienced and educated human eye, there is no evidence of contamination.

Visual observation is a first line of defense when a process deviates from normal. Any unusual observation should be a point of concern. We encountered an example where a production technician reported that a sample of solvent that normally is clear and colorless had a pink tint. The written certifications indicated that the solvent met the specifications. Upon investigation, it was determined that this material was sensitive to light exposure and that it may not have been stored properly. The pink coloration was a sign of breakdown that would have been missed were it not for visual observation by an alert technician.

Visual observation sometimes is aided by varying the lighting conditions (e.g., glancing angle illumination and UV [black] light) or by magnification. Visual observation also includes nonquantitative procedures such as the "white glove" test and the "water break" test. The water break test involves observing whether water beads up or channels on a surface. Such beading indicates the presence of oil-like contamination. When the drop sheets or flows rather than beads, it may mean that the surface is clean. However, the test is not foolproof. The presence of contaminants, like surfactants, that lower the surface tension also increases wetting; contaminants may be masked by surfactants.

Definitive Analysis

Generally, analytical testing refers to more than visual observation. In many cases, both qualitative and quantitative analyses are needed. There are a myriad of test methods. No one method can provide all the answers.

"Gee, we ran SEM/EDX, and we couldn't identify the organic contamination; so we added another cleaning step." We hear this comment on a regular basis (Kanegsberg and Kanegsberg, 2008). The reason is that most manufacturers have access to SEM/EDX, and it provides useful information. However, we sometimes expect more from SEM/EDX than the technique is capable of providing, such as molecular identification of a specific organic residue. Attempting to identify an organic compound by SEM/EDX is an exercise in futility. It is important to select an appropriate method to identify contaminants. Not only critical cleaning but also sample preparation (including extraction) and analysis have to be customized to the residue, surface, and required purity of the liquid or solid product. To customize, we have to first understand the contaminants.

Before running an analytical test, ask yourself, "Do I need to run a particular test?" Unnecessary testing diverts money and engineering resources that could instead be spent to grow the company. Part of "why" includes identifying the customer and the end use of the product (Kanegsberg, 2008).

Analytical testing involves more than specifying a general analytical technique. The specifics matter. Does the lab use a standard method? What is the background and experience of the analyst? How is the sample collected, shipped, and prepared within the analytical lab? What is the limit of detection of the method? What is the method variability? Are there interfering substances that would mask results? How variable are the results within a given lab? Among labs? Are the results analyst dependent? These are all relevant questions to be asked.

Elimination of contamination from the product is a common goal of process improvement efforts. The goal of zero contamination may be an edict from corporate management or a statement from a key customer, or may be you yourself fantasize about ridding your product of all contamination.

However, forget about achieving zero residue; it is an impossible goal. Instead, redirect your efforts to determine how clean your product needs to be for the application at hand. Invest in finding the most effective ways to minimize contaminants on the product and to minimize sources of contamination.

Acceptable levels of residue or contamination are set by a variety of factors, including customer requirements, industry-wide standards, benchmarking of historically acceptable products, and customary practice or habit. Are the acceptable levels protective of product quality? Are they rational?

Standards

A standard can be quantitative in that it can set quantitative levels (either to be exceeded or not to be exceeded). It can specify methods in that it can tell you how to conduct a particular activity. It can provide definitions (military or ISO standards). A standard can also be normative, which can be described as providing decision-making tools where there are complex series of operations with many possible variables. Standards are not the whole answer. For example, a standard adopted by most aerospace applications, IEST-STD-CC1246, "Product Cleanliness Levels and Contamination Control Program," (derived from MIL-STD-1246) defines and sets category levels of cleanliness. However, it neither specifies how to achieve these levels (either by cleaning or contamination control) nor how to test for them.

There are many sources of standards. A few of the largest organizations are the International Standards Organization (ISO), the American Society for Testing and Materials (ASTM), the Institute of Environmental Sciences and Technology (IEST), the Association for the Advancement of Medical Instrumentation (AAMI), the United States Pharmacopeia (USP), the Association Connecting Electronics Industries (IPC), and Surface Mount Technology Association (SMTA).

In some cases, companies create or amend a standard for their own processes. Sometimes, this is to build in more stringent requirements so as to increase the reliability of the final product. In some cases, this is for proprietary reasons.

Standards and Controls

Standards are not the same as controls; both are needed. Controls provide a benchmark of a process. Controls are customized to a particular application. Controls may be positive or negative. A positive control is known to work in the test or respond to the test; a negative control is known not to respond to the test. For example, when Barbara was monitoring the performance of a new pregnancy test, she included pregnant people as well as Carlos, Juan, and Bill as controls. At one point, test results indicated that Carlos, Juan, and Bill were all pregnant; immediate corrective action on the test was needed.

One negative control is a "known-to-be-clean" or "known-to-be clean enough" control. This is particularly important where contaminants must be measured at very low levels and/or where the samples may become inadvertently contaminated during shipping or even by air quality in the test lab. For this reason, such controls are sometimes referred to as environmental controls. They are shipped along with the unknown sample and tested at the same time as the unknown ones. If the results of the analytical test indicate that both the sample that is known to be clean and the unknown sample appear to have approximately the same levels of contaminants, there is probably no problem (Kanegsberg and Kanegsberg, 2006).

When you send controls to a test lab, it is important to identify the controls to the test lab as well as communicate what you want to happen to them. One client was trying to evaluate a new cleaning process at the facilities of an equipment supplier. They sent uncleaned samples for test cleaning along with controls that had been cleaned by their current process. They meant for the controls to provide a point of comparison so that when returned, they could have both the controls and the test samples analyzed to evaluate how well the proposed process worked. What they got back were the control samples along with a note, "Your controls looked dirty, so we cleaned them."

In Situ versus Extractive Analysis

It may be possible and desirable to observe the contamination directly on the surface. In situ measurement has the advantage of identifying exactly where the contaminant is; this can be a useful clue to where the contamination came from. However, not all situations lend themselves to in situ analysis. For one thing, the surface must be accessible; this may exclude contaminants in crevices or holes. In addition, without some method of concentrating the contaminant, it may not be possible to detect it; this is important where low levels of a contaminant dispersed over a surface are detrimental to product performance (see Chapter 12, Andrascik).

It is important to realize that the same approaches and limitations that apply to cleaning also apply to extraction. The goal of both cleaning and extraction is to remove the contaminant, be it particulate or thin film. In the case of extraction, the contaminant is recovered from the extraction medium so that it can be detected and/or quantified. The extraction technique, including the specific extraction chemistry, temperature, force, and time all contribute to what is removed from the part being tested. Issues of materials compatibility also apply to extraction; plasticizers and metal particles can be removed by unrealistically long or vigorous extraction methods. This leads to artifactual contaminants; the artifacts can mask true contamination.

We have seen extractive testing being used as a cleaning method. If the extraction test failed, it was repeated. Eventually, the extraction procedure removed enough contaminant from the surface for the product to "pass" the test. This method "worked" in that the part was now clean. However, it was a very time-consuming and costly cleaning method. In some instances, however, extraction is used as a cleaning and monitoring method in that on-line detection of contaminant in the process bath is used. When the level of contaminants in the cleaning or rinsing agent is sufficiently low, the product is deemed to be clean enough.

Airborne Molecular Contamination

Detecting airborne particles is a staple of clean room operation. The detection of airborne molecular contamination (AMC) should be just as routine. AMC is, as the name implies, carried by the ambient air and, since it is molecular in size, is not eliminated by many or most particulate filters. It therefore can be a silent poisoner of the process. AMC is any unwanted organic or inorganic molecule that is carried by the air to the surface of the product. The world of wafer fabrication categorizes AMC as acids, bases, condensables, and dopants (SEMI F21-1102 Standard). In a more global sense, AMC is productively thought of as any molecule carried by the air that masks the surface or reacts with it (Kanegsberg and Kanegsberg, 2009). AMC can come from many sources, both internal and external to your facility or clean room (Table 9.1).

If AMC adheres to the surface, it can be detected by extractive techniques or sensitive surface analytical techniques (Table 9.2) (see Chapter 10, Schiefelbein, and Chapter 11, Chawla). Collection tubes or witness samples, surfaces left to be exposed to the ambient air, are used to collect AMC. However, not all AMC adheres to the surface. For instance, an active molecule, like ammonia, does not adhere to the surface but can cause chemical changes to the surface when it impacts. In such cases, it is the ambient air that must be analyzed to detect the contaminant before it strikes.

TABLE 9.1   Sources of AMC

Source	AMC-Producing Activities/Conditions
Outside the clean room (or, before final assembly)	Proximal manufacturing processes within plant
	Proximal nonmanufacturing processes (e.g., food preparation)
	Process vented out of the plant
	Activities from neighboring plants
	General ambient outdoor air
	Supply chain activities (internal and external)
Transfer	Transfer conditions
	Transfer packaging
	Storage conditions/packaging
Inside the clean room (or, during final assembly)	Volatile process chemicals
	Plastics and other outgassing materials
	Personnel practices (volatile cosmetics)

Source: From Kanegsberg, B. and Kanegsberg, E., "AMC" (5-part series), *Controlled Environments Magazine*, June 2009.

TABLE 9.2 Examples: AMC Monitoring Techniques

AMC Sample Method	Test Methodology (Example)	Identity Information
Direct surface, witness sample	FTIR	Organics
	OSEE	None
Extraction from witness sample	Gravimetric	None
	FTIR	Organics
Air capture	TD-GCMS	Organics
	CRDS	Ammonia
	IMS	Acids, bases, organics

Source: Kanegsberg, B. and Kanegsberg, E., "AMC" (5-part series), *Controlled Environments Magazine,* July/August 2009.

Conclusions

Analytical tests can be of use not only for troubleshooting but also for determining what went *wrong*. With any process, a carefully designed series of tests can help benchmark what is *right*. Such a benchmark makes it much easier to return to normal if a process were to go awry. Consider a testing protocol to validate a process and to establish a benchmark. Consider also periodic monitoring of key parameters to keep the process on track.

References

Kanegsberg, B. How clean is clean? *Product Finishing Magazine,* March 2008.

Kanegsberg, B. and Kanegsberg, E. Navigating the forest of standards, standards philosophy. Part III: Standards and controls, *Controlled Environments Magazine,* July 2006.

Kanegsberg, B. and Kanegsberg, E. Find the contaminant by perturbing the surface: XPS and Auger (Part 1), *Controlled Environments Magazine,* October 2007.

Kanegsberg, B. and Kanegsberg, E. Thinking outside the box (or at least outside the SEM), *Controlled Environments Magazine,* November 2008.

Kanegsberg, B. and Kanegsberg, E. "AMC" (5-part series), *Controlled Environments Magazine,* June, July/August, September, October, November, 2009.

SEMI F21-1102 Standard, Classification of air-borne molecular contaminant levels in clean environments; available through ANSI.

10

Practical Aspects of Analyzing Surfaces

Introduction 135
Analytical Approach 136
Sample Handling and Packaging Issues 137
Analytical Methods 138
Sample Perturbation
Applicability of Different Analytical Approaches 140
Special Topics 141
Layer Characterization • Particle Analysis • Haze Identification
Summary 143
Bibliography 143

Ben Schiefelbein
RJ Lee Group

Introduction

Disciplines that specialize in the analysis of the surfaces of materials have evolved in response to the recognition of the importance of surface characteristics, with most of the development occurring since about 1970. Surface analysis has contributed to the understanding of materials, leading to process changes and control of surface traits that have resulted in improvements in a significant number of industries. The successes of many processes are dependent on engineering the characteristics of the surfaces, and analyzing the surfaces (accurately measuring these characteristics) is an integral part of the engineering process.

Surface characteristics that are often measured include, but are not limited to, the following:

Elemental chemical composition
Defect identification
Molecular structure determination
Surface binding states (determination of chemical state information)
Surface morphology
Surface roughness
The presence and identification of adsorbed surface layers
The presence and identification of particulate
The presence and identification of surface contaminants
Surface hardness
Crystal structure of surface layers and particulate
The presence and thickness and composition of surface layers
Dopant and impurity profiles
Surface homogeneity

Some traits that are dependent on the characteristics of the surfaces are as follows:

Adhesion
Tribology
Corrosion resistance
Wettability
Microanalytical characteristics
Wear characteristics
Electrostatic behavior
Electrical characteristics
Sample fractography
Biological compatibility
Bonding characteristics
Chemical activity

Some of the industries that make use of surface analysis are as follows:

Aircraft
Automobile
Chemical
Construction
Consumer products
Metals
Mechanical engineering
Mining
Paper products
Semiconductors
Utilities
Textiles
Optical devices
Telecommunications
Medical devices

Industries routinely use information about the characteristics of surfaces to affect a wide range of objectives, including

To enhance future performance of a part or component
To optimize manufacturing parameters
To improve or predict the reliability of a part or component
To carry out failure analysis on a part or component
To certify that a part or component meets specified requirements

Analytical Approach

The most affective analytical approach for a given project should be mandated by factors such as the nature of the sample, the relative thickness of any surface layers, the nature of the project, the availability of experience, and the availability of reference information and documented examples of similar projects. Often, several features that affect surface properties such as surface morphology, the presence of thin overlying and sub-layers, haze, particulate contamination, and surface contamination need to be evaluated. In too many cases, the methodology that is used to address a project is determined by the equipment that the analyst is familiar with or the equipment that is most readily available, often leading to wasted effort and delay in completion of the project and incorrect conclusions. Scanning electron microscopy (SEM), for example, is a poor choice for identification of a 100 Å thick surface layer, and auger electron spectroscopy (AES) and x-ray photoelectron spectroscopy (XPS) are poor choices for determination of the composition of micron-sized particulate. The wrong instrumentation is often applied to a project as a matter of familiarity rather than from consideration of the strengths and weaknesses of the methods.

A very important and too frequently overlooked aspect of analyzing surfaces involves a thorough examination of the sample, both prior and subsequent to the analysis. A preliminary evaluation using optical microscopy, for example, often reveals if the sample has excess surface particulate and/or oil contamination that may or may not be related to the project objective. The presence of oil contamination likely shows, at the very least, the need for evaluation of the sample handling and packaging procedures. It often proves prudent to evaluate the nature of contamination before carrying out a detailed analysis. Similarly, examination of the sample subsequent to analysis can prevent misinterpretation of the results. The results of XPS or Auger analysis of a stained material that shows, for example, the presence of a thin carbonaceous layer could be interpreted as showing that the cause of the stain is the carbonaceous layer. Optical examination of the analyzed sample subsequent to ion etching to a depth necessary to remove the carbonaceous layer may show, however, that the stain is still present. If the post-analysis examination showed that the stain had been removed by the etching process, it is likely that the carbonaceous layer was the source.

It is often helpful to characterize sample surfaces by more than one technique. Each of the methods that are commonly used has individual strengths, and it is imprudent to assume that a single technique can provide an effective characterization. A common practice is to examine each sample using optical microscopy to identify features such as excess surface contamination, surface particulate, and the location of areas of interest. For example, regions that optically show refraction colors may represent oil contamination, and circular groupings of particulate are likely deposited by evaporation of droplets (e.g., water spots). This practice would also show the presence of contamination such as tape fragments and other thick organic films that would make further analysis very difficult. Also, the results of an optical evaluation may show the need for additional sample preparation and/or indicate the need to use a method such as Fourier transform infrared spectroscopy (FTIR) to identify an organic component and/or use SEM and/or energy dispersive x-ray spectroscopy (EDX) to further characterize the sample before proceeding to techniques more sensitive to the composition of the near surface.

SEM/EDX analysis subsequent to optical examination represents an excellent screening technique. SEM/EDX has good sensitivity to a large number of elements, has excellent imaging capabilities, is relatively fast in comparison to XPS and AES, is appropriate for classifying materials as being organic or inorganic, and has flexible sample handling characteristics. Analysis by SEM/EDX often shows whether the next step should be, for example, analysis by FTIR to characterize the bond structure of an organic component.

Sample Handling and Packaging Issues

Proper sample handling, preparation, and packaging are very important aspects of surface analysis. The potential for and importance of sample contamination and consequent analytical errors cannot be overemphasized, due to the sensitivity of the techniques used for analyzing surfaces. Some of the techniques are designed to determine the composition of the first few atomic layers of the surface. Samples can be readily contaminated and/or altered during transfer, a sample preparation step, the packaging material, and inadequate protection during storage.

Precautions that should be considered mandatory for analysis of the near surface by methods such as XPS and Auger analysis include

- Cleaning of all tools on a routine basis, including implements such as sample cutting tools, mounting tools such as screw drivers, and sample mounts.
- Storage of implements in areas reserved only for surface analysis activities.
- Samples should never be touched by hand, not even with a gloved hand, but only with clean tweezers. Gloves should never be reused; gloves that are removed should be discarded immediately.

All work surfaces should be covered with a protective covering such as aluminum foil that is replaced on a regular basis

- Saws, vice gripes, etc., that are used for sample preparation should not be used for anything else, and should be wrapped in a protective covering such as aluminum foil when not in use.
- Samples should be covered or wrapped in aluminum foil or a non-contaminating material such as clean room nylon packaging at all times when not being analyzed or prepared. Even though aluminum foil is a convenient material that is frequently used and is, in many cases, an adequate sample packaging material, any contact between the sample surface and the foil is likely to result in contamination. Consequently, flat samples need to be packaged in such a way as to avoid contact between the sample surface and the packaging material or packaged using a non-contaminating material such as clean room nylon. Additionally, a common error is to reuse aluminum foil packaging material. Aluminum foil is subject to contamination—for example, during repeated handling or being allowed to lie on the bench for an extended period of time while the sample is being examined—and should be discarded.
- Ideally, the analytical equipment should be located in and sample preparation activities should be carried out in clean areas. At a minimum, an enclosed bench or glove box that has a flow of filtered air should be used.
- It is also often necessary to control the ambience of sample storage areas and containers due to the possible response of surfaces to ambient contaminants or stresses. The outer surfaces of materials are coated with a layer referred to as an environmental cap, which is in dynamic equilibrium, exchanging components with the atmosphere. Samples stored in areas that have ambient conditions that are hot or humid or are near a source of contamination such as salt from an ocean or oil from a machine shop will undergo changes in the composition of the surface layer. Proximity of an ocean, for example, will likely cause corrosion.

Analytical Methods

The following analytical techniques are often used for analyzing surfaces. For the purposes of this presentation, the term near surface analysis is used to refer to methods such as Auger analysis and XPS that can provide information about the first few atomic layers to differentiate them from other methods that are used to characterize other aspects of surfaces.

- Inductively coupled plasma (ICP) for both bulk and extracted sample including vapor phase (VP)/ICP
- Gas chromatography (GC) and GC/mass spectroscopy (MS) using extracted sample
- FTIR using both extracted samples and direct analysis
- Ion chromatography (IC)
- AES
- XPS
- SEM/EDX and wavelength dispersive x-ray (WDX) analysis
- Transmission Electron Microscopy (TEM)
- Secondary ion mass spectroscopy (SIMS), static SIMS and time of flight SIMS (TOFSIMS)
- X-ray diffraction (XRD)
- X-ray fluorescence (XRF)
- Total reflection XRF (TXRF)
- Atomic force microscopy (AFM)
- Raman spectroscopy

The analytical procedures generally employed for surface analysis can be subdivided into two categories: those that involve analysis of extracted samples and those that involve direct analysis. Each category has its strengths and its weaknesses, which define its primary realm of applicability. Analysis of extracted samples involves using a reagent or method to extract analytes from the surface of interest and subsequent off-line analysis of the extracted material. The methodology takes advantage of the sensitivity of methods such as gas chromatography, ion chromatography, and inductively coupled plasma to quantify the concentrations of analytes on the surface of the sample. The methods may have to rely on previous identification of the analytes of interests, are very sensitive, are often cost-effective, and are often ideal for quality assurance and quality control applications. These methods are limited by the amount of analyte that can be extracted and lack of real-time feedback. Analysis by gas chromatography has been further refined in the form of sample collection by chemical vapor deposition to improve sensitivity. Extraction of organic contaminants using a solvent followed by FTIR analysis is used to take advantage of unique infrared patterns to provide definitive analyte identification. Like GC and ICP, the sensitivity of FTIR can be enhanced by concentrating extracted material onto a small spot, improving the intensity of the infrared response.

Sample Perturbation

Most of the direct response analytical methods that are listed can be described as perturbation (excitation) of a sample by probing it with a beam of characteristic energy that elicits a response (or responses). This response is subsequently detected and processed to glean information about the sample (the adsorbing medium), based on the nature of the interaction between the sample and the beam.

The nature of the response is governed by the energy of the excitation and the characteristics of the sample. In general, the energy or nature of the beam is selected to interact in a specific manner, causing perturbations that can be measured and interpreted. Knowledge of the origin of a response is an invaluable aid in identifying and correcting artifacts that are caused by using inappropriate analytical parameters such as a high electron beam voltage (KV) that causes sample charging or x-ray energy that is too low to excite the analytes of interest. Some of the responses produced by electron beam, infrared, ion beam, and x-ray excitation are discussed below.

Electron Bean Excitation

- Electrons that have sufficient energy to penetrate thin samples and small particles can be used to provide images analogous to images produced by optical microscopes.
- Electrons can be diffracted by the periodic structure of crystalline solids such as a crystal lattice. Electron diffraction is used to determine the crystal structure of solids.
- Elastically scattered (reflected) incident electrons known as backscatter electrons can be used to provide an image of the sample surface. Additionally, the intensity of the backscatter signal is dependent on the average atomic number of the material providing compositional information about the sample.
- X-rays that have energies characteristic of the elements that produce them result from interorbital transitions caused by electron beam excitation. Subsequent characterization of the x-rays can be used to determine the composition of sample surfaces.
- Low-energy electrons known as secondary electrons result from inelastic interaction between the incident electron beam and the sample. The number of secondary electrons emitted is highly dependent on the surface topography producing dramatic pseudo three-dimensional images of sample surfaces.
- Auger electrons, low-energy electrons that are emitted from excited atoms, are complementary to the characteristic x-rays produced by interorbital transitions and are used to determine the composition of the near surface (first few atomic layers) of samples.

X-Ray Excitation

- X-ray excitation can result in inner orbital transitions and emission of characteristic x-rays similar to electron beam excitation and is used to determine sample composition.
- X-ray photo electrons are low-energy electrons of characteristic energy that are produced by irradiating surfaces using a well-characterized x-ray beam. The resulting spectra are very useful for determining the composition and chemical state of the near surface.
- X-rays diffracted by the periodic structure of crystalline solids such as a crystal lattice are used to determine the crystalline structure of solids.

Infrared

- Infrared excitation takes advantage of energy levels associated with specific frequencies at which functional groups within molecules rotate and vibrate. The frequencies can be related to specific functional groups, and provide a method for characterization of many molecular structures.

Ion Beam Sputtering

- Sputtering with a focused ion beam can be utilized to remove surface layers revealing the subsurface for analysis or to produce secondary ions that can be characterized using a mass spectrometer to determine the elemental, isotopic, and molecular composition of the surface.

Applicability of Different Analytical Approaches

Often, the most critical part of analyzing surfaces is determining which analytical procedure and/or piece of analytical equipment should be used to address a project. In many cases, reliance on previous experience or documented results obtained from similar projects will provide sufficient insight into the most appropriate approach.

Figure 10.1 provides a graphic representation of the applicability of surface analysis techniques based on feature size and concentrations. Figure 10.1 represents a good resource for selecting the appropriate methodology.

The sophistication of the analytical equipment used for surface analysis is astonishing with some of the instruments characterizing, for example, layers a few atoms thick, imaging atoms, and others providing data related to the number of contamination atoms per unit area. Each method has been engineered to maximize emission of a selected response (or responses) and optimize the detection of selected emissions. Some of the kinds of information that can be obtained from the most common analytical methods are described below.

Analysis by SEM, SEM/EDX, SEM/WDX, and FESEM makes use of secondary electrons, backscatter electrons, and characteristics x-rays to provide excellent images of the surface, along with a chemical analysis that is sensitive to most elements. Useful x-ray data are generated in the volume excited by the electron beam and by secondary x-ray penetration. Useful data are generally obtained from the first few microns of samples.

AES provides images generated from secondary electrons and backscatter electrons as well as near surface analysis using Auger electrons.

XPS provides near analysis as well as information about the chemical state of the near surface of materials as derived from evaluation of x-ray photoelectrons emitted from samples.

XRD is an invaluable tool for identification of the crystalline form of the sample.

XRF can provide excellent qualitative and quantitative analyses of materials. Although generally more useful for bulk analysis, a unique application, TXRF, makes use of extremely low-angle x-ray excitation to promote excitation of the outermost surface layers and can provide extremely low-level metal contamination on polished surfaces such as semiconductor wafers.

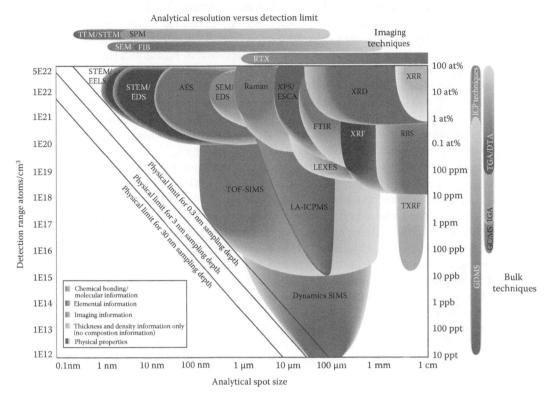

FIGURE 10.1 Graphic representation of the applicability of surface analysis techniques based on feature size and concentrations. (Courtesy of Evans Analytical Group LLC, Sunnyvale, CA. Copyright © 2010.)

FTIR spectroscopy provides a wealth of information about the nature of the chemical bonds of organic materials. The method is almost exclusively applicable to compounds with covalent bonds and is therefore particularly useful for characterization of organic components.

In practice, infrared spectra are compared with reference spectra of known compounds or reference energy band regions of spectra to reveal information about the structure of the sample. Identifications can be carried out on relatively thin films and particulate.

Transmission electron microscopy is a particularly versatile analytical tool that can utilize transmitted electrons, secondary electrons, backscatter electrons, diffracted electrons, and characteristics x-rays to provide a wealth of information about the surface of a sample.

Secondary ion mass spectroscopy relies on bombardment of surfaces using an ion beam, followed by mass spectrometry of the emitted ions. The strength of SIMS relies on its rapid profiling capability, with sensitivity in the parts per billion range.

Raman and scanning Raman spectroscopy rely on perturbation of low-frequency modes, and recent advancements are making them particularly attractive tools.

Special Topics

Layer Characterization

One of the important aspects of surface analysis is determination of the thickness and composition of the layers or thin films. In simplest terms, the surface of a material could be considered as the outer layer or layers of the sample, usually less than 1000 Å in thickness, along with an adventiceous outermost layer that is in dynamic equilibrium with the environment. Surface analysis projects, however, often involve characterization of the adventiceous layer, outer surface layer, and thin films to the substrate. In practical terms, it may be necessary to characterize a layer or layers and thin films totaling several microns in thickness to provide the information that is required to accomplish the desired goal of a project. The thickness of the surface layers acts as a constraint on the analytical procedures that can be employed to successfully address projects. SEM/EDX, for example, provides excellent information about the composition of samples to a depth of approximately 0.5–2 μm but provides very little information about the composition of the first few atomic layers, while AES and XPS provide excellent information about the composition of the outermost layers of the sample but determination of the composition to 0.5–2 μm depth requires an iterative process of etching and analysis.

Common protocols for layer characterization are SIMS, which uses mass spectroscopy to determine the composition of material as it is removed by ion etching; profilometry carried out by performing AES or XPS analysis of the composition of the near surface of a sample after each of a series of etch steps designed to remove a selected thicknesses of the sample surface until the desired depth has been reached; evaluation of cross sections by an appropriate microscopic method; and XRF. Each of these methods requires precise calibration and is subject to systematic errors. AES and XPS depth profiling, for example, require that the electron and ion beams have common focal planes. Additionally, the analyst must be eternally vigilant to prevent errors due to factors such as shadowing caused in the sample geometry. The determination of accurate thickness is also dependant on the use of the type of standards causing calibration problems.

Designing efficient profilometry protocols is further complicated by the lack of sufficient information about the composition and thickness of the surface, and it is often necessary to carry out multiple analyses to adequately characterize surface layers.

Particle Analysis

Particulate analysis provides special difficulties in surface analysis. Particulate represents one of the most pervasive forms of surface contamination. Particulate contamination ranges in size from submicron to hundreds of microns in size, often consisting of minerals, building materials, biological fragments, and textile and paper fragments. Although SEM/EDX can provide significant information about the size, composition, and morphology of particles larger than approximately 5 μm in size, the composition values will have systematic errors due to the morphology of the particle surface. Additionally, SEM/EDX analyses acquired on particles smaller than approximately 2 μm in size will often reflect the composition of the substrate supporting the particles as well as the particles and other particles in close proximity. TEM, field emission SEM, and AES provide more accurate information about the composition of particles smaller than 2 μm.

The advent of industrial uses of nanoparticles has further complicated particulate characterization. Nanoparticles are generally considered to be between 10 and 100 Å in size, often requiring TEM and/or FESEM for characterization.

Haze Identification

Identifications of stains described as a haze can present especially difficult analytical problems and are prone to misinterpretation. A reasonable reaction to the identification of stains described as a haze is to analyze the haze using SEM/EDX, AES, or XPS. The results and subsequent optical examination may show the presence of a thin contamination—for example, an oxide layer or a carbonaceous layer—and that ion etching had resulted in removal of the haze, leading to the reasonable interpretation that the haze is caused by the presence of the contamination. Unfortunately, a haze is often caused by the presence of very small (frequently smaller than a micron) carbonaceous droplets that are widely spaced so the analytical results are misleading. The wide spacing and droplet size result in low (often insignificant) concentration values of the contaminant. Furthermore, their presence is often over-looked even by SEM/EDX.

Summary

Analyzing the surface of a material is a discipline that evolves as more information about the sample, sample history, and performance requirements of the final surface are available. Cleaning a surface can be thought of as removing contaminants but in a broader sense often includes engineering the composition and thickness of the outer surface layers as well as other characteristics to conform to process and/or manufacturing requirements. Analysis of a surface and confirming that it corresponds to the requirements is an integral part of this process, making it possible to ensure that the surface will have the desired performance.

Designing a successful analytical methodology requires considering a significant number of factors including, but not limited to, confirmation that the sample has not been contaminated or altered by packaging materials, sample handling, sample transfer, sample storage, and sample preparation as well as identification of the appropriate analytical techniques and parameters, interpretation of the results, and relating the results to performance requirements. It is very difficult to set down a single set of rules that will provide positive results for every project, and it is almost impossible to anticipate every problem that might occur and provide a contingency to cover it. The process of designing a method is often an iterative process analogous to peeling an onion only to find another layer that needs to be addressed. Identification of and removing the source of a contaminant or surface deficiency often only makes it possible to identify additional surface irregularities that need to be evaluated.

There are obviously a significant number of precautions that have not been listed, and even more that are required for special circumstances.

Each of the analytical processes that are described here is explained in more detail in the cited references.

Bibliography

Azároff, L. V., Kaplow, R., Kato, N., Weiss, R. J., Wilson, A. J. C., and Young, R. A. *X-Ray Diffraction.* McGraw-Hill, New York, 1974.

Banwell, C. N. and McCash, E. *Fundamentals of Molecular Spectroscopy*, 4th edn. McGraw-Hill, London, U.K., 1994.

Benninghoven, A., Rüdenauer, F. G., and Werner, H. W. (eds). *Secondary Ion Mass Spectrometry: Basic Concepts, Instrumental Aspects, Applications, and Trends.* Wiley, New York, 1987, 1227pp.

Bertin, E. P. *Principles and Practice of X-Ray Spectrometric Analysis.* Kluwer Academic/Plenum Publishers, New York, 1970.

Briggs, D. and Seah, M. P. (eds). *Practical Surfaces Analysis by Auger and X-ray Photoelectron Spectroscopy.* John Wiley & Sons, New York, 1964.

Cormia, R. D., Schiefelbein, B., and Olsen, P. A. Electopolishing of stainless steel: Meeting the materials requirements for today's demanding applications. Paper Presented at Photo Chemical Machining Institute, East Dennis, MA, February 26, 1990.

Goldstein, G. I., Newbury, D. E., Echlin, P., Joy, D. C., Fiori, C., and Lifshin, E. *Scanning Electron Microscopy and X-Ray Microanalysis.* Plenum Press, New York, 1981.

Grant, J. T. and Briggs, D. *Surface Analysis by Auger and X-Ray Photoelectron Spectroscopy.* IM Publications, Chichester, U.K., 2003.

Jenkins, R. and De Vries, J. L. *Practical X-Ray Spectrometry.* Springer-Verlag, New York, 1973.

Kladnik, G., Schiefelbein, B., and Gill, M. Analyzing semiconductor process problems with beam probes. *Test and Measurement World*, November 1988.

Schiefelbein, B. and Strausser, Y. Put surface analysis to work in production. *Semiconductor International*, November 1988, p. 62.

Wagner, C. D., Riggs, W. M., Davis, L. E., Moulder, J. F., and Mullenberg, G. E. *Handbook of X-Ray Photoelectron Spectroscopy*. Perkin-Elmer Corp., Eden Prairie, MN, 1979.

Williams, D. and Carter, C. B. *Transmission Electron Microscopy. 1—Basics*. Plenum Press, New York, 1996. http://en.wikipedia.org/wiki/International_Standard_Book_Number

11

How Clean Is Clean? Measuring Surface Cleanliness and Defining Acceptable Levels of Cleanliness

Introduction 145
Definitions..146
Types of Contamination 146
Particle Contaminants • Thin-Film Contaminants • Microbial
Contaminants
Why Monitor Cleanliness? 147
Factors That Contribute to Inadequate Cleanliness................................148
Factors to Be Considered in the Selection of a Cleanliness
Monitoring/Measurement Method..148
Types of Cleanliness Measurement Methods 149
Indirect Methods • Direct Methods • Analytical Methods
Most Common Verification/Measurement Methods and Their
Principles of Operation 150
Indirect Methods • Direct Methods • Analytical Methods
Cost Impact of Cleanliness Levels..154
Methods for Defining an Acceptable ("Optimum") Level of
Cleanliness...155
Controlled Experiment • Production Testing
In-Process or On-Line Surface Cleanliness Monitoring..........................156
Conclusion 159
Bibliography 159

Mantosh K. Chawla
Photo Emission Tech., Inc.

Introduction

This section focuses primarily on surface contamination/cleanliness, as opposed to airborne or other types of contamination. Specifically, this section will cover the following:

- A brief description of the types of contamination encountered, the reasons for monitoring surface cleanliness and factors affecting surface cleanliness, followed by a discussion on various techniques for verifying/measuring surface cleanliness, their strengths, weaknesses, required operating skill level, and the approximate cost of available instruments. The most common analytical techniques are also listed.

- A discussion on ways to establish an acceptable level of cleanliness that can become a basis for evaluating alternative cleaning processes, optimizing existing cleaning processes, and ongoing monitoring of the cleaning process to ensure that an acceptable level of cleanliness is achieved.

Definitions

The following list of definitions has been elaborated in order to facilitate subsequent discussion and convey a consistent understanding of the information presented herein:

Contamination: Molecular and particulate surface material that has the potential to degrade the appearance or performance of a part, component, or assembly.

Molecular contamination: Nonparticulate contaminate material (film) without definite dimension; volatile species that may be physically or chemically absorbed on surfaces. This includes corrosive and noncorrosive films resulting from oil, greases, chemical residues, fingerprints, heat and vacuum applications, chemical action, and incompatible materials, such as films from outgassing.

Particulate contamination: Contaminate material with observable length, width, and thickness. In practice, an observable size will be approximately 0.1 μm.

Non-volatile residue: Soluble or suspended material and insoluble particulate matter remaining after controlled evaporation of a filtered volatile liquid.

Indirect methods: Any technique that gives an indirect indication of the level of surface cleanliness.

Direct methods: Any technique that provides a direct and relative measure of some surface characteristic that relates to the level of surface cleanliness.

Analytical methods: Any technique that provides information about the type/species and level of contamination in relative or absolute terms.

Optimum cleanliness level: A level of cleanliness that minimizes the total cost of cleaning and cost of nonconformance/failures due to poor surface cleanliness.

Types of Contamination

There are several types of contamination that can be present on the part surface and that may be undesirable for product performance. Some of the common types of contaminants are listed in the following.

Particle Contaminants

Contaminations present in the form of foreign particles on the surface, such as dust, hair, fibers, and metallic micro-fragments.

Thin-Film Contaminants

Contamination present in the form of a thin-film on the surface. This type of contamination includes both organic and inorganic thin-film contamination, such as skin oil, greases, processing fluids, surfactant/chemical residues, rinsing residues, oxides, and other unwanted thin-films on surfaces.

Microbial Contaminants

Contaminations present in the form of microbes on the surface, such as spores, bacilli, etc.

There are other types of contaminants, such as radioactive, heavy metal, etc. Discussion of these types of contaminants is beyond the scope of this chapter. The primary focus of this chapter is thin-film contaminants.

Why Monitor Cleanliness?

Have you been involved in discussions regarding the reasons for product nonconformance? Invariably, fingers are pointed at surface cleanliness as the culprit. The cleaning process people will swear that the process was working well and was not the cause for nonconformance. They may present records indicating quality control monitoring of the solvent, pH monitoring of the bath, or to a prearranged change-out schedule.

Do you identify with this scenario? If so, let us consider what is missing. It is obvious that, in this situation, no surface cleanliness verification is being done. The cleaning process people are going by their belief that the process was not exhibiting any abnormalities. Other folks blame surface cleanliness, because there is no verification that the process is working well. Monitoring the parameters of the cleaning process is important; however, this approach has its limitations. These are discussed in the next section, Factors that Contribute to Inadequate Cleaning.

In most applications, the presence of contamination can degrade the performance of parts, components, and systems. This results in nonconformance and, in the worst case, product failure. Molecular contamination of surfaces can drastically affect the performance of the parts.

Cleaning is part of many manufacturing operations. Parts may require cleaning before they can be electroplated or painted, before they can be soldered, or before they can be packaged and shipped for end use. Thus, cleaning is necessary for various reasons to ensure desirable product appearance or performance.

Since there is a need for cleaning parts for various reasons, it makes sense to monitor cleanliness on an ongoing basis to ensure consistent cleaning and part performance.

If a level of cleanliness is to be specified, then a method of verification must be specified at the same time for that level to have meaning. This generally leads to the question: What do we mean by "clean"? How clean is clean? Even so-called "clean" parts have a certain amount of contamination, even if it is at a microscopic level. With advances in technology, more and more applications are moving toward the need for higher levels of cleaning. Increasingly, precision cleaning cannot afford to rely on old methods of verifying or ensuring cleanliness.

Measuring cleanliness not only helps ensure product quality, it is an essential part of implementing pollution prevention approaches related to cleaning. Cleanliness measurement/verification methods can be utilized to (1) evaluate performance of any existing or alternative cleaning process, (2) optimize the cleaning process by analyzing parts during initial implementation of a new cleaning process, or (3) determine if better parts handling or other innovation may allow the cleaning process to be eliminated entirely. In addition, measuring cleanliness often prevents pollution by reducing rejects.

In order to specify a desired level of cleanliness, it is important to specify a method of measuring surface cleanliness that will help in ensuring the desired level of cleanliness. Once a method for measuring cleanliness has been selected, it can be used to establish the level of cleanliness achieved by any existing process. This level of cleanliness can be used as a benchmark to make changes to existing process to see if those changes improve the achieved level of cleanliness.

An established quantitative cleanliness can also be used to evaluate alternative cleaning processes. Evaluating alternative cleaning processes requires the answer to two questions: (1) Is the new cleaning process as effective as the one being replaced? (2) Does the new cleaning process leave behind any residue that could be detrimental to subsequent processing or performance? To avoid problems, answers to these questions are necessary before implementing a new cleaning process.

Factors That Contribute to Inadequate Cleanliness

In most cases, control of cleaning processes is achieved by specifying the operating parameters of the cleaning process, such as chemical concentration, temperature, water pressure, or the amount of time the parts are washed or rinsed. This approach defines how "clean" a part should be by specifying the process used to do the cleaning (i.e., dip part A in cleaning solution B at temperature C for X minutes), without regularly checking how clean parts actually are. This approach takes advantage of knowledge gained through experience with the cleaning process, or through measurements taken during initial testing of the cleaning process. This type of procedure generally also specifies the properties of the cleaning solution and replenishment of the chemicals on a periodic basis. This method, while practical and good most of the time, cannot be consistently relied upon for precision cleaning.

Even though the cleaning process variables are under control, there are several factors that can contribute to inadequate cleaning of parts.

First, the number of parts put through the cleaning process in a given period may vary. The more parts go through the cleaning process, the more contamination is removed from the parts, which is then mixed in the cleaning solution.

Second, from time to time, parts may have varying degrees of contamination present on them. The type and amount of contamination on each part varies from time to time and from vendor to vendor.

Third, the type of contamination (soil) on the parts may be different, and the cleaning process may not be effective in removing this type of soil. This is particularly true for the aqueous cleaning processes, which are generally less aggressive than the old organic solvent-based cleaning processes.

These changes can be related to a change in the product mix or to changes in processes by outside suppliers. Thus, this approach to ensuring cleanliness works only if the average number of parts and the average level of contamination on each part are consistent during a given period of time. If these conditions are not achieved, the cleanliness level will deteriorate below the acceptable level. Without the use of a surface cleanliness monitoring method, the lower level of contamination will not be detected until there are problems downstream.

It stands to reason that to truly control cleaning process, cleanliness monitoring must be an integral part of the cleaning process. Hence, in most cases, it is more effective for a level of cleanliness to be specified and monitored in production on a portion or on all of the parts cleaned. This is especially true in precision cleaning applications.

Factors to Be Considered in the Selection of a Cleanliness Monitoring/Measurement Method

There are a wide variety of cleanliness measurement methods. To determine which method is right for a given application, many issues must be considered. Some of the issues that affect the choice of method are

- *Type of contaminants to be monitored*: The method selected must be able to detect the types of contaminants of interest. For example, some methods will detect only organic contamination, and not inorganic contamination. If inorganic contamination is of concern, then such methods would not help. Some methods detect only certain types of contaminants. Such methods would be good if only certain types of contaminants are always expected to be on the surface. In general, it is better to have a method that can detect both organic and inorganic types of contamination, and that is not restricted to a certain type of contamination. This helps to ensure surface cleanliness, even if there is a change in any aspect of the production process upstream.
- *Type of substrate being checked*: If the part is being inspected directly, then the method must be compatible with the material being measured, without causing any damage. For example, certain

cleanliness measuring methods deposit some type of "measuring media" on the surface to measure the cleanliness. Care should be taken to make sure that the measuring media deposited on the surface is not going to affect the surface of the part. Care must also be taken to make sure that the measuring media does not contaminate the part surface.

- *Level of cleanliness that must be measured*: The method must be able to detect the contaminants at the minimum and maximum levels of interest. Each measurement method has a certain range of detection; in most cases, the minimum level of contaminant that can be detected is important for precision cleaning applications.
- *Accuracy and precision required*: This refers to how critical it is that the parts are cleaned to narrow specifications. Some methods provide gross estimates of contamination, even if they can detect contamination at very low levels, while others provide very precise measurement data for evaluation. The method selected must be appropriate for the application.
- *Features of the measurement method*: Some methods have certain features that may or may not be desirable. For example, some methods have to contact the surface or deposit something on the surface to make a measurement. It may not be desirable to contact the surface or deposit anything on the surface. Whether the method is noncontact, nondestructive, and/or noninvasive should be considered when selecting the right method.
- *Speed of measurements*: In most cases, it is not necessary to inspect every part. A representative sample at preset intervals is generally sufficient to track the performance of the cleaning process over time. Hence, the number of measurements that each method can complete per unit of time becomes important in selecting the right method. The method selected must be able to make analyses/measurements at the desired rate.
- *Acquisition and operating costs*: The more precise and automated measurement methods tend to be very expensive. In addition to acquisition cost, the operating costs, such as cost of any disposable supplies or costs of required operating skill, must also be considered. The total cost/benefit of the measurement method must be evaluated.
- *Skill level required*: The required skill level to utilize the technique and interpret the results varies a lot among various methods available. The ongoing cost of operating is higher for the more sophisticated techniques, particularly analytical techniques.

For a given cleaning process, it may be possible that more than one method is required to verify/measure all of the parameters of interest. There are many measurement methods that can be used to evaluate cleanliness in a manufacturing environment.

Types of Cleanliness Measurement Methods

The wide range of verification/measurement/analytical methods available can be differentiated in many ways. One simple way to categorize these methods is according to their mode of operation and the type of measurement yielded. The classification of various techniques based on these criteria is discussed in the following.

Indirect Methods

Most indirect methods of cleanliness measurement depend on a solvent of some type to dissolve any contaminants left on the part, after which the solvent is analyzed for contamination. This requires that the solvent used be stronger than the solvent that was originally used in the cleaning, to remove any residual the original cleaning solvent was not able to remove. Historically, these methods use solvents that are of the type many manufacturers are trying to eliminate from their cleaning processes. Recently, more environmentally benign alternatives have begun to be evaluated for this class of measurement methods.

Indirect methods that use solvents to extract contamination are usually only practical for small parts, due to the large volume of extraction solvent that would be needed for larger parts. Still, this method can analyze larger parts compared to some direct methods such as contact angle where very small parts must actually be able to fit in the equipment. Also, when extraction is used, none of the geometric limitations exist as they do for contact angle and some other direct methods.

Direct Methods

Direct methods actually measure cleanliness on the part of interest by analyzing the surface of the part directly. Direct methods, therefore, avoid many of the problems inherent in collecting contaminants off the part to be analyzed indirectly. However, since the part is being analyzed directly, there may be a limitation on the size or geometry of the parts that can be checked with some direct measurement equipment. Wherever possible, the preferred practice is to measure the cleanliness of the part directly, since the part surface is the one that is of direct interest.

Analytical Methods

Analytical methods are those that analyze the part surface or small piece of the part surface by studying the species of contaminants on the surface. Although these techniques are also direct techniques, they are classified separately because of their ability to identify the type of contaminants on the surface. Generally, this type of technique utilizes high vacuum for operation, and provides information about the type of contaminant on the surface, which cannot be provided by any of the indirect or direct methods. These systems generally involve laboratory-type instruments, and cannot be used on the shop floor/production area. The cost of this type of equipment is very high, generally in the range of $60,000 to high 6 figures. A very high skill level is required to operate the system and interpret the results. These techniques are very powerful and very useful in determining the type of contaminants present on the surface. The knowledge of the type of contaminants on the surface is of great use in locating and possibly eliminating the source of contamination.

Most Common Verification/Measurement Methods and Their Principles of Operation

Some of the most common indirect, direct, and analytical methods, with a brief discussion of their principles of operation, are presented in the following.

Indirect Methods

- *Net volatile residue (NVR)*: Also known as *gravimetric measurement*, this method requires a highly sensitive scale that can weigh parts to an accuracy of plus or minus 1 mg, or better. This method uses a volatile chemical, such as trichloroethylene, to flush the part. There are two ways to determine the level of contamination. One way is to weigh the parts after cleaning. Use a strong cleaning agent to flush the part, and weigh the part again after it is dry. The difference between the initial weight and post-flushing weight is attributed to the residual contamination that was left on the part by the regular cleaning process. If there is no difference in these two weights, the part is considered clean. This approach is good for small parts. If the area of the part is known, then the contamination/cleanliness level can be specified in some weight per unit of area.

    Another way, which is particularly good for large parts or surfaces, requires a container for collecting flushed fluids and a volatile chemical to flush the surface. The technique involves weighing the container, flushing a portion of the surface—preferably, a known area (i.e., a square foot)—with volatile chemicals, and collecting the flushed fluid in the container. After the volatile chemicals have evaporated, the container is weighed again. The difference in the weight of the container before and after collecting the flushed fluids is considered the amount of contamination left on the surface by the regular cleaning process. Once again, the level of cleanliness can be specified in weight per unit area (e.g., mg/ft^2).

This is a good gross measurement method when extremely high purity is not required. A small laboratory is needed to conduct these tests. Another variation of this method of gravimetric analysis is also used and is discussed in the following:

- *Ultraviolet (UV) spectroscopy*: This method has been used to measure flux residue left on printed circuit boards in the electronics industry, and has also been adapted to detect oils and greases on metal parts. This method requires the use of extraction equipment and an UV spectrometer, which are moderately expensive. In addition, the method requires that the contaminant to be analyzed have a unique absorption wavelength that can be identified in the UV spectrum. A calibration curve is then created by measuring samples of the solvent containing known concentrations of the contaminant at the unique wavelength. The method is only usable in the concentration ranges where the calibration curve is straight. Parts that are to be analyzed are extracted in a known amount of solvent to remove any of the contaminant. Typically, agitation or sonication is required during the extraction, which must be done in the same manner for each sample for the results to be meaningful. The solvent extract is then analyzed in the UV spectrometer at the unique wavelength. The absorbance is subsequently compared to the calibration curve to find the concentration of the contaminants in the extraction solvent. Based on the total volume of solvent and this concentration, the actual amount of contamination is derived. This method must be used in a laboratory, and requires a skilled operator.
- *Optical particle counter (OPC)*: An OPC gives both a count and size of particles in the solution measured, and can therefore be used to find out very specific information about the nature of the contaminants on the part. The method is typically useful when particle size is of interest, because, for example, particles below a certain size are acceptable as residual contamination, while those above the designated size are not. OPC requires extraction equipment to prepare a sample for analysis, and the OPC equipment itself. Two major techniques are available for particle counting: light extinction and light scattering. Light extinction (also called light blocking) uses a light source to shine a beam of light through a flow channel. Particles that pass through the beam block some of the light. The blockage of light creates an electrical pulse that is proportional to the particle size. A microprocessor counts and sorts the pulses according to size. Light extinction can measure particles as small as 1 μm. Light scattering is a more sensitive method that measures the light "scattered" by a particle as it passes through a light beam. Light scattering detects particles as small as 0.1 μm or even smaller, but does not work for particles bigger than approximately 25 μm. The sensitivity of OPC comes at a price. OPC is very expensive and requires skilled operators.

Direct Methods

- *Magnified visual inspection*: Visual inspection using a magnifying glass or low power microscope can be used to directly look at a part made of any material and to observe any gross contamination that may not be visible to the naked eye, but is still larger than the micron range. The method requires the part be removed from the production area and taken to the inspection area to be inspected. It is a pass/fail measurement method that may be used as a cross-check in precision cleaning applications that also use a more precise measurement method, or as a primary method in noncritical cleaning applications where only gross contamination need be removed. In practice, magnified visual inspection is only effective with smaller parts that can be handled by an operator and inspected in several short scans. It has the advantage of requiring minimum equipment. However, an area separate from production, such as a small laboratory, is almost a mandatory requirement, and inspectors must be well trained and thorough.

- *Black light*: This test requires a dark room and black light source for direct visual inspection of parts. This method is a pass/fail test that will work on any material with a contaminant that fluoresces under black light, provided the part itself does not fluoresce. The operator simply places the part under the black light and visually inspects the part. This method has most of the same application issues as magnified visual inspection, except that, since the contaminants fluoresce, they are even easier to notice if they are present. Although this method can be used on large surfaces, it is generally practical for testing smaller parts. Experience shows that the detection capability, even for the best operator, is limited to contamination of more than 100 mg/ft^2, which is approximately equal to 10,000 Å. This level of contamination is generally too much for precision cleaning requirements. Therefore, this method is only good for detecting gross fluorescing contaminants.

- *Water break test*: This simple method takes advantage of the fact that many contaminants of interest are hydrophobic. In this pass/fail test, which is typically used for metal surfaces, water is made to flow over the part. If it sheets off the surface evenly, the part is "clean." If the water channels or beads on certain areas, the part is rejected or sent for additional cleaning. This test can be conducted in production areas or as a batch test and is also usable on very large parts, such as airplane wings. To be effective, the water used in the test must be free of surfactants or other contamination that would cause the water to flow evenly even in the presence of contamination, and the parts must be of a geometry that allows water to flow across the surface of interest. This test will not detect inorganic contaminants and is not likely to detect small amounts of contamination or water-soluble contaminants.

- *Contact angle*: This method can be thought of as a more sophisticated water break test, as it also takes advantage of the fact that most contaminants cause water to bead up due to their hydrophobic nature. The test requires a small laboratory area and a highly skilled operator. To use this method, a contact angle goniometer (an instrument that can be used to measure contact angle) is required, and the part to be analyzed must be flat and of small size (approximately 3 in. × 3 in., or less). In addition, distilled water must be used, and other parameters, such as static electricity and humidity, must be carefully controlled. The test is performed by applying a distilled water droplet of reproducible size to the test surface. After waiting a couple of minutes for the drop to equilibrate, the operator examines the droplet using the goniometer and records the angle of contact the drop has with the surface. An idealized, perfectly clean metal surface would have a contact angle of 0°, which is impossible to obtain in laboratory air. A contaminated metal part would have a high contact angle, such as 90° or more. Some parts, such as plastics, have positive contact angles even when "clean," so the method is not typically used for cleanliness analysis for these materials. While a number is obtained from this test (the contact angle), the test still is nonquantitative in terms of the contaminants on the part. This method is very subjective and not capable of detecting scattered or small amounts of contamination. Water-soluble contaminants are not likely to be detected by this test. This method will not detect inorganic contamination.

- *Optically stimulated electron emission* (OSEE): The surface to be tested is illuminated with ultraviolet light of a particular wavelength. This illumination stimulates the emission of electrons from the surface. These electrons are collected and measured *as current* by the instrument. Contamination reduces the electron emissions and, therefore, the current measured. The inspected surface must emit electrons (i.e., the material must be photo-emissive) for the technique to work. However, most materials of engineering importance do emit electrons. The technique is simple to operate, fast, and relatively inexpensive. In addition, it is quantitative, nondestructive, and noncontact. The OSEE sensor can be hand-held and placed directly on the surface to measure cleanliness of the area of interest in a few seconds. This technique detects both organic and inorganic contamination, such as oxides, and can be used on any shape of parts as long as the geometry of the part is presented to the sensor in a consistent manner. This system lends itself to scanning small parts or large surface areas. This technique can be used in the production environment and on-line real time measurement of surface cleanliness. For example, an OSEE sensor has been mounted on-line in steel mills to measure the cleanliness of steel sheet moving as fast as 1000 ft/min prior to phosphate coating of the steel sheet.

- *Total organic carbon (TOC)*: Also known as *direct oxidation carbon coulometry (DOCC)*, this is a technique that works by oxidizing the sample surface to convert any carbon compounds present into carbon dioxide, and then detects and measures the carbon dioxide. The detection of carbon implies that there was some contamination that had carbon as its constituent. The level of TOC detected determines the level of cleanliness of any part. Since the TOC analyzers only detect carbon, the compound of interest must contain some carbon in a detectable quantity, in order for the analysis to function. This method works on a variety of materials and is surface-geometry independent. The method works only on small parts or pieces of larger parts. Due to the high temperature in the combustion chamber (more than 750°F), the method is not suitable for parts sensitive to high temperature. This method can be used in a laboratory, but is adaptable to the production environment.

- *Measurement and analysis of surfaces by evaporative analysis (MESREN)*: This technique utilizes radioactive decay to quantify organic contamination on a surface. MESREN can be used to directly analyze the surface, or to analyze the surface indirectly by using solvent extraction. The extent of radioactivity depends on the level of surface contamination present. This technique will not detect any inorganic contamination.

Analytical Methods

Analytical techniques are defined as those techniques that can analyze a surface to determine the type/species of contaminants and, in most cases, the absolute amount of contamination on the surface. These types of instruments/systems generally use high vacuum chambers, and are very costly to acquire and operate. In addition, a very high skill level is required to operate these types of systems and interpret the results. Most of these techniques can only examine small samples. This may require cutting a piece off the part to be inspected. Usually, careful sample preparation is required prior to analysis.

All these techniques are laboratory techniques and cannot be used on the shop floor. The samples have to be sent to the laboratory for analysis. In most cases, testing time is quite long, thus limiting the number of samples that can be tested. In some cases, it is important to know the species of contamination on the surface. For example, when developing a new cleaning process, the knowledge of the species of contamination helps in designing the most effective cleaning process. If the species is known, it is also possible to trace the source of contamination and minimize it or eliminate it. Due to the high cost and the high level of skill required, it is recommended that these techniques be used for development or investigative applications. These techniques very seldom provide real-time information to be used on a daily basis.

Some of the most common analytical techniques are as follows:

- *Electron spectroscopy for chemical analysis (ESCA)*: This technique is often called x-ray photo-electron spectroscopy, or *XPS*. This highly sophisticated and expensive measurement method uses special equipment to bombard the surface of interest with x-rays under vacuum conditions, causing electrons to be released from the surface. Since each type of element (e.g., carbon, oxygen) releases a unique amount of electrons under these conditions, the actual elemental composition of the surface can be quantified. The test requires a very small, flat surface and is not only expensive but is also lengthy. Its application is limited mostly to research and development, but it can be used to calibrate and evaluate other less sophisticated measurement methods. XPS equipment is typically only present in specialized university laboratories and larger industrial research laboratories.

- *Fourier transform infrared spectroscopy (FTIR)*: This technique uses infrared light focused on the part surface. Reflected light is detected and analyzed for the absorption of specific light frequencies. This information can be used to identify the type of organic materials that are on the surface.
- *Secondary ion mass spectrometry (SIMS—Static)*: This technique is often called time-of flight mass spectrometry or TOF-SIMS). An energized primary ion beam is used to bombard a surface in high vacuum. This causes the ejection of surface atoms as secondary ions. This technique identifies masses according to a gated arrival time at a detector. The mass of the various species present can be identified on the basis of the arrival time. The typical sampling depth ranges from 3 to 6 Å.
- *Secondary ion mass spectrometry (SIMS—Dynamic)*: Inherently a profiling technique, it uses O_2 or Cs ions to bombard a surface in high vacuum. This technique identifies masses according to a gated arrival time at a detector. The species of masses can be identified on the basis of the arrival time.
- *Auger electron spectroscopy (AES)*: Auger electrons, named after the discoverer of the process, are ejected when the surface is bombarded with electrons in a high vacuum. The energy level of Auger electrons provides information about the species of contamination.
- *Scanning electron microscopy (SEM)*: This is a surface imaging technique, performed in high vacuum. SEM utilizes a beam of electrons that is passed over a very small area of surface. This beam scatters when it strikes the surface. The "back-scattering" carried in by the return beam of electrons is measured with a microscope. This technique is well suited for identifying particulate and potentially nonuniform or thick films of contaminant. It is not very sensitive to very thin, uniform films, but is most useful for imaging particles. It also cannot test large areas.

Cost Impact of Cleanliness Levels

What are the costs one should look at if no cleanliness monitoring or verification is done?

First, consider the value-added costs associated with nonconforming parts if problems are discovered downstream instead of at the source (i.e., the cleaning process).

In one case, a customer returned 3000 parts for poor cleanliness. The vendor was not using any cleanliness monitoring or verification system. Instead, he was relying on control of cleaning process parameters. By using a cleanliness monitoring system, the vendor was able to test all of the parts and find approximately 30 parts that were not clean. These parts were re-cleaned, and the whole shipment was returned to the customer with full confidence that the parts were clean. The cost associated with shipping the parts back and forth, unpacking them for inspection, and repackaging them for shipping—not to mention lack of customer satisfaction—by far outweighed the cost of cleanliness monitoring.

In this case, there were no subsequent operations, so the cost of re-cleaning was minimal. When subsequent, value-added operations are performed on parts, it becomes even more critical to ensure that the cleaning process is controlled.

Other costs that should also be considered are those associated with time spent arguing about the source of any nonconformance encountered when surface cleanliness is not verified. These costs are nonproductive and can be minimized or eliminated by verifying cleanliness.

Cleaning cost is directly proportional to the level of cleanliness. The higher the cleanliness level, the higher the cost of achieving that level. Hence, costs associated with over-cleaning parts are nonproductive costs. Similarly, the level of nonconformance is indirectly proportional to cleanliness level; the lower the cleanliness level, the higher the nonconformance due to surface cleanliness. Costs associated with nonconformance due to poor cleanliness are also nonproductive costs. These two cost components can be combined to assess the "total cost" of cleaning. An acceptable level of cleanliness is the one that minimizes the total cost (Figure 11.1). That cost is associated with the "optimum" level of cleanliness.

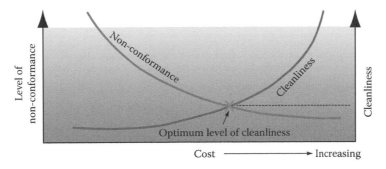

FIGURE 11.1   Optimum level of cleanliness.

Since all processes have some variation, there is bound to be some variation in the level of cleanliness achieved. An acceptable level of cleanliness would be defined as an acceptable variation around the optimum level of cleanliness at a minimum total cost.

Methods for Defining an Acceptable ("Optimum") Level of Cleanliness

Required cleanliness levels for desired performance can be determined by utilizing some means of quantifying the surface cleanliness/contamination testing to correlate product performance with contaminant level, thereby establishing the degree of contamination that can be tolerated. Let us call such level of cleanliness the "optimum" level of cleanliness. Once the optimum level of cleanliness is defined, a method is needed to quantitatively measure the level of cleanliness achieved during production to ensure that the surface meets the established cleanliness requirements.

In order to define the "optimum" level of cleanliness, the cleanliness level must be correlated with the failure/nonconformance rate of the subsequent process. A controlled experiment or production cleanliness monitoring can determine this correlation.

Controlled Experiment

One way to quantitatively define the level of cleanliness needed is to perform a controlled experiment. At a minimum, this experiment should involve preparing parts with different levels of cleanliness, measuring the surface cleanliness of each part, and correlating a measure of the "success" of the subsequent operation to the level of part surface cleanliness.

For example, if two parts are to be bonded together, the success of this operation is a good bond. The measure of bond strength can be correlated to surface cleanliness. If the parts are to be coated after cleaning, then the adhesion strength of the coating should be correlated to surface cleanliness. For the purpose of this discussion, let us assume that the parts are to be coated and that the adhesion of the coating is measured by the peel strength.

Parts with various degrees of cleanliness can be prepared by either altering some factors of the cleaning process or by applying contaminants. The contaminants can be applied by mixing known weights of contaminants with volatile chemicals and air brushing the mixture over the clean parts. The parts can be weighed when clean, and after the contaminants have been applied. The difference in part weight divided by the part surface area gives a measure of contamination in weight per unit area, such as milligram per square foot or similar units. Thus, by varying the weight of contaminants added to a given amount of volatile chemical, parts with various and known levels of contaminants can be prepared.

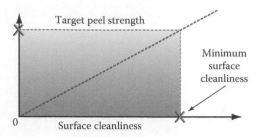

FIGURE 11.2   Surface cleanliness versus peel strength.

After the parts have been prepared, surface cleanliness measurements for each part should be taken and recorded. If possible, take several measurements per part. The next step is to apply the coating on to these parts. After the coating has been applied and cured/dried, take as many measurements of the peel strength as possible for each part, and record these values.

The mean surface cleanliness reading for these parts should be correlated with the mean measurement of peel strength. Figure 11.2 graphically depicts the typical result of correlating the peel strength of coating adhesion to the level of cleanliness of the surface. The cleanliness level that correlates with the target level of adhesion strength measurement then becomes the required *minimum* level of surface cleanliness.

Production Testing

Another way to establish an optimum level of surface cleanliness is to measure and monitor the surface cleanliness level achieved in production and correlate the achieved cleanliness level to the incident rate of product failure due to surface contamination. A cleanliness measuring method can be used to establish the level of cleanliness currently achieved by the cleaning process. It is also important to note the level of nonconformance due to this level of cleanliness. This level can be used as a benchmark by comparing the level of nonconformance resulting from surface contamination as the level of cleanliness changes from the benchmark level. Changes in the cleanliness level occur periodically. Thus, by monitoring the currently achieved level of surface cleanliness and correlating this current level of cleanliness to the nonconformance rate attributed to surface cleanliness can tell us if the total cost of cleaning has increased or decreased. For example, if the failure/nonconformance rate is too high, the surface cleanliness level will have to be improved in order to reduce the failure rate. On the other hand, a very low failure rate implies that the surface may be "over-cleaned." If no failures occur due to surface contamination, then it may be desirable to optimize the cleaning process by comparing the cost of nonconformance with the cost of reduced surface cleanliness.

It is also possible to change the surface cleanliness level up or down by altering cleaning process parameters in response to the actual failure/nonconformance rate. This way, the relation between the level of surface cleanliness and cost of nonconformance due to poor surface cleanliness can be established. Deliberately changing the level of surface cleanliness can help in establishing this relationship much faster than waiting for the cleaning process to give a different level of surface cleanliness.

In-Process or On-Line Surface Cleanliness Monitoring

Once the optimum cleanliness level has been established, a surface cleanliness measurement system can be used to monitor the process and to ensure that the desired cleanliness level is being achieved. By monitoring the surface cleanliness to an established level of cleanliness, the nonconformance due to surface contamination can be minimized or eliminated.

TABLE 11.1 Comparison of Various Direct/Indirect Methods of Cleanliness Verification/Measurement

Method	Type of Contaminant Detected	Relative Cost	Measurement Time	Quantitative	Part Geometry Limitation	Operator Skill Level	Noncontact	Nondestructive	Area Inspected	Limitations
NVR	Organic	Low	Few minutes	Yes	Some	Low	No	Yes	Limited	Generally small parts
UV spectroscopy	Some organic	High	Few minutes	Yes	Yes	High	Yes	Yes	No limit	Fluorescing contaminants only
Optical particle counter	Particulate	High	Few minutes	Yes	Yes	High	No	Yes	No limit	Large particle contamination only
Magnified visual inspection	Organic	Low	Few seconds	No	Yes	High	Yes	Yes	No limit	Only gross level of contamination detected
Black light	Some organic	Low	Few seconds	No	No	High	Yes	Yes	No limit	Only fluorescing gross contaminants
Water break test	Organic	Low	Few minutes	No	Some	Low	No	Yes	No limit	Only detects hydrophobic contaminants
Contact angle	Organic	Medium	Few minutes	Yes	Flat surface	Medium	No	Yes	Small	Only detects hydrophobic contaminants
OSEE	Organic and inorganic	Medium	Few seconds	Yes	No	Low	Yes	Yes	No limit	Does not detect particle contamination
MESREN	Organic	Medium	Few minutes	Yes	Flat surface	Medium	No	Yes	Limited	Does not detect inorganic contamination
Total organic carbon (TOC) or (DOCC)	Organic	Medium	Few minutes	Yes	No	High	Yes	Yes	Limited	Subjects the part to high temperature

TABLE 11.2 Comparison of Various Analytical Methods of Cleanliness Verification/Measurement

Feature	AES/SAM	EDX/WDX	ESCA (XPS)	DYNAMIC SIMS	STATIC SIMS	TOF SIMS	FTIR	ATR FTIR	RS	XRF
Bulk (<10 μm)	Yes	No	Yes	Yes	Yes	Yes	No	No	No	Some
Near surface (<2 μm)	No	Yes	No	No	No	Some	Yes	Yes	Yes	Yes
Bulk (>10 μm)	No	No	No	No	No	No	Yes	Some	Yes	Yes
Depth profiling	Yes	No	Yes	Yes	No	Yes	No	No	No	No
Identify interfaces	Yes	Some	Yes	Yes	No	Yes	Yes	Yes	Yes	Some
Quantitative measurement	Yes	Yes	Yes	Yes	No	With standards	Yes	Yes	Yes	Yes
Elemental	Yes	Yes	Yes	Yes	Some	Yes	No	No	No	Yes
Chemical	Some	No	Yes	Some	Yes	Yes	Yes	Yes	Yes	No
Used to identify organic species	No	No	Yes	No	Some	Yes	Yes	Yes	Yes	No
Used to identify inorganic species	Yes	Yes	Yes	Yes	Some	Yes	Some	Some	No	Yes
Area inspected	Small	Small	Small	Small	Small	Small	Small	Small	Small	Small
Limitations	Slow	Slow	Slow	Slow	Slow	Slow	Slow	Slow	Slow	Slow
Relative cost	Expensive	Expensive	Expensive	Expensive	Expensive	Expensive	Expensive	Expensive	Expensive	Expensive
Required skill level	High	High	High	High	High	High	High	High	High	High

Source: Cormia, R.D., Problem solving surface analysis techniques, *Advanced Materials & Processes*, December 1992.

Another advantage of in-process or on-line monitoring of surface cleanliness is that replenishment of chemicals or cleaning agents will only be done when needed, and not done according to a predetermined, somewhat arbitrary schedule. This replenishment schedule is usually time-dependent. In reality, the amount of contamination can vary considerably from part to part. Moreover, the number of parts being cleaned during a given time frame can also vary considerably. Thus, a time-dependent replenishment schedule is not the best way to control the cleaning process. The required level of chemical or cleaning agent concentration in the cleaning solution can be objectively determined and maintained by using a surface cleanliness verification system.

A surface cleanliness measuring system is not only very useful in helping establish an "acceptable cleanliness level," it is also very useful in evaluating the effectiveness of any new cleaning processes or changes to the existing processes. By measuring the cleanliness level after any changes to the cleaning processes are made, it is possible to assess the impact of these changes on the nonconformance/failure rate. The impact is determined by the previously established relationship between surface cleanliness level and the nonconformance/failure rate.

Conclusion

Evaluating alternative cleaning processes requires the answer to two questions: (1) Is the new cleaning process as effective as the one being replaced? (2) Does the new cleaning process leave behind any residue that could be detrimental to subsequent processing or performance?

There are many reasons for monitoring surface cleanliness, including quantitatively establishing the cleanliness level achieved by an existing cleaning process, measuring the cleanliness level achieved by the alternative cleaning processes to help select the most effective and cost efficient alternative, and monitoring the effectiveness of the cleaning process on an ongoing basis. Many factors affect the selection of a cleanliness monitoring method. Summaries of cleanliness verification methods are provided in Tables 11.1 and 11.2.

Bibliography

1. Gause, R. NASA Marshall Space Flight Center, A non-contacting scanning photo electron emission technique for bonding surface cleanliness inspection. NASA Marshall Space Flight Center. *Presented at the Fifth Annual NASA NDE Workshop*, Cocoa Beach, FL, 1987.
2. Chawla, M.K. How clean is clean? Measuring surface cleanliness. *Precision Cleaning*, 11–14, June 1997.
3. Mittal, K.L. *Treatise on Clean Surface Technology*, Vol. 1, Plenum Press, New York, 1987.
4. Cormia, R.D., Problem solving surface analysis technique. *Advanced Materials & Processes*, December 1992.

Conclusion

Bibliography

12

Cleaning Validations Using Extraction Techniques

Introduction 161
Validating a Cleaning Process.. 161
Defining the Validation Scope 162
Extractables versus Leachables • Reproducibility and
Effectiveness • Device Families • Defining the Surface • Revalidation
Identifying the Contaminants.. 164
Primary and Secondary Contaminants • Reaction Products • Volatile
Contaminants • Contaminant Families
Choosing the Test Method 165
Total Organic Carbon • Gravimetric Analysis • Ultraviolet/Visible
Spectroscopy • Particulate Analysis • Chromatography Methods
Choosing the Extraction Solvent... 167
Basic Solubility • Device Compatibility • Multiple Solvents
Setting the Extraction Parameters... 168
Technique • Temperature • Time • To Pool or Not to Pool
Validating the Extraction.. 169
Spike Recovery • Exhaustive Extraction • Cleaning Process Effectiveness
Establishing and Justifying Residue Limits... 170
Biocompatibility and Other Testing • Comparison • Risk-Based
Assessment
Conclusion 172
References... 172

Kierstan Andrascik
QVET Consulting

Introduction

Medical device manufacturers have always been under scrutiny to ensure that their product does no harm to the patient. For many years, the focus has been on biocompatibility and sterilization. While these still remain important factors in the medical device approval process, regulatory agencies have become increasingly aware of the potential dangers of residual manufacturing materials on medical devices. A thorough validation of the cleaning processes used to remove residual materials from newly manufactured medical devices is necessary to ensure patient safety.

Validating a Cleaning Process

According to the FDA, validation is defined as "establishing documented evidence which provides a high degree of assurance that a specific process will consistently produce a product meeting its

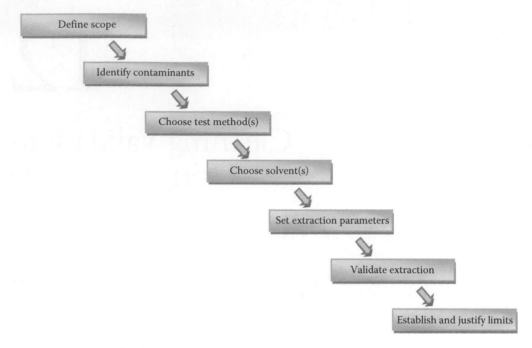

FIGURE 12.1 Design flow for cleaning validations.

predetermined specifications and quality attributes."[1] In terms of a cleaning process, the validation must demonstrate that a device can be effectively and consistently cleaned according to the acceptance criteria established.

The validation process includes design qualification (DQ), installation qualification (IQ), operational qualification (OQ), and performance qualification (PQ). When setting up a new cleaning process, each step in the validation process must be addressed. Validation plans are individualized based on the manufacturers' facility, device, and needs.

For many companies, the cleaning process has been utilized for many years, but a full cleaning validation has not been performed. Some people tend to skip over OQ when validating an established cleaning process, but this is a mistake. There are many variables in a cleaning process including exposure time to the cleaning solution or solvent, temperature, rinse time, and dry time. Cleaning processes should have acceptable tolerances for each variable (i.e., 50°C ± 5°C), and these tolerances need to be verified. In essence, the OQ provides the company with data showing that the high and low parameters are just as effective as the nominal parameters in cleaning the devices. The nominal parameters are validated during the PQ process to demonstrate reproducibility and effectiveness.

The basic flow for designing a cleaning process validation is shown in Figure 12.1.

Defining the Validation Scope

Cleaning validations can become overwhelming very quickly. By defining the validation scope, the purpose and goals of the validation come into focus. Normally, the scope of the validation specifies what cleaning process is being validated and what device types are included in the validation.

The scope also details what types of contamination the cleaning process is intended to remove. Usually, this consists of any material purposely placed on a device during the manufacturing process that is not part of the final device specification. In many cases, companies also need to consider materials that inadvertently contaminant a device. The main focus of a cleaning validation is usually on the removal of nonbiological contamination, but, sometimes, the cleaning process is intended to remove biological residues as well. This must be clearly defined in the scope.

Extractables versus Leachables

The validation of a cleaning process normally looks at extractables rather than leachables. Extractables are "compounds that can be extracted ... under exaggerated conditions (i.e., temperature, time, surface area) when in the presence of an appropriate solvent(s)" while leachables are "compounds that leach ... under normal processing conditions."[2] Since "leachables are a subset of extractables,"[3] a focus on extractables is a worst case.

Reproducibility and Effectiveness

The FDA requires that the reproducibility and effectiveness of a process be demonstrated. Reproducibility is twofold. Multiple devices from a single cleaning run must be analyzed to demonstrate uniformity within a cleaning run. Devices must also be analyzed from various cleaning runs to show consistency from run to run.

One way to determine how many samples to test from each run is using acceptable quality level (AQL) standards such as MIL-STD-1916[4] or ISO 2859-1.[5] The sample size is determined using a lot size, an inspection level, and an AQL. At times, following the AQL standards can be prohibitively expensive. At a minimum, three devices from each of three cleaning runs are recommended.

The effectiveness is determined by how well the cleaning process can remove the target contamination. Effectiveness is commonly indicated using a percent reduction or log reduction value.

Device Families

Many companies produce several different types of devices, and the cleaning process must be validated for each device that it is intended to clean. Rather than performing a complete validation for each device, companies can group the devices into "families" that have similar characteristics. These characteristics could include the following:

- Device material
- Size
- Configuration
- Surface treatments
- Intended use

Once device families are established, one device from each family is chosen for performing the validation. The chosen device must demonstrate worst case characteristics such as

- Largest surface area
- Blind holes
- Threads
- Textured surfaces

Justification of why this device was chosen must be included in the validation protocol.

Defining the Surface

Because target contaminants are found on the surface of a device, defining the surface is a necessary step in the validation process. At times, the "important" surfaces are those with patient contact or in the flow path of the device. A company may choose to prove the cleanliness of the "important" surfaces only.

Of course, there are risks with this approach, such as inadvertent patient contact. Also, depending on the device configuration, it may be very difficult to only test the "important" surfaces. If circumstances allow, it is best to validate the cleanliness of the entire device.

Once the surface has been defined, it is important to get an accurate representation of the surface area so that the residue amounts can be calculated in terms of the surface area. In most cases, the technical and engineering drawings are adequate, but, sometimes, the surface area is inconsistent from device to device, primarily due to coated, textured, or porous surfaces. In these situations, it may be best to calculate the surface area based on a theoretically smooth surface or use an average surface area calculation.

Revalidation

The scope should also list the circumstances under which a revalidation is required. A cleaning process should be assessed for revalidation when

- A change occurs in the process design.
- Any equipment is modified.
- Equipment is moved from one location to another.
- It is discovered that the process is no longer operating in a state of control.

Identifying the Contaminants

An important step in the validation process is identifying the contaminants, which the cleaning process is intended to remove. Contaminants usually fall into one of three categories: water-soluble residues, organic-soluble residues, and non-soluble debris. Ionic and polar compounds tend to be water soluble, while hydrocarbon compounds need an organic solvent to dissolve. For non-soluble debris, the solvent does not play as large of a role; it is removed primarily by mechanical means.

Primary and Secondary Contaminants

Manufacturing and cleaning processes have many obvious compounds that come in direct contact with the devices, but other contaminants may be less apparent. If a manufacturing line is used to produce several different devices, then cross-contamination between the devices may be an issue. Device 1 uses lubricant X and device 2 uses lubricant Y; you may need to indicate lubricant X as a primary contaminant for device 1 and a secondary contaminant for device 2, and vice versa. Other secondary contaminants may include hand soaps or lotions used by employees who handle the devices or cleaning agents used to clean the manufacturing lines. Some possible contaminant types are listed in Table 12.1.

Reaction Products

To complicate things further, some contaminants may be reaction products created during various steps in the manufacturing process. Passivation, anodization, and coatings are all common processes,

TABLE 12.1   Examples of Possible Types of Contaminants

Water Soluble	Organic Soluble	Insoluble Debris	Volatile
Detergents	Lubricants	Metallic particles	Degreasers
Acids	Cutting oils	Ceramics	Cleaning solvents
Coolants	Greases	Dust	
Hand lotions	Buffing compounds	Buffing compounds	
	Coolants	Polishing media	

which use chemicals to alter the surface of the device. If contaminants are not properly removed prior to these processes, the chemicals can react with the contaminants to create by-products. For example, nitric acid (commonly used in passivation) readily reacts with organic materials (including oils and greases). Even detergents can react with contaminants to create by-products. Reaction products are difficult to identify and, therefore, complicated to demonstrate efficiency of removal. Because these reaction products are unknown, it is advisable to use a general analysis method rather than looking for specific contaminants.

Volatile Contaminants

Some materials used during a manufacturing or cleaning process are volatile, such as vapor degreasers and isopropyl alcohol (IPA). Other contaminants are partially volatile. At times, volatility information is available on the Material Safety Data Sheet (MSDS).

Depending on the purity, use, and removal method of these volatile compounds, residue testing geared toward volatile contaminants may not be required. A high purity solvent, such as 99% IPA, is unlikely to leave any residue on the device. But if 70% IPA is used, there is a greater chance of residue (depending on what the remaining 30% consists of).

If the solvent is reused for several cleaning runs before it is replaced, then the issue is not with the solvent, but with the contaminants added to the solvent from previous cleaning runs. Ideally, these contaminants were already identified as primary or secondary contaminants for the devices.

The removal method for the solvent dictates the probability that it is still present during the next phase in the manufacturing or cleaning process. If the devices are heated to remove the solvent, then the solvent is less likely to still be there. Whereas, if the devices are only air-dried for a minute, the solvent could react with other materials to create reaction products. Solvents used on porous or textured surfaces are much more difficult to remove during the cleaning process.

For materials that are semi-volatile, the effect of that volatility depends on the method used to detect the residue. For some methods, it is important to obtain volatility information, experimentally if necessary, in order to get an accurate representation of residue levels.

Contaminant Families

Due to the sheer number of possible contaminants, some companies may choose to divide the contaminants into "families," similar to the device families discussed earlier. Contaminant families have similar characteristics such as

- Solubility
- Chemical composition
- Volatility
- Use

As with device families, a worst case contaminant is chosen from the group to use in the validation. For contaminants, worst case characteristics include the following:

- Most viscous
- Largest quantity used
- Lowest purity
- Hardest to remove

Choosing the Test Method

The most common validation method utilizes extraction techniques to quantify and characterize surface residuals that remain on a medical device following the cleaning process.[6] Appropriate test methods are chosen based on the characteristics of the target contaminants. Using multiple test methods may be beneficial due to variety in the target contaminants.

Total Organic Carbon

Total organic carbon (TOC) analysis is a nonspecific method, which measures the amount of organic carbon present in device extracts. It is a very sensitive method with detection capabilities in the parts-per-billion (ppb) range. Residuals are extracted from the device using purified water. The extraction ratio must be controlled for accurate analytical results. If too much water is used, the residue may not be detectable even though it is present. If too little water is used, the residue may not be adequately removed from the device.

Following extraction, the extract is analyzed on a TOC instrument. First, the carbon is oxidized to form carbon dioxide, and then the amount of carbon dioxide is measured. The resulting mass of carbon dioxide is proportional to the mass of TOC in the extract, which is interpreted as the TOC extracted from the device.

In order for TOC to be a suitable analysis technique, the carbon present on the device must be able to be extracted by purified water, and it must be oxidizable under the TOC test conditions. TOC does not identify or distinguish among different compounds containing oxidizable carbon; therefore, background carbon (carbon from sources other than the target compounds) should be limited. TOC will not detect inorganic contamination. By coupling TOC analysis with a conductivity/pH analysis, device extracts may also be analyzed for any ionized compounds such as acids, bases, or salts.

Gravimetric Analysis

A basic gravimetric analysis for medical devices is described in ASTM F2459.[7] This method quantifies any extractable residue that can be removed by the chosen solvent. One advantage of this method is that extraction solvents other than purified water may be used. However, the method excludes residues (or components of residues) that are more volatile than the extraction solvent.

The devices are normally extracted via refluxing or sonication. The extraction solution is evaporated, and the resulting residue is quantified gravimetrically. The sensitivity of the method is dependent on the sensitivity of the analytical balance used. The detection capabilities are usually between 0.01 and 0.1 mg. This method can detect soluble or insoluble extractables. If desired, a filtration step can be performed to separate and quantify the soluble and insoluble residues individually. This information is invaluable if non-soluble debris is a primary concern.

In some situations, it is advantageous to weigh the device before and after the extraction process. The cleanliness of the devices can then be gauged by comparing the pre- and post-weights. Once again, the sensitivity of this technique is dependent on the balance. Balances normally have a maximum weight associated with its sensitivity. For example, a balance may weigh to 0.1 mg only up to a weight of 200 g. Also, if a device weighs only 10 mg, the sensitivity of the balance to weight changes may be insufficient to be beneficial. Therefore, examining the device weight change may not be an effective technique for determining the cleanliness of very heavy or very light devices.

Residues from the gravimetric method can also be characterized by infrared (IR) spectroscopy. The spectrum of the residue can reveal the presence of certain compounds such as hydrocarbons, silicones, and amines. Identification can also be accomplished by comparing the residue spectrum to spectra of the target compounds.

Ultraviolet/Visible Spectroscopy

Ultraviolet/Visible (UV/Vis) spectroscopy can be used to quantify contaminants that readily absorb UV and/or visible light (i.e., detergents and dye penetrants). Devices are extracted in purified water, and the extracts are analyzed using a UV/Vis spectrophotometer. As with TOC, the extraction ratio must be controlled for accurate analytical results. If too much water is used, the residue may not be detectable even though it is present. If too little water is used, the residue may not be adequately removed from the device.

The concentration of the target compound is calculated using linear regression and a standard curve. Because each compound responds differently to this test, each analyte must be validated for method suitability. Some compounds are unsuitable for analysis using this method due to insufficient response to UV/Vis light. If several contaminants that absorb UV/Vis light are present, there is no way to distinguish one compound from another. Concentrations are calculated as worst case, assuming all resulting absorbance is due to a single contaminant. This method does not identify the absorbing substance(s) in the extract.

Particulate Analysis

When the main concern is insoluble debris, a particulate analysis is a useful test. Devices are extracted in purified water, and the extracts are analyzed for particulates. The two most common analysis techniques are found in USP <788>: the light obscuration method and the microscopic method. The light obscuration method uses a liquid particle counter that can determine the size of the particles as well as the number of particles of each size. For the microscopic method, the extract is filtered, and then the filter is examined microscopically.

Although these techniques are intended for use with pharmaceutical injections, they are commonly used to test medical devices. The acceptance criteria specified in USP <788> are also not intended for use with medical devices, but they can be used as a general guide.[8]

Chromatography Methods

Chromatography methods have the benefit of separation chemistry. The extracted residues are separated from each other, so they can be quantified and even identified individually. Devices can be extracted with various types of solvents, both polar and nonpolar. Gas chromatography (GC) is usually used for volatile or semi-volatile substances. Liquid chromatography (LC) is useful for semi-volatile and non-volatile residues. Both GC and LC methods can be coupled with mass spectroscopy (MS) to increase the identification power.

The biggest drawback of these methods is the expense, both in time and money. Methods must be developed and validated for each contaminant, and this is no easy task. Usually, chromatography methods are used when a specific contamination problem is identified (often via the methods previously discussed), and more detailed information is required.

Choosing the Extraction Solvent

Another step in the validation process is choosing the extraction solvent for the test method. For many test methods, the extraction solvent is predetermined by the method itself. But, in some cases, the solvent should be consciously chosen based on the solubility of the target contaminants.

Basic Solubility

Once the target contaminants are identified, basic solubility rules apply. In the case of organic compounds, the basic rule is "like dissolves like." Nonpolar materials, such as oils and greases, are much more soluble in organic solvents such as hexane or methylene chloride. Polar compounds, such as detergents and some types of coolants, are more soluble in water. Some common solvents are listed in Table 12.2.

TABLE 12.2 Examples of Possible Solvents

Polar Solvents	Nonpolar Solvents
Purified water	Methylene chloride
Isopropyl alcohol	Hexane
Methanol	Chloroform
Acetone	Carbon tetrachloride

Device Compatibility

Another factor in choosing the extraction solvent is that the solvent must be compatible with the device material(s). The extraction solvent must not dissolve or corrode the device itself. For example, if a device has polyethylene components, choosing methylene chloride for the extraction solvent would be inadvisable.

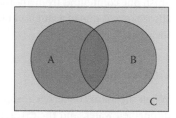

FIGURE 12.2 Venn diagram for multiple solvents.

Multiple Solvents

If the list of target compounds includes both polar and nonpolar materials, then a single solvent may not adequately remove all the contaminants. In these cases, multiple solvents are required.

A Venn diagram can be used to represent a simple two-solvent case (see Figure 12.2). Each solvent will remove all residues contained in its circle of solubility. The overlap represents residues that can be removed by both solvents. The surrounding area "C" represents residues that cannot be removed by either solvent. One significant goal in choosing the solvent(s) is to make "C" as small as possible.

There are two options when multiple solvents are chosen. The solvents may be used sequentially to extract the same device, or separate devices may be used for each solvent. If the solvents are used sequentially, those solvents must be used in the same order so that consistent, comparable results can be achieved.

Setting the Extraction Parameters

The next step in the validation process is choosing the appropriate extraction parameters for the test method. The three variables in the extraction are technique, temperature, and time.

Technique

Common extraction techniques are sonication, manual shaking, and reflux. Because ultrasonic baths are commonly used in cleaning processes, sonication is the most common extraction method. If a device is especially awkward in shape or size, it may not be able to be immersed in the extraction solvent, making manual shaking the technique of choice. A reflux extraction is gentler than ultrasonics, making it suited for fragile devices, but it cannot accommodate large devices.

Temperature

In general, solubility increases with temperature. Therefore, higher temperatures tend to produce higher efficiency in recovering the target contaminants. But higher temperatures can also chemically alter the contaminants. If contaminants are chemically altered, the recovered residue weight can also change. For safety reasons, the extraction temperature must not exceed the boiling point of the extraction solvent.

Time

The extraction time, especially with ultrasonics, must be chosen so that the target contaminants are efficiently removed without breaking down the device itself. Usually extraction times between 3 min and 1 h are sufficient.

To Pool or Not to Pool

When extracting test devices, it can be beneficial to pool multiple devices together for one analysis. This is especially true for small devices. For TOC and UV/Vis, pooling may be required in order to achieve the desired extraction ratio.

In general, pooling devices increases the sensitivity of the method because there are more residues that could be detected. For example, if a single device was extracted and the residue amount was below detection, say <0.01 mg, then that is the limit of the method's sensitivity at <0.01 mg/device. But, if five devices are pooled together, and the result is <0.01 mg, then the final result would be <0.002 mg/device.

The drawback of pooling is that there is an increased risk of adding non-soluble debris created through friction between the devices during extraction.

Validating the Extraction

In addition to validating the cleaning process, it is also important to validate the extraction technique being used. The extraction process must efficiently remove each target contaminant. For example, if your target contaminant is a heavy oil, then water would probably not be the solvent of choice, though it may be very useful when extracting detergent residuals.

There is no general criterion for what constitutes an efficient extraction, but, obviously, the target extraction efficiency is 100%. ASTM F2459 has established an extraction efficiency specification of greater than 75%. An efficiency of less than 50% usually would not be acceptable. The extraction validation must show that the extraction efficiency meets the chosen criteria. If it does not, the extraction parameters must be adjusted until the criteria are met. The two common methods for validating extractions are spike recovery and exhaustive extraction.

Spike Recovery

Spike recovery uses clean devices, which are spiked or doped with a known amount of contamination. The doping must be performed on-site at the testing facility to avoid losing portions of the contamination during packaging and shipping. Samples of each contaminant must be provided to the testing facility. To reduce background interferences, sometimes, it is advantageous to "pre-extract" the clean devices prior to doping. The "pre-extraction" is accomplished by subjecting the parts to the extraction process but not performing residue quantification.

Once the devices are "pre-extracted" and doped, a single extraction is performed on the doped device using the extraction parameters being validated. The recovery or extraction efficiency is simply the ratio of the amount of recovered residue to the amount of doped residue.

Sometimes, it is necessary to correct the extraction efficiency with the volatility of the contaminant. For example, if a contaminant is 30% volatile, one would only expect to recover 70% of the spiked amount.

The spike recovery technique has a high degree of specificity because it demonstrates the removal efficiency of each contaminant tested. Of course, this validation process becomes complicated as the number of contaminants and device types increase. For a comprehensive validation, each possible contaminant must be doped one at a time on each device type.

Exhaustive Extraction

The exhaustive extraction technique uses unclean devices, which contain an unknown level of contamination. The unclean devices are created by exposing devices to the normal manufacturing process but not the cleaning process being validated.

For target contaminants added during the cleaning process (i.e., detergents), the devices used for exhaustive extraction are prepared differently. Clean devices are cleaned again using the target compound but not rinsed. These devices must be pre-cleaned to reduce background interferences.

The devices are then extracted repetitively until the amount of detected residue is less than a certain percentage (i.e., 5% and 10%) of the initial residue or until there is no significant increase in the cumulative residue level.[9] The extraction efficiency is the ratio of the initial residue amount to the cumulative residue level. This technique provides the company with "real-life" data concerning the pre-cleaning (or pre-rinsing) levels of the target contaminants. When these data are compared to residue data from clean devices, the company can demonstrate the effectiveness of the cleaning process.

An important technical drawback of the exhaustive extraction method is that you may not be able to demonstrate adequate removal of every possible contaminant. For example, if the initial extraction produces a residue level of 20 mg, then after the final extraction, the device may still have almost 2 mg of residue on its surface. This remaining residue may comprise small amounts of several contaminants, or it may be a single contaminant, which could not be removed using the chosen extraction technique.

Cleaning Process Effectiveness

The easiest method to determine the effectiveness of a cleaning process is using an unclean or, in the case of detergent residuals, an un-rinsed device. The unclean or un-rinsed device would be prepared exactly as described for the exhaustive extraction, except that the device would only be extracted once. The effectiveness of the cleaning process is then calculated by comparing the amount of residue on the unclean or un-rinsed device (A) to the amount of residue on a clean device (B). The efficiency is calculated as follows:

$$\text{Percent reduction} = \frac{(A-B)}{B} \times 100\% \quad \text{or} \quad \text{log reduction} = \log_{10}\left(\frac{A}{B}\right)$$

Establishing and Justifying Residue Limits

There are no established regulatory limits for residual analysis. The primary reason for this is due to the number of variables involved. Every contaminant has differing degrees of toxicity, and that toxicity can change depending on where and how long it is exposed to the patient. Therefore, regulatory agencies have left it up to individual companies to establish and justify residue limits.

Residue limits are usually expressed in milligrams per square centimeter (mg/cm^2) or micrograms per square centimeter ($\mu g/cm^2$). By expressing the residue limit based on surface area, the same limit may be used for devices of differing sizes. If a surface area calculation is unavailable, then mg/device or μg/device is sufficient.

Limits need to be justifiable, but it is nice to have some wiggle room. Once a limit is established, it is much easier to justify lowering the limit than raising it.

Biocompatibility and Other Testing

Probably the most common technique used to justify residue limits is performing biocompatibility testing. Demonstrating the biocompatibility of the clean devices shows that any residues on that clean device are also biocompatible. Cytotoxicity is a very sensitive in vitro method because the device extracts are applied directly to the cells. The test is scored based on the degree of cellular destruction. Fortunately, this test has defined pass/fail criteria found in USP <87>.[10]

Many companies already have biocompatibility results on file, but these results can only be used if the cleaning process was exactly the same as the one being validated. Ideally, new testing should be performed so that the quantified residue amounts can be correlated to the biocompatibility results. Additional biocompatibility testing may also be needed based on where and how long the device is exposed to the patient.

Other tests that may be beneficial to perform are endotoxin and bioburden. These tests would only be necessary if you wish to claim that the cleaning process reduces endotoxin or bioburden levels.

Endotoxin is a substance released by gram-negative organisms such as *Escherichia coli*, *Salmonella*, and *Pseudomonas* when the cells die. Some common sources of endotoxin contamination are water, raw materials, equipment, and chemicals used to process a device. Endotoxin is not removed by sterilizing; it is usually removed by high-temperature baking or washing the devices with pyrogen-free water. Because endotoxin is so difficult to remove, the best course of action is to prevent or limit contamination during processing.[11] According to USP <161> and the 1987 FDA guideline, the endotoxin limit for devices used intrathecally (cerebrospinal fluid contact) is 2.15 EU/device. Other devices have a limit of 20.0 EU/device.

Bioburden testing provides information regarding the number of viable microorganisms on a test device. While sterilization is the primary method of bioburden reduction, the mechanical and chemical parameters of a cleaning process can also remove bioburden. No regulations establish an expected bioburden reduction for a cleaning process on newly manufactured devices, but a 3 log reduction (similar to a reprocessed medical device) could be used.

Comparison

Another common technique for establishing and justifying residue limits is through comparison. Once all the data have been collected concerning the residue amounts on the devices, various statistical calculations can be performed to establish residue limits. The most basic computations are calculating the average and standard deviation for each sample set. The sample sets can then be compared using statistical tests such as *t*-test or ANOVA. Confidence intervals can also be calculated. If the chosen test method has a high degree of variability, it is important to account for the uncertainty of the test method when establishing limits.

Clean samples from within a cleaning run can be compared to each other to confirm uniformity in a single cleaning run. Clean samples from multiple cleaning runs can be compared to demonstrate consistency from run to run.

When the cleaning process being validated has been established for a while, there is usually inventory available that was cleaned months or years before. These "off-the-shelf" devices can be analyzed to establish baseline data and compared to freshly cleaned devices. Since these devices have a history of nonproblematic patient use, an appropriate acceptance criterion is that the amount of residue on the freshly clean devices needs to be the same or less than the corresponding "off-the-shelf" devices.

Risk-Based Assessment

A risk-based assessment would include data concerning where patient exposure occurs and for how long. Obviously, a permanent device implant will carry a greater risk than a temporary implant. Devices used in the cerebral-spinal area put a patient at greater risk than a device used in the gastrointestinal area.

The risk of the detected residue amounts may be evaluated using available LD50 data for the target contaminants. "LD50 is the amount of a material, given all at once, which causes the death of 50% (one half) of a group of test animals."[12] The residue levels can be compared to the LD50, and the residue amounts can be expressed as a percentage of the LD50. Other available toxicity data can also be used (i.e., TDLO). Of course, these data may not be readily available due to the proprietary nature of many manufacturing and cleaning materials.

Conclusion

An effective cleaning validation is the result of a thoughtful and thorough validation plan. A successful validation plan consists of a practical design flow. The scope of the validation must be clearly defined and the possible contaminants identified. The test methods and solvent(s) used for the validation should be chosen carefully, and the extraction parameters must be consciously set and validated. Finally, residue limits should be established and justified using available data.

Though it may be an overwhelming process, this will benefit any manufacturer or supplier of newly manufactured medical devices. Regulatory agencies, such as the FDA, look for these validations. Above all, a better and safer product will be provided to the patient.

References

1. FDA. *Guideline on General Principles of Process Validation*. U.S. Food and Drug Administration, Rockville, MD, May 1987.
2. Extractables and leachables: A biopharmaceutical perspective. *American Pharmaceutical Review* <http://americanpharmaceuticalreview.com/ViewArticle.aspx?ContentID=266> (Accessed on August 2009).
3. Degrazio, F. The importance of leachables and extractables testing for a successful product launch. February 2007. *Controlled Environments Magazine* <http://www.cemag.us/articles.asp?pid=655>
4. MIL-STD-1916. Department of Defense test method standard: DOD preferred methods for acceptance of product. Department of Defense, Washington, DC, 1996.
5. ISO 2859-1. *Sample Procedures for Inspection by Attributes—Part 1: Sample Schemes Indexed by Acceptance Quality Limit (AQL) for Lot-by-Lot Inspection*. International Organization for Standardization, Geneva, Switzerland, 1999.
6. Andrascik, K. How to tell if a device is really clean. *Medical Design*, June 2008.
7. ASTM F2459-05. *Standard Test Method for Extracting Residue from Metallic Medical Components and Quantifying via Gravimetric Analysis*. ASTM International, West Conshohocken, PA, 2005.
8. Lunceford, R., Reynolds, S. *Analyzing Particulate Matter on Medical Devices*. Medical Device and Diagnostic Industry (MD&DI), May 2009.
9. ISO 10993-7. *Biological Evaluation of Medical Devices—Part 7: Ethylene Oxide Sterilization Residuals*. International Organization for Standardization. Geneva, Switzerland, 2008.
10. United States Pharmacopeia 32 & National Formulary 27, 2009. <87> *Biological Reactivity Tests, In Vitro*. United States Pharmacopeial Convention, Inc., Rockville, MD.
11. Bacterial Endotoxins/Pyrogens. U.S. Food and Drug Administration, March 1985 <http://www.fda.gov/ICECI/Inspections/InspectionGuides/InspectionTechnicalGuides/ucm072918.htm>
12. What is an LD50 and LC50. Canadian Centre for Occupational Health & Safety, June 2005. http://www.ccohs.ca/oshanswers/chemicals/ld50.html (Accessed on August 2009).

13

Biomedical Applications: Testing Methods for Verifying Medical Device Cleanliness

Introduction 173
Device Categories 174
Class I: Critical 174
Class II: Semicritical 174
Class III: Noncritical 174
Cleaning and Sterilization Residues 176
Analytical Testing 177
Chromatographic Analysis 178
Spectrophotometric Analysis 179
Infrared 179
Ultraviolet/Visible Spectroscopy 179
Total Organic Carbon 179
Biological Safety Assessment 179
Conclusion 181
References 181

David E. Albert
NAMSA

Introduction

Ensuring that medical devices and their component parts are free from contamination is exceptionally significant for the medical device industry. Inadequate or improperly cleaned medical device products can harm patients. Potential consequences include litigation and, ultimately, even business failure. Many devices are sterile products, and anything labeled as sterile is something that the United States Food and Drug Administration (USFDA) or any European regulatory agency is going to see as possessing the highest level of risk to the public. As a result, the control of potential sources of contamination becomes of primary importance.

Medical devices must be properly cleaned prior to manufacturing and packaging to ensure the removal of product residue, cleaning chemical residue, and microbes. Cleaning methods are developed and validated to minimize the risk of producing contaminated products by confirming that the cleaning process is sufficient. It is important to establish method limits and to select the proper cleaning techniques and the appropriate detection methods. Therefore, it is important to test parts once the cleaning process and sterilization are complete to ensure that they have been cleaned adequately. Validation and revalidation are key activities when establishing cleaning methods.

Once the cleaning, disinfecting process, or sterilization process has been designed and validated, it becomes necessary to demonstrate that chemicals used in the process are not left on the medical device or equipment at a level that may cause harm to medical personnel and/or to patients, either by a single exposure or as multiple exposures from repeated use of a medical device. One of the first steps in the development of an effective cleaning validation method is the determination of the necessary limits [1]. The FDA Guide to Inspections Validation of Cleaning Processes [2], for example, states, "The firm's rationale for the residue limits established should be logical based on the manufacturer's knowledge of the materials involved and be practical, achievable, and verifiable."

This chapter will examine various kind of contaminants associated with medical devices, and then explore various analytical techniques to measure their presence or absence. Another highlight or addition to this chapter is the use of toxicological risk assessment to evaluate the overall biological safety of residuals and contaminates.

Device Categories

For disinfection and sterilization purposes, reusable patient care equipment should be classified and processed according to recommendations of the Centers for Disease and the Association for Practitioners in Infection Control Guidelines on the Selection and Use of Disinfectants [3,4]. Before implementing any cleaning, disinfection, and sterilization of patient care equipment or device, the article to be cleaned, disinfected, or sterilized must be classified under the appropriate category [5].

Class I: Critical

These devices are in direct contact with blood or areas of the body not usually in contact with contaminants. Examples of such devices include surgical instruments, endocavity probes, implants, biopsy forceps/scissors, and ophthalmic irrigation devices. Sterilization is required for these devices.

Class II: Semicritical

These devices are noninvasive and normally contact intact mucous membranes. They include devices such as flexible endoscopes, endotracheal tubes, and breathing circuits. High-level disinfection is a minimum process requirement for products in this category.

Class III: Noncritical

Noncritical devices are not in contact with the patient. If contact is made, it is only with intact skin, for example, cuffs for measuring blood pressure, beds, and crutches. These devices rarely transmit disease. Intermediate or low-level disinfection is required, and products can be cleaned with a simple detergent. Examples of instruments, levels of disinfection, and the procedures recommended are provided in Table 13.1.

To ensure a successful cleaning and sterilization process, the following issues must be considered. All surfaces of the device must come into contact with the cleaning and disinfecting agents to adequately reduce bioburden. Bioburden is defined as the microbiological load (i.e., the number of viable organisms in or on the object or surface) or organic material on a surface or object prior to decontamination or sterilization. Bioburden, also known as "bioload" or "microbial load," is performed on pharmaceutical and medical products as a quality control measure. However, bioburden testing can serve a variety of purposes, and the relevance of the data can range from relatively insignificant to extremely critical. Bioburden tests can be used for general tracking or trending, for accepting incoming materials, for evaluating changes, or for establishing sterilization processes [6].

TABLE 13.1   Recommended Disinfection/Sterilization and Procedures for Class I, II, and III Devices

Instruments	Level of Disinfection/Sterilization	Procedure
Class I: critical Includes all invasive instruments (e.g., surgical instruments, IV catheters, implanted devices, etc.)	Sterility required	Moist heat, dry heat, or ethylene oxide
Class II: semicritical All instruments that contact mucous membranes (e.g., endotracheal tubes, endoscopes, airways, etc.)	High-level disinfection required. Must be disinfected between patients	Moist heat, 100°C for 30 min aqueous 2% glutaraldehyde for 20–30 min 1:10 dilution of bleach for 20 min
Class III: noncritical	Low-level disinfection required	Chemical disinfectants include ethyl or isopropyl alcohol (70%–90%), phenolic detergent solutions, iodophors for intermediate level. An exposure time of at least 10 min is required. For low-level disinfection, all the above plus quaternary ammonium compounds

Products or components used in the pharmaceutical and medical device industry require control of microbial levels during processing and handling. The device materials should be compatible, and unaffected by the process. After processing, the device must function as designed by the manufacturer. The cleaning, disinfection, and sterilization processes must allow for complete removal of the cleaning and disinfecting agents. Certain devices and configurations (see Figure 13.1) are known to be problematic [7].

The following are examples of design features that may cause challenges in designing and validating the cleaning procedures:

Narrow lumens of flexible endoscopes and stents
Crevices, hinges, and rough porous surfaces
Luer locks
Surfaces between insulating sheaths
Dead-end lumens

Reusable devices should be designed to withstand multiple sterilant or disinfectant exposures. The number of exposures to which a device can be subjected without loss of effective functioning will

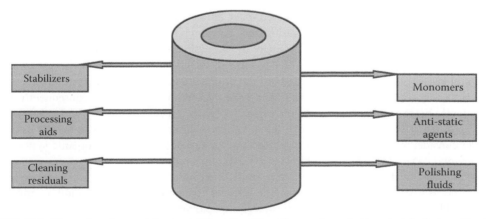

FIGURE 13.1   Examples of potential extractables and residues from polymeric biomaterials that could migrate into the surrounding environment.

help determine the useful life of the product. Most devices composed of sterilant-tolerant materials will withstand more than 100 cycles [5]. Testing should be designed to address not only the efficacy of the sterilization cycle, but also biocompatibility and the functional performance of the device. This can be accomplished by exposing a product to multiple cycles, including any cleaning steps between cycles, for the number of cycles that would be expected in the course of the projected maximum useful life of the device. Following these exposures, there should be a demonstration of functionality, physical integrity, and biocompatibility, as demonstrated through sufficient and appropriate testing.

When the device is a single-use device and the end user must sterilize the product, the manufacturer should provide the user with sterilization instructions. To provide these instructions, the manufacturer must have validation data to support the sterilization process.

A manufacturer should conduct studies in the same manner as recommended in recommended guidance documents, under Sterilization Efficacy Testing (ANSI/AMMI/ ISO 11737.1 and United States Pharmacopeia XXXII <1227>), to provide data to demonstrate that the recommended instructions provide the product with an equivalent Sterility Assurance Level (SAL) of 10^{-6} [8,9].

Cleaning and Sterilization Residues

To most patients, the cleanliness of medical devices begins and ends at whether or not the "things" about to enter their bodies are free from "germs." But to the medical device manufacturer, cleanliness is a far more complex issue. Not only is there concern about microbial contamination, but there is also the need to eliminate contaminants originating from manufacturing processes. These contaminants include, but are not necessarily limited to, cutting or polishing fluids used in machining, adhesives used to join polymer subassemblies, mold release agents, polymer processing aids, carbon residues from laser drilling and cutting, and airborne contamination from the factory environment. The presence of such contaminants on a finished medical device can negatively impact its biocompatibility, resulting in inflammation, infection, and systemic allergic responses in patients.

Fouling of devices with even low levels of any of the above-mentioned contaminants can affect subsequent manufacturing processes as well, resulting in problems of adhesion, bonding, or surface derivatization. The presence of surface films or particles, for example, interferes with the bonding of different parts in a medical device, and can lead to unacceptable levels of device failure [7]. Thus, cleanliness of medical devices is critical not only for patient safety, but also for manufacturability.

Conventional cleaning processes generally employ wet chemical techniques. Of these, the gentler approach is the use of aqueous surfactant solutions to dissolve organic and inorganic contaminants. Acids are also used to assist cleaning and to etch bonding surfaces. Unfortunately, despite the use of aggressive surfactants, highly purified deionized water, and several rinses, the solutions often leave residues in place of the original contamination. The new problem is to now remove the cleaner.

The use of organic solvents overcomes some of the limitations of aqueous systems, but introduces a different set of problems. Medical devices made from polymers that are sensitive to organic solvents can deform and lose the close tolerances needed for proper operation. While organic systems tend to leave fewer residues than aqueous systems, trace amounts of the solvent can cling and, consequently, lead to adverse device biocompatibility. In addition, organic solvents can be adsorbed by some composites. If these are not removed by a drying process, they can outgas, posing a potential risk to the host. While requiring considerably less energy to evaporate than its aqueous counterpart, organic systems pose environmental hazards that must be addressed at a prohibitive cost.

When wet chemical methods are ineffective in the removal of tenacious contaminants, sonication may be used to assist the cleaning process. The residual contamination, waste handling, and device damage problems associated with wet chemical techniques remain.

The potential impact of residue from sterilization processes must also be considered. Residues from steam and hydrogen peroxide processes are composed of nontoxic chemicals, i.e., water and/or oxygen.

For these processes, sterilant residues are not of concern. Other sterilization/disinfection processes, such as those using ethylene oxide (EO) and glutaraldehyde, may impart toxic residues. Additionally, reusable medical devices are sometimes made of polymeric or elastomeric materials, which may react with chemical sterilants (or, for that matter, with chemical cleaning agents) resulting in the formation of chemical residues or extractables. EO and EO by-products can be determined using the same procedures recommended for single-use devices [10]. Validation studies involving those processes for which residuals have not yet been characterized as toxic or nontoxic should include testing to determine whether toxic residues are present. If toxic residues are present, means to remove the residue (e.g., aeration or rinsing) or to reduce it to nontoxic levels before patient use must be included in the sterilization/disinfection instructions. Establishing a nontoxic level may mean the use of toxicological risk assessment to establish an allowable limit for each chemical residue. This topic is discussed in Biological Safety Assessment section.

Methods outlined in ISO 10993-7 recommend practices for determining residual EO in medical devices [10]. The International Standards Organization (ISO) has developed its EO Residual Standards, and the FDA guidelines have been revised. ISO EO residue requirements are different than current FDA standards or previous standards of the Association for the Advancement of Medical Instrumentation (AAMI), in that ISO will focus on EO limits based on expected patient exposure potential versus the current AAMI/FDA approaches that measure actual EO concentration values taken at particular points post-sterilization [7,10]. Presently, ISO will not require analysis of ethylene glycol (EG) residue. Since a reusable device will be subjected to multiple sterilant exposures, additional factors must be considered. Repeated processing may cause a buildup of EO or its byproducts ethylene chlorohydrin (ECH) and EG on the device. Increased aeration time may reduce residuals to low levels, or cleaning procedures employed between cycles may aid in the removal of residuals between resterilizations. It is important that residuals analysis be conducted on devices that have experienced the full complement of reprocessing procedures expected during their useful life.

Analytical Testing

Although the retention of functionality is recognized as one of the most pressing issues to deal with in a decision to sterilize and/or reuse, there is a scarcity of detailed scientific studies in the literature that have investigated the deterioration of materials and function in reused and re-sterilized medical devices.

Such studies would provide much-needed clarification. For one thing, medical devices are commonly made of polymers or plastics of varying density. Low-density plastics are less resistant to heat than high-density ones. However, the properties of plastics are also enhanced through the use of additives. Additives provide many benefits. They may stabilize the material to heat and light or reduce costs (fillers); enhance beneficial properties such as abrasion, resistance, and strength; or provide lubrication, flexibility, or antifungal properties. However, some of these additives are susceptible to reprocessing, sterilization, and reuse. They can be leached out, or their composition can be altered through exposure to light [10].

Mechanical testing can be used to help determine the functional integrity of devices before and after reprocessing and reuse, with the proviso that it may be hard to simulate in an experimental setting the various stresses a device may be exposed to in a clinical environment. Even if all the variables related to stress and strain were known, there is no one set of tests that is suitable for all devices.

However, a key mechanical test is the tensile test, which measures the force required to stretch a device through a range of extensions. The most widely used instrument for stress-strain measurement is the Instron Tensile Tester. This instrument is essentially a device in which a sample is clamped between grips and jaws, which are pulled apart at constant stress rates [11]. A variety of parameters are determined, such as elongation, elongation at break, breaking strength, and tensile modulus of elasticity. The tensile modulus is defined as the ratio of stress to strain, and is determined from the initial slope of the stress-strain curve. The modulus of a material is a measure of the specimen's ability to resist deformation. Tensile modulus is also referred to as Young's modulus. Some plastic materials can be weakened by the process of sterilization or disinfection, resulting in a decrease in the tensile strength of the material. Tensile strength data collected before and after cleaning, decontamination, or sterilization can help predict any changes in the strength of the material [12].

Hardness testing is another useful mechanical test used to determine the functional characteristics of material used in a medical device or instrument. Hardness is generally defined as an indication of the resistance to indentation, scratch resistance, and/or rebound resilience. ISO standards report three methods for measuring hardness: Shore hardness, the ball indentation method, and Rockwell hardness [13,14]. It is important to emphasize that hardness values obtained from one method in general cannot be compared with those derived from another, although data can be empirically compared. Generally, hardness is used as an end-performance property of material used in a device. The instrument that is used to obtain the Shore hardness and Rockwell hardness measurement is called a durometer.

TABLE 13.2   Thermal Analysis Tests for Certifying Product Quality

Thermal Analysis Techniques for Materials	
Test	Technique
Melting point	DSC
Degree of crystallinity	DSC
Glass transition temperature	DSC, TGA
Component quantification	TGA

Thermal analysis, which is the response of a polymer to controlled heating processes, is a family of techniques widely used in the development and characterization of materials, including plastics and elastomers. Some important techniques commonly used in thermal analysis are indicated in Table 13.2. Characterizing the glass transition temperature, melting point, and extent of crystallinity of a polymer is important, and is often used to produce materials whose properties are tailored to the product's ultimate application [15]. The primary thermal analysis techniques for certifying product quality are differential scanning calorimetry (DSC) and thermogravimetric analysis (TGA).

DSC results can give rapid measures of thermal characteristics, such as the polymer melting point and glass transition temperature. The nature of these transitions, in addition to identifying characteristics unique to each polymer, can also provide information about its phase structure, thermal history, and purity. The loss of polymer additives can have a dramatic effect on these thermal properties.

Chromatographic Analysis

The various chromatographic methods, such as gas, liquid, paper, and thin-layer, have become indispensable aids in the isolation, separation, and detection of chemicals. To detect the suspect residue, appropriate sample handling and extraction are essential. Furthermore, there are many system variables, so the power of any chromatographic system depends on the specifics, including the nature and sensitivity of the detector(s).

Gas and liquid chromatography, especially high-performance liquid chromatography (HPLC), have become powerful analytical tools used to characterize additives in polymeric materials used in medical devices, as well as in cleaning agents, disinfectants, and liquid sterilants adsorbed onto surfaces [12]. Gas chromatography (GC) and gas chromatography/mass spectroscopy (GC/MS) are commonly used for detection and quantitation of detergent residue. These methods provide separation, identification, and quantitation of results when an acceptable reference standard is used.

High-performance liquid chromatography is undeniably one of the fastest growing and one of the most useful of all the analytical separation techniques. The reason for this growth is attributable to the sensitivity of the method, its ready adaptability to accurate quantitative determinations, and its suitability for separating nonvolatile species or thermally fragile ones. Liquid chromatography, because of its great flexibility and widespread applicability, can be used for the analysis of more than 80% of all known organic compounds [11].

Spectrophotometric Analysis

Spectroscopic analysis can be used to identify classes of compounds (and sometimes, by inference, specific compounds). In some instances, it is possible to detect the compounds directly on the surface. More often, with the complex surfaces found in devices, analysis is performed after extraction in water and/or one or more suitable solvents.

Infrared

Infrared (IR) instruments measure the vibrational spectrum of a sample by passing infrared radiation through it, and recording which wavelengths have been absorbed and to what extent. Since the amount of energy absorbed is a function of the number of molecules present, the IR instrument provides both qualitative and quantitative information. Since the IR spectrum of a chemical compound is perhaps its most characteristic physical property, IR finds extensive application in identifying substances and their respective concentrations.

Many cleaning and disinfecting agents have distinctive IR spectra that can be used to identify residuals on the surface of devices and instruments [16]. For example, if residual glutaraldehyde is present on a medical device, then it can be extract in a solvent and the solvent analyzed for glutaraldehyde. Most detergents and enzyme (protein) solutions can be identified by this analytical method.

Ultraviolet/Visible Spectroscopy

Much like infrared, ultraviolet (UV) and visible spectrum of a chemical can be used to identify and determine the concentration of certain analytes in an extract. Enzymes, which are proteins, are very easily detected and their concentration determined by their distinct UV spectrum [12]. Most proteins absorb maximally at 280 nm.

The method provides quantitative results, offers fast spectral acquisition, and is not limited to water as the extraction solvent. However, it does lack peak separation and requires a chromophore for specificity.

Total Organic Carbon

Total organic carbon (TOC) analysis is specific to organic compounds and theoretically measures all the covalently bonded carbon in water. This method has been recognized and accepted as a valid way to detect residues of contaminants.

TOC analysis does have some limitations and drawbacks. Contaminants must be water soluble, and it must be established that a substantial amount of the contaminating material(s) is organic and contains carbon that can be oxidized under TOC conditions. Excellent water quality, such as United States Pharmacopeia (USP) "Purified Water" is needed for sensitivity and to reduce interfering substances.

Biological Safety Assessment

One of the greatest challenges in medical device chemical characterization is performing adequate assessment of biological or toxicological risks from extractables, including cleaning and sterilization residues that can compromise patient safety. ISO-10993-17 has clearly stated why and how risk assessments are an essential part of material biocompatibility and are necessary for the assurance of biological safety [17].

Toxicological hazard is a property of the chemical constituents of the materials from which a medical device is made, including the cleaning and processing chemicals, and should be considered in relation to the assurance of biological safety.

Therefore, for a biological safety assessment, the first step comprises chemical characterization of materials. Toxicological hazards can be identified from knowledge of the toxicity of materials or extracted chemicals.

A systematic analysis of biological risks is required using the general principles set out in clause 3 of ISO 10993-1 [18]. This standard states that "The selection and evaluation of any material or device intended for use in humans require a structured program of assessment" and that "The following should be considered for their relevance to the overall biological evaluation of the device":

1. The material(s) of manufacture
2. Intended additives, process contaminants, and residues
3. Leachable substances
4. Degradation products
5. Other components and their interactions in the final products
6. The properties and characteristics of the final product

Unfortunately, the matrix (a table categorizing devices and potential biological effects that need to be addressed) in ISO 10993-1 is often used as a checklist to perform a standard set of tests. The ISO materials biocompatibility matrix categorizes devices based on the type and duration of body contact. It also presents a list of potential biological effects. For each device category, certain effects must be considered and addressed in the regulatory submission for that device. What is actually needed is an appropriate scientific evaluation program based on the specifics of the device.

The results of all tests should be interpreted in the context of the overall risk assessment to know whether or not a specific residue level indicates acceptable risk. This means that a successful evaluation and assessment of risk must involve chemists, toxicologists, and risk assessors. This type of collaborative approach emphasizes the need for an overall scientifically valid risk assessment involving input from multiple scientific disciplines.

The suitability of a medical device for a particular use involves balancing any identified risks with the clinical benefit to the patient associated with its use. ISO 10993-17 states that "among the risks to be considered are those arising from exposure to leachable substances arising from medical devices." This standard provides a method for calculating maximum tolerable levels that may be used by "other standards-developing organizations, government agencies, and regulatory bodies. Manufacturers and processors may use the allowable limits derived to optimize processes and aid in the choice of materials in order to protect patient health."

The aim of the assessment should be to identify any biological hazards inherent in the materials used in the medical device and to estimate the risks resulting from these in light of the intended use. The goal is to develop a process that ultimately protects public health and establishes the safety of medical devices. This objective is supported by ISO 10993-17 in sub-clause 4.3 of the general principles for establishing allowable limits, which states that "the safety of medical devices requires an absence of unacceptable health risk."

The manufacturer of a medical device is responsible for assuring its biological safety, and for documenting the assessment of toxicological risks and establishing the effectiveness of the analysis. Evidence must be provided that an appropriate toxicological risk assessment has been carried out so as to ensure that public health is not endangered. ISO 10993-17 also adds that "where risk associated with exposure to particular leachable substances are unacceptable; this part of ISO 10993 can be used to qualify alternative materials or processes." This is another example of the way risk assessment can be used as a mechanism for critical decision processes.

Where significant risks arising from hazardous residues are identified by chemical characterization, their acceptance should be assessed in line with established toxicological principles. Biocompatibility tests identified in the ISO 10993 series of standards may be used to provide further assessment of risk.

The value of risk assessments has long been recognized by international organizations, and now by ISO 10993. In the risk assessment process, a decision must be made related to risk versus benefit.

Once risk assessment has been completed, the focus turns to risk management. Part 17 of ISO 10993 states that "Manufacturers and processors may use the allowable limits derived to optimize processes and aid in the choice of materials in order to protect patient health." Decisions should be made utilizing the results of risk assessment, biological safety testing, and safe clinical use of predicate devices as described in ISO 14971 [19]. When coupled or linked to biological safety testing, a successful biological safety assessment becomes a highly useful decision-making tool.

Conclusion

Successful cleaning, decontamination, and sterilization/disinfection of medical devices and instruments require both the careful selection of materials and consistent monitoring of the procedures used to process them. Chemical characterization of residues that may be adsorbed onto surfaces, and mechanical testing to ensure functionality, should provide sufficient information to evaluate the potential success of cleaned and disinfected/sterilized medical devices. By using a combination of chemical and mechanical analysis techniques, both manufacturing and decontamination processes can be optimized to ensure a safe and effective product. It is important to note that no test, however foolproof its design, can ever be considered a definitive predictor of clinical performance.

References

1. Brewer, R. Establishing residue limits for cleaning validation: In-depth examination of the factors and calculation that comprise a limit. *Institute of Validation Technology Cleaning Validation and Critical Cleaning Processes Conference*, Chicago, IL, July 24–27, 2007.
2. United States Food and Drug Administration. Guide to inspections validation of cleaning processes. FDA Available at: http://www.fda.gov/oral/inspect_ref/igs/valid.html (accessed June 5, 2009).
3. Spaulding, EM. Chemical disinfection of medical and surgical materials. In: *Disinfection, Sterilization and Preservation*. Lea & Febiger, Philadelphia, PA, 1986, pp. 517–531.
4. Alvarado, CJ. Revisiting the spaulding classification scheme. In: Rutala, WA (ed.). *Proceedings of the International Symposium on Chemical Germicides in Health Care*, Cincinnati, OH, May 1994, pp. 203–209.
5. Albert, DE. Biomedical applications: Analytical characterization for biocompatibility. In: Kanegsberg, B. and Konigsberg, E. (eds.). *Handbook of Critical Cleaning*. CRC Press LLC, Boca Raton, FL, 2001, pp. 563–574.
6. Bryans, T and Alexander, K. Using recovery tests to assess bioburden procedures. *Med. Dev. Diagn. Ind.* 2002, 10, 73–78.
7. Broad, J, Albert, D, and Nawrocki, L. Designing a cleaning efficacy program for reusable devices. *Med. Dev. Tech.* 2006, 16(5), 17–22.
8. ANSI/AMMI/ISO 11737-1. Sterilization of medical devices—Microbiological methods—Part 1: Estimation of population of microorganisms on product. ISO, Geneva, Switzerland, p. 150, 1995.
9. United States Pharmacopeia XXV. Validation of microbial recovery from pharmacopeial articles. U.S. Pharmaceutical Convention, pp. 2259–2261.
10. ISO 10993-7. Biological evaluation of medical devices—Part 7: Ethylene oxide sterilization residuals, ISO, Geneva, Switzerland.
11. Albert, DE. Materials characterization as an integral part of global biocompatibility. *Med. Plast. Biomater.* 1997, 4, 16–23.
12. Albert, DE and Wallin, RF. A practical guide to ISO 10993-14: Materials characterization. *Med. Dev. Diagn. Ind.* 1998, 20, 96–99.

13. ASTM D2238, D2583. *Annual Book of ASTM Standards.* American Society for Testing and Materials, Philadelphia, PA, 1991: 08.02—Plastics (II).

14. ISO. International Standard 868. 2003. Plastics and ebonite—Determination of indentation hardness by means of a durometer (shore hardness). ISO, Geneva, Switzerland.

15. DiVito, MP, Fielder, KJ, Curran, GH, and Feder, MS. Recent advance in routine thermal analysis instrumentation. *Am. Lab.* 24(1), 30–36.

16. Palley, I and Signorelli, AJ. Physical testing. In: Sibilia, JP (eds.). *A Guide to Materials Characterization and Chemical Analysis.* VCH Publishers, Inc. New York, 1988, pp. 273–284.

17. ISO 10993-17. Biological evaluation of medical devices—Part 17: Establishment of allowable limits for leachable substances. ISO, Geneva, Switzerland.

18. EN ISO 10993-1. 2003. Biological evaluation of medical devices—Part 1: Evaluation and testing. ISO, Geneva, Switzerland.

19. ISO 14971. 2007. Medical devices—Application of risk management to medical devices. Annex I: Guidance on risk analysis procedures for toxicological hazards. ISO, Geneva, Switzerland.

14

Material Compatibility

Overview 183
What Is Meant by Compatibility? 183
Small Investments in Compatibility Can Eliminate Unwanted
Problems 184
Preventing Damage to the Hardware • Preventing Effects That Alter
Hardware Performance • Protect the Processing Equipment and Preserve
the Processing Fluid • Identify Optimum Materials and Processes
Who Should Consider Compatibility? 186
Compatibility Data Are Most Useful When Considering a Change 187
Changing the Cleaning Process (Not Including the Cleaning
Fluid) • Changing the Cleaning Fluid • Changing the Parts
Finding and Interpreting Good Compatibility Data Can Be a
Challenge 188
Finding the Data • Interpreting the Data • Examples of Common
Compatibility Test Data
Compatibility Testing Is Easy 190
Getting Good Test Material • Setting Up a Good Test
Example of a Successful Compatibility Test 192
Take-Home Lessons 193
References 193

Eric Eichinger
Boeing

Overview

Compatibility is the ability of two or more things to remain in harmonious contact over a period of time. It is a subject that is very important for those working with sensitive materials or with configurations that can entrap cleaning fluids. Compatibility becomes especially important when making changes to a process. This chapter will summarize a variety of compatibility concerns, as well as provide guidance on how to assess and generate compatibility data.

What Is Meant by Compatibility?

Compatibility is a concept that everyone is familiar with at some level, but it is exactly this familiarity that leads to a wide range of precision cleaning problems. The compatibility road is filled with pitfalls stemming from both what we do not know and, perhaps more importantly, what we think we already know. There is a very real need to understand compatibility because of the many forces of change impacting the critical cleaning industry. Changes motivated by factors such as new environmental requirements or improved technology can frequently have unintended consequences. The aerospace industry often relies upon unique or extreme requirements, so even small problems related to compatibility can have catastrophic consequences. These consequences are not only costly but can also compromise the quality of the finished product. Therefore, knowing when and how to assess compatibility will improve quality and save money.

For the purposes of this discussion, compatibility is the ability of two or more things to combine or remain together with no aftereffects. Although the definition is simple, the "things" in question must be specifically identified. Subsequent sections will describe what is meant by "combine" and what exactly an "aftereffect" is.

In precision cleaning, the "things" involved are usually one of two types: first, the things that we can control; second, the things that are, for whatever reason, beyond our control. Examples of things that can be controlled include cleaning equipment, cleaning solvents, cleaning processes, and the cleaning environment. Examples of things that may be beyond control include the hardware being cleaned and its service environment. Therefore, the trick to solving compatibility problems is to manipulate the "things" that can be controlled to eliminate potential "aftereffects."

Small Investments in Compatibility Can Eliminate Unwanted Problems

So what are these "aftereffects," and why do we try so hard to prevent them? There are at least three answers to this question, and each must be considered independently.

Preventing Damage to the Hardware

The amount and variety of damage to hardware due to poor compatibility are surprising. Nonmetallic seals and other parts may swell, soften, gain or lose weight, embrittle, craze, or otherwise degrade when exposed to a cleaning compound. Metallic parts may rust/corrode or become more susceptible to stress fracture. Coatings and lubricants are likely to raise compatibility concerns due to their ability to dissolve, soften, debond, discolor, and generally degrade in performance. In addition, these effects are dependent on temperature, contact time, and other factors such as the pressure and presence of oxygen. Higher temperature will generally amplify the effects, as shown in Figure 14.1. Extended contact time will also amplify effects but will generally do so according to Figure 14.2.

In addition to the cleaning solvent selected, there are other process variables that can damage hardware. The best example of this is the use of ultrasonic energy to enhance cleaning. Those same cavitation bubbles that do such a great job at removing tenacious soils also can erode soft metals such as aluminum, or can remove coatings. Lower ultrasonic frequencies, such as 27 kHz, generally cause more

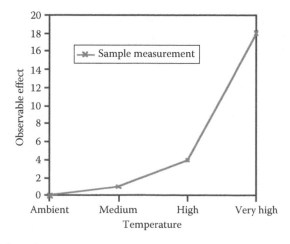

FIGURE 14.1   Compatibility effect versus temperature.

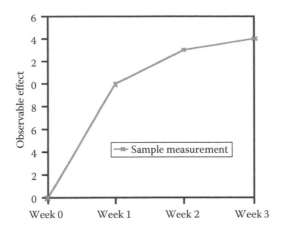

FIGURE 14.2 Compatibility effect versus time.

damage than the higher frequencies (40 kHz and above). Many ultrasonic baths have so-called hot spots, where standing wave patterns inside the tank increase the threat to parts. Certain sensitive part configurations, such as small bellows, may not be compatible with many ultrasonic processes.

Other process variables that can damage hardware include drying temperature, which, if too hot, can degrade chemical conversion coatings on aluminum among other things.

Preventing Effects That Alter Hardware Performance

There are some cases in which the precision cleaning process does not damage a part, yet the aftereffects can be severe. It is important to understand how the part operates, as well as where it operates. Here are a few examples that are associated with complex valves. A valve cleaned with a water-based solvent is vacuum dried. However, rather than evaporate away, the water entrapped in the valve freezes during the vacuum dry process and forms a block of ice. Now the valve will no longer function and, even if the ice melts, the part will still be wet.

Now, in a similar example, the valve is instead cleaned with a flammable solvent like isopropyl alcohol (IPA). After drying, a small amount of IPA remains trapped behind a secondary seal. The valve is then placed in service with liquid oxygen. The combination of the service fluid and an incompatible flammable solvent creates an explosion risk that will likely jeopardize the entire system.

In the example given in the preceding text, liquid oxygen is a known compatibility concern, but milder service fluids should also be considered. Take the case explained in the preceding text, but switch the service fluid from liquid oxygen to hydraulic fluid (Mil-PRF-5606H, to be specific). This relatively unreactive fluid has an acrylic constituent that can precipitate out in the presence of IPA. The precipitate can clog fluid passageways and jeopardize the entire system.

Protect the Processing Equipment and Preserve the Processing Fluid

The last kinds of aftereffects worthy of concern are those that damage the cleaning system. Compatibility should be a top consideration when changing cleaning equipment or the cleaning solvent. When the new wonder solvent dissolves the cleaning system seals and ends up on the floor, the high cost of downtime will arrive with the high cost of cleaning solvents and high cost of equipment repair. What can be done about this?

Start by selecting either a cleaning solvent or a cleaning device. If the solvent is selected first, evaluate cleaning devices that can handle the solvent. Remember the story of the inventor who brought forward a flask of the universal solvent? The king recognized the false claim because the solvent did not dissolve (read was compatible with) the flask. A great solvent will usually be incompatible with many things—that is often why it is such a great solvent. Keep in mind all the concerns mentioned in the preceding text. This may be a case where exposure time really makes a difference. Ask the cleaning device manufacturer if that view port is really glass or is it Plexiglas?

Compatibility problems cannot always be avoided by selecting water-based cleaners. High pH detergents (>10) can damage aluminum, and even certain nonmetallics such as Vespel rubber.

When the cleaning device is selected before the solvent, the compatibility review works in reverse. List the materials of construction used in the device, and then find the cleaning solvent that is most compatible.

Identify Optimum Materials and Processes

The three sections above can all be summarized by saying that initial ignorance of compatibility can lead to high-cost headaches and irritated customers. However, compatibility consideration is also an effective way to save money. A good compatibility assessment as well as some performance data will identify all the potential cleaning materials and processes available. Then, a selection can be based on implementation and operation costs. Usually, the more tools there are in the toolbox, the better the final result will be.

Who Should Consider Compatibility?

We can just as easily consider the question of who should avoid fatty foods because the answers are the same. Everybody! Yet, as with fatty foods, there are special "at-risk" groups that must be the most vigilant. The two at-risk groups in terms of compatibility are those working with sensitive materials, and those working with things that can entrap cleaning solvents.

There is no compendium on what a sensitive part is, but if there were, it would include some metallics, such as plated surfaces. There would be lots of nonmetallics, perhaps with polycarbonate at the top. A good rule of thumb with nonmetallics is "guilty until proven innocent." Parts with coatings will likely have special needs. Even if the coating does not visibly change during cleaning, there is no guarantee that its performance will be unaffected. For example, corrosion-inhibiting chromates will leach out of coatings into pure water with no visible change. This may, however, upset the customer when the part corrodes, not to mention the facilities folks who now must dispose of pure water with carcinogenic chrome in it! Then there is the at-risk group that "cleans" parts with lubrication requirements. Bearing assemblies are good examples of this type of hardware. In terms of compatibility, this type of cleaning is more of an art than a science, and is way beyond the scope of this book.

The other at-risk group is those working with entrapment areas. Whoever said, "What goes in must come out," never saw this type of hardware. Hardware with a high potential for incompatibility may contain faying surfaces, blind holes, seals, threaded inserts, or seat assemblies, which can trap or soak up and retain cleaning solvents.

Just because you clean at the "piece part" level does not mean that there is no potential for entrapment. A piece part may just be the lowest logical level of disassembly. Only when you are working with a noncomplex, mono-material item do the fears of entrapment fade.

So, to summarize, you are at risk and should definitely consider compatibility if you

1. Are cleaning plated surfaces
2. Are cleaning nonmetallic surfaces
3. Are cleaning coated surfaces
4. Are cleaning hardware with lubricity properties
5. Are cleaning hardware with entrapment areas
6. Are changing your cleaning process (see next section)

Compatibility Data Are Most Useful When Considering a Change

One nice thing about compatibility worries is that they only need to be dealt with during changes associated with the precision cleaning process, cleaning fluid, or parts. Sometimes, however, recognizing the change can be the most difficult part. The next three sections summarize some common changes that trigger the need for evaluating compatibility, as well as some less obvious changes.

Changing the Cleaning Process (Not Including the Cleaning Fluid)

These are usually the easiest changes to cope with. Process engineers have the ability to control what the new process will be like. The supplier usually has a large amount of literature to assist in decision making, and can often demonstrate a similar process in operation. Compatibility data is a great tool for distinguishing between two otherwise acceptable processes. Make sure that the supplier has performed all the necessary tests with the appropriate materials. If they have not, it may be wise to augment some of the testing (see later sections).

Compatibility should be considered even when replacing one piece of equipment with something almost identical. Certain types of precision cleaning equipment tend to have their own personality. For example, two 27 kHz ultrasonic tanks with the same power rating will not be exactly the same from a compatibility standpoint. Each tank will have its own unique hot spots where part damage potential is highest. If sensitive parts are used in this tank, the hot spots should be mapped out and avoided.

Pay close attention to the "incidental" process materials that would not obviously contribute to compatibility problems. For example, many cleaned parts are sealed in polyethylene bags. Polyethylene is compatible with almost everything and is usually factored out from compatibility considerations. However, certain additives to the polyethylene bags can cause compatibility issues. In one case, silver-plated parts tarnished when an antistatic agent from the bag contacted the part surface. In another case, a company received a new shipment of bags that contained an elevated quantity of plasticizer. Plasticizer is a compound added to keep the bags flexible and easy to work with. The extra plasticizer deposited out onto the hardware during weeks of storage (storage could be considered a change), however, forcing extensive and expensive recleaning and delaying final assembly. Changes like this are very difficult to identify, but the consequences can be severe.

Changing the Cleaning Fluid

Whoever coined the phrase "nothing in life is easy" could very well have been changing their cleaning fluid. As described in the preceding text, cleaning performance and compatibility are necessarily at odds with one another. Thus, performance and compatibility evaluations are best performed together. Remember to evaluate compatibility under actual use conditions. Also remember to consider compatibility with the people who have to work with the cleaner (toxicity). Frequently, the best cleaner is a compromise between performance, compatibility, and toxicity.

Just as with the changes in cleaning processes, there are some subtle changes with cleaners to be aware of. For example, if two cleaners, each with great compatibility, are utilized in series, what happens when the cleaners are mixed? There really is no way to predict what will happen, so a good rule of thumb is to avoid letting cleaning fluids mix. One notable exception to this rule is to use deionized water to "chase" out IPA.

Although intentionally mixing cleaners can usually be avoided, what do you do when the cleaner itself is a mixture of materials? Be aware of the following three types of cleaners: simple mixtures, azeotropes, and isomers. Simple mixtures, unlike the other two, are clearly blended products. A review of the product data as well as the material safety data sheet (MSDS) will usually indicate what has been mixed together. Compatibility of the mixture as a whole should be assessed, and not as a sum of its parts; for example, compatibility of chlorine gas and compatibility of water do not combine to define the compatibility of hydrochloric acid.

A second type of mixture is the azeotrope. An azeotrope is a blend of two or more liquids that keep their composition after distillation. Unlike simple mixtures, azeotropes can be used in vapor degreasers because their composition in the vapor phase is identical to that in the liquid phase. Even though these cleaners may seem to be a singe compound, it is important to remember that they are really a special type of mixture.

The last common type of mixture is the isomer. Isomers are two compounds that have identical chemical formulas (same elements in the same proportions). However, the atoms feature different arrangements and isomers have different properties. How different are the properties? Most people can easily distinguish between spearmint (L-carvone) and caraway (D-carvone). The "D" or "L" that precedes the compound name indicates you are dealing with an isomer. One common isomer used for aerospace cleaning is D-limonene (derived from oranges). The properties of L-limonene (derived from pine) are different. The *trans* and *cis* designations also indicate an isomer, as in the commonly used *trans*-dichloroethylene.

So what is the big deal with all of these mixtures? First, it important to know exactly what is in a cleaner. The solvent HCFC 225 often contains two isomers, the toxic "ca" and the much less toxic "cb." Also, it is important to recognize that there may be potentially proprietary additives. Often, cleaning solvents require trace amounts of stabilizers, coupling agents, or anti-oxidizers. These "secret ingredients" can be manipulated by the manufacturer, and the user will be the last to know.

A perfect example of "secret ingredient" manipulation involves a cleaner considered for use on space hardware that had to be compatible with liquid oxygen. The cleaner demonstrated oxygen compatibility and was near implementation. Then, the vendor decided to address complaints received from other market sectors by changing a stabilizer. The product name did not change, nor were the users aware of the change. Yet, the cleaner was no longer compatible with oxygen. Fortunately, a test identified this change and the cleaner was not implemented. In the case of mixed solvents used for critical processes, compatibility testing may be required on a lot-by-lot basis.

Changing the Parts

The most difficult types of changes to respond to are those associated with the part or with its service environment. Any design changes in configuration or material selection should prompt a compatibility assessment. These changes are often difficult to respond to because they are not usually made by the group that does the cleaning. The key again is awareness, and awareness relies upon good channels of communication between the redesign group and the cleaning group.

Compatibility between the precision cleaning process and the part's ultimate service environment is essential. Consider the potential for cleaning fluid entrapment and other compatibility effects. In one case, a logistics depot decided it would be more efficient if all their valves were cleaned to the strictest cleanliness level. This would allow flexibility when the valves were needed by a particular subsystem since they would always be clean enough. However, this decision was short-lived, because the cleaning fluid in the strict process was not compatible with all of the subsystem service fluids.

Finding and Interpreting Good Compatibility Data Can Be a Challenge

A good compatibility assessment may cost more time and money than is available, so the best thing to do first is try to find existing data. The plethora of compatibility data available seems, at first, a blessing, which can turn to a curse as you wade hopelessly through a mountain of data that does not address your specialty application.

Finding the Data

Vendors often have compatibility literature that is applicable. However, it will not have the unbiased flavor of data generated by a "third party." The "third-party" data can be found, with a little library or Internet searching, in books and periodicals. The majority of these data cover cleaner–material interactions (especially at room temperature). Thus, if you know the materials involved, these books should help you select a cleaner. They will also assist in the selection of things like plastic gloves and storage containers. However, there is less information available on compatibility with ultrasonic equipment and other process-related factors.

Interpreting the Data

Generally, compatibility data is a measurement of a change in properties over a period of time. The time may be very short, as in an impact test, or the time can be much longer, as in immersion compatibility tests. Specifically, the data must be representative to be meaningful. Representative data usually means the materials tested are the same or very similar to the material in question. Also, the exposure time should represent a worst-case scenario (remember to consider entrapment potential). Temperature, pressure, and other environmental factors should also be factored in. Subtle differences can potentially determine whether the data from a given study are useful in evaluating the specific application. For example, the amount of headspace in the sample container will affect the amount of oxygen available to immersion test specimens. Thus, there is almost always a degree of subjectivity in compatibility data because it is not common to find directly representative data.

As if compatibility data were not subjective enough, much of the published data is qualitative. Rather than list quantitative data, such as a linear swelling of 5%, publications often assign a letter rating and then, somewhere in the fine print, explain an "A" is fine for long-term storage, "B" is fine for short-term, "C" materials should be avoided, and "D" means "run for your life" or some other threatening note. It is best to allow only the amount of subjectivity merited by the criticality of the hardware. If the criticality is high, compatibility testing may be in order.

Examples of Common Compatibility Test Data

The different compatibility tests (described in more detail in the next section) measure a variety of properties. Often, multiple types of data will be required to make a thorough assessment. The next section briefly summarizes the most common compatibility tests used to quantify degradation (yes, the data are usually quantitative and then reported in a qualitative form). Note that there are many specific types of compatibility tests, such as breakthrough time for gloves, which are not described in this chapter but may be distant cousins to some of the tests that are described.

Metals are the most straightforward. Immersion testing is often performed either with single materials or with galvanic couples (two metals in contact with each other). Immersion frequently assumes covering the sample to some degree with a liquid but can also include immersion in a controlled environment such as an oven or an atomic oxygen chamber (which simulates an earth orbit environment). Metals sensitive to stress corrosion, such as titanium, usually require a more complex (stressed) specimen and may require elevated temperatures. Important factors include exposure time, the percentage of the specimen that is immersed, and the amount of oxygen available. Data taken from these specimens include the difference between the starting and ending weights (pay attention here to the specimen size), amount of oxide formed, and depth of corrosion (frequently reported in the rather odd units of mils per year—1 mil = 0.001 in.).

Elastomers, like metals, are frequently immersion tested. Testing can be done on "chunks" of material, actual parts (such as "O-rings"), or parts in a service environment (under compression, specific temperature, etc.). Weight loss data are again common, but there are two other performance-related

TABLE 14.1   Some Common Tests for Compatibility

Material Type	Test	Important Variables	Result Reported
Metallic	Immersion	Time, oxygen availability	Weight loss/gain, amount of corrosion
Elastomers	Immersion	Time, specimen configuration	Weight loss/gain, shore hardness, % swell
Coatings	Immersion	Time, coating age	Visual change, adhesion loss, pencil hardness change
	Wipe	Coating age	Visual change
Liquid	Solid immersion	Time	Purity change, NVR
	Mixing	Relative amounts, time	Temperature change, pressure change

measurements that are also common. "Shore hardness" is measured by pressing a blunt pin against the elastomer before and after exposure. Frequently, solvents will effect a noticeable drop in shore hardness of elastomers. The shore hardness recovery may also be important. Over time, after the immersion compatibility test is over, an elastomer usually will gradually recover some or all of its original shore hardness.

A second frequently reported elastomer compatibility result is "percent swell." Not surprisingly, this is the dimensional change associated with a period of immersion. The dimensions of an elastomer can begin to change within the first 5 min and may take an hour to a month to stabilize. There is no universal "acceptable amount" of swelling used to generate the qualitative rankings published, but a good rule of thumb is that >10% swell is a problem. Just as in the hardness test described in the preceding text, an elastomeric compatibility specimen will usually try to return to its original dimension. This is not always a good thing, however, because if swell is induced by a high vapor pressure solvent (very volatile), uneven shrinkage is likely to occur as the edges shrink faster than the center. Elastomers can crack and break due to uneven shrinkage.

Coatings compatibility tests are usually performed either by immersion or wiping. Data are gathered visually, or by using tape to measure changes in adhesion (ASTM D3359), or by measuring the ability of a series of sharpened pencils to break the coating (ASTM D3363).

It is sometimes important to assess whether a particular liquid will be degraded upon contact with another liquid or a solid. Liquid compatibility is assessed by immersion of solid specimens or by mixing of two liquids. Immersion results will demonstrate whether there are changes in purity, or if the liquid extracts something from the solid. Extraction is frequently quantified as non-volatile residue (NVR). The NVR is what is left behind after the liquid is evaporated away. A good thing about NVR is it can often be identified through a variety of instrumental methods.

Liquid mixing is an excellent way of quantifying reactive compatibility effects. In this test, two liquids are placed in a sealed pressure vessel. Temperature and pressure changes can be recorded to assess whether the liquid mixing is a catastrophic threat or a potential nuisance. Again, at the conclusion of the test, chemical breakdown products and NVR can be quantified and identified. This is a very useful test to evaluate the effect of an entrapped processing media on the ultimate service media.

Table 14.1 summarizes the common types of compatibility tests.

These tests will be described in more detail in the next section.

Compatibility Testing Is Easy

So, after spending some time in the library and on the phone, it is clear that there is not enough relevant compatibility data to say whether the new wonder solvent will be a future workhorse or a Trojan horse. Perhaps the solvent is too new or too obscure. Whatever the case, the best bet is to collect enough data to ensure there are no unwelcome surprises.

The following description of common compatibility tests is not intended to provide the "how to" details, as that can be left to the ASTMs. However, the quality and interpretation of the data obtained from a compatibility test can be heavily influenced by the detailed approach. Thus, the following is intended to point out key considerations during test set up, or good questions to ask when reviewing existing data.

Getting Good Test Material

Since many compatibility tests are performed before implementation, the need for a test sample frequently arises. Pay careful attention to the sample! Where did it come from? How similar is it to what will be used? Answer these questions first, or the test may be doomed before it is even started. For example, it is not uncommon for small samples to originate in a laboratory rather than a manufacturing line. Laboratory samples are usually more pristine with less impurities and no traceability. These pristine samples may not represent the production material very well. The sample should represent what will actually be used on the actual product to be cleaned. Following this logic, materials are often procured based upon one requirement, and then discarded when they no longer meet a second-use requirement. A good worst-case compatibility test should consider material at the low end of the use specification as well as the high end.

Setting Up a Good Test

"Never ask a question for which you are not prepared for the answer." In the world of compatibility, this platitude suggests that Pass/Fail criteria always be established. The difference between passing and failing may hinge on how the test is set up.

Immersion Testing

Immersion tests are often the simplest meaningful compatibility test available. The variables to measure have already been covered in the previous section, but the test conditions are every bit as important. The exposure to air is a good example of an important condition. Fully immersed specimens will be exposed to dissolved air. To control dissolved air, the sample container should be sealed with a minimum of headspace. For partially immersed specimens, the air–liquid interface will often behave differently than the fully immersed portion. Finally, if wet–dry cycles are likely, an alternate immersion test should be considered. Alternate immersion tests are especially well suited to measuring stress corrosion.

In addition to simulating environmental factors, immersion tests can also simulate exposure to service fluids. When there is the potential for entrapment, it may be worth considering if any reaction will occur between the processing material and the service material. Changes in pressure, temperature, or color will indicate whether a reaction occurs. The ratios of the two test fluids are subjective, but should be based on worst-case entrapment.

Since reactivity usually increases with temperature, run the test at the highest anticipated fluid temperature. However, since the amount of dissolved air decreases at higher temperatures, it may also be necessary to repeat the test at a lower temperature if the effects of dissolved air could be important.

Other Types of Compatibility Tests

Although immersion testing is the single most common test, there are many other tests to consider. These tests can be selected based on the specific configurations of the parts and the processing equipment involved. For example, if there are entrapment locations or capillary passageways, a "sandwich corrosion" test (ASTM F1110) may be in order. A small amount of sample is placed on a flat sheet with another placed on top to make the "sandwich." The sandwich is put into some sort of hostile environment, most commonly a salt fog chamber, and then checked for damage. This test has been used effectively to identify aqueous products that are safe to use on aluminum.

Coatings can be tested by a simple wipe test, but the age of the coating as well as the number of wipes are important. Metals can be stressed by bending into "U" or "C" type specimens and then placed in a variety of environments (including simple immersion). In this case, the alloy is very important and should be representative or one that is sensitive to stress corrosion cracking.

In summary, the difference between "passing" or "failing" a compatibility test can be determined by how the test was done. The tests can be simple, but should always be as representative as possible.

Example of a Successful Compatibility Test

Let us look at the example of Clevis McKleen's new degreasing system. He would like to replace an old vapor degreaser with a new pre-cleaning process. To better understand his options, Clevis first examines the things he cannot control about his process. He knows he must remove light hydrocarbon and particulate contamination from oxygen regulators for diving equipment. The regulators have the potential to entrap his cleaner. Furthermore, he has several Teflon O-rings and an anodized aluminum housing that are assembled into the oxygen regulators. Clevis must select an appropriate cleaning agent and process to safely clean the regulators.

The first step, Clevis decides, is to set some boundaries based on the things he has no control over. Since the parts are used in a breathing air system and can entrap solvent, he decides that he must use a cleaner that is relatively nontoxic. The entrapment potential also precludes the use of highly acidic or basic cleaners. He can consider flammable materials if he takes some safety precautions, and can use a volatile organic compound (VOC) if he can keep the emissions low. Water-based products and ultrasonic immersion baths are attractive options.

Given the conditions described in the preceding text, Clevis selects a mild, water-based cleaner he already uses for another process, and a 40 kHz ultrasonic bath for consideration. Clevis knows he needs compatibility data before implementing this process, and contacts the appropriate vendors. The vendor for the cleaning agent has compatibility data that indicate the cleaner will not damage any of Clevis' materials, even at the proposed use temperature of 150°F and concentration of 10%. However, all of the data were taken from flat immersion panels. Since Clevis is worried about entrapment, he decides to augment the vendor data with a sandwich corrosion test of his own.

The ultrasonic bath vendor also has compatibility data, but these data suggest that the 40 kHz bath may damage the anodized aluminum housing. The vendor offers an 80 kHz bath as a safer alternative. Clevis takes a scrapped regulator to the vendor for trial processing and finds that the anodic film is not harmed by the 80 kHz process, but the higher frequency is unable to remove his particulate contamination.

To make matters worse, Clevis gets his sandwich corrosion specimens back and the data do not look good. Even the mild cleaner caused corrosion to the aluminum specimens. Clevis decides to consider an organic solvent, perhaps in an immersion cleaning bath. He steers away from the ozone depleters and the ones with poor solvency. He must also consider his facility's VOC concern and rules out solvents with a vapor pressure over 50 mm of mercury. IPA appears to be a good candidate.

A brief review of the published compatibility data on IPA indicates no problems with any materials. Clevis decides to run a simple immersion test with the Teflon O-rings, just to be sure. After 2 days exposure, he observes a 2% weight gain and no change in shore hardness. However, during his literature review, Clevis finds that IPA can fuel an explosion in high-pressure oxygen systems. This makes the complete removal of the IPA essential, so Clevis decides to place the regulators in a vacuum drier at the end of the process.

Clevis learns that his immersion bath is too high in VOCs, but that a recirculating flush with IPA will reduce the VOCs to an acceptable level and minimize the fire hazard. Knowing that he plans to use IPA, Clevis reviews all of the materials in the recirculation system with the equipment vendor for compatibility. No issues are identified, and Clevis confidently implements his IPA recirculation system for cleaning his regulators.

Take-Home Lessons

Compatibility is clearly an important consideration, which can lead to many types of problems when ignored. Pretty much any change to a process, whether intentional or accidental, should trigger a compatibility review. Reviews should generally leverage existing data, but often there is not enough relevant data to make a conclusion. Testing may be the only way to gain the confidence needed to safely resolve a compatibility concern. This chapter should help in identifying those concerns and addressing them in a safe and effective way.

References

ASTM D3359. Test methods for measuring adhesion by tape test, American Society of Testing and Materials.

ASTM D3363, Standard test method for film hardness by pencil test, American Society of Testing and Materials.

ASTM F1110. Test method for sandwich corrosion test, American Society of Testing and Materials.

Mil-PRF-5606H with amendment 3. Hydraulic fluid, petroleum base; aircraft, missile, and ordinance, September 7, 2006. Available at http://www.everyspec.com/MIL-PRF/MIL-PRF+%28000100+-+09999%29/MIL-PRF-5606H%28AMENDMENT3%29_6003/

II

Applications

15 **Clean Critically: An Overview of Cleaning Applications** *Barbara Kanegsberg*........197
Overview • Why Clean? • What Is Critical Cleaning? • Cleaning Options • Electronics
Cleaning/Defluxing • Aerospace • Biomedical Applications • Industrial
Metal Cleaning • Nanotechnology • Selecting and Improving the Cleaning
Process • References

16 **Cleaning Validation of Reusable Medical Devices: An Overview of Issues in
Designing, Testing, and Labeling of Reusable Devices** *John J. Broad and
David A.B. Smith*..229
Introduction • Guidance • Issues and Test Programs • Testing Methodology
and Validation Study Design • Appendix A.1: Product Families for Cleaning
Validation • Appendix A.2: Instruments Representative of the Six Product Families
Utilized for Validation Testing • Appendix A.3: Cleaning Methods • Results—Cleaning
Validation • Compliance with Acceptance Criteria • References

17 **Critical Cleaning for Pharmaceutical Applications** *Paul Lopolito* 255
Overview • Regulations • Selecting Cleaning Agents • Selecting Cleaning
Agent Suppliers • Cleaning Methods • Critical Parameters • Equipment
Design Considerations • Cleaning Method Design • Product and Equipment
Grouping • Sampling Methods (Direct and Indirect) • Sample Site Selection • Recovery
Studies • Analytical Detection Methods • Residue Limits and Acceptance
Criteria • Cleaning Validation Master Plan and Validation Protocol Design • Cleaning
Standard Operating Procedures • Change Control • Biological Contamination • QbD
and PAT Considerations • Conclusion • References

18 **Cleaning in the Food Processing Industry** *Hein A. Timmerman* 271
Chemistry of Soils • Films on Equipment • Standards of
Cleanliness • Summary • References • Bibliography

19 **Electronic Assembly Cleaning Process Considerations** *Mike Bixenman* 283
Introduction • Why Clean Electronic Assembly and Advanced
Packages? • Electrochemical Migration Risk • Cleaning Process
Design • Miniaturization • Miniaturization and Lead-Free Drive Solder Flux
Requirements • Science of Electronic Assembly Cleaning Agents • Matching the
Cleaning Agent to the Flux Residue • Materials Compatibility • Thermodynamics versus
Kinetics • Process Validation • Conclusion • Acknowledgments • References

20   **Precision Cleaning in the Electronics Industry: Surfactant-Free Aqueous
      Chemistries**   *Harald Wack* 319
      Surfactants • Critical Cleaning in the Electronics Industry • Surfactant-Free Aqueous
      Cleaning Products • Overall Cost of Ownership of the Cleaning Process • Conclusions:
      The Most Important Characteristics of Surfactant-Free Aqueous Cleaning
      Agents • Acknowledgment • References

21   **Contamination-Induced Failure of Electronic Assemblies**   *Helmut Schweigart* 333
      Contamination-Induced Climatic Failure • Humidity and Pollution Effects on Electronic
      Assemblies • Conclusions • Acknowledgment • Reference • Bibliography

22   **Surface Cleaning: Particle Removal**   *Ahmed A. Busnaina* 345
      Introduction • Particle Removal • References

23   **Cleaning Processes for Semiconductor Wafer Manufacturing (Aluminum
      Interconnect)**   *Shawn Sahbari, Mahmood Toofan, and John Chu* 359
      Introduction • Basic Operations in Wafer Fabrication • General Wafer Cleaning
      Techniques • FEOL Cleaning Processes • BEOL Cleaning Processes • Chemistry of
      Positive Photoresist Strippers • Chemistry of Negative Photoresist Strippers • Chemistry
      of Post-Plasma Etch Polymer Removers • Challenges of Future Technology • References

24   **Advanced Cleaning Processes for Electronic Device Fabrication (Copper
      Interconnect and Particle Cleaning)**   *Shawn Sahbari and Mahmood Toofan* 379
      Chemical Mechanical Planarization • Back End of Line Cleaning • Wafer
      Backside and Bevel Cleaning • Particle Cleaning: An Introduction • Cleaning
      Stationary Particles • Cleaning Mobile Particles • Cleaning and
      Rinsing • Conclusion • References

25   **The Cleaning of Paintings**   *Richard C. Wolbers and Chris Stavroudis* 399
      Aqueous Cleaning • Cleaning Paintings

26   **Road Map for Cleaning Product Selection for Pollution
      Prevention**   *Jason Marshall* 411
      Overview • Introduction • Goals • Alternative Product Selection • Metal Cleaning
      Summary • Conclusions and Lessons • Acknowledgments • References

27   **Wax Removal in the Aerospace Industry**   *Bill Breault, Jay Soma, and
      Christine Fouts* 429
      Introduction • Selective Plating and Masking • Wax Removal Technologies • Aerospace
      Specifications/Industrial Practices • General Process Guidelines for Wax
      Removal • Summary • References

28   **Implementation of Environmentally Preferable Cleaning Processes for Military
      Applications**   *Wayne Ziegler and Tom Torres* 439
      Overview • DoD Alternatives Implementation Road Map • Barriers to
      Implementation • Keys to a Successful Alternative Technology Implementation
      Effort • DoD Solution • Conclusion • References

15

Clean Critically: An Overview of Cleaning Applications*

Overview 197
Why Clean? 198
What Is Critical Cleaning? 198
Cleaning Options...199
Avoiding Cleaning • Cleaning Processes: A Brief Overview
Electronics Cleaning/Defluxing 200
Why NOT Cleaning Can Be an Option • Why Cleaning Electronics
Assemblies Should Be an Option
Aerospace 206
Keys to Success in Aerospace and Elsewhere • Conservatism • Case
Study: Cleaning in an Airless/Vacuum System Prior to Applying Engineered
Coatings • Inertial Navigation Systems: Complexity and the Team
Approach • Optics
Biomedical Applications...221
Case Studies: Medical Applications
Industrial Metal Cleaning 226
Nanotechnology 226
Selecting and Improving the Cleaning Process.................226
References..227

Barbara Kanegsberg
BFK Solutions

Overview

This section of the book addresses specific applications of critical cleaning. Even in a book as comprehensive as this, it is impossible to address all applications. We have chosen a number of representative areas and invite the reader to look at all the various approaches, not just those related to their own field. For example, a professional engaged in building components for an aerospace application may find insights from the chapters into art restoration or food processing.

In this chapter, we inspect the concept of critical cleaning as a process by focusing on specific case study applications in electronics, aerospace, and medical devices. "Why clean?" is a question that must be answered by anyone with the responsibility of manufacturing any object. If the answer is "to clean," there are a myriad of process and product choices that must be addressed to effectively and reliably produce the product.

---

* With comments by Bev Christian. *Note:* This chapter contains material adapted from B. Kanegsberg, Very high performance, complex applications, in *Handbook for Critical Cleaning*, 1st edition, CRC Press, Boca Raton, FL, 2001.

Why Clean?

Soil is matter out of place. Cleaning is the removal of matter out of place.

Do not clean because Mom told you to, unless Mom is your boss (Kanegsberg, 2009a). While cleaning should be a value-added process, there is the tacit assumption that we have to clean all surfaces at all steps in the manufacturing process and that the more we clean, the better. The opposite is true. Any cleaning step has the potential to modify the surface; undesirable surface modification can compromise the product. In addition, cleaning incurs costs; at any step, it must be determined that the investment (capital, labor, consumables, time, etc.) is worthwhile.

There are certainly cases where not cleaning is an option; Bev Christian presents a logical argument as to why at least certain electronics assemblies should not be cleaned (please see section "Why Not Cleaning Can Be an Option," in this chapter). There are many instances where cleaning is essential for proper functioning of the product. However, we spend a significant amount of time advising clients and colleagues to begin by evaluating their assumptions about the need to clean.

Critical cleaning includes designing the appropriate steps to set up a cleaning process and also the steps at which not to clean. As we have indicated in the preface to this edition, the critical cleaning step may not be the final cleaning step or the most sophisticated cleaning process. It may occur in a clean room; it may occur during initial fabrication. The reason for cleaning is to achieve the appropriate surface qualities, the appropriate surface attributes. Appropriate is the operative word. A complete absence of matter out of place is impractical and impossible. Cleaning for the sake of cleaning may be counterproductive; you may need to gently question authority, be it Mom, Dad, your boss, or even a regulatory person.

Having said this, the most manufactured objects require the removal of soil at one or more steps in the process. The issue is surface quality, and as products become smaller and more complex, the surface-to-mass ratio decreases. For nanodevices, the surface is the product, so surface properties, attributes, and appropriate cleaning are everything.

What Is Critical Cleaning?

After some consideration (a few decades), the term "critical cleaning" seems a reasonable, descriptive term. Precision cleaning has been defined as cleaning of products of perceived high value, submicron level particulate removal, or cleaning products where the results of improper cleaning could be catastrophic. Precision cleaning has also been described as cleaning something that did not look particularly soiled in the first place (Le Blanc, ca. 1999). On the other hand, those involved in processing optics could argue that optics may be embedded in heavy pitch and wax, which then have to be completely removed. While embedded in pitch, those parts do not look visually clean.

Back in the early 1990s, this author explained that everyone knew the difference between high-precision cleaning and ordinary cleaning. High-precision cleaning was simple to define: that would be *my* manufacturing process. Ordinary cleaning, on the other hand, encompasses everyone else's manufacturing process (Kanegsberg, 1993). I went on to explain that understanding as much as possible about the manufacturing process in question and then collaborating with others who were also faced with stringent cleaning requirements is crucial to advancing the science and art of contamination control.

What we typically look on as achieving the appropriate level of cleanliness includes cleaning (removal of surface contaminants), contamination control, and surface quality or properties. While these aspects can, in theory, be separated, for all practical purposes, they become enmeshed. The critical cleaning steps may be during final assembly in a clean room, or they might take place very early in the process in an environment that more resembles an automotive repair facility.

Cleaning Options

Avoiding Cleaning

You have options in setting up a critical cleaning process. The first is don't clean. Determine if cleaning is necessary. Only work with clean parts; if you have a choice, perhaps selecting components suppliers that provide you with cleaner parts is a more economical option. Some suppliers of components offer various grades of cleanliness; you have to ask. There is, of course, a proviso. We have occasionally observed large companies setting policies that compel their smaller, less powerful suppliers to do enough cleaning that the large, final assembler can assert that they use no dangerous chemicals. Since the smaller suppliers may not have the in-house expertise in industrial hygiene and environmental regulations, this is a highly inappropriate approach, and it is not recommended. Increasingly, fortunately, we are more likely to find larger companies encouraging their suppliers toward better chemical handling and stewardship. Some companies avoid cleaning by outsourcing the process. While this may be reasonable for companies that required cleaning processes that they cannot control, and while the company doing the outsourcing is technically not doing the cleaning, they are still ultimately causing cleaning to be performed and are responsible for the process performance.

You can also avoid cleaning by keeping the part or component clean before, during, and after the cleaning process. For example, it is a waste of effort and money to use a substandard drying system that redeposits soil on the part or to allow parts to air-dry in an area where airborne contamination is a problem.

Another way to avoid cleaning involves redesigning. Design the product or component so that it is less complex, so that it is readily manufactured. Design for manufacturability is not necessarily part of a CAD program. Educate your designers about manufacturing and cleaning issues. Of course, such efforts are more likely to impact new programs; with current programs, designs may be contractually specified. Sometimes, the soil can also be changed. Low-residue, "no-clean" flux is one notable example.

Cleaning Processes: A Brief Overview

Cleaning techniques are discussed in detail in the other book in this series, *Cleaning Agents and Systems*. Cleaning is a process, not a chemical. A cleaning process consists of wash, rinse, and drying steps. Each has a separate function and must be considered and planned carefully. Cleaning processes can be aqueous-based, solvent-based, or so-called solvent-free or chemical-free cleaning. Examples of "chemical-free" cleaning include dry abrasive blast, plasma, and steam. Of course, steam is a chemical, and, in plasma cleaning, the cleaning chemistries are generated in situ. Aqueous or water-based processes include water alone, water with salts (e.g., bicarbonate) or with a mixture of organic and/or inorganic additives. The proportion of water to additives in the "as-used" (diluted) cleaning solution is variable. The organic additives may or may not be volatile organic compounds (VOCs), and they may have varying degrees of impact on workers or on the environment. Solvents in the world of critical cleaning mean organic solvents. Organic solvents can be used alone or in blends. The aggressiveness of organic solvents on materials of construction, on workers, and on the environment is variable. Debates about what constitutes a safe cleaning agent continue, and regulatory agencies still find it easier to regulate on a chemical by chemical basis. Of course, you have to comply with applicable regulations. At the same time, please be aware that cleaning is a process, and process performance, worker safety, and environmental impact are related to the process, not to the chemical alone. Some considerations and resources needed in redesigning processes are indicated in Figure 15.1 (Kanegsberg, 2008b).

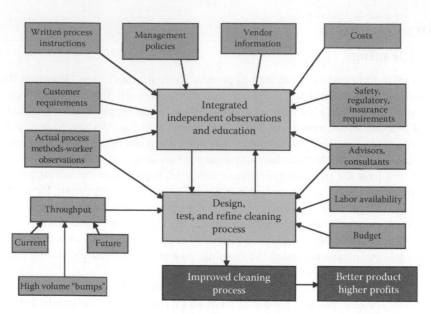

FIGURE 15.1   Examples of considerations/resources in selecting and improving critical cleaning processes.

Electronics Cleaning/Defluxing

Why NOT Cleaning Can Be an Option

Bev Christian, Research in Motion

The one question that must be asked right at the beginning is, "Do I really need to clean my circuit assembly?" This book gives many reasons why one should, so they are not discussed here yet again. This section, bluntly, deals with "Why bother?"

Since we are talking about active electronic circuits, at least one of the three legs of the electrochemical corrosion triumvirate—electric potential, moisture, and significant ionic contamination—must not be present to avoid cleaning. The first leg cannot be avoided by the very nature of the objects under discussion.

For the second leg, moisture, electronic products used in the high arctic, dry deserts, outer space, and highly air-conditioned rooms are not likely to be subjected to condensing moisture in any kind of regular fashion. However, it must be admitted that except for the latter man-made spaces, inexcessibility may dictate, along with the need for ultrahigh reliability for space and for most electronic military and medical items, the rigors of cleaning. This still leaves a significant portion of the pantheon of electronic devices that may not need cleaning—most IPC Class I and Class II electronics in dry and air conditioned environments. Note that the operative term here is "may." Examples of Class I and Class II products could be an inexpensive portable radio and a high-end smart phone, respectively. It will depend on what market segment the manufacturer wants to aim for and the level of expense the philosophy of the company will support.

The real key is the ionic cleanliness of the electronic assembly. Since the objective is to not clean, this means that the circuit board, electronic components and the assembly materials must be ionically clean. The easiest way, although appearing on first blush not the least expensive, is to buy products from top-tier suppliers who you audit on a regular basis.

Circuit boards bought from a premier circuit board manufacturer may never show boards with high ionic contamination over literally a period of years. Suppliers of this caliber should be able to supply you with data on the values of various control parameters throughout the manufacturing process and the final board cleanliness and solderability characteristics. It is no good to have a perfectly clean board that you can't solder because, for instance, the final cleaning step was so harsh that it oxidized your metal finish on your pads and in your plated through holes.

Components bought from a myriad of suppliers are more problematic, especially if an occasional unusual part must be bought from a "specialty" supply house. However, if your company is doing business with suppliers that it can trust, examines manufacturing information from the supplier (run charts, etc.) on a regular basis, and carries out regular on-site audits, the likelihood of cleanliness problems is not banished but made very small.

Laboratory examinations in the author's lab have shown that most modern SMT components are ionically clean. Analog and digital components, capacitors, resistors, diodes, LEDs, buzzers, switches, crystals, transistors, and connectors have been tested. So far, fluoride, chloride, bromide, nitrate, sulfate, nitrite, and phosphate have been the inorganic ions tested for. Organic anions tested for include formate, malate, acetate, phthalate, and methane sulfinate. Except for one diode that had $10.6\,\mu g/cm^2$ of malate, the highest concentration of any organic ion found on all of the components tested was $0.61\,\mu g/cm^2$. Excluding connectors, the highest value for an inorganic ion surface contamination concentration was $0.41\,\mu g/cm^2$. One connector had a bromine level of $14.0\,\mu g/cm^2$, while three different connectors had nitrate levels of 6.5, 9.7, and $10.8\,\mu g/cm^2$, respectively.

What was advocated for bare boards and components above, can also be said for liquid fluxes, paste fluxes, cored wire, and solder paste. If it is financially possible, it is best to carry out your own testing. This would include such tests are halide ion, copper mirror, fluoride ion, the British corrosion test, surface insulation resistance, and electrochemical migration. They are listed more or less in increasing order of a combination of complexity, time, and cost.

Issues with cleaning electronic circuit packs include compatibility of the solder masks, components, and flux residues with the cleaning chemistry and conditions (temperature, spray pressure); increasing inability to clean under low-standoff components; effect of solvents on labels, paints, and markings; the floor space needed for cleaning equipment; the consumption of utilities (electricity, water, air pressure); maintenance of the equipment; safe handling and storage of the cleaning chemicals; and processing, storage, and disposal of spent cleaning fluids.

Unfortunately, some cleaning companies use scare tactics to try and sell their point of view about cleaning. However, companies like Lucent and Nortel Networks started selling in the late 1980s/early 1990s high-end, no-clean, Class II telecommunications switches with expected 15–25 year lifetimes. These pieces of equipment were backed by warranties that included severe monetary penalties that would be imposed on them by carriers like AT&T and Bell Canada for even seconds of downtime over the course of a year. These two companies are just two examples of many companies that carried out the supplier selection, observations, and materials testing discussed above. No cases of large-scale crashes of whole systems made by these and other reputable telecommunication equipment suppliers who made no-clean product are known.

Why Cleaning Electronics Assemblies Should Be an Option

I have a somewhat different view than Bev Christian. Recently, design and environmental challenges have caused many manufacturers to look at cleaning options for electronics assemblies. No-clean fluxes are an approach for many applications; we find increasing evidence that defluxing (cleaning) of many electronics assemblies is necessary. For the better part of a generation, post-solder defluxing has been relatively straightforward. Manufacturers used no-clean flux, and they either did not remove it or removed it with water (Kanegsberg, 2008a).

History

For decades, cleaning electronics assemblies was desirable, even required, and approaches to cleaning were rather set in stone. Rosin fluxes, traditionally used in military and other high-reliability applications, had to be cleaned with CFC-113 or trichloroethane. Ultrasonic cleaning was largely not acceptable.

Eliminating ozone depleting chemicals (ODCs) required replacing chlorofluorocarbons (CFCs). The replacement efforts, beginning in the late 1980s, were arduous and challenging. The author was involved in those efforts; implementing replacements for ODCs in electronics was part of the basis for her EPA Stratospheric Ozone Protection Award. In retrospect, it was an incredible experience that benefited industry and the environment. Options increased. Aqueous, semi-aqueous, and advanced solvent processes grew in popularity. Rosin mildly activated (RMA) flux was supplanted by water-soluble organic acid (OA) flux or low-residue (no-clean) flux, even for many high-reliability applications. Manufacturers who had been in the electronic assembly field for many years saw cleaning problems (or at least cleaning requirements) go away. Even if no-clean flux required a bit of cleaning, water or dilute aqueous cleaners did the job.

New Challenges

Within the last 3 years, defluxing has again become an area of interest. Drivers toward defluxing include increased design complexity, higher performance requirements, economic pressures, and U.S. and international environmental regulations that impact the makeup of solders and the availability of cleaning agents.

Fortunately, cleaning of electronics assemblies is a relatively well-studied area, which for generations has been the topic of books (Tautscher, 1976). Aqueous cleaning has been a viable option for some time (Cala, 1996). Professional organizations, notably the IPC (Association Connecting Electronics Industries), and Surface Mount Technology Association (SMTA) are good resources for discussions of issues in electronics assembly. The IPC, in recognition of the reemergence of cleaning issues, is developing a consolidated defluxing handbook. In addition, there is a new effort to develop a handbook and/or guidelines regarding potting and encapsulation, which are distinguished from conformal coating. The documents provide good guidance; you have to assess your own situation.

Design Requirements

Design for manufacturability is a wonderful goal; design that favors manufacturability generally involves simplicity. If the components of a circuit board assembly could be compared to a city, it is desirable to have open spaces and to avoid an overly dense population. However, given performance requirements, design for manufacturability is a relative term. Many design engineers, often with good reason, take the approach of automatically attempting to pack 5 lb of stuff in a 10 lb bag. There may be good reasons, such as some technical requirement for a compact or miniaturized product. However, the more closely spaced are the components, the more difficult it is to clean. The architecture becomes complex, with tightly spaced components and low standoff. In critical cleaning applications, the physical properties of the assembly at the micro level meet the impasse of the physical properties of water, water with additives, and even some solvents. Important physical properties of the cleaning agent include the surface tension, density, and viscosity. For components with low standoff, roughly speaking below 5 mil, aqueous cleaning is a challenge. The challenge can often be met by prolonged force; this translates into slower speeds and longer in-line systems.

This effect of component density on ability of a particular cleaning system to work effectively was illustrated two decades ago by W. Machotka, C., Knapp, and B. Kanegsberg in a study at Litton Industries. While the study was performed with RMA flux and while water-soluble and low-residue fluxes are easier to remove, the principle that densely populated assemblies are difficult to clean is being rediscovered today. In the study, leadless, one inch square component simulators were placed at varying standoff distances (Kanegsberg, 1998). That is, components were spaced at 0.003, 0.005, 0.008, and 0.015 in. from the surface of the board. The component simulators were dipped in RMA flux. Parts were charred by heating immediately before cleaning to simulate soldering. They were then cleaned in actual production batch and in-line cleaning systems using a variety of cleaning agents. Efficiency of removal of RMA flux was estimated gravimetrically. Cleaning equipment and cleaning agents are identified only by code. This experiment was not completely balanced in that the cleaning agents tested in a particular

TABLE 15.1 Percent Flux Removed Relative to Standoff

Equipment	Cleaning Agent	0.003 in.	0.005 in.	0.008 in.	0.015 in.
In-line a	1,1,1-Trichloroethane azeotrope	98	100	100	100
In-line b	HCFC	88	100	100	100
In-line c	D-Limonene A, semi-aqueous	79	100	100	100
Batch a	D-Limonene A, semi-aqueous	69	100	100	100
In-line d	Hydrocarbon blend, semi-aqueous	56	100	100	100
In-line e	Aqueous/saponifier A	79	93	100	100
Batch b	Aqueous/saponifier A	65	96	94	100
In-line f	Aqueous saponifier B	71	100	100	100
In-line g (older model)	Aqueous/saponifier B	26	38	63	73

Notes: Five samples were run per test. Batch systems were manually operated; in-line systems were automated.

piece of equipment depended on what was being used at the time in a particular production situation. In that sense, the options tested were considered acceptable, but all available options had not been tested by the production facility.

As indicated in Table 15.1, below a 5 mil (0.005 in.) standoff, it is difficult to clean with less aggressive solvents, semi-aqueous, and aqueous methods. In addition, automated systems tend to be more effective in soil removal. Results provide additional evidence that system design can influence cleaning; the very old cleaning system was not nearly as effective as the newer designs. This author has empirically observed that 5 mil tends to be the borderline level between relatively straightforward and relatively difficult cleaning applications for all manner of assemblies, not just electronics.

Properly motivated design engineers, on being presented with such cleaning information, might modify the assembly to allow easier fluxing and defluxing. It is typically helpful to involve the design engineers in process modification plans, keeping them on any teams or at least keeping them up to date on proposed changes. Conversely, those involved in assembly process development would do well to look at the next generation of assemblies to determine if they can be cleaned using the methods under consideration.

However, miniaturization is a reality; this is one reason why electronics assemblies are becoming difficult to clean. The term "densely populated" board is used frequently to describe assemblies that present challenges with flux residue. The issues of new electronics design and of cleaning densely populated assemblies in a timely manner and with adequate soil removal in in-line aqueous cleaning systems seem to arise during all current programs and seminars (Kanegsberg, 2009b). Water does not wet surfaces as well as do most solvents, it does not "creep" under tightly spaced components, so cleaning agents and soils can remain trapped. Water and those aqueous formulations do not have as high a wetting index as do many organic solvents. Not all organic solvents, including isopropyl alcohol, provide adequate wetting for all electronics applications. Aqueous formulations are being refined to address the issue. In order to assure adequate removal of cleaning agent residue, cleaning agents must be readily removed during the rinse step. If an in-line aqueous system is selected, adequate force, careful design of nozzle, and appropriate fixturing are essential for good soil removal (Woody, 2008). This may mean that the conveyor belt is operated 15.2 cm (6 in.)/min; the conveyor belt may need to be very long. In your minds eye, visualize something moving that slowly.

Soils (Fluxes et al.)

In electronics assembly, solder flux is the primary soil. Flux is, in a sense, a cleaning/surface modification agent in that it facilitates soldering by preventing the buildup of oxides. However, once used, it must itself be cleaned. Many manufacturers have made significant strides in manufacturing by modifying the soil, in this case, the flux. Choices in flux include

- Rosin based
- Water soluble
- Low solids, the so-called no-clean fluxes

All have their advantages and difficulties.

Rosin-based flux is based on pine tree sap with additives, including activators, some of which are acids. RMA flux is classically used in military and other high-end applications. Rosin-based fluxes had their problems. As naturally occurring materials, they can show wide variations in soil from lot to lot, because, for one thing, groves of pine trees vary in composition of the sap. Flux residue can be a problem, particularly if the flux cures after soldering and prior to cleaning. This most often happens with significant delay prior to the cleaning process. In addition, many of the polar additives are difficult to remove; they may leave complex mixtures known as the infamous white residue. Because flux is a mixture of polar and nonpolar components, cleaning agents with a wide solvency spectrum are preferred, or azeotropes, such as IPA/cyclohexane azeotrope, have proven effective.

Flux residue of any sort is not considered desirable. Residues can form crystals (dendritic growth) that can impair product function or even result in product failure (please refer to Chapter 21, Schweigart). Dendritic growth is a sort of microscopic stalactite or stalagmite, which appears on assemblies after cleaning, often after aging in a humid atmosphere. Acid or salt residue can be corrosive, damaging materials of construction. While manufacturers claim that they want no detectable residue, it has become widely accepted that rosin fluxes may leave a residue that is less damaging to the product than are some other types. Rosin flux, when heated in the soldering operation and cured due to delayed cleaning or further heating with inadequate cleaning, may leave trace residues under components. However, these residues are often hard and jewel-like, and they appear to be nonreactive. One would suspect that such flux residues form a sort of artificial amber. Amber is petrified tree sap, and, in fact, several jewelers have quietly confided that realistic, artificial amber can be produced by heat-curing rosin flux.

Other fluxes are widely used because of pressures to avoid organic solvent cleaning or to avoid cleaning all together. OA fluxes or water-washable fluxes can be very effective and can obviate the need for solvent cleaning. They are meant to be cleaned with water-based cleaning agents, not with solvents.

While some OA fluxes are synthetic, others are based on lemon juice or apple juice. This has caused some confusion among manufacturers who confuse lemon-based flux with D-limonene (orange-peel derived) cleaner. Changes in the soldering procedure are needed, and care must be taken in the cleaning process to avoid residues that can produce dendritic growth or other interfering or corrosive residue.

Low-solids or no-clean fluxes are designed with minimal solids and minimal residue to avoid the need for cleaning altogether. Control of the overall assembly and soldering process is typically more exacting with no-clean fluxes. An inert atmosphere may be required, and there is generally a much narrower process window. Because of the high level of process control, the most successful initial implementation of low solids fluxes occurred in very large-scale manufacturing facilities where ongoing training and process control could occur.

Because solder joints may not be as aesthetically pleasing with no-clean processes, it is often necessary to modify in-house or customer requirements from visual requirements to functional requirements. In addition, there are applications where even the small amount of residue left by low-solids fluxes is unacceptable. Sometimes, only a very dilute aqueous-based cleaning agent is needed, or water without additives may be acceptable. It should be pointed out that some low residue fluxes are more amenable to being cleaned with water than are others. Some no-clean flux residue can result in product degradation in a humid atmosphere, a distinct disadvantage if the product is to be used in New Jersey or Texas in July. It is the preference of this author to use a cleanable no-clean flux where possible.

The "et al." issue cannot be ignored. While most manufacturers think of flux residue as the primary contaminant, other contaminants, including nonionic contaminants, can cause problems. Often, contamination is introduced by a components supplier and may not be detected until some problem arises after final assembly. Other residues include oils, sulfur-based compounds from machining fluids, and metal or ceramic particles. Determining residue of flux and other contaminants can involve a host of visual, microscopic, and analytical testing such as surface insulation resistivity and ion chromatography. However, to this day, the conventional wisdom is to test for ionic contamination; if the assembly passes the standard tests for ionics, the assembly should operate appropriately. While ionic testing is logical in terms of immediate functionality, we are seeing situations where it is not sufficient, other testing would be beneficial. For example, the initial and long-term adherence of paralyne coating and of epoxies and potting compounds depends on having a surface that is free of contaminants that could leach or outgas, and this includes nonionic organic compounds that are not likely to be detected via standard test methods. In addition, as Schweigart points out (Chapter 21), capacitive coupling from nonionic residues can degrade performance at high frequencies.

Product Performance Requirements

High-reliability electronics in aerospace, military, medical, and other critical requirements require long-term performance. In fact, for products associated with medical devices, expectations of long-term reliability are increasing. The performance of the cleaning process is important for both short-term and long-term reliability. This means careful selection, validation, and monitoring of all new processes. Conventional wisdom is that consumer electronics do not necessarily require careful cleaning. The product may not have a long expected lifetime; inexpensive may trump quality. However, given global competition and movements toward sustainability, the paradigm may again shift.

Cleaning and contamination removal challenges are not limited to flux removal. The entire fabrication process must be considered in terms of residual contamination, surface properties, and product integrity Circuit board and component materials can have compatibility issues with cleaning chemicals, both aqueous and solvent.

Lead-Free Solder and Other Regulatory Issues

The move to lead-free solders was spurred by the European Union (EU) directive Restriction of the use of Certain Hazardous Substances in Electrical and Electronic Equipment (RoHS). The RoHS directive affects electronic components sold in the EU, but lead free is also being mandated in other areas of the world, including California in the United States. The impact of lead free on electronic cleaning is primarily associated with the higher temperatures required with lead-free solders. High temperatures can bake residues on a surface; baked-on residues are more adherent and more difficult to clean.

Lead free tends to be the "fall guy" for assembly problems. Even if some of the product line can be produced with more traditional solders and with fluxes that are readily removed, there is still the issue of multiple assembly and cleaning processes.

In addition, other current regulatory issues like limitations on VOCs and toxic air pollutants can restrict uses of cleaning agents with more effective cleaning and wetting capabilities. At the same time, other restrictions on cleaning agents are increasing, and these restrictions may impact solvent cleaning plans. Most cleaning and defluxing agents contain at least some VOCs. There is pressure to reduce VOCs (smog producers), and not just in California. At the DoD level, the joint services solvent substitution working group (mercifully referred to as JS3WG or JS3; refer to Chapter 28, Ziegler and Torres) is tasked with eliminating hazardous air pollutants (HAPs) and minimizing or eliminating VOCs.

The impact of greenhouse gas emissions is very likely to affect all manufacturing, especially cleaning processes. California recently passed advanced legislation in the area. Activities in California cannot be dismissed elsewhere. What happens in California often becomes the norm throughout the United States. In addition, the DoD is issuing policy on climate change.

Energy use may be restricted by regulation or through economic pressure. Many critical cleaning processes need large amounts of energy, notably when large quantities of heated water are used and for drying operations.

Water is a valuable, increasingly scarce resource. Water use may be subject to restrictions through direct regulation or economic pressure, especially in drier climate regions such as the American Southwest. Solvent processes that declined in use after the CFC phaseout may experience a resurgence when energy and water costs increase. Well-designed solvent equipment can now better contain emissive vapors, decreasing the risk to workers and the environment.

As with the replacement of ODCs—which we sometimes refer to as "The Great Freon® Phaseout—these restrictions can be recast and in fact exploited to achieve more effective, reliable electronics defluxing and critical cleaning. In the course of studies performed for University of Massachusetts, Lowell, we unearthed studies indicating the performance and economic advantages of insulation. Initial investments in well-insulated equipment, adoption of less-emissive cleaning equipment, recycling, and closed-loop processes are likely to yield benefits not only in lower costs but also in superior process performance.

Choosing a Cleaning Process

The type of cleaning process depends on multiple factors including the type of flux, the design of the assembly, expected product end use, customer requirements, worker preferences, cleaning agent/cleaning equipment costs, and the local mix of safety and environmental regulations. It should be noted that it may also be possible to modify the build process. Alternative solders and soldering techniques are being developed. Epoxies may be used, or laser ablation may replace soldering.

It should also be remembered that, for many high-end applications, electronics assembly is often much more than defluxing. An assembly may contain mixed OA and rosin fluxes, machining, oils, and lapping compounds. Sub-vendors may change the process, resulting in changes in soil residue and in cleaning agent residue.

An array of cleaning process have been used with electronics assemblies ranging including no-clean, water, aqueous/saponifier, water with large amounts of organic additives including nonlinear alcohols and unidentified proprietary additives, semi-aqueous, co-solvent, classic chlorinated solvents, brominated solvents, engineered solvents alone and as blends, and flammable solvents.

For every successful application of a given flux and cleaning process, an unsuccessful one could be cited. For all the standardization, cooperative testing, and understanding of materials compatibility, cleaning electronics assemblies is a very site-specific effort. It should also be remembered that eliminating electrostatic discharge (ESD) is important not only to avoid assembly failure but also to achieve contamination control.

Each defluxing application has specific challenges. For example, where no-clean flux was successful in a hand-soldering operation, the manufacturing engineer noted a number of critical factors that basically add up to understanding what is being soldered (Miller, 1997):

- Condition and materials compatibility of the components and bare boards
- Cleanliness of alloys and base materials (essential for adequate wetting)
- End use of the product
- Signal-to-noise requirements
- Component and design requirements

With no-clean fluxes, there is not as large a process window as with rosin or OA fluxes. The company worked both with clients and materials suppliers to choose the proper materials. Issues of the composition of various alloys and thickness of application over the base metal can affect solderability and shelf life. The engineer emphasized that it is a constant educational process to make vendors and customers knowledgeable. The take-home lessons from this study can be applied to other situations where the acceptable process window may be narrowed, such as with lead-free solders.

Aerospace

Because the products absolutely must perform reliably in exacting situations over a long period of time, critical cleaning in aerospace applications presents exciting challenges. Engineers must address requirements and restrictions from private clients, military clients, and an assortment of regulatory agencies. There is often an overlay of secrecy that makes process change difficult and that requires each aerospace company to reinvent solutions to problems. In terms of customer requirements, there can be hundreds of pages of specifications and documents, and these can be outdated, contradictory, and may not address current issues. This is a particular problem for smaller companies, farther down in the supply chain, which may be faced with reconciling cleaning requirements from multiple customers while still attempting to produce a reliable product and to make a profit. Older legacy systems are a particular challenge in that they may specify cleaning agents that are no longer available. We have even seen consumer, household cleaning agents listed in specs. These may have morphed from the time when they were first "called-out," and, as indicated in Chapter 1 (Kanegsberg) of *Handbook for Critical Cleaning: Cleaning Agents and Systems*, household cleaners are generally suboptimal for industrial applications.

Given that there are issues of product reliability, military requirements, competition, and costly testing, which may take years to complete, aerospace has been understandably conservative in adopting new processes. At the same time, aerospace has been fearless and inventive in evaluating and eventually choosing from among the range of new solvents, mixtures, and cleaning techniques. Cleaning techniques adopted include supercritical CO_2, aqueous cleaning, co-solvent systems, proprietary blends, and acetone and other low flash-point cleaning systems. In LOX systems (systems that will be exposed to liquid oxygen), the issue has been to find cleaning agents that pass stringent tests to prevent fires or explosions during use. For LOX cleaning, a range of cleaning agents may be allowed for initial cleaning; then, at the final stage, a level of defined but noninterfering residue is tolerated.

A number of authors discuss approaches to removal of soil in aerospace applications. For example, please refer to the discussion of dewaxing by Breault, Soma, and Fouts (Chapter 27). In addition, two case studies are provided as examples of approaches used to improve cleaning processes. Both involve performance requirements, economics, and regulatory issues. In addition, please peruse approaches used in other critical applications outside of aerospace; you may find that some can be adapted.

Keys to Success in Aerospace and Elsewhere

In the course of approximately a quarter century of interaction with aerospace groups, we have observed three keys to success in process change or optimization. These three keys to success hold for other critical applications as well. The first key to success is to achieve strong, collaborative company teams. Examples of groups who should be represented in a process change team include designers (yes, some designers actually are concerned that what they dream up can be cleaned and assembled successfully), engineers, assemblers/technicians, maintenance people, quality control representatives, safety/environmental professionals, management, appropriate advisors, and key people in the supply chain. The size of the team depends on the nature of the company; teams can range from three people to several dozen people. Such teams can be successful because they brainstorm (sometimes pointedly), but they do not engage in paralysis by analysis, and they actually reach conclusions and *do* something. When I first entered the workforce, I was frankly skeptical of teams, because, as a woman of my era (the Paleozoic Era), I was constantly told to be a "team player" and was then marginalized as the taker of notes and supplier of cookies (actually, cookies make meetings run more smoothly). I was wrong. Teams work; they help make better product, and they save money. If you shy away from teams, please get over it.

The second is communication and ongoing education. Successful teams actually talk to each other. They also continue to educate themselves about technical and regulatory issues; they attempt to find cohesive solutions rather than to wear blinders, solve problems in a vacuum, and pass a nonworkable solution over the fence. They educate themselves proactively, sometimes inviting selected speakers to in-house seminars, combining education with brainstorming.

The third is testing. There is either too much or not enough. I have observed that individuals and committees can make the mistake of specifying a piece of cleaning equipment with all sorts of bells and whistles, getting it on the floor, adding the cleaning agent they thought would work, and then being a bit disappointed that efficacy of cleaning was not as described in the colorful brochures. They then scramble to find an appropriate cleaning agent or they to reconfigure the cleaning equipment. Sometimes, they may end up paying many times the original investment in capital equipment. There are many variables to consider, including variables in materials of construction. There are so many variables that cleaning process selection becomes both art and science. Process qualification is not a list of specs. Test first.

Involve the production people in testing. In aerospace and in the manufacture of other high-value product, process change and process improvement will not occur without the input, cooperation, and collaboration of the production assemblers. For many high-precision processes, the end product is produced on a very small scale and may be based on hundreds of assembly and cleaning steps. Automated cleaning processes may be impractical. In such cases, build and cleaning processes are highly specialized, involving the input of skilled and experienced assemblers (Kanegsberg, 1993). Particularly in situations with repeated hand-cleaning processes involving skill and judgment, it is important to involve those who will work with the alternative cleaning processes on a day-to-day basis. Often, the assemblers themselves are able to detect potential problems that are not picked even with sophisticated analytical testing. For example, in flux removal, a pilot project was undertaken in which individuals performing overhaul and rework and hand cleaning operations evaluated several hydrocarbon blends and orange terpenes. While initial laboratory-scale evaluation indicated that all of the products provided for pilot test cleaned equivalently, the technicians reported subtle differences. Some said they could see differences in cleanliness, and saw problems with certain of the cleaning agents. More detailed, costly testing including surface analysis and outgassing confirmed that, indeed, certain cleaning agents were leaving a previously undetected residue. Attempting to conduct all possible analytical testing would have been time-consuming, costly, and probably unproductive. In general, where the production people have been involved at various stages in new process development, a much more robust process has resulted.

Conservatism

Over a period of perhaps a quarter of a century, aerospace applications were highly dependent on solvents, notably the ODCs, CFC 113 (commonly referred to by the trade name Freon), and 1,1,1-trichloroethane (TCA). Given the time required to validate new processes, many high-precision, high-value processes changed only to develop more and more steps. Over the years, if there were problems at a particular stage, which could conceivably be traced to contamination, the engineer in charge might recommend additional cleaning with CFC 113, TCA, or isopropyl alcohol (IPA), with some IPA and perhaps acetone for drying.

Many aerospace applications use classic solvents, or HCFC 225 (scheduled for phaseout in 2015), or blends of hydrofluorocarbons (HFCs) and hydrofluoroethers (HFEs). Others use a variety of aqueous and/or solvent blends. This is not surprising. Manufacturing may be global, but regulatory restrictions are local, and they are confusing. Since the mid- to late-1980s, increasingly stringent environmental regulations covering VOCs, air toxics, and climate change (not to mention ODCs) have driven many process changes. The process of change and the regulatory drivers has resulted in a number of false starts and inefficient processes. On the bright side, where process development has been approached with an eye to better performance rather than simply coping with the regulation of the moment, more efficient processes with fewer steps and lower usage of cleaning agents have been adopted.

Case Study: Cleaning in an Airless/Vacuum System
Prior to Applying Engineered Coatings

Engineered coatings are many and varied. When applied to metal surfaces using techniques such as thermal spray, physical vapor deposition (PVD), or plasma vapor deposition, the coatings can impart desirable performance characteristics. First, however, the surface must be clean. The experience of Plasma

Technology Inc (PTI) illustrates how a high-tech company achieved the surface requirements needed for an array of aerospace customers (Dowell, 2003). Teams, education, and testing were all employed.

PTI, headquartered in Torrance CA, applies some 300 coatings to varied substrates using eight techniques including plasma vapor deposition, PVD, and ion vapor deposition. The customer requirements are exacting. They had been removing organic contaminants primarily by vapor phase cleaning in perchloroethylene (PCE, tetrachloroethylene) using a vapor degreaser that met Federal NESHAP requirements. However, because perchloroethylene (PCE) is a toxic air contaminant in California, the south coast air quality management district (SCAQMD) required that open-top degreasing be discontinued as of January, 2003.

PTI had to change the cleaning process. PTI set up a team (I became part of that team) that included engineering, safety/environmental, and management. The team set goals and requirements for the new cleaning process. First, the process had to work. They had to have an effective cleaning process, one that removed organic soils and left the surface amenable to PVD and related processes. It had to be cost-effective. It had to be environmentally sound. It had to be able to be managed by workers in a safe and responsible manner. It had to be acceptable to SCAQMD. PTI analyzed the cleaning options. Not cleaning was not an option; the coatings would not adhere acceptably. They could not change the product of the soil; as a job shop, their customers controlled those factors. Based on testing over the years, they knew they could use aqueous processes for some but not all applications; while they do some aqueous cleaning, they determined that their major cleaning system could not be aqueous. "Solvent-free cleaning" would not be effective at the soil levels they encountered. Impingement cleaning was not enough; aluminum oxide blast was needed, but if they did not first remove the organic soils, the blast process would just push the soil around. They could use a VOC-exempt solvent or one with low levels of VOC additives, which provided limited options. They could apply to SCAQMD for an exception or adopt alternative approved controls; both options were deemed to be difficult and costly. They could use a classic chlorinate solvent in an airless system or a solvent containing VOCs; this would at least provide cleaning characteristics that they were accustomed to. They could outsource the cleaning, but they wanted to retain the control of performance and control over their process flow. They could move the plant, but they risked losing at least some of their educated employees.

The team educated themselves on the options relative to their understandable concerns about process change. The solvent had to be effective in removing heavy oils and mixed soils, many of them applied by the customer. There had to be good materials compatibility with a wide range of metal substrates. They were concerned about changes in the product line and in end-use requirements. There are customer requirements and constraints. There are product reliability issues. They determined that there is no perfect cleaning agent, that no approach works for everything, and that all cleaning agents have regulatory baggage in terms of impact on the environment and on workers (worker exposure to the chemical and flammability). It was decided that the general approach would be solvent cleaning, because new cleaning equipment is costly. They determined that the solvent approach was known to give reliable performance over a broad range of soils; they knew it was accepted by their customers.

Initially, they evaluated HCFC 225. It is VOC exempt, and they could avoid the cost and process constraints of an airless system (please see Chapter 24, Sahbari and Toofan). The phaseout date of 2015 seemed far away at the start of the project. However, while HCFC 225 showed reasonable performance in immersion cleaning with ultrasonics, the solvent was not sufficiently aggressive in vapor phase cleaning. Immersion with ultrasonics was not practical for some of the very large parts; a large capital investment would be required and, given the physical characteristics, significant solvent loss would be expected. HCFC 225 is available with additives, but most blends have a VOC level that would result in their not being VOC compliant.

PTI looked at solvent and equipment choices. They needed an aggressive, stable solvent with a wide solvency range. The primary choices were PCE (their ongoing solvent of choice, the known quantity) and acetone, a VOC-exempt solvent. PCE would require an airless system. Acetone is flammable; they needed to investigate potential materials compatibility issues. They looked at five vendors supplying airless (vacuum) vapor degreasers and one vendor supplying low-flash-point systems.

They did initial evaluations with equipment vendors. They invited vendors to the plant, interviewed them, and evaluated how well the vendors appeared to understand process issues and regulatory issues. They received initial bids; timeliness and responsiveness counted. They looked at the details of equipment design and vendor experience with similar processes. They looked at cost estimates, and those estimates showed over $100,000 in variability. Vendors were asked to provide customer references; we checked those references. Most important, since SCAQMD had to accept the process, the team assessed vendor understanding of SCAQMD issues and willingness to work with SCAQMD. A few seemed reluctant to work with the regulators or, surprisingly, refused to do so. Vendors who refused to work collaboratively with the regulatory agency were eliminated from consideration. We also worked with people at SCAQMD to assess their acceptance of specific iterations of airless/vacuum technology.

The details of the cleaning equipment can make or break the process; and two points were deemed crucial in this application. The first was the use of stainless steel construction as opposed to carbon steel. For airless systems, some were convinced that carbon steel was adequate. Theoretically, if there is no air, no moisture, corrosion will be forestalled. In actual practice, based on performance of many systems in the field, we decided that stainless steel construction was a wise investment. The second was the importance of flexibility; we insisted on a system that could be readily adapted to use a number of different solvents. Given the regulatory scrutiny and constraints on using PCE, the system would have to use trichloroethylene (TCE), methylene chloride, *n*-propyl bromide (*n*PB), and azeotropic blends of HFCs or HFEs with high levels of *trans*-1,2-dichloroethylene. Materials compatibility of the various solvents and blends are different; in this application, while compatibility with metals had to be considered, the primary issue was compatibility of plastics and elastomers used in seals. Because some of the azeotropes wet very effectively, they sort of "creep" out of cleaning systems, so pump design was also an issue.

Asking for customer references is important; following up on those references is even more important. We contacted references via telephone. In some cases, the equipment and process bore no relation to the proposed PTI cleaning process. While the vendor(s) had supplied excellent aqueous cleaning equipment and/or standard vapor degreasers, the PTI project was their initial foray into the world of airless/vacuum systems. We asked about durability, initial training, ongoing support (including availability of replacement components for the system), and about the speed and rapidity with which the supplier responded to problems. We asked how often the system needed to be repaired. The answers were illuminating. We conducted site visits wherever possible. We looked at the cleaning equipment to see how well it held up after leaving the showroom floor. We checked the design of the robotics and the design of the software interface. Any laments of sobbing technicians were carefully considered. Always ask for references, and always check the references.

After weeding out equipment designs that did not seem promising, a selection of representative parts were tested in the cleaning equipment at the vendors' facilities. The parts were soiled with representative soils at expected contamination levels. The solvents evaluated were PCE, TCE, and acetone. For each combination of parts and soils, a control or benchmark wash shipped with the parts. The control was a part that had been contaminated with soils to be tested and then cleaned with PCE using the then-current open top degreaser. The control was used to provide clues to the equipment manufacturer as to what a clean part should look like. The control part would be shipped back to PTI along with the vendor-cleaned parts to correct for artifactual contamination. In this case, some inorganic contamination was acceptable, but the parts had to be thoroughly degreased. Replicates were tested where possible. In this situation, actual hardware was used. The parts were shipped in clean aluminum foil rather than plastic to avoid artifactual contamination with plasticizers. We spoke with the engineer in each vendor test lab prior to shipping samples. The required testing protocol was provided to the analyst in writing, after confirming that the test could be run in a timely manner without wearing out the welcome of the test facility. Clean parts were examined by the vendor and then returned for visual examination for the absence of interfering soils, signs of materials compatibility issues, and corrosion.

Responses of the vendor test laboratories differed, and these responses weighed into the decision. They were taken as an indication of expected vendor support after the sale of the equipment.

Even with a clear go-ahead from the lab personnel at the equipment vendors, response time ranged from the expected weeks to the unacceptable months. One equipment vendor re-cleaned the control. In some cases, immersion with ultrasonic cleaning was attempted for large parts; this was not acceptable because the capital equipment costs would have increased significantly. Finally, some test reports were more complete than others. A test report consisting of pages of boiler plate ending with "it's clean, buy our equipment" was not considered impressive. Acetone was eliminated from consideration primarily because of materials compatibility problems. While it had been used successfully in wipe cleaning at ambient temperature, in a heated system (one designed specifically for low-flash-point solvents), some product became pitted, and visible materials compatibility with magnesium (a visible "bloom") was observed. There was also the concern with process consistency because acetone and water are miscible, so the composition of the cleaning agent could vary. Chlorinated solvents can break down in the presence of water to form undesirable acids. However, the solubility of water in chlorinated solvents is very low, and chlorinated solvents for vapor degreasing are formulated with stabilizers.

In addition to the considerations indicated above, costs and warranty were considered, and the safety/environmental advisor checked things out with a fine-tooth comb. In fact, a certified industrial hygienist was an active member of the process change team. Some considerations included worker exposure to the solvent during equipment setup, routine operation (very low, because it is an airless system), and maintenance. This includes potential exposure during solvent addition, cleaning of solvent sumps, and filter changes. There were also issues of physical safety including electrical, ergonomic, and the mechanics of parts handling; some of the parts are very large. Waste management was addressed. Because a steam boiler was to be used, appropriate employee training was needed.

The matrices in the decision-making process were complex. In the end, an airless system using PCE was selected. The system is primarily vapor and spray but also contains a small ultrasonic tank for immersion cleaning of smaller, complex parts. However, acceptance was contingent on meeting the regulatory requirements. The PTI process change team, a representative of the equipment vendor, and representatives of the SCAQMD met to discuss the specifics of equipment design, to agree on the use of several alternative solvents, and to develop and agree on a permitting strategy, including allowable solvent emissions for each solvent under consideration. As it turned out, restrictions on VOCs, even VOCs that are generally considered to be relatively safe for surrounding neighbors, are subject to extremely stringent emissions allowances. Therefore, a vacuum/airless system was preferable to a low-flash-point system in terms of solvent options. For all options, a final carbon trap or filter is required to trap any residual solvent, and the filter had to be permitted. The permits provide for use of not only PCE but also TCE, *n*PB, and an azeotrope consisting of HFE blended with high levels of *trans*-1,2-dichloroethylene.

Was equipment setup smooth and trouble free? Absolutely not. The vendor was compelled to modify the equipment design to reinforce the structure. The heating system did not initially mesh with the cleaning equipment. The ultrasonic tank was not correctly affixed, and it had to be replaced. Employee education/training was critical. This was a new process, and, in contrast with an older vapor degreaser, operating vacuum system requires that you tell it exactly what you want it to do; the cleaning technician could not make the changes on the spot. Of course, this requirement resulted in very well-defined, consistent processes.

The cleaning processes required refinement, and the adjustment of the ultrasonic system was critical (Kanegsberg, 2008c). Even after even after the ultrasonic tank was reinstalled, cavitation, as measured by the aluminum foil test, was not adequate (Figures 15.2 through 15.4). The reason proved to be that, while a so-called vacuum system does not really pull a vacuum, the pressure in the cleaning chamber is reduced; PCE was too close to the boiling point at the reduced pressure. Reducing the temperature slightly still provided adequate solvency, and, at the same time, cavitation was achieved. The net effect was that process flow improved because parts could be handled with less cooldown time.

FIGURE 15.2   Ultrasonic cavitation erosion pattern of aluminum foil in PCE, at atmospheric pressure and ambient temperature, indicating effective cavitation.

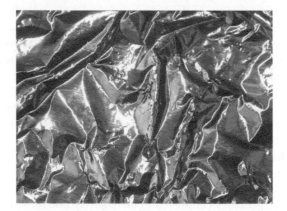

FIGURE 15.3   Ultrasonic cavitation erosion pattern of aluminum foil in PCE, reduced pressure, 218°F, indicating suboptimal cavitation.

FIGURE 15.4   Ultrasonic cavitation erosion pattern of aluminum foil in PCE, reduced pressure, 190°F, indicating effective cavitation.

TABLE 15.2 Annual PCE Emissions to Air in an
Airless Cleaning System with Carbon Trap

Cleaning System	Timeframe	Emissions PCE in Pounds (Avg.)
Open-top degreaser, NESHAP Compliant	2001, 2002	1500
Airless system	2003–2009	<1 (est. 1 oz)

There were additional costs, and PTI had budgeted for extra costs, including the engineering costs associated with process development, costs associated with delivery, facilities modification, and other permitting costs.

The outcome was that the process is now routine, and it has been in place for over 7 years. PTI achieved reliable critical cleaning in a stringent regulatory environment. As indicated in Table 15.2, solvent losses to the environment are miniscule (Kanegsberg, 2009c). PTI achieved a solution that is well suited to their customer requirements; they achieved a degree of process and permitting flexibility, so, even if one solvent came under severe regulatory distress, others could be used.

Inertial Navigation Systems: Complexity and the Team Approach

Litton Guidance and Control Systems Division (now part of Northrop) has conducted several modifications in processes for a beryllium-based instrument. As with many such applications, process modification was required for environmental/regulatory considerations, notably the phaseout of ODC's. Process modification began in the mid-1980s. Because new cleaning agents and processes have been developed, process modification has been multistep and is ongoing. A particular application was modified first by adopting a co-solvent system (solvent cleaning followed by rinsing in another solvent), and then by evaluating a solvent. The study is provided to show the logic of the approach used in a given situation, not to indicate that this is the desirable solution for all aerospace-related applications.

Navigation systems consist of gyroscopes and/or accelerometers along with the surrounding electronics. While cleaning problems associated with electronics may seem daunting, cleaning is a much more complex and diverse problem for high-accuracy instruments than for typical electronics assemblies. In electronics assembly, there are relatively limited materials of construction and configurations, and there are widely accepted industry standards.

Fluxes, including traditional fluxes, are used in inertial navigation systems. However, the build process and end-use requirements make qualification of the cleaning process more complex and exacting (Kanegsberg, 1993). Definition and control of residue, residue of soil, and of the cleaning agent itself, is a major issue. The RMA flux was known to produce an inert residue; very small amounts of this residue, should they occur, would not be acceptable. While low-solids fluxes have been implemented in some hand-cleaning operations in other high-precision builds, process control and reliability testing were judged impractical for this application. One reason is that in some cases, the sealed system is in an atmosphere of a poly-halogenated flotation fluid; as such, there must be no residue of cleaning agent. Use of water is a controversial issue in this and in some other high-precision applications. Many of the components are water sensitive and difficult to dry, so aqueous cleaning is unacceptable at many stages.

Even under clean room conditions, building precise instrumentation inherently generates testing, rebuild, and, doing so, generates an array of soils including

- Greases
- Oils
- RMA flux

- Water-soluble flux
- Fingerprints
- Particles
- Poly-halogenated flotation fluids

In assembling classic gyroscopes and accelerometers, one finds a much larger range of materials of construction than are found in the typical electronics assembly. For inertial navigation systems, an array of materials of construction is used in instrument build. An entire list of materials of construction would require half a dozen pages. Some representative materials of construction are indicated in Table 15.3, along with potential cleaning and contamination problems. It is important to note that, for Litton's applications, it is not enough to consider each individual material separately. The entire assembly or subassembly must be evaluated to avoid potential issues of cleaning agent residue, galvanic interaction, outgassing, and product deformation. For this reason, the feedback of experienced instrument assemblers is crucial.

The other issue is determining how "clean is clean," or "how clean is clean enough." When the product may be expected to perform continuously and reliably for a quarter of a century, addressing these ultimately unanswerable questions becomes increasingly important. Whereas in the electronics world, there are established cleaning standards, for precision cleaning, the goals are often pragmatic, based on expected use of the product; a combination of analytical testing of residue with performance is often used. The goal is to minimize contamination and residue to its lowest level. Those actually using inertial navigation systems, particularly aircraft pilots, have been enthusiastic support for aggressive contamination control.

Process modification was multistep and involved a team approach by in-house engineering and production staff as well as ongoing collaboration with manufacturers of cleaning agents and cleaning equipment. Initially, an array of cleaning agents and cleaning sequences were evaluated in-house. Some more promising cleaning sequences including various hydrocarbon and D-limonene (orange terpene) blends followed by self-rinsing or rinsing in IPA or perfluorinated materials were then tested using the facilities and personnel of a major cleaning agent manufacturer, in cooperation with a number of other cleaning agent manufacturers. Results were evaluated visually and by analytical techniques (FTIR and ESCA). Some in-house rare gas analysis (RGA) analysis was also incorporated in the study. Surprisingly, given the stringent process requirements, at least half a dozen promising cleaning sequences were found. The most promising cleaning sequences were then tested against the most exacting critics of all: experienced production assemblers.

TABLE 15.3   Representative Materials of Construction of Gyroscopes and the Process Concerns

Materials of Construction	Primary Concerns
1. Beryllium, Aluminum	Oxidation, erosion, cleaning agent residue
2. Stainless steel, other ferrous metals	Corrosion erosion, cleaning agent residue
3. Complex assemblies containing 1 and/or, 2, and combinations of gold, magnesium, tungsten carbide, copper, Hy-MU-80 (nickel alloy)	Same as 1,2, plus galvanic interactions
4. Sapphire, specialized glasses	Cleaning agent residue, subtle surface changes
5. Kapton, plastic coatings, coated and uncoated epoxies, composites	Solid cleaning agent residue, outgassing of vapors, softening, deformation, solubilization of material of construction
6. Flotation fluid	Becomes a soil under test, rework conditions, residue undesirable, especially if reacts with cleaning agent (nucleophilic substitution reaction, SN2, with alcohols)
7. Complex assemblies, many materials of construction (1–5) Complex soils	Same as 1–5, plus entrapment of cleaning agent; reactivity of cleaning agents with nonmetals and residue of soils; situation exacerbated in sealed systems

Based on assembler input, the cleaning sequences were refined and implemented. The approach adopted was a co-solvent system consisting of initial cleaning with a hydrocarbon blend containing various alcohols followed by two to three rinses with IPA. This new process allowed elimination of TCA cleaning; some perfluorinated material continued to be used as a final rinse to assure thorough removal of fluorolube. Overall, the number of process steps was reduced; often an 18-step process was reduced to four to six steps. Even the process introduced initially required refinement, because IPA was found to react with beryllium; intermittent residues were found. Eventually, IPA was replaced with volatile methyl siloxanes. At the time, the new processes represented the best option to replace ODCs.

The co-solvent processes, while allowing replacement of TCA, were far from optimal. The subassemblies are very complex, with close tolerances and blind holes. While the cleaning agent can be removed with careful process control, extreme and constant vigilance is required to assure that no cleaning agent residue is left. The hydrocarbon blends are costly, and some of the operators found the odor to be disagreeable. The search for a more reliable process continued.

After extensive testing, a specific *n*PB formulation was adopted to replace some of the multistep cleaning. The Abzol VG (VG) cleaning agent has been implemented for cleaning during instrument build where cleaning activity similar to TCA is required. The cleaning agent was chosen for a number of reasons including solubility parameters, low residue, liquid and vapor phase cleaning characteristics, and compatibility with materials of construction.

Operations engineers conducted the primary evaluation of the VG cleaning agent. The materials and processes group conducted some preliminary evaluations and found it to be very effective for removal of a wide range of soils, including flux, without the requirement for rinsing with another cleaning agent, and it has properties similar to 1,1,1-trichloroethane (Carter, 1999). In addition to cleaning capability and reliability, the process time was reduced by over 40% and cleaning agent usage was reduced to essentially one-third.

While *n*PB was adopted because of its aggressive solvency, which is similar to 1,1,1-trichloroethane, there is always a balance between solvency and compatibility. Beryllium metal, used in some legacy inertial navigation systems because of its relatively high ratio of strength to weight, is also a relatively reactive metal, forming beryllium oxide. To put it into perspective with more commonly used metals, beryllium can be thought of as rather temperamental aluminum. Beryllium coupons supplied by Litton were tested by the cleaning agent manufacturer. After 24 h of exposure at the boiling point, beryllium submerged in the solvent showed no sign of discoloration or tarnish. Extensive tests were conducted by the cleaning agent manufacturer (Shubkin, 1997). It should be emphasized that the results of these studies should be used to indicate the approach to be used in evaluation; no process of product is a universal solution.

The interim process eliminated Class I ODCs but was cumbersome to use. Adding a more aggressive cleaning agent with a higher evaporation rate along with reevaluating the necessity of various cleaning and rinsing steps has significantly simplified the process. Because of changing regulations and given the ongoing development of new products and cleaning methods, process modification and process improvement is often best achieved in a multistep manner.

In all of the applications described above including ultrasonic cleaning, spray and flushing systems, the measured employee inhalation exposure to *n*PB was below 10 parts per million (ppm). Area samples, used in some cases to measure the concentration of *n*PB at specific locations in the workroom to predict "worst case" potential exposures, were also below 10 ppm. The observed exposure depends on the type of cleaning, cleaning action, ventilation, and worker education and awareness. At the time, staying below 10 ppm provided a very comfortable margin. In the interim, American Conference of Industrial Hygienists (ACGIH) established a TLV of 10 ppm. While there is a common convention among industrial hygienists calling for worker exposure to be kept below half of the established permissible exposure limit (PEL) or threshold limit value (TLV), a comfortable margin could be established without additional engineering and process controls. However, in December, 2009, the California OSHA Standards Board adopted a PEL for *n*PB (generally referred to as 1-bromopropane by Cal/OSHA) of 5 ppm, with a skin notation. This is a legally enforceable standard throughout California, and it would be expected to be implemented in 2010. This would mean that if such processes were to be run in California, safety professionals would expect the level of *n*PB to be below 2.5 ppm, a challenging goal in most industrial applications.

Optics

Optics cover a range of applications from eyeglasses and contact lenses to an array of specialized sensors and components of specialized devices. The substrate is sometimes a plastic but is more often than not a specialized glass. This section deals primarily with high-precision glass optics. Some of the considerations apply to plastics. With plastics, one must also factor in issues of compatibility with cleaning agents.

Processing of optics is shrouded in mystery. There are a number of reasons. For one thing, with biomedical applications, there may be regulatory concerns with the Food and Drug Administration (FDA). In addition, competition-sensitive issues and military concerns result in secrecy, vagueness, and lack of communication. While many of these concerns are no doubt justified, the result has been an array of complex, byzantine processes that are difficult to control and troubleshoot. Because of the many fabrication steps involved, processes are often performed for traditional reasons; it is often impractical to take the time and effort to justify a particular operation. We know certain components have to be very clean, and we often perform complex laboratory analysis. In the end, it comes down to maintaining the process so that the next fabrication step can be successfully completed and that the final device operates acceptably. Optics processing inherently develops into a tightly woven sequence of processes. A supposedly minor change to one process, perhaps mandated by some regulatory requirement, may impact other parts of fabrication such that extensive modification and reevaluation is needed.

In processing optics, the substrate is first machined and grossly shaped or sliced. Various surfaces are then polished with specific slurries, and the surfaces may be etched and/or acid treated. Finally, specific coatings are applied, often by vacuum deposition. At various stages, cleaning has to occur in such a manner to remove the soils without damaging the surface that has just been carefully polished, treated, or coated. Cleaning must be accomplished without redepositing soil, adding residue of cleaning agent, depositing particles, or changing the surface in some undesirable manner.

With optics and with other sensitive materials, it should be remembered that water quality and composition can impact surface properties. Contaminants can be deposited on and react with the surface. In addition, ultrapure DI H_2O can leach materials out of surfaces and alter the surface.

A vast, often fanciful assortment of materials, which will be referred to as blocking compounds, are used to hold the substrate in place for polishing, or in order to protect certain portions of the optical system from treatment. Blocking compounds may include such materials as

- Pitch (asphaltum)
- Soft wax
- Bees wax
- Rosins
- Nail polish
- Epoxies
- Thermoplastics
- Mixtures of plastics, rosin, and waxes of variable composition

The materials are chosen for a variety of reasons, often relating to the forces needed for polishing. In some cases, optics may be set into a 3 mm or more thick base of an organic-based blocking compound on a larger base plate. This mosaic-like object is then polished in a specific slurry. At the end, the blocking compound and polishing compound must be removed. Classically, 1,1,1-trichloroethane (TCA) was used for many optics operations. The optics could be soaked in TCA, either at ambient or at elevated temperature; the blocking compound would dissolve, leaving a light residue. Final cleaning would then be accomplished by vapor phase cleaning. With the advent of the ODC problem and the TCA phaseout, a number of other chemicals and processes have been used with varying success. Issues involve

- Inadequate solvency
- Cleaning agent residue
- Scratching of the substrate
- Etching of the substrate

Deblocking Agents

Deblocking processes are used in a number of aerospace, aeronautics, and military applications (please refer Chapter 27, Breault, Soma, and Fouts). Some blocking agents currently in use are summarized in Table 15.4. Details regarding some deblocking processes follow. In addition, it might be pointed out that, in general, particularly for rinsing and final cleaning of optics, there is no substitute for high-quality, low-particulate cleaning agents. Where possible, electronics grade, HPLC grade, or the equivalent should be used. Certification of desired properties by the supplier and some assurance that the formulation will not be changed without notification are crucial.

Chlorinated and Brominated Solvents

PCE and other chlorinated and brominated solvents have been tested, with varying success. PCE has an inherent limitation in that the higher boiling point may produce bubbling of the blocking compound. The bubbling, boiling action may jar the optics, effectively undoing the previous polishing process. *n*PB has also been used effectively in some applications, and it has the advantage of allowing for final cleaning in the vapor phase. Many of the chlorinated solvents are covered by a Federal National Emissions Standard for Hazardous Air Pollutants (NESHAP), so that careful record keeping and reporting are required. *n*PB is acceptable for use in vapor degreasing by the U.S. EPA SNAP group. It is a VOC, and, the low allowable worker exposure level poses challenges, especially in California.

Acetone, Alcohols, and Low-Flash-Point Blends

Acetone, IPA, and methyl alcohol are widely used in fabrication of optics. Acetone has a certain appeal, particularly in areas of poor air quality, in that is VOC exempt. In addition, many low-flash-point solvents are of low cost and can be obtained in high quality. I cannot stress highly enough that low flash-point solvents can be used safely and with confidence in appropriately designed equipment. There have been semi-disastrous attempts to heat low-flash-point solvents or to adapt existing equipment in-house. Even though equipment for low-flash-point solvents is costly, the decrease in solvent usage and in disposal costs may result in a rapid payback period.

Hot Water or Cold Shock

Either approach depends on physical rather than chemical removal of the blocking agent. In the case of hot water, one is simply melting the wax or rosin. While heat is important in many cleaning processes both to boost solvency and to promote melting, one must be aware of the inherent limitations of heat in the absence of solvency. Cold shock, either by freezing the optics or by using a CO_2 snow gun has had limited success in releasing the optics via changes in thermal expansion or by physically cracking the blocking material. More often than not, however, particularly in production situations, the process has resulted in damaged substrate.

Aqueous/Surfactant

Water-based cleaning is becoming increasingly popular because of safety and environmental concerns. Adopting water-based cleaning agents often requires a change in the blocking agent. Heavy pitches and very hard waxes are often difficult if not impossible to remove with water-based cleaning agents. In addition, aqueous/surfactant blends require increased cleaning action such as mechanical agitation

TABLE 15.4   Summary: Some Considerations and Examples of Deblocking Compounds

Deblocking Agent	General Characteristics	Performance Concerns	Regulatory/Safety Issues
Chlorinated solvents	Liquid and vapor phase cleaning Self-rinsing, Aggressive solvency Rapid process	High boiling point (PCE) may be result in components damage Must be tested with individual application to assure no surface changes	Engineering controls required for personnel protection Air toxics Solvent containment, reporting required
*n*PB	Liquid and vapor phase cleaning Similar performance to TCA	Must be tested with individual application to assure no surface changes	VOC Low ODP Engineering controls required for personnel protection Safety/employee
D-Limonene (orange terpene)	Good solvency for many blocking compounds Liquid phase cleaning Some formulations can be rinse with water	Must be tested with individual application to assure no surface changes Leaves significant residue for most high-end applications Residue from additives may interfere with subsequent applications Rinsing with solvent or water required Some blends can oxidize	VOC Distinct odor can be an issue
Esters (e.g.,: ethyl lactate, di-basic esters) Alone or in blends	Good solvency for many blocking compounds Liquid phase cleaning Formulations available for aqueous rinsing	Must be rinsed	VOC Distinct odor can be an issue
NMP	Moderate to good solvency for many blocking compounds Liquid phase cleaning Often less prone to leaving residue than D-limonene or ester blends	Must be rinsed Often somewhat longer processing time	VOC
Proprietary blends, water soluble organics	Moderate to good solvency for many blocking compounds	Require rinsing	May contain significant amounts of VOCs Appropriate mixed waste stream handling needed
Acetone	Good to aggressive solvency Can be used as heated liquid vapor phase (with proper equipment) Useful in combination with other deblocking agents, including aqueous	Exceedingly rapid evaporation	VOC-exempt Not an air toxic Very low flash point If heated, must be used in specially designed equipment
Alcohols (isopropyl alcohol, methyl alcohol)	Good to aggressive solvency Cold cleaning Can be used as heated liquid vapor phase (with proper equipment) Useful in combination with other deblocking agents, including aqueous	Solvency often limited for blocking agents of interest	VOCs Low flash point If heated, must be used in specially designed equipment
Hot water	Low solvency, acts by melting Useful with soft wax Rinse agent	Limited solvency May require multiple rinsing Must control water quality	Disposal of waste streams

TABLE 15.4 (continued) Summary: Some Considerations and Examples of Deblocking Compounds

Deblocking Agent	General Characteristics	Performance Concerns	Regulatory/Safety Issues
Cold shock	Nonchemical	May damage substrate	
Aqueous/surfactant	Moderate to good solvency for many blocking compounds Liquid phase cleaning	May require multiple rinsing Additives may produce subtle surface changes, not immediately evident	Disposal of waste streams Additive packages may have significant VOCs or require special handling
Solvent cleaning sequences (co-solvent, bi-solvent, includes HFCs, HFEs)	Multiple solvency properties Often treated as low VOC processes	New cleaning equipment may be required Equipment may be restricted in terms of utilizable solvents Thorough removal of cleaning solution required	Process control essential Waste management may be an issue

or spray, ultrasonics, additional heat, and careful rinsing. Where agitation is used, it is important to protect the substrate.

Rather than struggle through the evaluation of succession of aqueous cleaning agents and processes, it is often more productive to change the blocking agent to one more readily removed with water-based cleaners.

Co-Solvent and Bi-Solvent Processes

An assortment of such processes are available; some include biobased cleaning agents like D-limonene or methyl soyate that are mixed with a displacement/rinsing agent such as an HFC or HFE. The displacement or rinsing agent generally does little to remove the blocking agent; the importance is in removing the original cleaning agent; in some cases, it acts to suppress the flammability properties of the cleaning agent. Depending on the byzantine specifics of the local regulatory situation, some of these processes have found favor in areas of poor air quality. Such processes often require purchase of specific, dedicated equipment and use of a limited array of cleaning and displacement chemistries. This may limit process flexibility.

Effect of Additives

Glass is soft. In modifying older, solvent-based processes, one might be aware that additives both in aqueous cleaning agents and water-soluble organics have the potential to produce subtle surface changes. These changes may not be immediately apparent but may manifest as some other problem many steps later in the process.

Additional Issues and Suggestions

Cleaning versus Surface Modification

Many optics preparation processes are old and steeped in the rich tradition of the company. (Translation: no one knows why they work, but if we attempt to change anything, the process fails, and YOU are blamed). Part of the problem relates to subtle changes in the surface of the substrate. This means that cleaning problems may become entangled with problems of surface preparation. For example, optics are often cleaned in mixtures of very strong acids and salts. In one instance, in an attempt to reduce usage of a chromic acid mixture, solvents were tested for cleaning. The solvents were judged unacceptable on the grounds of a change in contact angle measurement. One might suspect, however, that the acid was modifying the surface and so changing the contact angle. The problem was not one of cleaning but probably rather one of surface modification.

Troubleshooting

Optics processing is notorious for intermittent, inexplicable problems. With so many steps in the fabrication process, precise control of every factor in every operation is difficult. If you are in the oh, so fortunate position to be put in charge of troubleshooting an optics fabrication process, it is often difficult to determine the source of the problem. You have to become a detective. Let us suppose that a coating process is suddenly no longer successful. One would, of course, look at the coating operation itself and factors in the immediate vicinity of the suspect problem including

- Analytical testing to determine the nature of the contaminant (coated surface, surface prior to coating)
- Coating equipment operation
- Personnel changes or dissatisfaction
- Chemical changes
- Water purity, if applicable
- Clean room conditions
- Surface preparation
- Storage conditions prior to coating

If careful examination of these factors does not yield the source of the problem, one needs to venture further afield, considering factors mentioned above as they apply to earlier operations and also including

- Earlier cleaning, polishing, and etching steps
- Changes in any chemical
- Change in any process
- Equipment condition
- Recent equipment repair or overhaul
- New lot or new source of substrate
- Modification in initial slicing or polishing operations

Above all, see if you can turn the investigation into a team effort, involving all of the people who might be influenced or who might be the source of the problem. It can be a real challenge to keep communication open and maintain a no-fault atmosphere, even within the facility. Of course, it is always possible that someone is purposely breaking clean room discipline (I find the term *discipline* to be unproductive and unprofessional). It is more likely that someone or several people have made changes that they honestly felt were the equivalent of the status quo, or that they just weren't aware of. It can be a real challenge to persuade people to communicate. Many people who become involved in high-technology, clean room operations like optics fabrication do so because they would prefer to work quietly, out of the public eye. Some new idea may have been quietly implemented which, while wonderful in isolation, does not quite mesh with the whole operation. The important thing is to keep the atmosphere open, productive, and as free as possible from finger pointing. This is a real art; it is an art I myself am still developing. Side benefits include more up-to-date, well-documented processes in that production workers may confide what they are actually doing. They may also have some wonderful ideas for improvement. Quality circles and teams may seem a generation out of date, but honest communication and respect for the ideas of others can really pay off.

Teaming ideally involves the vendor community as well. If subcontractors are involved, you need to find out as tactfully as possible whether or not they have made changes. The same holds for cleaning agent and cleaning equipment vendors and for any outsourced items such as maintenance of the clean room or of the water system.

Maintaining and Improving the Process

With optics fabrication, as with all multistep high-precision processes, maintaining the process is crucial. This involves

- Using the highest quality (chemicals, equipment, water, disposables, substrates)
- Documentation of chemical quality through on-site testing and/or vendor certification
- Control of disposables, storage conditions
- Maintaining records of product and chemical lots
- Documentation of water quality
- Clean room control
- General thorough record keeping
- Employee education
- Value the workers, listen to their ideas

In essence: invest in quality, trust nothing, strive for consistency, but don't become a tyrant. Someday, we'll all get this right!

Biomedical Applications

Sterilization is not the same as cleaning. Dirt that has undergone terminal sterilization is still dirt. A product with sterile contaminants will not perform acceptably; this includes biomedical devices. Implantable medical devices face problems from soils, even if there are no apparent biocompatibility issues. Soil can interfere with mechanical functioning of the product. Minute amounts of soil can change surface quality and surface properties. Minute amounts of soils can leach off of the device and damage the host (the person who has received the implantable device). The importance of surface preparation has long been recognized, "All prosthetic implants should be scrupulously free of contaminating overlayers at the instant of their biological placement" (Meyer, 1988). One of the biggest positive changes we have seen is that the prevalent attitude has changed from "we worry about sterilization not about cleaning" to "we are concerned with leachable residue" (translation: leachable residue is soil).

Cleaning and manufacturing for biomedical applications, in the context of this discussion, includes

- Implants (metals and plastics) for long-term use in humans
- Catheters and other devices for use in animal experimentation
- Surgical instrumentation, which may include long and extremely fine-bore tubing
- Assorted plastics and metal disposable and non-disposable materials
- Subassemblies for clinical instrumentation, including automated equipment for clinical laboratory testing (blood tests, etc.)

The biomedical community is justifiably concerned about quality processes, but each group tends to work in isolation and secrecy that surpasses that of many military applications. There is an overwhelming reluctance to discuss either successes or difficulties in a public forum.

A number of factors promote secrecy. It is difficult and costly to bring a new product to the marketplace, or to change the manufacturing process for an existing product. The manufacturer faces not just the hurdles of coping with strict and often conflicting local and national air quality, water quality, and employee safety regulations; there are also the issues of biocompatibility, pyrogens, bacterial growth, and other harmful residues.

Gaining FDA approval is often a major hurdle. Some companies reportedly fear that emerging FDA guidelines will render their complex, costly testing inadequate. Therefore, even groups doing careful, thoughtful testing may be fearful of calling attention to themselves by the FDA or by any other regulatory agency.

The FDA does not regulate or specify cleaning per se. However, the FDA does regulate the effect of the manufacturing process on the finished device. This would then be part of the 510k premarket notification. One FDA spokesperson notes that while the 510k format is standard, because there must be tens of thousands of different devices, across-the-board testing requirements would be difficult to define.

In designing any new medical device, it is a reasonable presumption that surface attributes, including surface cleanliness, impacts performance (Kanegsberg, 2006a). Device manufacturers, regulatory agencies like the FDA, and clinicians are concerned with the cleaning process, both in terms of the impact of residue on the host and in terms of the impact of cleaning, both positive and negative, on surface attributes. Problems can occur if cleaning is not considered to be integral to the build process. There can be changes during prototype and production, during plant relocation, and in response to safety/environmental regulatory constraints. This can be a major issue for devices involving complex, far-flung supply chains. It must be reiterated that while cleaning, contamination control, and surface quality or properties are technically separate issues, in practice, they must be considered holistically.

In some instances, medical devices are manufactured using cleaning chemistries that have been commonly accepted in hospitals, notably IPA and aqueous cleaning solutions. As devices become smaller and more complex, with tight spaces and blind holes, and as the recipients of the devices live longer, more exacting performance is required. IPA and aqueous chemistries are excellent for polar and mixed soils. Both have limited wetting capabilities. Aqueous cleaning agents must be rinsed to remove leachable residue. In fact, effective rinsing and drying are important for most cleaning agents used in medical devices. Leachable residue can include outgassing. Increasingly, device manufacturers consider fine-tuning the cleaning sequences to include cleaning agents with properties that target the specific soils of concern.

The cleaning process impacts the surface quality, and more is not necessarily better. There is a perhaps understandable tendency among medical device manufacturers to do repeated ultrasonic cleaning so as to extract leachable residue from the device. However, ultrasonic cleaning works by erosion. In fact, the classic test for functionality of an ultrasonic process is via cavitation erosion of aluminum foil after it is subject to the ultrasonic process. While aluminum is much more susceptible to ultrasonic damage than are some classic implantable devices, there is the potential to extract parts of the device from the device and/or to change important surface properties in ways we did not intend to. Cavitation damage has been observed on silicon wafers by electron microscopy (Busnaina, 1991). Documented ultrasonic damage includes delamination of gold coating from a quartz crystal used in a quartz crystal microbalance. While we cannot discuss specific medical device applications, judicious extrapolation is reasonable. Control the process, including the use of ultrasonic cleaning (Kanegsberg, 2006b).

Some issues of importance to the biomedical community in modifying their cleaning process are indicated in Table 15.5. Many of these factors are also important to other critical cleaning applications.

Case Studies: Medical Applications*

Because companies are fearful of incurring increased inspection by either the competition or by government agencies, they keep both their successes and failures to themselves. In fact, some individuals declined to discuss issues involved in manufacturing biomedical instrumentation, even anonymously, on the grounds that their phraseology would be recognized. The author would like to thank those colleagues, clients, and associates, throughout the United States, who shared their concerns and success stories. It should be noted that in the following examples and case studies, the names of companies and of individuals have been pointedly omitted to assure candor and to avoid competition-sensitive issues (Kanegsberg, 1997).

---

* This subsection is based on a paper presented at CleanTech '97, a conference sponsored by the Cleaning Technology Group, Witter Publishing Corp., April 1997, Cincinnati, OH.

TABLE 15.5   Examples of Potential Contamination Problems in Biomedical Applications

Potential Contamination	Potential Problems	Possible Solutions
Soils (greases, oils, polishing compounds)	Toxicity, biocompatibility Increase in bioburden Blockages Mechanical malfunctions	More aggressive cleaning agent Enhance cleaning action (turbulation, ultrasonics) Monitor soil loading
Cleaning agent residue from solvating agent or aqueous/ saponifier	Toxicity Variable changes in bioburden Lot to lot variability Mechanical malfunctions	Avoid cleaning agents with significant nonvolatile residue Provide more rinsing than is typically used Monitor the process to avoid excess carryover of cleaning agent

Solvent or rinse agent residue	Variable residue on component
	Outgassing leading to
	    Toxicity
	    Damage to components in sealed devices
	Materials compatibility issues
		Choose a solvent not readily adsorbed by materials of construction
		Provide adequate drying

Microbial contamination	Unsuitability for use in sterile situations Mechanical breakdown of clinical equipment	Choose a more aggressive solvent More aggressive cleaning, longer cleaning Clean at a higher temperature

One company, which produces a variety of plastic and metal components for biomedical applications, switched from TCA to PCE in a contained solvent system. The selection process involved many months of comparison studies and testing, including bioburden testing. Determination of bioburden involves contaminating the product with microorganisms, cleaning, and then testing to determine the amount of remaining microorganisms relative to the control or reference cleaning method.

For this particular group, given the configuration and spacing of components, entrapment of solvent is an issue. While cleaning in IPA or with IPA azeotropes has been used for many biomedical applications; in this case, low-flash-point solvents were not removed rapidly or completely enough to allow adoption of a flammable liquid/vapor phase system. Even using high-temperature cleaning with PCE, specially designed fixtures were crucial to assuring adequate cleaning and drying.

Aqueous cleaning has always been a popular option, particularly where compatibility issues make solvent cleaning unwise. Aqueous cleaning can be successfully adopted where careful attention is paid to process design (Cala and Winston, 1996). One company successfully implemented both solvent cleaning to replace TCA and aqueous cleaning to replace CFC 113. The group found that initially, it was more difficult to ramp up the new aqueous process because blind holes in some of the disposable plastics components made drying a limiting part of the process. Their drying technique had to be improved.

Another group makes sealed implantables for relatively long-term application in humans. The device must of necessity be extremely compact, so all components are very closely spaced. They use surface mount electronics, and have been replacing CFC 113 for cleaning rosin flux. They also have an array of bonding problems, including bonding of plastics to metals. If there is any flaw in surface preparation, subsequent bonding may not be successful. Perhaps even more potentially distressing, any flaw in the surface can result in subsequent adhesive failure in a warm, saline environment. One of the engineers notes that "like everyone else, we tried orange terpene alone, we weren't successful."

This maker of implantables has adopted cold cleaning with IPA cleaning, and aqueous cleaning prior to coating as an interim measure. IPA alone is not efficient for removing most rosin fluxes. The process is admittedly inefficient and labor intensive. Their cleanliness standards are

- Ionograph measurements
- Coating adhesion at a subsequent step (i.e., the parylene sticks)

The group would like to adopt aqueous cleaning for environmental reasons. They expect that some solvent usage may be required. However, they are concerned about the capital outlay and chemical handling issues associated with use of low-flash-point solvents.

The group wished to replace cleaning processes utilizing HCFC-141b and TCE. A manufacturer of titanium bone replacement implants and ultrahigh-molecular-weight polyethylene (UHMWPE) cartilage replacement implants, found that a stabilized *n*PB formulation met the company cleaning requirements, including

- Acceptable removal of buffing compound
- Low concentration of retained solvent in the UHMWPE
- Reduction of bacterial spore count in the UHMWPE by 50% or greater

Retained solvent was measured by gas chromatography/mass spectrometry. Results for *n*PB were compared with HCFC 141b and TCE. After 24 h of drying at ambient temperature with good air-flow, 30 ppm of *n*PB were detected. By comparison, the level of retained solvent was over five times as high for HCFC 141b and over 15 times as high for PCE. After 96 h of drying, the level of retained *n*PB decreased to 3.7 ppm. Measurements made at 106 h were 4.4 ppm for HCFC 141b and 27 ppm for PCE. Bioburden studies indicated spore count reduction of 68% and 73% after cleaning with *n*PB. Details of these studies are reported in another paper (Shubkin, 1997).

Yet another group has had good success in using an HFC blend to clean plastic and metal surgical devices. HFC has low surface tension and provides increased wettability along with compatibility with a wide range of materials. The blend was chosen because it has a somewhat higher polarity, and therefore a greater solvency range than the HFC alone. The engineer in charge of the project notes that one of the major challenges is that cleanliness is not defined for the medical industry; FDA has no information as to how clean is clean. The process had to be developed and justified to company management based on what would be logical standards considering the end-use application. In this case, residual particulate material of greater than 5 μm is considered crucial. HFC's (as well as HFE's and many volatile methyl siloxane (VMS)) are particularly effective in removing particles.

Ongoing Concerns and Challenges in Device Manufacture

Safety Note: Ultrasonics and Flammables

It is not appropriate to fill an ultrasonic tank with and/or heat IPA or acetone. While we see the problem in many fields, there seems to be a particular trend in this practice among those developing new medical devices. When the process is scaled-up, often after the method has been validated and assuming that you have not managed to set fire to the facility, it becomes technically onerous and costly to purchase large-scale production equipment.

Regulatory Constraints

Safety and environmental regulatory concerns limit process choices. This can impact both final assembly and the supply chain. A process may be deemed noncritical, for example, cleaning after machining. A low VOC process may be substituted in the interest of environmental compliance on the grounds that cleaning after machining could not possibly be critical. If that low VOC process leaves a residue and if that residue is adequately addressed at other steps, an undesirable leachable residue may result. This residue could compromise product performance, resulting in performance and liability issues. Cleaning after machining could be the critical process.

There are also issues with regulating product safety. There is an approach among regulatory agencies of requiring device manufacturers to "prove" that no residue exists. It is impossible to achieve zero residue. In addition, agencies may issue warning letters about a given chemical or set of chemicals where problematic residue has been found. Just as with environmental regulatory agencies, there are campaigns against chemical culprits, and these campaigns may be conducted on a chemical by chemical basis. Many chemicals, used improperly, can leave a leachable residue or be adsorbed into plastics and then outgas. Given the number of materials of construction, the assortment of sizes and configuration of devices, and the number of cleaning steps required, one overall document mandating specific requirements for cleaning, cleanliness, and residue is impossible. It would be productive if instead guidance about the process was provided by agencies.

Validation and Pre-Validation

When we are called in to validate a cleaning process, we often find an eclectic mix of cleaning steps using a fanciful array of cleaning agents with little or no rationale regarding the selection process. We still find consumer household products in use (always a risky idea), and there may also be complex blends that have not been adequately characterized as to the potential level and toxicity of residues. In such instances, I am tempted to say something like "there, there little cleaning processes, you certainly *are* valid." Let's find a more reasonable approach.

Design the cleaning process before you validate (Kanegsberg, 2010). Pre-validation is a prudent plan and sets your company up to do an economical, defensible validation. This means evaluating cleaning agents, cleaning processes, determining likely problematic residues, obtaining cleaning agent information from vendors, and perhaps doing preliminary residue testing. The MSDS is one resource, but it does not indicate all potential residue, so you may need to work directly with the suppliers of cleaning agents to determine all of the delicious ingredients that might impact the device, signing confidentially agreements where appropriate. Explaining to regulatory agencies that you use "a proprietary blend" ought not to be adequate.

Validating processes to control viable and nonviable contamination is a challenge (Broad, 2007). Because adequate cleaning or "absence" of residue has to be demonstrated, ISO 10993-17 is a normative standard that can be used as part of the cleaning validation to identify potential contaminants and residues along with their associated risks. The decision tree at the end of the standard is particularly helpful, and, in fact, the approach can be extrapolated for use in other situations where risk analysis is required. Depending on the product, speciating the residue can be crucial. For many new devices, information about bulk residue like total organic carbon (TOC) and gravimetric analysis provide necessary but not sufficient information. Extraction can be an issue, and ASTM F2459-05 "Standard test method for extracting residue for metallic medical components and quantifying via gravimetric analysis" can be applied where appropriate.

Quantifying how clean is clean enough is difficult. If a new device can be compared with an existing, successful one, a benchmark approach can be used. Completely new designs may require a combination of extrapolation, documentation, and a logic tree. Simply showing a TOC level and asserting cleanliness ought not be sufficient.

Prototype to Type

Many advanced devices are created by small companies that are producing on a very small scale. They then have to decide whether to mass-produce the product or to license it. We continuously run into lab-scale processes that are conducted at the bench, where cleaning is accomplished in beakers, with stirring rods, or magnetic stirrers. If the anticipated scale-up is from dozens per day to millions per day, it is wise to design cleaning processes that can be adapted to large pieces of equipment. *Cleaning Agents and Systems*, the other book of the *Handbook for Critical Cleaning*, describes options in cleaning agents, cleaning equipment, and cleaning processes. As you design the process, consider what will happen on a larger scale. Whether your company produces the product or licenses it, it is unlikely that technicians will be toiling over large vats of IPA and stirring with 2 m long glass stirring rods.

Combination Devices

Combination devices combine traditional devices, pharmaceuticals, and/or biologicals. This means regulatory complexity and major challenges in design and cleaning (Kanegsberg, 2008a,d). Residue of a cleaning agent must not compromise the active pharmaceutical ingredients (API). The philosophies of device manufacturers must be meshed with the philosophies and approaches of pharmaceutical manufacturers. The languages and meanings of the communities do not always coincide. In our experience, planning and communication are essential.

Paralysis through Analysis

It is possible to become so concerned with what might go wrong that a decision is deferred, to the potential detriment of the company, the clinician, and the transplant recipient. Analytical tests are not definitive. Judgment must be exercised. If you don't, funding for product development may run out and a decision may be made by someone with less sensitivity to the problem than you or your team. Engineers, chemists, analysts: please take responsibility.

Industrial Metal Cleaning

Metal cleaning includes process issues discussed in *Handbook for Critical Cleaning*, but also specialized issues relating to anodic and cathodic cleaning and electropolishing. Well-established processes have been developed. Cleaning agents and equipment, particularly technologies geared to anodic and cathodic cleaning are very specialized. Keep in mind that electropolishing may be primarily surface modification rather than cleaning. For detailed information, the reader is referred to a classic, concise, readable, and relevant reference on the subject (Peterson, 1997). The book covers process options, design of experiment, and process monitoring, geared specifically toward the metal finishing community.

Nanotechnology

Nanotechnology has moved from a marketing buzzword to viable production of product. At one point, I erroneously thought that nanotechnology would be the ultimate solution to the need to clean. Nanotechnology may eventually result in self-cleaning surfaces and may obviate the need for traditional cleaning. However, in our experience, the manufacture of nanodevices requires not only controlled environments but also judicious implementation of cleaning processes. Planning the process, having an interdisciplinary cleaning team, and involving management and the assemblers is, if anything, even more crucial in nanodevice assembly.

Selecting and Improving the Cleaning Process

For decades, cleaning has been described as fragmented, as a large group of niche applications. In the following sections, experts in a number of fields teach valid and appropriate cleaning methods. Which one should you choose for your application? Perhaps one of the approaches can be adopted or adapted with little or no change. However, for many if not most applications, the typical approach may not work for one reason or another. I suggest you do what we do as consultants: browse through approaches and cleaning techniques used in a number of fields, then design a cleaning process that is best for your company.

We often find that two seemingly similar manufacturing facilities may adopt vastly different cleaning processes, and this holds true for critical cleaning processes ranging from hydraulic pump repair to implantable medical devices. The logic tree is company specific and includes technical, budgetary, regulatory, personality, and political considerations.

The world is demanding increasingly high standards for high-value processes. Increased miniaturization and expected longevity and reliability (e.g., in pacemakers) implies designs that will be more difficult to assemble, increasingly difficult to clean, and where residual soils can be catastrophic. We need standards: in clean rooms, chemicals, and process performance. We set standards, without necessarily knowing whether those standards are relevant, sufficient, or overkill.

Fear is probably one of the biggest deterrents to progress in precision cleaning. Fear results in unrealistic, dogmatic processes that experienced assemblers simply ignore. Fear results in ever-increasing process steps, which are not only inefficient but may actually produce contamination through excess product handling. Fear blocks communication, including communication with the work force, the engineers, with subcontractors and vendors, and with those involved in other, often seemingly unrelated applications.

Some general keys to successful process implementation include

- Testing the actual process in the proposed equipment
- Comparing results with the control method
- Providing appropriate automation
- Involving the production team (including management and assemblers) in process development and decision making
- Providing for more than one cleaning option
- Assuring thorough rinsing
- Providing for rapid drying
- Educating (not just training) the production people
- Understanding that process optimization is an ongoing issue, one of continuous improvement, with potential high production rewards (Kanegsberg, 1997)

High-value components may involve processes that are so specialized that standards may be nearly impossible to define. Certainly, understanding the performance of related processes and even seemingly unrelated product lines can be helpful in avoiding problems. In all, the best guidelines remain

- Logic
- Choice of quality cleaning agents and processes
- Documentation of ongoing processes
- An educated production force
- Good communication

References

Broad, J. and Kanegsberg, B., Minimizing viable and non-viable contamination: Standards and guidelines for medical device manufacturers. *Controlled Environments Magazine*, September 2007.

Busnaina, A., Ultrasonic cleaning of surfaces: An overview. In *Particles on Surfaces*, ed. K. Mittal, pp. 217–237. Plenum Press, New York, 1991.

Cala, F.R. and Winston, A.E., *Handbook of Aqueous Cleaning Technology for Electronic Assemblies*. Electrochemical Publications Ltd., Port Erin, 1996.

Carter, M., Andersen, M.E., Chang, S., Sanders, P.J., and Kanegsberg, B., Cleaning high precision inertial navigation systems, a case study and panel discussion. In *CleanTech '99 Presentation and Proceedings*. Rosemont IL: Witter Publishing, pp. 294–301, 1999.

Dowell, R., Norris, S., Unmack, J., and Kanegsberg, B., Case study: Cleaning process prior to PVD of critical metal substrates. In *CleanTech 2003*, Chicago, IL, March 2003.

Kanegsberg, B. Cleaning options in the high precision cleaning industry: Overview of contamination control working group XIII. In *International CFC and Halon Alternatives Conference*, Washington, DC, p. 943, 1993.

Kanegsberg, B., Cleaning for biomedical applications. In *Precision Cleaning '97*. Cincinnati, OH: Witter Publications, 1997.

Kanegsberg, B., Successful cleaning/assembly processes for small to medium electronics manufacturers. In *Nepcon West '98*, Anaheim, CA, March 1998.

Kanegsberg, B., Osseointegration, precision cleaning, and surface quality. USC Graduate School of Periodontics, Los Angles, CA, June 2006.

Kanegsberg, B., Selecting an aqueous cleaning process, June 19, 2008a. http://www.pfonline.com/articles/pfdkanegsberg01.html (accessed March 18, 2010).

Kanegsberg, B., Electronics in 2010+: Cleaning challenges. *Clean Source Newsletter*, November 2008b. http://www.bfksolutions.com/Newsletter%20Archives/V5-Issue5/Electronics_In_2010.html (accessed March 18, 2010).

Kanegsberg, B., Cost-effective cleaning for quality thermal spray coating. International Thermal Spray Association Presentation, Fairport Harbor, OH, April 2009a.

Kanegsberg, B., Cleaning processes for high-quality plating. In *Presentation and Proceedings SUR/FIN 2009*, Louisville, KT, June 2009b.

Kanegsberg, B., Abbink, B., Dishart, K.T., Kenyon, W.G., and Knapp, C.W., Development and implementation of non-ozone depleting, non-aqueous, high precision cleaning protocols for inertial navigation subassemblies. In *MICROCONTAMINATION '93 Proceedings*, San Jose, CA, 1993.

Kanegsberg, B. and Kanegsberg, E., Parameters in ultrasonic cleaning for implants and other critical devices. *Journal of ASTM International* 3(4) (April 2006).

Kanegsberg, B. and Kanegsberg, E., Cleaning validation and contamination control in manufacture of high technology products, medical devices, and precision instruments, one day workshop. In *UCLA ERC*, California, CA, March 2008a.

Kanegsberg, B. and Kanegsberg, E., Two challenges in contamination control of combination devices. *Controlled Environments Magazine* 11(3) (March 2008b): 42.

Kanegsberg, B. and Kanegsberg, E., Design, the key to cleaning, defluxing, and productivity. In *Presentation, LA/OC SMTA*, Irvine, CA, November 2009.

Kanegsberg, B. and Kanegsberg, E. Validation readiness, parts 1 and 2. *Controlled Environments Magazine*, January and February 2010.

Kanegsberg, E., Kanegsberg, B., and Phillips, J., Cleaning validation issues for combination devices. *Controlled Environments Magazine* 11(3) (March 2008): 10.

Le Blanc, C., Office of the deputy under secretary of defense. Personal communication. ca. 1999.

Meyer, R.E. and Baier, A.E., Implant surface preparation. *Intl. J. Oral Maxillofac. Implants* 3 (1988): 9–20.

Miller, M., Manu-tronics case study. *Nepcon West 1997*, February 1997.

Peterson, D.S., *Practical Guide to Industrial Metal Cleaning*. Cincinnati, OH: Hanser Gardner Publications, 1997.

Shubkin, R., A highly effective solvent/cleaner with low ozone depletion potential. In *Precision Cleaning '97 Presentation and Proceedings*. Cincinnati: Witter Publishing Corp., 1997, pp. 27–43.

Tautscher, C.J., *The Contamination of Printed Wiring Boards and Assemblies*. Bothell, WA: Omega Scientific Services, 1976.

Woody, L., Nozzle design and implementation for cleaning residues from high density, low profile component assemblies. In *IPC High Performance Electronics Assembly Cleaning Symposium*. Rosemont, IL: IPC, 2008.

16

Cleaning Validation of Reusable Medical Devices: An Overview of Issues in Designing, Testing, and Labeling of Reusable Devices

John J. Broad
NAMSA

David A.B. Smith
Tissue Banks International

Introduction 230
Guidance...230
Designing, Testing, and Labeling • Sterilization of Medical
Devices • Sterilization of Reusable Devices • Compendium
Issues and Test Programs...233
Design Considerations • Decontamination and the Removal of Endotoxins
and Exotoxins • Decontamination of Devices Exposed to Transmissible
Spongiform Encephalopathy
Testing Methodology and Validation Study Design 237
Phenol–Sulfuric Acid Test for Carbohydrates • BCA Protein Assay and
Limulus Amebocyte Lysate Endotoxin Assay • Acceptance Criteria •
Regulatory Considerations and Labeling • Summary and the Future
Appendix A.1: Product Families for Cleaning Validation 245
Appendix A.2: Instruments Representative of the Six Product
Families Utilized for Validation Testing 246
Distractor—Group 1 • ALIF Insertion Instrument—Group 2 • Lollypop-
Rasp—Group 3 • Trial Sizer—Group 4 • Trial Inserter—Group 5 •
T- Handle—Group 6
Appendix A.3: Cleaning Methods 248
Corrosion Prevention • Stain Prevention • Treatment of New
Instruments • Simple and Complex Instruments • Manual Cleaning
(All Instruments—Simple and Complex) • Simple Instrument
Cleaning Steps • Complex Instrument Cleaning Steps • Lubrication
of Instruments • Inspection of Instruments • Autoclave
Sterilization • References
Results: Cleaning Validation 251
Bioburden Recovery: Uncleaned/Cleaned Devices
Compliance with Acceptance Criteria..253
References..253

Introduction

Reusable medical devices have been utilized in the delivery of medical care for well over a hundred years. Especially over the last century, devices have become increasingly more complex because advances in diagnostic and therapeutic medicine have led to more sophisticated designs. As designs become more complex, the process of adequately cleaning, disinfecting, and sterilizing these instruments has become more complex as well. The focus of this chapter is to address the validation of cleaning processes for reusable medical devices.

As early as the nineteenth century, there was an understanding within the surgical community that the use of silver cerclage wire for open reduction and internal fixation of bony fractures would reduce the chance of or prevent a surgical wound from developing pus. This was not based on an understanding of sterility and microbial bioburden, but rather the antibiotic property of silver. Its usage, however, does indicate an appreciation of the consequences of postoperative infections in that era.[1,2] Even in the present day, the use of silver as a biocidal in ointments[3] and also in nanoparticles for its antibiotic properties is not uncommon.[4,5] With the advent of anesthesia in the mid-nineteenth century, the surgical treatment options were greatly expanded; however, this did nothing to stop the spread of infection. The postoperative death rates in some hospitals were sometimes as high as 90%.[6] It was Joseph Lister who was largely credited with being the first to clean and sterilize surgical instruments and silver wire suture. He also developed sterilized catgut for use as resorbable sutures.[7]

Even though the value of cleaning and sterilizing surgical instruments is not a new concept, designing and executing a thorough cleaning validation to meet specific standards is a recent development in the healthcare products industry in the last several decades. Manufacturers of reusable medical devices are responsible for supporting their product claims for reprocessing reusable devices in healthcare facilities. Furthermore, the standard of care has dramatically improved over the years and the public and the healthcare community have become increasingly aware of the societal costs (monetary and nonmonetary) of postoperative nosocomial infections resulting from surgery. In many examples, what would have been considered reasonable methods of designing, manufacturing, validating, and processing surgical instruments several decades ago (as well as many treatment modalities) would now be considered malpractice or negligence.

Meeting the responsibility of developing and validating surgical instruments that can be effectively cleaned and sterilized not only involves appropriate design criteria but also developing instructions for preparing the device for reuse and conducting testing that validates these instructions. Proving cleaning sterilization and/or disinfection efficacy is the primary consideration involved in device reuse, but other issues related to device function, physical integrity, and biocompatibility must also be addressed. The reprocessing methods of medical devices must be included in the initial and ongoing development of the design criteria for the device.

Guidance

Scientific guidelines have been developed to provide assistance to device manufacturers in designing, testing, and labeling of devices intended for reuse in healthcare facilities. These are discussed in new and existing standards. Other local, national, and international guidelines may be applicable in a particular situation. Given the constant impetus to revise, modify, and harmonize standards, the reader is advised to use the following as guidelines and to utilize the most recent updates.

When choosing a validation process for reusable medical devices, U.S. manufacturers rely on guidance from Food and Drug Administration (FDA)[8] and validation methods developed by the Association for the Advancements of Medical Instrumentation (AAMI),[9] as well as international standards. The AAMI has been working with the International Organization for Standardization (ISO)[10] and the American National Standard Institute (ANSI)[11] on the development of guidance for cleaning and sterilization methods for reusable medical devices. ANSI is the only U.S. representative of ISO. Playing an active role in ISO governance, ANSI also participates in a large part of the technical programs, and administers many important committees and subgroups (approximately 20% of all ISO technical committees and subcommittees).[11] The American Society for Testing and Materials (ASTM)[12] provides guidance on the design, materials selection, surface treatments, and standards of performance (of certain instruments) in addition to some cleaning information.

The criteria set forth by the standards from these organizations and the applicable regulations contribute to acceptable methods and validations of cleaning and sterilization of the instrument as well as design factors that affect the ability to clean, disinfect, and/or sterilize the instrument. Because there have been a number of changes in standards in the last decade, questions have been raised as to how manufacturers should design cleaning and sterilization validation protocols in order to be in compliance with the many new standards.

Designing, Testing, and Labeling

The updated Technical Information Reports (AAMITIR12:2004,[13] AAMITIR30:2003[14]) cover a number of general topics:

- Design considerations (physical, material, total system, and user design)
- Decontamination (cleaning agents, water quality supplies, equipment, and methods used in healthcare facilities and cleaning validation programs)
- Disinfection with liquid chemicals (classification of medical devices, based on degree of risk, low intermediate and high level, disinfectant selection, toxicity, and material compatibility)
- Sterilization (available processes, design considerations, efficacy testing methods, sterilant residue, and documentation requirements of hospital information)
- Device/sterilant/equipment compatibility
- Regulatory considerations (FDA regulatory classification and requirements)

The American Society for Testing and Materials (ASTM) Standard Specifications[12] provide guidance on many issues relating to surgical instruments and implants. These include but are not limited to:

- Design
- Materials selection
- Surface treatments
- Standards of performance (of certain instruments)
- Decontamination
- The prevention of corrosion

ISO[10] and, in the United States, the ANSI[11] also have similar standards. The United States FDA[8] recognizes many of these as "recognized consensus standards," examples of which include:

- Recognition Number 11-167: ASTM F1089-02, Standard Test Method for Corrosion of Surgical Instruments. (Orthopedic Surgery)
- Recognition Number 8-67: ISO 7153-1:1991/Amd. 1:1999, Surgical Instruments—Metallic Materials—Part 1: Stainless Steel. (Materials)
- It is further recommended that vascular graft manufacturers utilize relevant provisions of ANSI/ AAMI VP20-1994, Cardiovascular Implants—Vascular Prostheses, where appropriate.

The FDA also provides guidance for reusable medical devices with cleaning, sterilization criteria as well as labeling instructions in other documents.[8,15]

Sterilization of Medical Devices

The 2004 standard (ANSI/AAMI/ISO ST 81:2004[16]) provides requirements for the information provided by medical device manufacturers, so that medical devices can be reprocessed in a safe manner. The standard covers:

- Preparation at the point of use
- Preparation, cleaning, and disinfection
- Drying
- Inspection, maintenance, and testing
- Packaging
- Sterilization
- Storage

This standard also provides examples of processes (decontamination and sterilization) for various types of reusable devices.[16]

Sterilization of Reusable Devices

This ISO standard (ISO 17664:2004[17]) provides requirements for the information provided by medical device manufacturers, so that medical devices can be reprocessed in a safe manner. The standard covers same topics as mentioned in ANSI/AAMI/ISO ST81:2004[16]:

Compendium

The AAMI document, AAMI TIR30:2003,[14] provides information regarding test protocols, test soils, and acceptance criteria in order for manufacturers to validate the cleaning process for reusable medical devices.

A manufacturer should conduct validation studies in the same manner as recommended in guidance documents (e.g., Designing, testing, and labeling reusable medical devices for reprocessing in health care facilities: A guide for medical device manufacturers, Technical Information Report, AAMI TIR12:2004,[13] under Sterilization Efficacy Testing) to provide data to demonstrate that the recommended instructions provide the product with an equivalent sterility assurance level (SAL) of 10^{-6}.

Device Categories and CDC Recommendations

It is important to understand the intended use of the device in order to select the appropriate cleaning method. The Center for Disease Control and Prevention (CDC) recommends disinfection or sterilization based on three end-use categories, critical, semi-critical, and noncritical.[18]

Critical Items

These devices are in direct contact with blood or areas of the body not usually in contact with contaminants. Examples of such devices include surgical instruments, endocavity probes, implants, biopsy forceps/scissors, and ophthalmic irrigation devices. Sterilization is required for these devices. In most cases, the device in this category has significant blood contact, which will contaminate not only the external surfaces of the device but also the internal spaces of the device. Routine cleaning using hot water rinses and brushing the device are usually not effective. Cleaning operations should employ the use of enzymatic cleaning solutions and/or mechanical cleaning methods using sonication and mechanical cleaning equipment.

Semi-Critical Items

These devices are noninvasive and normally contact intact mucous membranes. They include devices such as flexible endoscopes, endotracheal tubes, and breathing circuits. High-level disinfection is a minimum process requirement for products in this category. The effective cleaning of devices contacting mucous membrane can be as complicated as cleaning critical devices. The majority of applications for semi-critical items will have mucous membrane fluid including blood contact. In these cases, the use of enzymatic cleaning solutions and/or mechanical cleaning methods using sonication and mechanical cleaning equipment will be necessitated.

Noncritical Items

These devices have only minimal contact with the patient and rarely transmit disease. If contact is made, it is only with intact skin. Examples include cuffs for measuring blood pressure, beds, and crutches. High- or low-level disinfection is required and products can be cleaned with a simple detergent. Because such devices rarely have contact with blood, effective cleaning methods may be less aggressive and enzymatic cleaning may not be required.

Issues and Test Programs

Design considerations, decontamination, and labeling requirements are issues that must be addressed by manufacturers considering test programs for new reusable devices. The issues are also of concern to those needing to support changes in product design or disinfection/sterilization processing instructions.

Most devices composed of sterilant-tolerant materials will withstand more than 100 cycles. Testing should be designed to address the efficacy of the sterilization cycle, and functional performance of the device. In addition, biocompatibility in terms of any residuals remaining from the cleaning process should also be assessed. This testing can be accomplished by exposing a product to multiple cycles, including any cleaning steps between cycles, for the number of cycle's equivalent to the projected maximum useful life. Following these exposures, there should be a demonstration of functionality, physical integrity, and biocompatibility demonstrated through adequate testing. Implementing an inspection procedure based on the number of surgeries or a period of months, will provide the instrument manufacturer with valuable information about wear, corrosion, and functionality of the device over time.

Whether a single-use or multiple-use device and the end user must clean and sterilize the product, the manufacturer must provide the user with cleaning and sterilization instructions. To provide these instructions, the manufacturer must also have validation data to support the cleaning and sterilization processes.

Design Considerations

A reusable medical device intended for reprocessing must be designed to function safely and effectively following sterilization in a healthcare setting. Reusable devices must be designed to withstand multiple sterilant or disinfectant exposures. Furthermore, the number of exposures to which a device can be subjected without loss of effective functioning is an important factor in determining its useful life.

Within the scope of this discussion of cleaning issues, several engineering aspects each can make substantial contributions to the final ability to clean and sterilize the instrument. These include material selection, heat treatments, surface treatments, labeling, design of the instrument, design of hinges and joints, and other sliding surfaces. Careful selection of the materials and materials treatments can be based on many factors, which include strength, appropriate hardness, flexibility, weight (density), friction coefficients (between tissue and the instrument, the surgical glove and the instrument, etc.), surface finishes, and color (nonreflective, anodized, etc.) laser etching.[19]

It is important to note that some materials may be acceptable for use as surgical instruments but are not acceptable for implants. A manufacturer may make the decision to use an implantable material in an instrument in order to reduce some potential risks. For example, in the case that an oscillating bone saw might come in contact with a second instrument and create some debris that would drop into the wound, even after irrigation some particulates may still remain in the patient. Using an implantable material for the second instrument would reduce the risk of the patient having a reaction to any remaining fine particulates.

It could be argued that the most important factor is surface texture. This does not necessarily need to be a smooth polished finish but it should not have small intrusions in the material where soil may not be removed by cleaning chemicals and processes. These small surface flaws can also contribute to fatigue failure or other types of failure of the device.[19] Furthermore, if corrosion occurs over time, the corrosion products may also provide regions where soil can build up and may not be effectively removed by cleaning operations. If not fully removed, autoclaving can then bake the soil onto the surface, making removal even more difficult. If the use of abrasive cleaners becomes necessary to remove baked-on soil, then the instrument may need to be refinished, passivated, or coated again in order to place it back into service. Corrosion can result from oxidation (e.g., pitting, sensitization of stainless steels, or other processes), or other chemical exposure (e.g., hydrogen embrittlement) or corrosion influenced by the presence of mechanical stress with or without the influence of chemicals (stress corrosion cracking, fretting, galling, and spalling).[19] Furthermore, the persistent presence of soil on the device can trap water and ions against the device, which can potentially accelerate corrosion processes.

The following are some design features that may cause potential challenges in the cleaning procedures along with suggested approaches and solutions:

- Narrow lumens of flexible endoscopes: where possible, lumens should be accessible to cleaning agents (e.g., enzymatic cleaners and brushes). Luer Locks—Luer lock assemblies should be designed to disassemble for easy cleaning.
- Surfaces between insulating sheaths: insulating sheaths should be avoided. Contaminants can become trapped between the sheath and device. In some cases, the sheathing material can crack and collect soil.
- Dead-end lumens: this design should be avoided. If dead-end lumens are necessary for the proper use of the device, outlets for adequate irrigation during cleaning should be included.
- Smooth surfaces: smooth surfaces are easier to clean (Figure 16.1), but they are prone to scratching and corroding. So, cleaning protocols should be carefully specified in the reprocessing instructions. For example, in the case of a stainless steel instrument, the instructions might read, "The use of stiff nylon cleaning brushes is recommended, as nylon will not scratch the passive oxide layer on the surface (any metal brushes, abrasive cleaners, etc., can damage the surface of the instrument, which can lead to corrosion)."
- *Crevices hinges and rough porous surfaces*: Designs with recesses, channels, etc., are difficult to clean. Such designs should be avoided. Where possible, hinged areas should be designed so that they can be disassembled for ease of cleaning (Figure 16.2).
- *Hinges and grasping surfaces*: Designs with features such as recesses and grasping areas should be cleaned with brushes where possible (Figure 16.3). Hinged area should be disassembled for easy cleaning where possible. Alternatively, the hinge can be tight, i.e., made to tight tolerances, which prevents soil from penetrating. By machining the surfaces with a high degree of

FIGURE 16.1 Smooth surfaces are more readily cleaned but may become scratched or corroded.

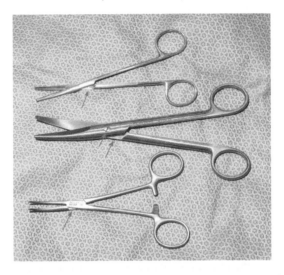

FIGURE 16.2 Hinged instruments should be designed to be disassembled for accessibility during cleaning.

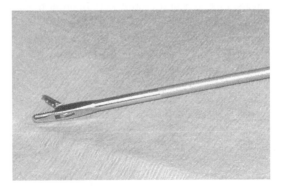

FIGURE 16.3 Hinges and grasping surfaces may require cleaning with brushes.

precision, to tight specifications of both flatness and a tight contact, penetration of blood and other fluids to which the instrument is exposed can be minimized or even eliminated. One common approach is to combine a soft and hard material together, such as with a bushing. At sufficiently high stresses, the softer material can exclude air and liquid between the two surfaces. In some cases, a softer metal may shear and yield instead of sliding. A typical example would be pairing a harder stainless steel (e.g., grade 440 C stainless) with a softer one, 304 stainless steel.

Decontamination and the Removal of Endotoxins and Exotoxins

The cleaning process may serve several purposes; removing blood, protein, fats, carbohydrates, and other potential contaminants from surfaces, crevices, joints, and lumens of a device and reducing the quantity of particles, microorganisms, and endotoxins. Endotoxins are high-molecular-weight complexes associated with the cell walls of Gram-negative bacteria that are pyrogenic (fever causing) in humans. Aside from having their own toxicity, the organic material of endotoxins can also provide a protective coating for some microbes, which may impair the effectiveness of the cleaning agents and methods (liquid chemical disinfectants, washing, and sonication) and the sterilization process, e.g., steam.

Pyrogens are considered any substance that is capable of producing a fever response in a living host. In the healthcare industry, validating the cleaning of pyrogens with devices that directly or indirectly contact the cardiovascular, lymphatic system, or the cerebral spinal tissue, e.g., administration sets, catheters, implants, and infusion sets is a requirement prior to product release. Pyrogens are classified into two groups:

- Microbial, e.g., bacteria, fungi, and viruses
- Nonmicrobial, e.g., drugs and device material

The most significant pyrogens are the endotoxin from Gram-negative bacteria. Other forms of microbes, e.g., fungi and viruses can also cause a fever response although to a lesser degree than the Gram-negative bacteria. Endotoxins are high-molecular-weight lipopolysaccharides (LPS) components of the outer bacterial cell wall. They can cause fever, fluctuation in blood pressure, and meningitis when introduced into the blood or tissues of the body. The cell walls, which are composed of proteins, phospholipids, and LPS, are released from the Gram-negative bacteria during division, apoptosis (programmed cell death), or cell lysis. In manufacturing processes that use water for cleaning (and/or for testing the device), endotoxin contamination is difficult to prevent since endotoxins are very small in size and can easily pass through filters.

Endotoxins have a strong resistance to heat (dry and wet). In order to effectively inactivate endotoxins, a dry heat process of 250°C for 30 min is recommended. Reusable device materials must be heat tolerant in order to withstand these high temperatures. There have been reports that lower temperatures have been effective. Endotoxins can be inactivated using soft-hydrothermal processing in liquids at 130°C–140°C.[20] This is significant with devices that cannot withstand the higher heat required for dry heat processes.

Gram-negative endotoxins responsible for infection include *Escherichia coli*, *Pseudomonas*, and *Salmonella sp.* There are, however, endotoxins other than LPS. The delta toxin of *Bacillus thuringiensis* endotoxin can be harmful to insects but not humans. In some cases, high levels of endotoxins from Cyanobacteria in tap water sources have been demonstrated.

Decontamination of Devices Exposed to Transmissible Spongiform Encephalopathy

Prion-related diseases such as transmissible spongiform encephalopathies (TSE), primarily Creutzfeldt–Jakob disease (CJD), are heat stable and can bind strongly to stainless steel and other materials used in reusable medical devices making them very difficult to remove and inactivate by normal cleaning and sterilization procedures.[21,22] Currently, in the United States, most manufacturers of reusable medical devices include a statement in their reprocessing instructions along the lines of,

> If instruments are or were used in a patient with, or suspected of having Creutzfeldt-Jakob Disease (CJD), the instruments cannot be reused and must be destroyed due to the inability to clean and sterilize to eliminate risk of cross-contamination.

The United Kingdom has a similar policy to minimize risk of iatrogenic infection.[22] These measures have been taken not only due to the fact that prions are difficult to inactivate, but also due to difficulty in quantifying the efficacy of the methods[23] and in validating such procedures.[21-23] In many countries, however, the prevalence of these slow insidious diseases is higher than initially thought and that such diseases may have been spread through blood products.[22]

There are several accepted methods to inactivate prions that involve the use of hypochlorite or strong alkali solutions, but both of these are highly corrosive to many materials used in surgical instruments and other reusable medical devices. Other solutions are detergent proteases but they may only cleave the proteins and will have different efficacy for different strains of TSEs.[22]

A team at the University of Edinburgh led by Baxter has extensively described the use of radio-frequency gas-plasma, which not only destroys the proteins but also removes the remaining material. The combustion products are typically simple gasses such as CO_2, SO_2, NO_2, and water; however, the combustion takes place at a relatively low temperature (room temperature) and, thus, the surfaces being treated are not damaged by heat.[22,23] Validation of these new technologies for deactivation of prions is being developed but the animal/prion strain model may not be suitable to properly validate the process for human TSEs. Furthermore, the scientific community's knowledge of the adherence of prions to the various materials utilized in reusable medical devices is still largely uncharacterized. Critical validation questions include maximal study duration, and utilizing an animal study population significantly large to attain "sufficient statistical validity." Log reduction of infectivity should be as large as possible.[21] Rather than utilizing an animal model as the endpoint for whether an instrument was satisfactorily cleaned (for a validation study), a new approach to quantify the efficacy of protein removal from solid surfaces using fluorescein-labeled bovine serum albumin and epifluorescence scanning (EFSCAN) has been recently described.[23]

As there are few technologies available in this area and these are still being developed, scientists working in this area caution that a multistep approach to the removal of these difficult-to-eradicate molecules is probably the most effective way to deal with TSEs.[22] A limitation of the radio-frequency gas-plasma devices is the inability to penetrate long, tight passageways and into mating surfaces. It is likely that the radio-frequency gas-plasma methods will also be utilized in the deactivation of endotoxins and exotoxins in the future.

Testing Methodology and Validation Study Design

The information provided in AAMI TIR30:2003[14] lists methods that have been developed by several countries to test the efficacy of cleaning of a variety of reusable medical devices. Furthermore, this technical information report includes testing methods for the detection of soil proteins, fats, carbohydrates, endotoxins, and viable microorganisms. These tests can be used to provide data to support verification that the manufacturer's recommended cleaning procedure using a specific cleaning agent is effective for a particular device.

Cleaning efficacy studies employ the use of soil markers to determine the efficacy of the cleaning method. The following markers have been used:

- Protein
- Carbohydrates

TABLE 16.1   Examples of Soil Markers Used in Evaluating Efficacy of Cleaning of Various Devices

References	Soil Marker	Device (Examples)
TIR12: 2004[13], Rutala[24]	Whole blood, serum, milk powder, and saline	Surgical instruments
TIR30:2003[14]	Peanut butter, evaporated milk, butter, flour, lard, dehydrated egg yolk, saline, and printer's ink blood, Huckers artificial soil (fecal equivalent)	Colonoscopes
British standard 2745: part 3[25]	Glycerol, dehydrated hog mucin, horse serum, unbleached plain flour, aqueous safranine solution water	Anesthesia accessories
Kramer[26]	Blood	Endoscopes

- Lipids
- Hemoglobin
- Endotoxin
- Bacterial spores in artificial soil

Depending on the device under consideration, one or more test soils are appropriate for cleaning efficacy studies. The cleaning procedure developed should be based on the type of contamination expected on the device, design features, and the potential for the patient to come in contact with pathogens. For example, the test soil(s) that is appropriate to verify cleanliness for devices that enter the sterile body cavity will be different than that for a device that contacts the mucosa and/or intact skin. The case study included in this chapter provides an example of an effective approach to demonstrate the effectiveness of a manual cleaning process of surgical instruments. Table 16.1 provides examples of soils for use in evaluating various instruments.

Verification is usually performed in a laboratory setting, selecting an artificial soil mixture that closely simulates the level of contamination expected in clinical use. For example, simulated blood or body fluids may be formulated using a mixture of calf serum, dry milk powder, and a 1:1 rabbit blood/saline mixture.[24] Protein and carbohydrates are other markers that can be used to evaluate the effectiveness of cleaning a medical device.

A method proposed by AAMI TIR12:2004[13] for evaluating the effectiveness of the cleaning process is to add *Bacillus stearothermophilus* spores at approximately 10^4–10^5 organisms per device. The spores are used as a tag to test the efficacy of cleaning. The soiled device is then cleaned according to the manufacturer's instructions and the remaining spores are recovered and enumerated. It is important that the recovery technique be validated prior to determining the efficacy of cleaning. The efficacy of cleaning can then be determined by subtracting the spores recovered from the device after cleaning, from the spores recovered from a control sample (soiled without cleaning).

Replication of testing should be sufficient to ensure that the cleaning procedure can be duplicated in healthcare facilities. A benchmark bioburden reduction level by cleaning should show at least a 3 log (1000-fold) reduction. Published data show that this is possible for flexible endoscopes[27] that are considered relatively challenging to clean.

Viable count determination can be useful, but viable counts should not be used as the only marker for cleaning efficacy. While loss of viability may demonstrate a reduction in viable organisms, it does not indicate the removal of soil residues. Two methods of evaluating cleaning efficacy are the phenol–sulfuric acid test for carbohydrates, bicinchoninic acid (BCA), protein assay, and limulus amebocyte lysate (LAL) endotoxin assay.

Phenol–Sulfuric Acid Test for Carbohydrates

The phenol–sulfuric acid assay[28,29] is a broad spectrum method for the detection of carbohydrates, measuring both mono- and polysaccharides. Calibration curves of micrograms sugar versus 490 nm for mannose, glucose, and galactose are shown to provide reliable determinations (typically +/− 3%–4% variability) of corresponding methyl glycosides and linear and branched-chain oligosaccharides containing the corresponding reactive hexose residue.[14,28,29]

BCA Protein Assay and Limulus Amebocyte Lysate Endotoxin Assay

The BCA protein assay utilizes a highly sensitive reagent for the spectrophotometric determination of protein concentration in solution. It is currently considered to be the most sensitive of the colorimetric protein methods available. Many of the detergents (Triton X-100, SDS, Brij, Lubrol, Chaps, Tween 20 and 80, etc.), buffers (sodium phosphate, hepes buffer, sodium acetate, etc.), and salts (sodium and potassium chloride) commonly used to study and remove proteins are compatible with the reagent.[30,31] The endotoxin levels eluted from cleaned medical devices can be measured by LAL assays as described in ANSI/AAMI ST 72:1998.[32] LAL assays are a family of tests utilized to detect and quantify endotoxins (high-molecular-weight complexes associated with the cell wall of Gram-negative bacteria that is pyrogenic in humans). The tests are all based on extracts of blood cells from the horseshoe crab (*Limulus polyphemus*).

Acceptance Criteria

A major challenge faced by instrument manufacturers is determining for how clean is "clean." The AAMI TIR30:2003 Compendium[14] has provided a number of references that address issues regarding acceptance criteria for cleaning effectiveness. Based on a review of published data for various types of reusable devices,[27] the average levels of the various markers after cleaning of reusable devices were

- Protein: $<6.4\,\mu g/cm^2$
- Carbohydrate: $<1.8\,\mu g/cm^2$
- Hemoglobin: $<2.2\,\mu g/cm^2$
- Endotoxin: $<2.2\,EU/cm^2$
- Bacterial spore reduction: 3 logs

Regulatory Considerations and Labeling

The U.S. FDA has requirements to be addressed by manufacturers for reusable devices, which include labeling, premarket clearance, and quality systems. In 1996, the FDA's Office of Device Evaluation issued a draft guideline entitled "Labeling Reusable Medical Devices for Reprocessing in Health Care Facilities: FDA Reviewer Guidance."[15] This document was intended for use as a checklist to evaluate a reusable device's labeling content for conformance with all applicable requirements to ensure an adequately prepared device for repeated, multi-patient use.[13,15] The guidance document also covers some single-use-only devices. This document identifies seven criteria for reprocessing, which must be addressed in the instructions.[15]

As was covered in the introduction to this chapter, many practices that were considered acceptable decades ago are no longer sufficient. The labeling practice instructing users to "follow routine hospital reprocessing procedures for this device" is considered unacceptable, as there are many devices for which "routine hospital reprocessing procedures" do not exist.

The FDA outlines seven criteria for device reprocessing instructions that are utilized for evaluation by the FDA reviewer. In the context of this discussion, these are

1. Labeling for a reusable device must include reprocessing instructions and instructions on how to make the device ready for use (even if not reusable).

2. A statement that the device must be thoroughly cleaned (even though there may be other steps involved in readying the device for clinical or surgical use).

3. The appropriate microbiocidal process that would also indicate sterilization or high-, medium- or low-level disinfection.

4. The cleaning, disinfection, and/or sterilization process must be feasible at the healthcare provider facility (i.e., surgery center, physician clinic, etc.). All locations (e.g., hospitals, outpatient surgery centers, etc.) where the devices may be used and the equipment available at those facilities should be carefully considered.

5. The instructions must be understandable to the intended audience and be presented in a clear and concise format and presented in a logical sequence.

6. The instructions must be comprehensive to provide a precise understanding of the steps and must include (if needed for proper cleaning):

   a. Any special accessories.

   b. Special preprocessing handling steps.

   c. Disassembly/reassembly instructions.

   d. Cleaning method. If a specific method is not identified, then the cleaning qualification must include a representative selection of commonly used cleaning methods.

   e. Cleaning and/or lubricating agents. If a specific agent or a class of agents are not identified, then the cleaning qualification must include a representative selection of commonly used cleaning agents.

   f. Adequate rinsing must also be described. Rinsing must also be qualified to demonstrate that residual levels of any cleaning agents are low enough as to not interfere with further reprocessing steps and that they are removed to a safe level.

   g. Disinfection or sterilization method. If only a generic method is identified, then the qualification must cover all forms of the generic method.

   h. Special postprocessing handling.

   i. Reuse life, which includes how many times the device can be safely used.

   j. Special warnings and precautions. Careful consideration should be given to any condition (of the device or relating to other factors) that could affect the performance, cleaning, disinfection, and/or sterilization of the device.

   k. Lay use. The instructions should be targeted for the audience that will be using the device.

   l. Reference to guidance documents or to labeling of accessory devices. The manufacturer may refer to instructions for standard commercially available equipment, but is still obligated to qualify the reprocessing of the device with that equipment.

   m. Telephone number to request information.

   n. Statement directing the user to qualify any deviations from the recommended method.

7. The instructions must include only devices and accessories that are legally marketed.

The reader is encouraged to obtain the full document (Labeling Reusable Medical Devices for Reprocessing in Health Care Facilities: FDA Reviewer Guidance[15]) from the FDA website[8] for the full criteria. It is recommended that the technicians who will perform the cleaning validation, review, and provide feedback to the proposed cleaning instructions before the validation studies are initiated.

The responsibility for the labeling is dual. The manufacturer of the device is responsible for supporting the claim for reuse, providing procedures that can be reasonably executed by the user. The user of the device (healthcare facility) is responsible for having the proper facilities and equipment to execute and follow the instructions.

In the early stages of designing a device, these factors should be considered and design criteria should be developed to make the device reprocessable in an acceptable fashion. Careful up-front planning will help to ensure a smooth process for validation studies.

CASE STUDY—TRANZGRAFT ALIF (SPINAL) SURGICAL INSTRUMENT SET WITH INSTRUMENTS USING BACTERIAL SPORES AND ENDOTOXIN AS SOIL MARKERS

Editor's note: While the following case study is one example of an approach to a cleaning validation for manufacturers of reusable medical devices, it is not meant to be a "cookbook recipe." Instead, it is suggested that manufacturers set testing and acceptance criteria that are appropriate to the configuration, materials of construction, and end-use requirements of each specific device under consideration. To achieve this, the rationale for testing and the rationale for acceptance criteria must be carefully considered and documented for the device in question.

PURPOSE

The purpose of this protocol is to define the rationale, test methodologies, and acceptance criteria for the cleaning validation of the TranZgraft Anterior Lumbar Interbody Fusion (ALIF) surgical instrument set for spinal fusion. This study is based on the guidelines in AAMI TIR12:2004,[13] and AAMI TIR30:2003 (4).[14]

SCOPE

Cleaning of a medical device is performed to remove organic material, microorganisms, and endotoxins after use to ensure that the device does not present an unreasonable challenge to the terminal sterilization method. This study provides the test parameters to determine the effectiveness of the recommended cleaning procedure. It should be noted that a cleaning validation covers the efficacy of a cleaning process to reduce bioburden by a set level, while a sterilization validation covers the sterilization of one entire unit, often a surgical instrument tray and all instruments in their appropriate bays for autoclaving (or other sterilization method).

It is the responsibility of the manufacturer (or specification developer) to select appropriate cleaning methods for the devices. In this case study, conservative cleaning and sterilization methods were selected as a "worst-case" scenario. The other issue was the possibility that the surgical instruments might be utilized in a smaller hospital in a medically underserved community where more sophisticated cleaning equipment might not be available. Unfortunately, increasingly more aggressive cleaning methods (longer soak times and then eventually sonication) needed to be utilized in later studies in order to accomplish a 3 log reduction in bioburden with some instruments.

The basic study consists of device inoculation with an artificial soil medium (ASM), followed by subsequent cleaning per sponsor's instructions and then quantitative assessment of the remaining bioburden. The resistance of the indicator organism (*Geobacillus stearothermophilus*) to the sponsor-proposed cleaning procedure is evaluated based on log reduction values. The ability to reduce presence of endotoxin after performing the sponsor-proposed cleaning procedure is based on meeting an endotoxin limit of 2.2 EU/cm² according to AAMI TIR30:2003.[14]

PRODUCT DESCRIPTION

The Tissue Banks International TranZgraft ALIF surgical instrument set for spinal fusion used for testing was composed of 33 stainless steel devices that would contact the blood. The product families were designated such that each instrument represented other instruments that had similar shapes, geometries, moving parts, and size and mass. Six devices from each of the six product families designated by the sponsor were required for testing (Table 16.2). Furthermore, future instruments can also be justified as "validated as cleanable" with the cleaning methods validated if a justification can be made that the new instrument can be categorized in an already defined (and validated) product family (Appendix A.1).

TEST PROCEDURES

BIOBURDEN REDUCTION STUDIES

Bioburden Reduction: Soil Preparation

ASM was prepared by mixing the following:

Five milliliters of rabbit blood saline mixture (1:1)
Five milliliters of fetal bovine serum
Three grams of dry milk powder

The prepared ASM was inoculated with an appropriate volume of *Geobacillus stearothermophilus* (ATCC 7953) spores so that each inoculum aliquot will contain 10^4–10^5 CFUs. A population verification was performed on the spore suspension to verify the amount of *Geobacillus stearothermophilus* spores that were delivered to the ASM.

Bioburden Reduction: Inoculation of Devices

All devices including positive controls were cleaned and steam sterilized to remove any remaining spores after the completion of the bioburden study and prior to initiating the endotoxin reduction studies.

TABLE 16.2   Device Requirements. Six Instruments Representative of the Product Families (see Appendix A.2)

	Cleaning Method	
Device Type	Bioburden Reduction	Endotoxin Reduction
Distractor Group—1	6	6[a]
ALIF insertion instrument Group—2	6	6[a]
Lollypop Rasp Group—3	6	6[a]
Trial sizer Group—4	6	6[a]
Cannulated Instruments Group—5	6	6[a]
T—Handle Group—6	6	6[a]

[a] All devices including positive controls were cleaned and steam sterilized to remove any remaining spores after the completion of the bioburden reduction study and prior to initiating the endotoxin reduction studies. All products used for testing are representative of normal production quality as produced under the current good manufacturing practice (CGMP) requirements (Code of Federal Regulations, Title 21, Chapter I, Subchapter H: Medical Devices, Part 820 Quality System Regulation).[8]

Six sterile devices of each type were inoculated on the area of the device deemed most difficult to clean with inoculated ASM to yield 10^4–10^5 *Geobacillus stearothermophilus* spores per device. All inoculated devices were allowed to dry under a laminar airflow (LAF) hood at ambient conditions for 30 min. Three replicates of each device type were used to determine the cleaning efficacy; and three replicates were used as the positive controls (i.e., not cleaned). Immediately after drying, the positive controls were processed using the methods described for the bioburden recovery validation of the uncleaned controls, i.e., there was no cleaning intervention with this group. The remaining inoculated devices were cleaned using the sponsor-specified method (see below).

Bioburden Reduction: Cleaning Procedure

Three inoculated devices of each type following the sponsor-specified method (Appendix A.3) were cleaned to assure that all surfaces and cavities were free of ASM. Each device was visually inspected for remaining debris.

Bioburden Reduction: Determination of Log Reduction Value

A population determination was performed to verify the amount of *Geobacillus stearothermophilus* spores present in the ASM aliquot used for inoculation of devices. Three were simply inoculated and the other three were inoculated then cleaned. Using the results of this test, log reduction values (LRV) are provided in the results section entitled, "Interpretation," to verify that the cleaning methods met the three log limit criteria according to AAMI TIR30:2003.[14]

ENDOTOXIN REDUCTION STUDIES

Endotoxin Reduction: Soil Preparation

ASM were prepared by mixing the following:

Five milliliters of rabbit blood saline mixture (1:1)
Five milliliters of fetal bovine serum
Three grams of dry milk powder

The prepared soil was inoculated with an appropriate volume of *Escherichia coli* control standard endotoxin to achieve a final endotoxin concentration of $(10^4$–$10^5)$/0.1 mL.

The ASM inoculum containing endotoxin was enumerated by using LAL kinetic chromogenic methods as described above in the section, BCA protein assay, and the LAL endotoxin assay.

An aliquot of the inoculated ASM was tested at 1:100,000 and the recovery determined to attain between 10^4 and 10^5/0.1 mL of ASM inoculum.

Endotoxin Reduction: ASM Inoculation of Devices

Six sterile devices of each type were inoculated in areas deemed most difficult to clean with 0.1 mL of the artificial soil medium/endotoxin per device. All inoculated devices were allowed to dry as described in the inoculation test specification prepared prior to the initiation of the study. Three replicates of each device type were used to determine the cleaning efficacy and three replicates will be used as the positive controls (i.e., not cleaned). All inoculated devices were dried under a LAF hood at ambient temperature for 30 min. Immediately after drying, the positive controls were processed using the methods described in the test specification created during the bioburden recovery validation of the uncleaned controls. The remaining three inoculated devices were cleaned using the sponsor-specified method (see below). Inoculum-containing endotoxin was enumerated by using LAL kinetic chromogenic method per the NAMSA standard operating procedure. Aliquots of the inoculated ASM were tested at 1:10,000 and 1:100,000.

Endotoxin Reduction: Cleaning Procedure

Three inoculated devices of each type were cleaned following the sponsor-specified method (Appendix A.3) to assure that all surfaces and cavities are free from ASM. Each device was visually inspected for residual debris.

Endotoxin Reduction: Endotoxin Determination

An endotoxin assay was performed using the LAL method to detect residual inoculated soil on the cleaned and uncleaned devices. For each type of device, six tests were performed (three cleaned and three uncleaned positive controls). The surface area of each type of device was provided, and the results of the endotoxin assay were reported as EU/cm^2.

Using the results of this test for each device, the quantity of recoverable endotoxin was determined to verify the sponsor's cleaning methods met the $2.2\,EU/cm^2$ limit according to AAMI TIR30:2003.[14]

INTERPRETATION

The effectiveness of the cleaning procedure to remove bioburden from device surfaces as calculated as a $\log_{10}$ reduction recovery value (LRV):

$$LRV = Log_{10} \frac{\text{Average Initial Population Recovery (three uncleaned devices)}}{\text{Average Final Population Recovery (three cleaned devices)}}$$

The inoculum verification plates confirmed that the population of spores delivered onto each device was 10^4–10^5 CFUs (colony forming units). Results of the endotoxin assay after cleaning should be comparable to benchmark levels of $<2.2\,EU/cm^2$ (4). Inoculum-containing endotoxin verification was used to confirm the EU delivered onto each device was $(10^4$–$10^5)/0.1\,mL$. If the inoculum were outside the targeted concentration, the test would be repeated.

ACCEPTANCE CRITERIA

The effectiveness of the cleaning method was reported as a $\log_{10}$ reduction of spores following the prescribed cleaning procedure. A $3\,\log_{10}$ reduction or greater was considered acceptable according to AAMI TIR30:2003.[14] Results of the endotoxin assay were comparable to benchmark levels of $<2.2\,EU/cm^2$.[14] Such results would indicate that the recommended cleaning process is effective in removing the simulated soil from all designated surfaces of the device that would come in contact with the patient or are accessible to tissue, blood, or body fluids. Testing should be adequate to ensure that the process can be successfully duplicated in the hospital environment. In the event that the results do not meet acceptance criteria, a new cleaning procedure is to be used and the validation study is to be repeated.

Summary and the Future

The major challenges faced in the reprocessing of reusable medical devices are the removal of bacterial endotoxins and other pyrogens and the decontamination of prions as well as the validation of these cleaning and disinfection processes. There are a number of technologies that are currently being developed for reusable medical devices to address these challenges such as soft-hydrothermal processing, which has been shown to inactivate endotoxins. Radio-frequency gas-plasma has been shown to largely reduce the presence of transmissible spongiform encephalopathies such as CJD. It is likely this technology will also be applied to the removal of endotoxins and exotoxins in the near future.

The objective of a cleaning validation should be to successfully demonstrate the appropriate log reduction for the cleaning or disinfection methods of the devices being assessed. Numerous factors can contribute to the ultimate success of a cleaning validation for a reusable medical device but, most importantly, selecting the appropriate guidance documents, cleaning methods, and endpoints is critical. By considering the effective cleaning of reusable medical devices and the validation of the cleaning throughout the development of the product, one can minimize the possibility of having to repeat cleaning validation studies with increasingly more rigorous methods.

Cleaning endpoints are not 100% defined as described in TIR 30. The FDA has recently provided new expectations of the cleaning endpoints. The agency has recommended that the industry moves away from artificial soil markers, e.g., spore and consider the use of protein markers that can be analyzed quantitatively to endpoint values at below $4\,\mu g/cm^2$ following manual and/or mechanical cleaning procedures.[33] Endotoxin markers should be restricted to use with products with pyrogen-free claims when the device is exposed to cleaning or rinsing agents that may be contaminated with Gram-negative endotoxin sources. The FDA believes the cleaning endpoints should be at minimum detectable levels of organic soil. Submissions should contain a three test program: (1) visual examination, (2) protein with an endpoint at $6\,\mu g/cm^2$, and (3) clinically relevant tests, e.g., TOC, endotoxin (for pyrogen-free claims), and/or hemoglobin.

Furthermore, developing a basic understanding of cleaning issues such as device materials, surface finishes, cleaning equipment available, and the end user of the specific device in the early stages of the device development will help the medical device manufacturer to generate successful cleaning/disinfection methodology and the accompanying device reprocessing instructions. In doing so, the resulting methodology and the associated instructions will not only be satisfactory from a regulatory perspective but will also assure that the cleaning methods as executed by the end user will effectively clean or disinfect the device when in clinical use and thus maintain patient safety. As reusable device design becomes more sophisticated, the combined efforts of manufacturers, healthcare practitioners, and regulatory agencies will be needed in order to develop safe and effective instructions for use in reprocessing reusable medical devices in healthcare facilities.

Appendix A.1: Product Families for Cleaning Validation

TRANZGRAFT ALIF SURGICAL INSTRUMENTS—INSTRUMENTS/INSTRUMENT FAMILIES FOR CLEANING VALIDATION

INSTRUMENT #		Part name to be etched on part
1	*DISTRACTOR*	ALIF-DISTRACTOR
2	*ALIF INSERTION INSTRUMENT*	ALIF-INSERTER
3		
	REPRESENTATIVE INSTRUMENT:	
	LOLLYPOP-RASP	ALIF-RASP
	OTHER INSTRUMENTS IN FAMILY	
	TAMP	ALIF-TAMP
	CURETTES	ALIF-LEFT
	PADDLE DISTRACTORS	ALIF-DISTRACTOR-8MM
		ALIF-DISTRACTOR-10MM
		ALIF-DISTRACTOR-12MM

		ALIF-DISTRACTOR-14MM
		ALIF-DISTRACTOR-16MM
		ALIF-DISTRACTOR-18MM
	PADDLE REAMERS	ALIF-REAMER-8MM
		ALIF-REAMER-10MM
		ALIF-REAMER-12MM
		ALIF-REAMER-14MM
		ALIF-REAMER-16MM
		ALIF-REAMER-18MM
4	**FAMILY—TRIAL SIZERS**	
	REPRESENTATIVE INSTRUMENT:	
	TRIAL SIZERS—25 mm DIAMETER × 18 mm TALL	TRIAL-25MM × 18MM
	OTHER INSTRUMENTS IN SIMPLE INSTRUMENT FAMILY	
	TRIAL SIZERS—25 mm DIAMETER	TRIAL-25 MM × 10 MM
		TRIAL-25 MM × 12 MM
		TRIAL-25 MM × 14 MM
		TRIAL-25 MM × 16 MM
	TRIAL SIZERS—28 mm DIAMETER	TRIAL-28 MM × 10 MM
		TRIAL-28 MM × 12 MM
		TRIAL-28 MM × 14 MM
		TRIAL-28 MM × 16 MM
		TRIAL-28 MM × 18 MM
5	**FAMILY—CANNULATED INSTRUMENTS**	
	REPRESENTATIVE INSTRUMENT:	
	TRIAL INSERTION TOOL	ALIF-TRIAL-TOOL
	OTHER INSTRUMENTS IN FAMILY	
	SLAP HAMMER	ALIF-SLAP HAMMER
6	*PADDLE T-HANDLE*	ALIF-T-HANDLE

Appendix A.2: Instruments Representative of the Six Product Families Utilized for Validation Testing

Distractor—Group 1

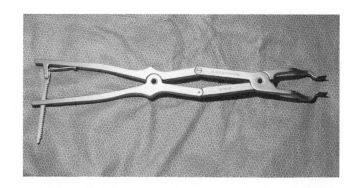

ALIF Insertion Instrument—Group 2

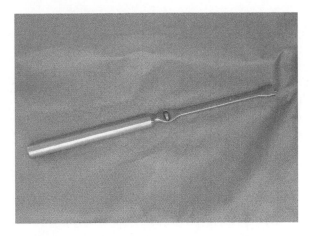

Lollypop-Rasp—Group 3

Trial Sizer—Group 4

Trial Inserter—Group 5

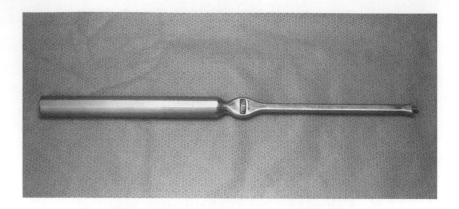

T- Handle—Group 6

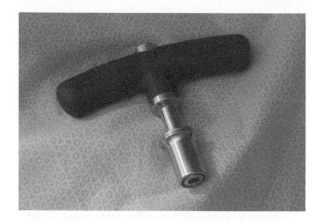

Appendix A.3: Cleaning Methods

TBI/Tissue Banks International
TranZgraft® ALIF Surgical Instrument Set
Cleaning and General Maintenance Instructions

Corrosion Prevention

TBI's surgical instruments are manufactured from high-grade stainless steel. A passive oxide layer on the surface protects the steel from spotting, staining, and adds corrosion resistance. Although this stainless steel has corrosion resistance, exposure to caustic compounds can attack and damage the surface of the instruments. To reduce or eliminate this damage, minimize or avoid the exposure to strong acids, salt solutions, bleach, and chloride-containing solutions.

Stain Prevention

To avoid staining, use demineralized or distilled water in the care and maintenance of surgical instruments. If only tap water is available, dry instruments immediately after the final rinse with filtered, compressed air to prevent spotting. Avoid processing instruments of different metallic compositions together.

Treatment of New Instruments

All instruments are provided non-sterile, and require cleaning, lubrication, and sterilization before surgical use.

Simple and Complex Instruments

Simple instruments are defined as surgical instruments that are one single piece that do not have small holes, cannulae, or threaded holes.

Complex instruments are surgical instruments that have more than one piece (metal, rubber) and/or have small holes, cannulae, or threaded holes. They may or may not have moving parts (hinges, mating surfaces or rods that turn, etc.). Some instruments such as the curettes with rubber handles should be disassembled before the enzymatic (Enzol) soak and cleaning procedures (as labeled on the instrument).

Manual Cleaning (All Instruments—Simple and Complex)

After each use, the instruments should be cleaned as soon as possible to prevent fluid and debris from drying. If the instruments cannot be cleaned soon after surgical use, they should be soaked in enzol enzymatic detergent to prevent drying of soil. These cleaning procedures were validated as described in AAMI TIR30:2003, using enzol enzymatic detergent at a concentration of 1 oz/gal enzol for simple instruments and 2 oz/gal enzol for complex instruments (ordering information at the end of this document).

Simple Instrument Cleaning Steps

1. Rinse in cold running water to remove most fluid and debris from the instruments as soon as possible after surgical use. Flush out any holes or passageways with a standard pressurized spray nozzle.
2. Brush the instruments during the rinse with a stiff nylon cleaning brush to break up any soil. With any holes, use a tight-fitting plastic cleaning brush to clean the inner area and to remove debris. Force enzymatic cleaning solution into holes and other difficult-to-reach areas of the instruments with a syringe or other device. Check the instruments visually and clear any soil remaining. The use of stiff nylon cleaning brushes is recommended as nylon will not scratch the passive oxide layer on the surface (any metal brushes, abrasive cleaners, etc., can damage the surface of the instrument, which can lead to corrosion).
3. Soak the instruments for 10 min in enzol enzymatic detergent at a concentration of 1 oz/gal solution. Agitate the instruments to remove air bubbles from any tight spaces.
4. Brush the instruments with clean brushes after the soak to break up any remaining residual soils. Attention must be paid to tight areas. Check the instruments visually to ensure the removal of soil before proceeding.
5. Rinse the instruments thoroughly on all surfaces and any holes or tight geometry with pressurized water.
6. If, on visual examination, any soil remains, reclean those instruments.
7. Dry the instruments with filtered, compressed air.

Complex Instrument Cleaning Steps

1. Rinse in cold running water to remove most fluid and debris from the instruments as soon as possible after surgical use. Flush out any holes or passageways with a standard pressurized spray nozzle.
2. Submerge and agitate the instruments in enzol enzymatic detergent at a concentration of 2 oz/gal solution. Move instruments through their full range of motion. Open and close releases and turn nuts back and forth. Soak for a minimum of 5 min.
3. Brush the instruments during the rinse with a stiff nylon cleaning brush to break up any soil. With any cannula or holes, use a tight-fitting plastic cleaning brush to clean the inner area and to remove debris. Force enzymatic cleaning solution into cannula and holes and other difficult-to-reach areas of the instruments with a syringe or other device. Check the instruments visually and clear any soil remaining in hinges, mating joints, or rough surfaces. The use of stiff nylon cleaning brushes is recommended as nylon will not scratch the passive oxide layer on the surface (any metal brushes, abrasive cleaners, etc., can damage the surface of the instrument that can lead to corrosion).
4. Rinse the devices in water again. Agitate the devices by hand for at least 1 min. Repeat this process two additional times.
5. Ultrasonically clean (sonicate) the instruments for 15 min in a fresh solution of enzol enzymatic detergent at a concentration of 2 oz/gal solution.
6. Brush the instruments with clean brushes after the soak to break up any remaining residual soils. Attention must be paid to tight areas. For instruments with moving parts (T-Handles, Inserters, etc), turn nuts and actuate any other movements by opening and closing the instrument through the full range of motion. Check the instruments visually to ensure the removal of soil before proceeding.
7. Rinse the instruments thoroughly on the outside surfaces and the internal passageways with pressurized water. For instruments with moving parts, open and close or move the instrument through its range of motion while rinsing.
8. If, on visual examination, any soil remains, reclean those instruments.
9. Dry the instruments with filtered, compressed air.

Lubrication of Instruments

Stainless steel surgical instruments with moving parts and hinges should be lubricated with a water-soluble surgical instrument lubricant after cleaning. This will protect the instruments from rusting and staining.

Inspection of Instruments

After cleaning and lubrication, all instruments must be visually examined for any signs of corrosion and/or damage. Sharp instruments should have uninterrupted cutting surfaces. Instruments with moving parts should be briefly tested (opened and closed, etc.) to ensure proper function for the next surgical use. The full tray should be checked to verify all instruments are present.

Instrument sets should be returned to Tissue Banks International at the address below every 3 months for a thorough inspection to ensure optimal performance during surgery.

Autoclave Sterilization

Sterilization does not replace proper cleaning procedures. Only clean instruments can be properly sterilized for surgical use.

All instruments require autoclave sterilization before use in surgery. After cleaning, drying, and lubrication (and reassembly if applicable), place each instrument back into its designated location within the instrument tray for autoclave sterilization. Wrap the tray as per hospital procedures. Following this autoclave sterilization and drying cycle will help to maintain sterility and prolong the life of the instruments. These instruments are validated for the following autoclave sterilization with a gravity cycle as described in AAMI TIR12:2004.

Item:	Exposure Time at 132°C (270°F) Gravity Cycle	Drying Time
TranZgraft ALIF surgical instrument tray with instruments, *wrapped*	15 min	15–30 min

References

AAMI TIR30. 2003. A compendium of processes, materials, test methods, and acceptance criteria for cleaning reusable medical devices.

AAMI TIR12. 2004. Designing, testing, and labeling reusable medical devices for reprocessing in health care facilities: A guide for medical device manufacturers.

ANSI/AAMI ST81. 2004. Sterilization of medical devices—Information to be provided by the manufacturer for the processing of resterilizable medical devices.

ANSI/AAMI ST79. 2006. Comprehensive guide to steam sterilization and sterility assurance in health care facilities.

ASTM F 1744—96 (Reapproved 2002) Standard guide for care and handling of stainless steel surgical instruments.

Information from Tissue Banks International, Document Number: MI.0002.00. (*Note*: This is not the full document but only the text relevant to the material presented in this chapter.)

ISO 17664:2004. Sterilization of medical devices—Information to be provided by the manufacturer for the processing of resterilizable medical devices.

Results: Cleaning Validation

Bioburden Recovery: Uncleaned/Cleaned Devices

Geobacillus stearothermophilus spores prepared with the soil: Refer to Table 16.3.

Bioburden recovery of uncleaned devices: Refer to Table 16.4.

TABLE 16.3 Bioburden Reduction Study—Population of Inoculum

Sample	Population Per Device
Initial aliquot	3.5×10^5
Final aliquot	4.1×10^5
Mean population per Device	3.8×10^5

TABLE 16.4 Bioburden Reduction Study—Bioburden Recovery of Uncleaned Devices

Sample No.	Average CFUs/Plate	Dilution Factor	Total CFUs Per Device
1	61	1:5000	3.1×10^5
2	57	1:5000	2.9×10^5
3	59	1:5000	3.0×10^5
Avg. per device	59	1:5000	3.0×10^5

Bioburden recovery of cleaned devices: Refer to Table 16.5.
Initial Log Reduction Value (LRV): Refer to Table 16.6.

TABLE 16.5   Bioburden Reduction Study—Bioburden Recovery of Cleaned Devices

Sample No.	Average CFUs/ Plate	Dilution Factor	Total CFUs Per Device
1	2	1:50	100
2	3	1:50	150
3	1	1:50	50
Avg. per device	2	1:50	100

The $\log^{10}$ reduction of spores (cleaning effectiveness) was calculated:

$$\text{Log}_{10} \text{ reduction} = \frac{\text{Log}_{10} \text{ average initial population recovery (uncleaned devices)}}{\text{average final population recovery (cleaned device)}}.$$

TABLE 16.6   Bioburden Reduction Study Summary

Sample	Avg. CFUs Recovered Uncleaned	Avg. CFUs Recovered Cleaned	Average Log Reduction Value	Acceptance Level; LRV = 3
Distractor	3.0×10^5	100	3.5	Pass

Endotoxin Recovery Results: EU recovered from cleaned and uncleaned devices (each device was covered with 500 mL) and EU recovered per cm^2 of device: Refer to Tables 16.7, 16.8, and 16.9.

TABLE 16.7   Endotoxin Reduction Study—Endotoxin Recovered from Uncleaned Devices

Sample No.	EU/Device Recovered Uncleaned
1	5.75×10^5
2	7.65×10^4
3	7.30×10^4
Avg. per device	2.4×10^5

TABLE 16.8   Endotoxin Reduction Study—Endotoxin Recovered from Cleaned Devices

Sample No.	EU/Cleaned Device Recovered
1	3.5
2	<25[a]
3	3.0
Avg. EU per device	10.5

[a] 25 EU/Cleaned Device was used as a whole value to calculate the average EU per device.

TABLE 16.9   EU Recovered Per cm^2 for a 1571 cm^2 Reusable Distractor

Sample No.	Average EU/Cleaned Device Recovered from Table 16.8	EU/cm^2	Acceptance level 2.2 EU/cm^2
Avg. per device	10.5	0.007	Pass

Average EU recovered per device/cm^2 per device = 10.5 EU/1571 cm^2 = 0.007 EU/cm^2.

TABLE 16.10 Summary of the Cleaning Effectiveness for this Study

Sample	Avg. CFUs Recovered: Uncleaned	Avg. CFUs Recovered: Cleaned	Log Reduction Value	Acceptance Level: LRV = 3	Avg EU/Cleaned Device Recovered (Table 16.9)	EU/cm²	Acceptance Level: 2.2 EU/cm²
Distractor	3.0×10^5	100	3.5	Pass	10.5	0.007	Pass

Compliance with Acceptance Criteria

Bioburden and reduction study: The bioburden LRV values for the reusable distractor met the acceptance level of the specified 3.0 $\log_{10}$ reduction in accordance with AAMITIR30:2003.[14] These cleaned devices were macroscopically examined for visible remaining soil, and soil residues were not observed.

Endotoxin reduction study: The endotoxin reduction study showed the average level of 10.5 endotoxins present on each device after cleaning was less than the acceptable maximum level of 2.2 EU/cm² established by AAMI TIR30:2003.[14]

In summary, the bioburden reduction study devices and the endotoxin reduction study devices passed all acceptance criteria. Thus, the cleaning procedure proposed by Tissue Banks International (Appendix A.3 refer to Table 16.10) for the reusable distractor will safely reprocess the devices in a healthcare facility.

References

1. Sommé, D.R. A case of ununited fracture of the thigh-bone. *Medico-Chirurgical Transactions*, 1831, 16 (Pt 1), 36–45.
2. Hinton, J. Case of recent fracture of patella, Treated by Wire Suture. *The British Medical Journal*, Mar. 7, 1885, 1(1262), 480–481.
3. Agarwal, A., Weis, T.L., Schurr, M.J., Faith, N.G., Czuprynski, C.J., McAnulty, J.F., Murphy, C.J., and Abbott, N.L. Surfaces modified with nanometer-thick silver-impregnated polymeric films that kill bacteria but support growth of mammalian cells. *Biomaterials*, Feb. 2010, 31(4), 680–90. [Epub 2009 Oct. 28].
4. Samuel, U. and Guggenbichler, J.P. Prevention of catheter-related infections: The potential of a new nano-silver impregnated catheter. *International Journal of Antimicrobial Agents*, Mar. 23, 2004 (Suppl 1), S75–S78.
5. Furno, F., Morley, K.S., Wong, B., Sharp, B.L., Arnold, P.L., Howdle, S.M., Bayston, R, Brown, P.D., Winship, P.D., and Reid, H.J. Silver nanoparticles and polymeric medical devices: A new approach to prevention of infection? *Journal of Antimicrobial Chemotherapy*, Dec. 2004, 54(6), 1019–1024. Epub 2004 Nov. 10.
6. Atlas, R.M. *Microbiology, Fundamentals and Applications*, Macmillan Publishing Company, New York, 1988.
7. Harding-Rains, A.J. *Joseph Lister and Antisepsis*, Priory Press, Hove, Sussex, 1977, pp. 72–73.
8. The United States Food and Drug Administration (FDA) website, WWW.FDA.GOV.
9. Association for the Advancement of Medical Instrumentation (AAMI) website, WWW.AAMI.ORG.
10. International Organization for Standardization (ISO) website, WWW.ISO.ORG.
11. American National Standards Institute (ANSI) website, WWW.ANSI.ORG.
12. American Society for Testing and Materials (ASTM) website, WWW.ASTM.ORG.
13. AAMI. Designing, testing, and labeling reusable medical devices for reprocessing in health care facilities, A guide for medical device manufacturers, Technical Information Report, AAMI TIR12:2004, AAMI, Arlington, VA, 2004.
14. AAMI. A compendium of processes, materials, test methods, and acceptance criteria for cleaning reusable medical devices, AAMI TIR30:2003, AAMI, Arlington, VA, 2003.

15. FDA Office of Device Evaluation. Labeling reusable medical devices for reprocessing in health care facilities: FDA Reviewer Guidance, Office of Device Evaluation, Silver Spring, MD, April 1996.
16. AAMI. Sterilization of medical devices, information to be provided by the manufacturer for the processing of resterilizable medical devices, ANSI/AAMI/ISO ST81:2004, AAMI, Arlington, VA.
17. ISO. Sterilization of medical devices—Information to be provided by the manufacturer for the processing of resterilizable medical devices, ISO 17664:2004, ISO, Geneva, Switzerland, 2004.
18. Rutala, W.R. et al. Guideline for Disinfection and Sterilization in Healthcare Facilities, CDC, 2008 (update of 1985 document.)
19. Smith, D. Metallurgical engineering considerations in spinal reconstruction, *SPINE: State of the Art Reviews*, Hanley & Belfus, Philadelphia, 10(2), 199–230, May 1996.
20. Miyamoto, T. Okano, S., and Kasai, N. Inactivation of *Escherichia coli* endotoxin by soft hydrothermal processing. *Applied and Environmental Microbiology*, 2009, 75(15), 5058–5063.
21. Murphey, S.A., Mayhall, E.S., Russek-Cohen, E., Brown, R., and Hidderley, A. Scientific issues in evaluating products intended to decontaminate surgical instruments exposed to TSE agents. Center for Devices and Radiological Health (CDRH) Discussions of a Recent FDA Device Advisory Panel. October 31, 2005.
22. Baxter, H.C., Jones, A.C., and Baxter, R.L. Prion deactivation and reprocessing of Surgical Instruments: An Update. Hospital Healthcare Europe, Theatre and Surgery, 2009; T20–23. www.hospitalhealthcare.com
23. Baxter, H.C., Richardson, P.R., Campbell, G.A., Kovalev, V.I., Maier, R., Barton, J.S., Jones, A.C., DeLarge, G., Casey, M., and Baxter, R.L., Application of epifluorescence scanning for monitoring the efficacy of protein removal by RF gas-plasma decontamination. *New Journal of Physics*, 2009, 11, 115028 (13pp). http://www.njp.org
24. Miles, R.S. What standards should we use for the disinfection of large equipment? *Journal of Hospital Infection*, 1991, 18(A), 264–272.
25. British Standard. Washer disinfectors for medical purposes, Part 3—Specification for Washer—disinfectors except those used for processing human-waste containers and laundry, British Standard 2745: Part 3, BSI, London, U.K., 1993.
26. Kramer, A. German Society of Hospital Hygiene, DGKH Executive Committee, Prufung und Bewertung der Reinigungs und Disifektion Swirkung Von Endoskop—Dekontaminator—sowio—Desinfektion Sautomaten. *Hygiene und Madizin*, 1995, 20, 40–47.
27. Alfa, M.J., Olson, N., Degagne, P., and Jackson, M. A survey of reprocessing methods, residual viable bioburden and soil levels in patient-ready endoscopic retrograde choliangiopancreatography duodenoscopes used in Canadian Centers. *Infection Control Hospital and Epidemiology*, 2002, 23, 198–206.
28. Dubois, M., Gilles, K.A., Hamilton, J.K., Rebers, P.A., and Smith, F. Colorimetric method for determination of sugars and related substances. *Analytical Chemistry*, 1956, 28, 350.
29. Saha, S.K. and Brewer, C.F. Determination of the concentration of oligosaccharides, complex type carbohydrates, and glycoproteins using the phenol-sulfuric acid method. *Carbohydrate Research*, 1994, 254, 157–167.
30. Smith, P.K., Krohn, R.I., Hermanson, G.T. et al. Measurement of protein using bicinchoninic acid. *Analytical Biochemistry*, 1985, 150, 76–85.
31. Sapan, C.V., Lundblad, R.L., and Price, N.C. A review: Colorimetric protein assay techniques. *Biotechnology and Applied Biochemistry*, 1999, 29, 99–108.
32. AAMI. Bacterial endotoxin—Test methodology, routine monitoring, and alternatives to batch testing. ANSI/AAMI ST 72: 2002. AAMI Arlington, VA, 2002. American National Standard.
33. Lappalainen, S.K., Gomatam, S.V., and Hitchins, V.M. Residual total protein and total organic carbon levels on reprocessed gastrointestinal (GI) biopsy forceps: Center for Devices and Radiological Health, Food and Drug Administration, Rockville, Maryland 20850. Published online September 5, 2008 in Wiley InterScience (www.interscience.wiley.com). DOI: 10.1002/jbm.b.31202.

17

Critical Cleaning for Pharmaceutical Applications

Overview 255
Regulations 256
Selecting Cleaning Agents 256
Selecting Cleaning Agent Suppliers 258
Cleaning Methods 259
Critical Parameters 259
Equipment Design Considerations 260
Cleaning Method Design 261
Product and Equipment Grouping 262
Sampling Methods (Direct and Indirect) 263
Sample Site Selection 263
Recovery Studies 263
Analytical Detection Methods 264
Residue Limits and Acceptance Criteria 265
Cleaning Validation Master Plan and Validation Protocol Design 266
Cleaning Standard Operating Procedures 267
Change Control 267
Biological Contamination 268
QbD and PAT Considerations 269
Conclusion 269
References 269

Paul Lopolito
STERIS Corporation

Overview

This chapter focuses on critical cleaning of pharmaceutical process equipment. The guiding principles described here apply to other government-regulated industries such as biopharmaceuticals, medical devices, dietary supplements, and cosmetic products that contain active ingredients. Government regulations require a high level of assurance that manufactured products are not adulterated with contaminants or residue from products previously made in the equipment. That level of assurance can be achieved by following good manufacturing practices (GMPs) and by verifying, validating, or continuously monitoring the cleaning process:

- Cleaning verification confirms that the surface has been cleaned to pre-approved acceptance criteria after each product is manufactured.

- Cleaning validation is done through repeated documented trials. Validation ensures that the cleaning procedure meets predefined specifications, which then eliminates the need to verify cleaning. Validation generally requires a minimum of three successful trials.
- Continuously monitoring critical cleaning parameters via quality by design (QbD) and process analytical technology (PAT) principles allows equipment to be rapidly released.

Regulations

The U.S. Food and Drug Administration (FDA) regulates the manufacture of pharmaceuticals, biological products, food products, radiation-emitting products, medical devices, cosmetics, and dietary supplements sold in the United States. The FDA consists of seven centers and offices:

- Office of the Commissioner Organization
- Center for Biologics Evaluation and Research Organization (CBER)
- Center for Devices and Radiological Health Organization (CDR)
- Center for Drug Evaluation and Research Organization (CDER)
- Center for Food Safety and Applied Nutritional Organization (CFSAN)
- Center for Veterinary Medicine Organization (CVM)
- Office of Regulatory Affairs Organization (ORA)

The FDA is responsible for protecting public health by setting laws that govern manufacturing practices and by enforcing these laws through site inspections and pre-market approvals. An FDA document, 21CFR Part 211, Current Good Manufacturing Practices for Finished Pharmaceuticals (cGMP), defines acceptable manufacturing practices and includes facility and equipment cleaning requirements. Table 17.1 shows citations within the cGMP requirements that address cleaning and microbial control in facilities.

Defined GMPs are not limited to the United States. There are separate, specific GMP requirements for the European Union (EU), Health Canada, Therapeutic Goods Administration (TGA), Medicine and Healthcare Products Regulatory Authority (MHRA), World Health Organization (WHO), Pharmaceutical Inspection Co-operation Scheme (PICS), and other countries and organizations.

Harmonizing manufacturing practices and test procedures is an ongoing process led by the International Conference on Harmonization of Technical Requirements for Registration of Pharmaceuticals for Human Use (ICH).

The ICH consists of regulators and industry representatives, and includes six members, three observers, and the International Federation of Pharmaceutical Manufacturers Association (IFPMA). Members include the EU; the European Federation of Pharmaceutical Industries and Associates (EFPIA); Ministry of Health, Labor and Welfare, Japan (MHLW); Japan Pharmaceutical Manufacturers Association (JPMA); the U.S. FDA; and the Pharmaceutical Research Manufacturers of America (PhRMA). The three observers are the WHO, Health Canada, and the European Free Trade Association (EFTA). Examples of ICH guidelines include Q7: Good Manufacturing Practices for Active Pharmaceutical Ingredients and Q2(R1): Validation of Analytical Procedures: Test and Methodology.

While regulations and guidelines vary from country to country, they are designed to ensure the safety, efficacy, and security of the medicinal product being manufactured. Cleaning and sanitizing practices for facilities and equipment are critical elements of GMPs. These practices should be clearly defined in standard operating procedures (SOPs). Employees should be well trained and practices should be verified, validated, or continuously monitored.

Selecting Cleaning Agents

Cleaning agents used in pharmaceutical settings include organic solvents, water, commodity chemicals, and formulated cleaners. Organic solvents such as acetone and methanol are commonly used to clean bulk active pharmaceutical ingredients (API). Due to the environmental need to reduce volatile organic

TABLE 17.1 21 CFR Part 211 Citations Regarding Cleaning and Microbial Control

Subpart	Section
Subpart C: Buildings and Facilities	Sec. 211.42 (10) Aseptic processing, which includes as appropriate:
	(i) Floors, walls, and ceilings of smooth, hard surfaces that are easily cleanable
	(v) A system for cleaning and disinfecting the room and equipment to produce aseptic conditions
Subpart D: Equipment	Sec. 211.67 Equipment cleaning and maintenance
	(a) Equipment and utensils shall be cleaned, maintained, and sanitized at appropriate intervals to prevent malfunctions or contamination that would alter the safety, identity, strength, quality, or purity of the drug product beyond the official or other established requirements
	(b) Written procedures shall be established and followed for cleaning and maintenance of equipment, including utensils, used in the manufacture, processing, packing, or holding of a drug product. These procedures shall include, but are not necessarily limited to, the following:
	(1) Assignment of responsibility for cleaning and maintaining equipment
	(2) Maintenance and cleaning schedules, including, where appropriate, sanitizing schedules
	(3) A description in sufficient detail of the methods, equipment, and materials used in cleaning and maintenance operations, and the methods of disassembling and reassembling equipment as necessary to assure proper cleaning and maintenance
	(4) Removal or obliteration of previous batch identification
	(5) Protection of clean equipment from contamination prior to use
	(6) Inspection of equipment for cleanliness immediately before use
	(c) Records shall be kept of maintenance, cleaning, sanitizing, and inspection as specified in 211.180 and 211.182
Subpart E: Control of Components and Drug Product Containers and Closures	Sec. 211.94 (c) Drug product containers and closures shall be clean and, where indicated by the nature of the drug, sterilized and processed to remove pyrogenic properties to assure that they are suitable for their intended use
	(d) Standards or specifications, methods of testing, and, where indicated, methods of cleaning, sterilizing, and processing to remove pyrogenic properties shall be written and followed for drug product containers and closures
Subpart F: Production and Process Controls	Sec. 211.113 (a) Appropriate written procedures, designed to prevent objectionable microorganisms in drug products not required to be sterile, shall be established and followed
	(b) Appropriate written procedures, designed to prevent microbiological contamination of drug products purporting to be sterile, shall be established and followed. Such procedures shall include validation of any sterilization process
Subpart I: Laboratory Controls	Sec. 211.165 (b) There shall be appropriate laboratory testing, as necessary, of each batch of drug product required to be free of objectionable microorganisms
	Sec. 211.165 (e) The accuracy, sensitivity, specificity, and reproducibility of test methods employed by the firm shall be established and documented. Such validation and documentation may be accomplished in accordance with 211.194(a)(2)
Subpart J: Records and Reports	Sec. 211.182 Equipment cleaning and use log
	A written record of major equipment cleaning … of each batch processed. … In cases where dedicated equipment is employed, the records of cleaning… shall be part of the batch record

carbon (VOC) emissions, newer bulk active manufacturing equipment is being designed to use traditional solvent reflux cleaning or aqueous cleaning. Water is a widely used solvent and can be used alone or in preparing commodity and formulated cleaning agents. Aqueous commodity cleaners include sodium hydroxide, citric acid, and phosphoric acid. Formulated aqueous cleaners can include neutral, alkaline, or acid cleaners as well as detergent additives or sanitizing agents. Adding components to a formulated cleaner can improve cleaning. Mechanisms that influence cleaning are solubility, solubilization, wetting, dispersion, emulsification, hydrolysis, oxidation, and antimicrobial activity (Verghese, 1998). Table 17.2 contains a list of common components and their functions within formulated aqueous cleaners.

The expression "clean smarter, not harder" holds true when selecting the right cleaning agent. Often, the simple addition of a surfactant or oxidizing agent can drastically reduce the time or effort required

TABLE 17.2   Common Formulation Components, Aqueous Cleaners

Component	Examples	Function
Water	Distilled (DI), water for Injection (WFI), USP purified, softened, potable, etc.	Solvent for salts, polar materials; carrier for additives
Surfactant	Nonionic, cationic, anionic, or amphoteric	Wet, solubilize, emulsify, disperse
Chelants	Ethylenediamine-tetraacetic acid (EDTA), nitrilo triacetic acid (NTA), gluconates or polyphosphates	Tie up calcium, iron, and other metals
Solvent	Ethanol, methanol, acetone, D-limonene	Solubilize
Bases	Potassium hydroxide, sodium hydroxide, sodium bicarbonate	Alkalinity source, hydrolysis
Acids	Organic (citric, glycolic) or inorganic acids (phosphoric, nitric)	Acidity source, hydrolysis
Builders	Phosphates, silicates, or carbonates	Assist in detergency; multifunctional
Dispersants	Polyacrylates	Suspend solids
Antimicrobials	Quaternary ammonium, phenols	Kill, reduce microbes
Oxidant	Hydrogen peroxide, sodium hypochlorite	Oxidize, kill microbes

to clean a surface. For example, thickening agents such as methylcellulose, microcrystalline cellulose, or carbopols are traditionally hard to clean, but using an alkaline cleaner and adding an oxidant makes them easier to remove.

In most cases, a single cleaning agent is not sufficient for every application within a site. Thus, it is common to find a combination of cleaners in use. A neutral cleaner works well for manual cleaning applications. A caustic or alkaline cleaner works to clean hard-to-remove organic soils and for automated cleaning applications. An acid cleaner is good for removing hard water scale or for routine derouging and passivation. Finally, a sanitizing agent can be used to control bioburden, if necessary.

Selecting Cleaning Agent Suppliers

In the past 10 years, pharmaceutical companies have paid more attention to how they source and monitor raw materials used in manufacturing. The quality of raw materials is just as critical as availability and cost. For most pharmaceutical companies, cleaning agents are treated as raw materials.

As part of the supplier approval process, most companies set requirements and may request specific documents from cleaning agent suppliers. Requirements might include proof or completion of some or all of the following:

- International Standards Organization (ISO) certifications
- Proof of GMP compliance
- Quality questionnaire
- Supplier questionnaire
- Quality agreement
- Supplier agreement
- Raw material specifications
- Change notification policy
- Contingency planning policy
- Expiration dating
- Availability for audit
- Formulation disclosure policy
- Statements that products are free of questionable materials such as certain dyes, perfumes, and animal-derived materials

Some of the documents listed above may require confidentiality or other agreements between the chemical supplier and end users. Such agreements can be challenging to achieve. For example, a request for sourcing raw materials only from North America can be restrictive for the supplier. It could lead to inflated costs or to a higher risk of shortages. Another example is how some change notification policies require that customers receive prior notice and approve all changes to cleaning agents. This may seem reasonable on the surface, but imagine seeking approval from hundreds or thousands of customers before making an insignificant change in the process or product package.

The supplier should be able to provide sufficient documentation to use the product for your specific application. For GMP applications, the cleaning agent supplier should be able to supply the following supporting documents: material safety data sheet (MSDS), certificate of analysis, technical data sheet, substrate compatibility, product stability, rinsability data, toxicity information, and analytical methods for detection of residual cleaning agents.

Choosing the right supplier is almost as important as choosing the right cleaning agent and process. The supplier must have the technical expertise and experience to provide technical support and quality assurance documentation to support cGMP requirements.

Cleaning Methods

The methods used for pharmaceutical cleaning include manual (brush, wipe, foam, or spray application with low or high pressure), ultrasonic, solvent reflux, agitated immersion, automated parts washing, and clean-in-place (CIP) applications. The cleaning method is typically selected based on equipment size and the degree of automation desired (Table 17.3). Examples of equipment with small parts might include filling equipment and assemblies (e.g., filling needles, stopper hopper, and guide plates) and tablet manufacturing equipment (e.g., tablet tooling, turret, and slats).

The cleaning method directly impacts the level of action on the surface. In a manual cleaning method, the operator can apply a lot of force in scrubbing the surface, but operator safety considerations may limit time and temperature options.

The solvent reflux cleaning method typically results in a cascading flow: boiled solvent condenses on the reactor side walls, the condenser, and the header before cascading down into the kettle. CIP systems mainly rely on a combination of spray and cascading flow. Non-manual cleaning methods such as agitated immersion, ultrasonic, automated washers, and CIP systems may provide less action on the surface, but time and temperature parameters can be increased to enhance chemical reaction and activity.

Critical Parameters

Understanding requirements and defining critical parameters for each cleaning process is the key to controlling and reproducing the procedure. Critical parameters include cleaning agent, method of cleaning, concentration, time, and temperature. In most cases, these parameters can be adjusted to achieve a desired effect. If plant efficiency (time) is critical, then increasing the cleaning agent concentration or application temperature might reduce the time. Laboratory testing is an ideal way to gauge the

TABLE 17.3 Typical Cleaning Methods for Small and Large Process Equipment

Small Parts and Equipment	Large Process Equipment
Manual	Manual
Wipe/brush	Wipe/brush
Soak tanks	Pressure spraying
Recirculation washers or tanks (COP)	Foam cleaning
	Agitated Immersion
Spray washers	Solvent reflux
Ultrasonic	Clean-In-Place (CIP) system

effect of changing critical parameters within a cleaning process. Other parameters such as water quality, surface properties, environmental conditions, and soil type and conditioning (e.g., drying) also influence cleaning. These parameters can be defined in the SOPs for manual cleaning applications, or they can be detailed with automation specifications for a CIP system.

For manual cleaner applications, the SOPs should specify the cleaning agent and the use concentration, water source and temperature, cleaning tools (brush, wiper, spray, or foam applicator), and the technique used for applying the agent. For example, the cleaning method might include instructions such as "Soak part A for 10–15 min in 2% v/v cleaning agent B at 45°C, and then scrub for 20–30 s with a nylon bristle brush. After scrubbing, rinse part A for 20–30 s using ambient temperature purified water." In this example, the cleaning agent, concentration, time, temperature, cleaning device, water quality, and rinse time are clearly defined.

For manual cleaning processes, operator training should be documented and routinely audited to ensure that all operators are following the procedure. For automated processes, these parameters can be controlled with sensors and switches. For example, some processes use conductivity to verify the cleaning agent concentration and to verify that the final rinse water meets specification. Other sensors might include thermometers or resistance temperature devices (RTDs) to verify wash temperatures, pressure sensors/transducers to insure flow rate, and timers to verify duration of wash and rinse times.

Equipment Design Considerations

Equipment design and configuration are important factors in cleaning pharmaceutical equipment and associated piping. The key design issues include liquid coverage in tanks, piping and equipment, liquid flow rates, surface finish, material of construction, and component selection and drainability (Voss, 1996).

Coverage in tanks and equipment can be verified using riboflavin (0.2 g/L water) and an ultraviolet (UV) light. The maximum fluorescence for riboflavin (vitamin B2) under UV light is 565 nm, but a standard 365 nm UV lamp will do. Thoroughly apply the riboflavin solution to all surfaces and use a UV light to verify coverage. Perform a single rinse and then re-inspect using the UV light. Riboflavin, an extremely water-soluble compound, will fluoresce in the presence of UV light, indicating incomplete coverage. Typical problem areas include under spray balls, the backsides of baffle blades, behind inlet ports, and the undersides of mixer blades. Spray balls are hollow fixed spray devices with small holes, allowing cleaning solutions under pressure to pass through them. The holes of the spray balls are typically positioned to allow the spray pattern to cover the top of the vessel and cascade down the sides. Baffle blades are thin strips mounted to the walls of vessels to prevent gross vortexing, especially in low viscosity systems. Results of the coverage test should be included in factory acceptance testing (FAT) or site acceptance testing (SAT).

When using spray balls, typical flow rates for vertical cylindrical tanks are 2.5–3 gallons per minute (gpm) per foot of vessel circumference (see Table 17.4). Typical recommended flow rates for horizontal tanks are 0.1–0.3 gpm/ft^2 of total internal tank surface (4–12 Lpm/m^2).

The recommended flow velocity in piping should be greater than 5 ft/s. Refer to Table 17.5 for recommended flow rates to achieve the proper flow velocity. Minimize dead legs within piping and equipment. The length-to-diameter ratio of dead legs should be 1.5 or less. To allow effective drainage, position dead legs horizontally and at a slight angle. Positioning dead legs horizontally or vertically without any angle can trap air or particulates, respectively. Select and place valves (preferably diaphragm valves) to allow full cleaning. Angle inlet and outlet ports for best cleaning and drainage.

Surface roughness and materials used for construction can influence cleaning parameters. In general, the rougher the surface, the more difficult it is to clean. Eliminate surface irregularities such as corrosion, weld defects, scratches, etches, crevices, obvious pits, and any engraving or embossing. These allow soil to penetrate and adhere to the surface, making it harder to clean. For optimal operation and cleanability, make sure product contact surfaces and interior surfaces of equipment pipes are smooth, and surface irregularities are removed by polishing or repair. A surface roughness of less than 25Ra (roughness average) is common for pharmaceutical and biopharmaceutical manufacturing.

TABLE 17.4 Recommended Flow Rates for Cleaning Vertical Cylindrical Vessels Having Dished Heads

Vessel Internal Diameter (ID)		Flow Rate	
Feet (ft)	Millimeters (mm)	Gallons/minute (gpm)	Liters/minute (Lpm)
1.5	457	12–14	45–53
2	610	16–19	60–72
3	914	24–28	90–106
4	1219	31–38	117–144
5	1524	39–47	148–178

Source: Reprinted from ASME, Table SD-6, ASME BPE-2007, ASME, New York, 2007, by permission of The American Society of Mechanical Engineers. With permission.

TABLE 17.5 Recommended Flow Rates to Achieve 5 ft/s (1.52 m/s)

Tube Size					
Outer Diameter (OD)		Inner Diameter (ID)		Flow Rate	
Inches (in.)	Millimeters (mm)	Inches (in.)	Millimeters (mm)	Gallons/minute (gpm)	Liters/minute (Lpm)
0.5	12.7	0.37	9.4	1.7	6.5
0.75	19.1	0.625	15.9	4.8	18
1.0	25.4	0.875	22.2	9.4	35
1.5	38.1	1.375	34.9	24	90
2.0	50.8	1.850	47	42.8	162
3.0	76.2	2.875	73	102.0	386

Source: Reprinted from ASME, Table SD-5, ASME BPE-2007, ASME, New York, 2007, by permission of The American Society of Mechanical Engineers. With permission.

Common substrates for the pharmaceutical, biopharmaceutical, cosmetic, and dietary supplement industries are 316L and 304 grade stainless steels. Hardened steels (S-7, D-3, and D-2) are common for tablet tooling. Glass-lined or hastelloy reactors are used to manufacture small-molecule API products. Nonmetallic surfaces such as silicone, Teflon®, polypropylene, polyethylene, polysulfone, and other thermoplastics and elastomers are common but can degrade or wear over time. Periodically inspect and replace them if needed as part of a sound preventive maintenance program.

The equipment, components, and associated piping of any system should be designed for full draining. Eliminate horizontal surfaces, angle piping, and slope inlet ports; and minimize dead legs and dome vessel tops and bottoms to allow complete drainage. Use the recommended pitch for piping of 1/8 in./ft and limit wide variations in pipe diameter so as not to interfere with flow rates.

Cleaning Method Design

Developing a robust, efficient, and easily validated cleaning process requires an understanding of the combined role and influence of the following: the cleaning parameters, the nature of product residue, and the complexities of surface and equipment design issues. Table 17.6 provides a list of typical variables to consider for cleaning processes.

Using a sound cleaning process is critical since it can have a significant influence on regulatory compliance and overall manufacturing costs. Laboratory testing can efficiently evaluate multiple variables

TABLE 17.6　Typical Variables to Consider for Cleaning Processes

Residue and Soil Type	Surface and Equipment	Cleaning Parameters
Wet	Material of construction	Cleaning agent
Dry	Irregularities	Cleaner concentration
Baked	Roughness	Time (wash, rinse, etc.)
Steamed	Flow rates	Temperature (wash, rinse, etc.)
Compressed	Coverage	Water quality
Dirty hold time	Drainability	Environmental factors (e.g., humidity)
Composition	Minimize dead legs	
Amount	Orientation of dead legs	Action
	Valve selection	Rinsing
	Insert selection	

and, when properly designed, can save valuable time and resources prior to investing in field trials. It can also be performed in parallel with equipment and manufacturing process design. When possible, the variables should be similar to what you would experience during manufacturing. Knowing the cleaning variables is also important in trouble-shooting a cleaning failure or microbial contamination issue.

Product and Equipment Grouping

By grouping products or equipment, you can reduce the amount of testing needed to support a validated cleaning process. Group products of similar types (coated tablet, lotion, liquid, etc.) that are cleaned by the same process (i.e., that use the same cleaning agent, concentration, time, and temperature). Ideally, the products will be manufactured on similar equipment and have similar toxicity.

Group products by type and then choose a worst-case product within the type. Select the worst-case product using one or more of the following approaches:

1. Pilot scale or lab cleaning studies
2. Operators' experience
3. Activity and toxicity of the active component
4. Solubility of the active ingredient

Group products by the same cleaning process; if one product requires a different cleaning concentration, then the higher concentration should be used for all products grouped. Time or temperature can be similarly adjusted for multiple products as long as the longer time or higher temperature is used. The cost of raw materials, use of the facilities, and efficiency of the plant are all affected when adjusting the cleaning procedure, so factor these in when making grouping decisions.

You can group equipment that is identical. You can also group equipment of the same configuration but of different sizes. In this case, it's typical to test both the largest and the smallest sizes to support validation. As with product grouping, the same procedure must be used to clean the equipment. Small parts of large equipment can be grouped as long as the cleaning procedure is the same. Typically, you will test what is determined by operator experience or configuration to be the most complicated or hardest-to-clean part. For example, a screened gasket is more difficult to clean than a standard gasket.

Grouping strategies for automated parts washers can be difficult, because there may be multiple parts and multiple types of soils. But you can use the same guidelines as above to group these parts and types of soils. With an automated parts washer, load configuration must be considered. In some situations, the parts are loaded in a custom- designed rack, for example, with a filling line assembly.

There are also spindle design racks, in which a part is positioned over one or more spindles. For these style racks, it is important to either (1) define the loading pattern, or (2) perform additional testing during FAT or SAT. If you can demonstrate spindle equivalency, you can then allow non-defined loading. You can place parts of various sizes on the spindle, and evaluate the smallest and largest sizes. When baskets are used for small parts, you define either the loading configuration or the quantity of parts allowed in a basket. Also, be sure to define the position of the basket on the rack.

Sampling Methods (Direct and Indirect)

There are two types of surface sampling: direct and indirect. Direct surface sampling utilizes a swab, wipe, or scanning instrument to measure the residue on the surface. With this approach, analysts can sample specific areas—generally hard-to-clean spots, such as the backside of a baffle, an inlet port, impeller shaft, etc. It's also important to sample the largest representative area, such as the walls of the tank, and thus provide a baseline level of residue on the equipment. The second type of sampling, indirect, is often called rinse sampling. This approach allows a larger and relatively inaccessible area to be sampled, such as a length of piping or a mixing vessel and drain line.

While designing the cleaning process, it is also important to correlate the results from direct surface sampling with indirect surface sampling. This is generally done by combining results from surface sampling and rinse sampling and performing rinse and swab recovery studies.

Sample Site Selection

Selecting sample sites is important for demonstrating overall cleanliness. The sample sites should represent not only the hardest-to-clean areas, but also the largest surface areas. Cleaning operators should be able to assist in identifying hard-to-clean areas. Typical spots might be near gaskets, at the liquid/air interface on a tank wall, the backsides of baffles, the bottoms of spray balls, under impeller blades, and inside inserts. Make sure to clearly define swab sites and to define quality control sampling procedures in the cleaning validation protocol, discussed later in the chapter.

Cleaning validation protocols might recommend more sampling sites than needed for routine monitoring or re-validation of protocols. Including photographs along with a general description is an excellent way to define sampling sites within a particular protocol or procedure. If, due to the size of the part, a sample area is less than the standard swab area, you might sample several parts, use a correction factor, or define a unique specification for that sample.

Recovery Studies

Surface and rinse sampling are methods used to determine the amount of residue remaining on a surface. Because these sampling methods may not remove 100% of residue, analysts perform recovery studies to determine the percent of residue actually recovered. That number will be used to correct sampling values to reflect the actual amount of residue on the surface. Recovery studies are generally performed in the laboratory. They are intended to simulate actual conditions in the field, and often take a worst-case approach to designing processes.

The residue (process soil, contaminant, or cleaning agent) is applied to the surface at or below predetermined acceptance criteria. The residue should simulate the characteristics of the actual residue on the surface of the equipment (e.g., duplicate hold time, temperature, humidity, and number of repeated applications). For direct surface sampling, the same procedure used in the field should be used for the lab studies (e.g., duplicate surface material, swab material, wetting solution if needed, area swabbed, and swabbing technique). For rinse recovery studies, it may be a little more difficult to simulate the technique used in the field. In the design of the laboratory rinse technique, be sure to consider the following rinse sampling conditions: the balance of rinse-volume to surface-area, flow pattern, rinse time, and temperature.

The test coupons used for the laboratory recovery studies should be constructed of the same materials as the surfaces of the equipment in the field. Typically, stainless steel represents the greatest surface area, and has been used as a representative surface for swab recovery values (Forsyth et al., 2007).

Analytical Detection Methods

There are two types of detection methods: specific and non-specific. Specific detection methods measure the concentration of a specific component. An example of a specific method is high performance liquid chromatography (HPLC).

Non-specific methods measure a general property that may have contributions from different components. Total organic carbon (TOC), pH, and conductivity would be considered non-specific methods. You might consider a non-specific detection method to be specific for a target residue if you assume the target residue is the only contributor to the quantity detected (LeBlanc, 2000). Table 17.7 includes a listing of analytical methods and specificity of the method in detecting residues from cleaning agents.

For detection of process residues, the analytical methods commonly used vary based on active ingredients and industry type. Small-dose API uses UV spectroscopy and HPLC; dietary supplements use HPLC and ATP. Biopharmaceuticals use conductivity, TOC, microbial testing, endotoxin testing, and protein assays. Solid and liquid dose facilities commonly use HPLC.

You must validate the analytical method to ensure accuracy, precision, specificity, linearity, and robustness.

- Accuracy: how close the analytical method is to detecting a known value
- Precision: how reproducible the measured value is over multiple tests
- Specificity: determining that the method is specific for the analyte with no interfering substances

TABLE 17.7　List of Analytical Methods Potentially Available to Detect Residual Formulated Cleaning Agents

Method	Rinse	Swab	Specificity
pH	P	P	N
Conductivity	P	P	N
Alkalinity	P	P	N
Acidity	P	P	N
TOC	P	P	N
ATP (adenosine triphosphate)[a]	P	P	N
AA (atomic absorption)	P	P	N
FTIR (Fourier transfer infrared spectroscopy)	P	P	Y
UV (ultraviolet spectroscopy)	P	P	Y
IC (ion chromatography)	P	P	Y
HPLC (high performance liquid chromatography)	P	P	Y
IMS (ion mobility spectroscopy)[b]	P	P	Y

Note:　P, potentially available; N, non-specific method; Y, specific method.

[a] ATP or adenosine triphosphate is a chemical compound in which energy is stored. ATP is present in all animal, vegetable, bacteria, yeast, and mold cells. The bioluminescence test for ATP was developed using the firefly enzyme luciferase. The substrate luciferin in the presence of oxygen and magnesium ion catalyzes the conversion of ATP to light.

[b] IMS or ion mobility spectroscopy is capable of detecting low concentrations of specific compounds by ionizing a sample and measuring how fast the ions move through air at atmospheric pressure in a homogenous electric field. IMS is an alternative test method to HPLC for measuring active components during cleaning verification or validation (Walia et al. 2003).

- Linearity: how proportional the measured value is to the amount of analyte present over a specified assay range
- Robustness: determining that the values are reproducible by using different instruments or by performing with different analysts

Residue Limits and Acceptance Criteria

Residues or contaminants from previous products should not impact the strength, identity, appearance, or quality of the next product produced. Guidelines indicate that, after cleaning, surfaces must be visually clean and "free" (within suitable limits) of detergent residues. There are no established acceptance criteria for product, contaminant, or cleaning agent residues. The preference is to have no residue present, though that is generally not achievable or measurable. It is also not recommended to set acceptance values below the limit of detection (LOD) of an analytical method. The FDA guide to inspecting validation of cleaning processes references an article on setting acceptance criteria (Fourman and Mullen, 1993). The following criteria are noted:

1. Surfaces are to be visually clean.
2. Any active agent is present at less than 10 ppm in any subsequent product.
3. Any active agent present in a subsequent product is less than 1/1000 of the minimum daily dose of the active in the maximum daily dose of the subsequent product.

The organoleptic specification of "visually clean" is significant in that if there is visible soil on the surface, then the equipment is not considered clean. Generally, though not necessarily for every soil, the limit of detection of a visually clean surface is in the range of $4\,\mu g/cm^2$. This depends on many factors: residue, surface, lighting, and the viewing angle and distance of the inspector. The level of visible detection can be experimentally determined by deliberately adding a known concentration of residue to a surface and having analysts determine if the residue is visible or not. The analysts should follow the visible inspection procedure used for the specific equipment. If the level of detection is determined to be less than the calculated acceptance criteria for that cleaning agent and process residue, then it might be justifiable to use "visually clean" as the sole acceptance criterion. If the calculated acceptance criteria are less than the visible limit, then visual inspection can demonstrate a dirty surface but should not be used as the sole acceptance criterion.

Setting limits can apply to (1) the limit in subsequent product, (2) the limit per surface area, or (3) the limit in an analyzed surface area (LeBlanc, 2000). The following items should typically have limits set:

1. The concentration limit of product A in the next product manufactured
2. The absolute limit of product A in the next manufactured batch
3. The limit per surface area of the cleaned equipment
4. The limit per analytical method (swab or rinse volume)

Calculations that follow will address setting limits for items 1–4 above.

The first step is to calculate the acceptable level of actives of one product in the second product. Based on the Fourman and Mullen approach, the calculation is 1/1000 of the minimum daily dose of active in Product A divided by the maximum daily dose of Product B (Fourman and Mullen, 1993 and FDA Guide to Inspecting Validation of Cleaning Process, 1993). The safety factor is generally 1/1000 or 0.001.

$$\text{Limit in } \mu g/mL \text{ or ppm} = \frac{(\text{Safety factor})(\text{Minimum daily dose of active in product A in } \mu g/day)}{(\text{Maximum daily dose of product B in mL/day})}$$

Next, calculate the absolute limit in the next manufactured batch, also commonly referred to as the maximum allowable carry-over (MACO). The MACO is calculated by multiplying the limit in subsequent product by the batch size of the subsequent product.

$$\text{Limit in } \mu g = 1000 g/kg \times (\text{limit in subsequent product in ppm}) \times (\text{batch size in kg})$$

The limit per surface area of the cleaned equipment is calculated by dividing the total limit in μg by the total product contact area in cm². The limit per surface area is commonly expressed in μg/cm². The surface area of the equipment should include all wetted parts. For example, a mixing tank would include the vessel interior, the piping, and any inserts or attachments (such as mixing blades, scrapper inserts, baffles, and agitator shaft).

$$\text{Limit per surface area in } \mu g/cm^2 = \frac{\text{Limit in } \mu g}{\text{Surface area in } cm^2}$$

For swab sampling, limit per swab diluent is calculated by multiplying the limit per surface area by the swab sample area, divided by the swab diluent volume.

$$\text{Limit per swab diluent } (\mu g/g \text{ or } \mu g/mL) = \frac{\text{Limit per surface area in } \mu g/cm^2 \times \text{swabbed area } (cm^2)}{\text{Swab diluent volume in mL or g}}$$

For rinse sampling, limit per rinse is calculated by multiplying the limit per surface area (μg/cm²) by the surface area (cm²), divided by the rinse volume used (mL). The final units for the rinse residue would be in μg/mL or ppm. The calculations for the analytical method (swab or rinse) limits should be adjusted by the recovery factor as described earlier.

$$\text{Limit per rinse } (\mu g/mL \text{ or ppm}) = \frac{\text{Limit per surface area in } \mu g/cm^2 \times \text{surface area } (cm^2)}{\text{Rinse volume in mL}}$$

The calculations noted above are based on the assumption that residue build-up and distribution are uniform into the next subsequent batch. For certain types of equipment, such as filling lines and tablet presses, this may not be the case. The residue on the tablet press and tablet tools, as well as in the filling line and needles, will typically contaminate the first few tablets or vials of the manufacturing batch. In these cases, define a specified number of tablets or vials to be discarded at the beginning. Use calculated or experimental data to estimate the number of tablets or vials to be discarded.

Cleaning Validation Master Plan and Validation Protocol Design

A master plan for cleaning validation addresses an overall approach to ensure that residue from the previous product and contaminants do not adulterate the next product manufactured. The plan might be area- or facility-specific, or it might cover multiple facilities. The master plan document should contain the following:

- Rationale for product and equipment grouping
- Analytical method selection and validation practices
- Determination of specification and acceptance criteria calculations
- Responsibilities
- Validation schedule
- Revalidation practices

- Change control policy
- The overall cleaning approach

This document should also include an approved cleaning agent list, along with analytical methods and acceptance criteria for carry-over of each cleaning agent.

Cleaning validation protocols are a subset of the cleaning validation master plan. Their scope can vary from a single item to a whole manufacturing process. Cleaning validation protocols should include the following:

- Objective
- Scope
- Definitions
- References
- Related SOPs
- Safety rules (if required by facility)
- Responsibilities
- Equipment and supplies
- Test methods
- Test results
- Acceptance criteria
- Data collection forms
- Revision history
- Deviation log

Prior to execution, cleaning validation protocols should be reviewed and approved by at least the owning department, validation services, quality control, and quality assurance. Upon completion, analysts should issue a validation protocol report that summarizes the results, with specific reference to the acceptance criteria. The report should explain and bring closure to any deviations reported while executing the protocol. The final report should be approved by at least the same groups that approved the validation protocol prior to execution.

Cleaning Standard Operating Procedures

The keys to reproducible and effective cleaning are detailed SOPs and effective training for all operators who execute them. The SOPs should be understandable, specific, and repeatable, and they should contain the following sections: Purpose, Scope, Objectives, Responsibilities, Safety, References, Equipment and Materials, Procedure, and Revision History.

The Procedure section within the cleaning SOPs should address specific issues: labeling or pre-inspecting equipment, dismantling or assembling equipment prior to cleaning, preparing the cleaning agents if needed, and washing the equipment (pre-rinse, wash, or wash-and-rinse procedure). The washing procedure should specify the critical cleaning parameters such as wash/rinse times, temperature, water quality, and cleaning agent concentrations. In every case, be sure to carefully document equipment cleaning in log books, equipment tags, or worksheets.

Change Control

As with a validated process, you must document every change to a procedure, whether intentional or unintentional. An example of an intentional change might be increasing the rinse time, adding an additional spray ball, or replacing a pump. An instance of an unintentional change might be using a cleaning agent past the expiration date or using an uncalibrated conductivity meter. Regardless of whether a change was intentional or not, you must assess the impact each change might have on the cleaning process as well as its possible effect on the subsequent manufactured product.

Biological Contamination

GMPs require biological contamination control for the facility, equipment, and containers. Cleaning and disinfection practices for the facility should include disinfectants with proven efficacy against bacteria, fungi, viruses, and spore-forming organisms. In the United States, disinfectants, sanitizers, and sterilants are classified as pesticides, and are registered and approved for use by the Environmental Protection Agency (EPA). The United States Pharmacopeia-National Formulary (USP-NF) 32 General Chapter <1072>, Disinfectants and Antiseptics, lists information on disinfectants—selecting, applying, and rotating them—as well as information about their mechanism of action and appropriate validation practices. It also contains information about using in vitro and in situ testing to establish a facility cleaning and disinfection program.

Equipment can be biologically contaminated with bacteria, fungi/molds, viruses, endotoxins (bacterial pyrogens), prions, bacteriaphage, or biofilms. You can control biological contamination by designing appropriate cleaning procedures or by using separate sanitizing or sterilizing procedures. Be sure to create cleaning or process validation protocols with defined specifications for biological contamination. While specification limits for biological contaminants should be scientifically justified, many companies default to limits set for the quality of the water used for the final rinse step.

Standard biological methods include enumeration, isolation, identification, antimicrobial effectiveness, endotoxin testing, microbial limits, and sterility testing. Table 17.8 provides a summary of biological tests and the corresponding USP test methods. The ICH (Q6), European Pharmacopeia (Ph. Eur.) and Japanese Pharmacopoeia (JP) also have test methods available for the biological test methods included in Table 17.8.

Containers and closures used to package final dosage forms must be cleaned and sterilized to control microbial and pyrogenic contaminants. Pyrogens are fever-inducing contaminants; endotoxins are a pyrogenic component of gram-negative bacterial cell walls. Endotoxins can be effectively removed or inactivated chemically or with dry heat. You can verify the efficacy of the inactivating process with the rabbit pyrogen test, the gel clot method, or by using chromogenic or kinematic test methods.

In addition to the biological test methods indicated in Table 17.8, the USP-NF also contains guidance for establishing microbial limits for non-sterile pharmacopeial products, non-sterile nutritional and dietary articles, and controlled environments. The USP-NF guidance documents are USP <1111>, USP <2023>, and USP <1116>, respectively. Biological contamination control is a critical part of GMP requirements and should be considered an integral part of the cleaning, sanitizing, and manufacturing process.

TABLE 17.8 Biological Test Methods and USP-NF Validation Resources

Test	USP-NF
Antimicrobial effectiveness	<51>
Sterility	<71>
Microbial limits	<61>
Bacterial endotoxins	<85>
Rapid micro methods	<1223>

QbD and PAT Considerations

QbD and PAT are relatively new concepts to GMP. Guidance documents on both concepts have been issued by the ICH and the FDA. QbD identifies the critical parameters within the process and defines the design space and limits of these critical parameters. Within a cleaning process, the critical parameters may include cleaning agent concentration, wash temperature, and wash time. To achieve the most effective cleaning, you can use laboratory studies to set critical limits for concentration, time, and temperature. PAT helps you monitor these critical parameters. When monitoring shows these critical parameters are running within the acceptable range of the design space, you may be able to reduce or eliminate traditional cleaning validation or verification.

Conclusion

Critical cleaning for pharmaceutical applications must begin early in the design stages. Such early consideration allows time for developing effective cleaning cycles using laboratory or field studies, selecting cleaning agent suppliers, and (if needed) modifying equipment design. When you evaluate the cleaning process early, you have more time for analytical method development and validation. This approach results in a robust and efficient cleaning process that consistently meets predetermined specifications and regulatory requirements.

References

ASME (2007). Bioprocessing equipment, ASME-BPE 2007. ASME, New York, pp. 8–47.

Food and Drug Administration (FDA) (1993). Guide to inspections of validation of cleaning process. FDA, Silver Spring, MD, July 1993.

Forsyth, R., O'Neill, J. C., and Hartman, J. L. (2007). Materials of construction based on recovery data for cleaning validation. *Pharmaceutical Technology*, October 2, pp. 103–116.

Fourman, G. L. and Mullen, M. V. (1993). Determining cleaning validation acceptance limits for pharmaceutical manufacturing operations. *Pharmaceutical Technology*, 17(4); 54–60.

LeBlanc, D. A. (2000). *Validated Cleaning Technologies for Pharmaceutical Manufacturing*, Buffalo Grove, IL: Interpharm Press. pp. 135–168.

Verghese, G. (1998). Selection of cleaning agents and parameters for cGMP processes, *Proceedings of the INTERPHEX Conference*, Philadelphia, PA, Reed Exhibition Co, Norwalk, CT, pp. 89–99.

Voss, J. (Ed.) (1996). *Cleaning and Cleaning Validation: A Biotechnology Perspective*, Bethesda, MA: PDA, pp. 1–39.

Walia, G. et al. (2003). Ion mobility spectrometry speeds cleaning verification, *Pharmaceutical Processing*, September.

18

Cleaning in the Food Processing Industry

Chemistry of Soils 271
    Soil Attachment • Dairy Soils • Brewery Soils • Food Processing Soils
Films on Equipment...278
Standards of Cleanliness...278
Summary 281
References...282
Bibliography 282

Hein A.
Timmerman
Diversey Inc.

Chemistry of Soils

Soils encountered in food and beverage processing facilities vary greatly depending on the point in processing and the surrounding environment. On the interior of processing and storage equipment, one will find the food product along with some micro-organisms. In other areas, including packing and storage areas, soils will be heterogeneous mixtures of food deposits, minerals, micro-organisms, lubricating greases and oils, and other gross dirt and debris. It is necessary to understand the composition of soils in order to choose the correct cleaning agents and cleaning program. Table 18.1 lists the types of soils found in food processing facilities and their characteristics. Regardless of the soil type, the sooner cleaning takes place after deposition, the easier it will be to remove. While many chemicals could be efficacious in removing food soils, there is considerable limitation when taking into account all the issues concerning food safety. The chemicals intended for cleaning in the food processing industry should pass a strict clearing policy toward all food safety risks, such as chemical residues in the end food products, chemical reactions with the soil, as well as operator and environmental safety.

Food soils themselves can be organic or inorganic in nature. The most common soils, the carbohydrates, are also the easiest to remove. Carbohydrates include sugars, starches, and cellulose, which are all water-soluble. Heating will alter the structure of each, making them more difficult to remove, but still relatively easy compared to other soils. Examples include syrups, potatoes, rice, grain, pastas, fruit, and most soft drinks. All of these soil types can easily be removed using an alkaline cleaner. Warm to hot solutions are generally used, at a pH of 7–12.

Proteins are probably the most difficult food soils to remove from surfaces. They are very large molecules, which are folded into the specific structures that make them functional. Changes in their environment, including heat and acidic pH, cause the destruction of many of the bonds that hold them together, leading to the unfolding of the molecule (denaturation) and the exposure of binding sites to other molecules (Figure 18.1). The proteins are now long instead of compact, and the binding of other molecules, including proteins, leads to an irreversible coagulation of the large proteins. Their tenacity is increased when they combine with minerals.

271

TABLE 18.1   Characteristics of Food Soils

Type of Soil	Example	Water-Soluble?	Effect of Heat	Cleaner	Removal
Organic					
Carbohydrate	Sugar, starch	Yes	Caramelize, sticky	Alkaline	Easy
Protein	Casein	Some	Denature, combine with salts, more tenacious	Chlorinated alkaline or strong alkaline	Very difficult
Fat	Tallow, lard, corn oil	No	Polymerize, making it more tenacious	Alkaline	Difficult
Petroleum	Lubrication oils, greases	No	None	Solvent	Difficult
Inorganic					
Monovalent salts	Sodium chloride	Yes	None	Acid	Easy
Water stone	Calcium Nitrate	No	Precipitation	Acid	Easy to difficult
Food stone	Calcium oxalate (beer stone); tricalcium phosphate (milk stone)	No	Makes it more tenacious	Acid or very high levels of sequestrants.	Easy to difficult
Metallic deposits	Rust, oxides	No	None	Acid	Easy

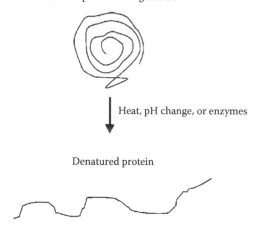

Native protein configuration

Heat, pH change, or enzymes

Denatured protein

FIGURE 18.1   Protein structures.

Foods high in protein include eggs, milk, meat, and beans. Some proteins are water soluble and some are not. Naturally, insoluble proteins are those structural proteins that make up muscle, hair, animal hoofs, nails, and other tissue. Those that are soluble become insoluble when changed or denatured by heat or acid. Cooking an egg is an example of denaturing a protein. The heat causes the soluble proteins called albumins to denature and precipitate out as the white solid material we know as egg white. Most proteins bind more tightly to surfaces at temperatures above 85°C (185°F), where they are denatured. That is why overcooked eggs or milk is so hard to remove from vessels or utensils.

Strongly alkaline or chlorinated alkaline cleaners (at pHs 10–13) are usually used to clean protein-aceous soils because of their ability to destroy bonds that hold proteins together. This is also called peptizing because the bonds that hold proteins together are peptide bonds. A chlorinated alkaline cleaner is often a mixture of sodium hydroxide, potassium hydroxide, and sodiumhypochlorite, where the sodiumhypochlorite is used for its oxidative character or bleaching properties. Acidic cleaners denature proteins, but do not break them into smaller pieces like chlorinated alkaline cleaners do, so they are not effective cleaners. Hot solutions are used, but they are generally 85°C (185°F) or less to avoid further denaturation.

Fatty soils (triglycerides) are not water soluble, and are thus more difficult to clean than carbohydrates, but probably not as tough as proteins. These types of soils also provide an excellent medium to trap other soils on surfaces, including proteins and minerals. Heat during processing can polymerize fats, making them more difficult to remove. Vegetable oils and animal fats are commonly encountered in food plants. Alkaline cleaners are used because of their ability to saponify the fatty soils and create soluble soaps that are easily washed away. Surfactants are usually added to help wet the surfaces and soils. Hot temperatures, ~5°C (~15°F) above the melting point of the fat in question, are generally used for the most efficient cleaning.

Petroleum-based soils do not originate from food products, but rather from the processing machinery. Materials used to keep motors, pumps, and conveying equipment running smoothly can build up and trap other soils. Since these organic lubricating greases and oils are not water or alkali soluble, a solvent cleaner is needed to remove them.

Mineral salts account for the inorganic food soils. Salts from the water or food products can precipitate onto equipment, especially heated surfaces. Water stone is common in areas with hard water supplies. If cleaners do not have the ability to control hard water, or if insufficient concentrations are used, scales can result. Some vegetables like spinach and celery contain large amounts of oxalic acid that deposit out as "stone" on equipment when combined with calcium or magnesium (calcium oxalate). The calcium or magnesium can come from the food itself, the processing water, or other foods being processed simultaneously. "Beer stone" is a result of the high oxalic acid content of barley and calcium (from other beer components and the water). Celery and spinach stone are also calcium oxalate. Milk stone is calcium phosphate with or without calcium casienates and phospho-lipids. Milk stone is formed when minerals in milk precipitate out after heating. Mineral salts are easily removed with acid cleaners. For the high oxalate containing deposits, sequestrants such as EDTA, in moderately alkaline cleaners give best results. These stones, however, are not only inorganic, but are a complex matrix of organic and inorganic soil, requiring a well-focused approach.

Metallic deposits, rust, and metal oxides are the final class of inorganic soil. While these soils do not originate solely from the food, the type of food, the processing environment, and the cleaning program influence their formation. Rust and oxide formation can occur on exterior or interior surfaces. Chemical-to-surface incompatibility and poor sanitation can be responsible for these soils. As with all inorganic soils, acid cleaners are used to remove these deposits.

Soil Attachment

Soils can be *associated with* or *adsorbed onto* surfaces. Association can be via simple attachment by ordinary attractive forces like wetting, or it can be via entrapment. Pits and crevices, and other areas where surfaces are not completely smooth are sites where soil particles can lodge. Layers of previously deposited soil also provide an excellent "trap" for fresh soil. The intimacy of any association is affected by contact time, temperature, surface characteristics, particle size, and water content. Some soils that are very loosely associated with a surface can be removed merely by wiping or rinsing the surface, while others require more energy and detergents to remove them.

Adsorption on the other hand involves a chemical reaction between a polar molecule and a polar surface. To remove soils that are adsorbed onto surfaces requires a surfactant to lower surface activity and break the chemical bonds between soil and surface.

Dairy Soils

Dairy soils can vary greatly in composition depending on the product, the processing conditions, and even the animal from which the milk comes. The major components in milk are sugar (water soluble), fat, protein (both water soluble and insoluble types) and minerals or ash. The composition of the product itself and the temperature of processing play the largest roles in determining the type of soil deposited. Table 18.2 is provided to illustrate the differences between hot and cold soils compared to milk itself, but should not be taken as the definitive source for soil compositions.

Cold milk soil contains a large amount of sugar (mostly lactose) and fat, while hot soils contain only traces if any. Sugar and fat both become more soluble with increasing temperature, which is why they are found in larger proportions in cold soils as opposed to hot soils. Fats melt at higher temperatures, making them less likely to come out of solution onto surfaces, and easier to wash away if they do. Proteins are a major component of both hot and cold milk soils because of their large size and ability to combine with minerals, sugars, and fats. There are two major proteins in milk. Whey proteins are water soluble, but begin to denature when heated above 65°C. Casein proteins are not water soluble and are easily precipitated with minerals, enzymes, or heat. (Alfa Laval Corp. 1986) Minerals are an important component of both hot and cold soils because of their relative insolubility. Many researchers have proposed that they are the first component to deposit onto surfaces, forming the foundation for other deposits (Burton 1968).

Hot milk soils differ greatly depending on the processing temperature and the holding time. There is generally more soil in the hotter portion of the processing equipment than in the nonheated parts. As processing temperature increases, the mineral content of soil also increases (stone) while the fat content decreases. Minerals tend to precipitate out of solution at higher temperatures becoming more of a soiling concern. They can also combine with protein, making them more tenacious. As previously stated, proteins unfold when heated, exposing more sites that are capable of binding to surfaces. This results in more tenacious protein-based soils under heated conditions. Soils found at the hottest part of the pasteurizer are mostly mineral in nature with some protein, and they are more difficult to remove than soils in other areas of the equipment. It is obvious that soils in raw milk holding tanks are quite different than those found on pasteurizer plates, and their susceptibility to cleaning programs will differ as well.

Whatever the soil composition, it appears that the mineral components (calcium, phosphate, etc.) and proteins are responsible for the tenacity of soil. The changing structure of proteins upon heating, and their ability to interact with the surface and with minerals are responsible for their influence. An initial layer of minerals deposit, upon which proteins attach. The structure of the protein and mineral portion of the soil helps determine the amount of fat that will be trapped within (Figure 18.2).

The soils resulting from other dairy products will be different than simple milk soils because of the additional processing involved. Butter is actually milk fat, so removal from equipment will require melting and a great deal of emulsification. Cheeses undergo enzymatic reactions to denature proteins and

TABLE 18.2   Composition of Cows' Milk and Milk Soils

Soils Are % Dry Weight Basis	Whole Milk	Cold Milk Soil	Hot Milk Soil[a]	Milk Stone	
				Min	Max
Protein	3.6	26.6	30.3	4.1	43.8
Sugar	5.0	38.11	Trace	None	Trace
Fat	3.7	29.9	23.1	3.6	17.66
Minerals	0.7	5.3	46.6	42.3	67.3

Sources: Adapted from Harper, W.J., *Food Sanitation*, 2nd edn., AVI Publishing Company Inc., Westport, CT, 1980; Belitz, H.D. and Grosch, W., *Food Chemistry*, Springer Verlag, Berlin, Germany, 1987.

[a] From a tubular heater.

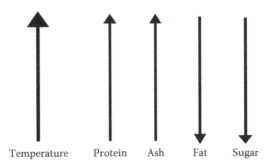

FIGURE 18.2 Effect of heat on milk soil deposition.

produce the curds, while soluble whey proteins are removed. These soils are much more tenacious than simple milk soils, so they are more difficult to remove, especially if allowed to dry. Cultured products like sour cream and yogurt contain bacteria and their by-products which changes the soil composition as well. Fruits, nuts, and flavorings added to frozen desserts and yogurts add to particulate debris and increased sugar content.

Since most of the components of dairy soils are organic in nature, alkaline and chlorinated alkaline cleaners are used. Alkalis are able to saponify the fat portion and emulsify all the organic portions of the soil, including the proteins. Chlorinated alkaline cleaners are even better able to remove proteinaceous soils. A minimum of 60 ppm chlorine in use solution is needed to clean effectively, and the efficacy improves with an increasing chlorine concentration up to about 200 ppm. Residual protein forms a bluish iridescent film on metal equipment and is easily removed with a chlorinated alkaline cleaner. To be effective on the mineral portions of the soil, and to combat the effects of hard water, alkaline cleaners must contain sequestering agents like phosphates or EDTA. The alkaline cleaners are generally used at 54°C–80°C (130°F–178°F). These warm temperatures help melt the fat (milk fat melts at −32°C or 90°F) and allow the cleaner to clean most effectively.

Sometimes, acidic cleaners are used *after* the alkali to remove any mineral deposits left behind from the milk (milk stone) or hard water (water stone). Milk and water stones are similar in appearance, usually a chalky white or grayish residue easily visible on stainless equipment. Alternatively, acidic cleaners are used *before* the alkali on burnt on soil to loosen the minerals that increase the tenacity of other soil components. Acid washes are generally done at ambient temperature. This acid wash would be followed by an alkaline wash to remove the soil completely from the surface.

Brewery Soils

Beer is a yeast catalyzed fermentation product of sugar containing liquids. Sugars are obtained from grains like barley, corn, and rice. The characteristic flavor of beer is due to essential oils and resins found in the hops plant and barley. The resulting product is a naturally carbonated, alcohol containing beverage. Brewing is both a science and an art that has been generations in the making. Clean equipment and environment are essential for successful brewing.

The composition of the average finished beer is as follows:

Water	93.0%
CO_2	0.4%
Alcohol	3.2%
Carbohydrate	3.0%
Protein	0.3%
Mineral	0.1%

Malted barley, hops, yeast, and water are the four major ingredients in beer and are responsible for resultant soils. There are no fatty soils, but lots of sugars and starches, proteins, polyphenolic compounds, resins, oils, and minerals. The barley, hops, and yeast as well as extraneous bacteria and mold all contribute protein content to the finished product and soil. Hops also contribute essential oils and resins that are the primary sources of the flavor and foaming characteristics of beer, and hops act as a natural defense against bacterial spoilage. Both hops and malted barley are sources of tannins (polyphenolic compounds), which contribute to the flavor and color of beer. Barley is naturally high in oxalic acid content, which precipitates out as calcium oxalate or "beer stone."

Carbohydrate components of beer and brewing soil (sugars and starches) are water soluble. They are easily removed with water alone or with alkaline cleaners. The proteinaceous soils are more difficult to remove and may require chlorinated alkaline cleaners. Proteins in beer often combine with the tannins resulting in a more tenacious soil. The acidic pH of beer (partially due to acidic hop resins) also contributes to the protein-tannin soil tenacity. When formed during boiling of the wort (the sugary liquid prior to fermentation), the deposit is called "hot break"; when formed during cooling of the wort, it is called "cold break." These tannin-protein complexes and resins and oils are all difficult to remove from surfaces, and improper cleaning will result in visible build up. In addition to the protein tannin complexes, fermentation tanks often have a light brown foamy residue, which is proteinaceous in nature and due to the yeast accumulating on either the top or bottom of the tank. Chlorinated alkalis are often used to remove all of the above deposits. Beerstone, which is a mineral deposit may be white or gray brown and is easily removed with acidic cleaners.

Alkaline or chlorinated alkaline cleaners are used for the majority of brewery soils. A concentration giving 1%–2% or even up to 5% causticity as NaOH will adequately neutralize soils. Chlorine at 400–800 ppm will remove proteins. The problem with alkaline cleaners is their incompatibility with the carbon dioxide (CO_2) present in tanks and lines. When caustic and CO_2 are mixed, carbonate and bicarbonate are produced, reducing the amount of NaOH available to clean. While carbonate and bicarbonate are both cleaning agents, they do not contribute as much alkalinity as caustic. The formation of carbonate also produces scales, which can block lines and must be removed with an acid cleaner. Besides reducing the efficacy of the caustic, valuable CO_2 is consumed in the reaction. When cleaning with caustic, care must be taken in CO_2 environments.

$$\underset{\text{caustic}}{2NaOH} + \underset{\text{carbon dioxide}}{CO_2} \rightarrow \underset{\text{sodium carbonate}}{Na_2CO_3} + \underset{\text{water}}{H_2O}$$

$$\underset{\text{caustic}}{2NaOH} + \underset{\text{carbon dioxide}}{CO_2} + H_2O \rightarrow \underset{\text{sodium bicarbonate}}{NaHCO_3}$$

or

$$NaOH + CO_2 \rightarrow NaHCO_3$$

and

$$NaHCO_3 + NaOH \rightarrow Na_2CO_3 + H_2O$$

Cleaning temperatures are generally 50°C–65°C for stainless surfaces and lower temperatures for bright beer tanks In many cellars, where the temperature is controlled, the tanks will be cleaned at the cellar temperature; this may be as low as 5°C (40°F).

Acidic cleaners are used on a periodic or regular basis to remove mineral deposits like beer stone. These acid washes are generally done after most of the soil has been removed with alkaline cleaners (Salisbury 1983).

Recently, there have been acidic cleaners developed for brewery use that are designed to replace alkaline cleaners. The special blend of acids and surfactants are able to remove the soils traditionally cleaned with alkaline cleaners.

When using acids, there is no concern about carbon dioxide incompatibility. Total acidic cleaning is still a relatively new concept and is increasingly accepted.

Food Processing Soils

A variety of soils are found in food processing plants ranging from simple sugars to fats and proteins. Often, a cleaning regime is needed that will be effective on a complicated soil matrix. Since most food soils are organic in nature, alkaline cleaners are generally used. The particular formulation is chosen based on the soil tenacity, surface type, and method of cleaning.

Heavy duty alkaline or chlorinated alkaline cleaners are often used on meat and poultry soils that are fatty, protein based, and generally heavy. For cooked or burnt on soils of any type, including soups, sauces, and gravies, heavy duty alkaline cleaners are used. If soiling is severe in equipment like fryers and kettles, the cleaners may be heated to 85°C–95°C (180°F–200°F) to increase effectiveness on carbonized soils. As much as 1%–2% causticity as NaOH is used for these applications. For fatty soils and all other proteinaceous soils including dairy soils, egg, and pasta-based soils, chlorinated alkalis are usually used. Phosphoric or organic acid cleaners are often used as a follow up to alkaline cleaners if mineral deposits are a problem. Heavy duty acid cleaners are used where severe scaling is a problem due to water quality, food stone, or improper rinsing of alkaline cleaners. Greases and oils from processing machinery are removed from floors and equipment with solvent cleaners.

Foods containing eggs or egg products are generally the most difficult to clean in processed food plants. Because of the desirable properties eggs impart to foods, they are used in many foods including pastas, breads, sauces, dressings, and baked goods. Egg whites are basically soluble proteins that become insoluble (denatured) upon heating. They begin to coagulate at 62°C. Yolks are fat in water emulsions that contain approximately one-third protein and two-thirds fat. The protein in egg yolks begins to denature or coagulate at 65°C. Both whites and yolks also contain trace amounts of minerals.

While from a cleaning standpoint the coagulation of eggs is undesirable, that is, the single property that makes eggs so useful in so many foods. When cooked, the proteins denature and help hold food together. The denaturation of egg proteins upon agitation also makes for foaming properties that improves the texture of many whipped foods like puddings, batters, and dressings. Finally, the ability of eggs to help emulsify other ingredients makes them valuable to the food processor. Mayonnaise is an oil in water emulsion held together by eggs. Heavy duty alkaline cleaners and chlorinated alkaline cleaners are needed to clean the egg protein deposits on surfaces. The removal of egg-based foods is much easier if the soils are not allowed to dry or be excessively cooked onto surfaces.

Meats have a high moisture content(~76%) and a high protein content (~21%). The fat content varies with the cut of meat and the animal. Soluble proteins only account for 20%–30% of the total protein content, while the rest is insoluble (the flesh itself). Fish flesh contains similar levels of protein with varying fat and water content. Both protein and fat are difficult to remove from surfaces. Heated alkaline solutions are needed to melt and saponify fat, but excessive heat can cause fats to polymerize and become more difficult to remove. A temperature of ~5°C above the melting point of the fat is desirable. Since animal fats are mixtures of fatty acids, they have wide melting ranges. Beef tallow has the highest melting range at 40°C–60°C. Cleaning temperatures of 50°C–60°C adequately melt most animal fats. Alkaline cleaners are more effective on proteins at higher temperatures, but again excessive temperatures can cause proteins to denature and adhere more tightly to surfaces. Temperatures in the 40°C–60°C (100°F–140°F) range are used. It is important to thoroughly rinse gross particles of blood, meat, and fish products from surfaces before cleaning with hot solutions to avoid "baking" the soils onto the surface. Chlorinated alkaline foam or spray cleaners are used for most processing areas and equipment.

Doughs, batters, and pasta all contain eggs, which is why these food soils are difficult to remove. Cooked and dried soils of these types are even more tightly bound to surfaces because of protein denaturation and sugar caramelization. Pastas contain an average of 14.5% protein. Heavy duty alkaline or chlorinated alkaline cleaners at 50°C–60°C (120°F–140°F) are usually able to remove even baked or burnt residues, but acid cleaners are sometimes needed to loosen soils before the alkali wash. Mineral residues are not a big problem in processing areas dealing with these types of foods, unless hard water is used for processing and sanitation, or if the rinsing of alkaline cleaners is incomplete.

Rice and other grains leave very starchy soils with some protein. Alkaline or chlorinated alkaline cleaners are used at temperatures of 60°C–70°C (140°F–160°F).

Since vegetables are mostly composed of sugars and starches, their soils are generally easy to remove with alkaline cleaners at 40°C–60°C (100°F–140°F). The protein and fat content of vegetables is so small that they contribute little to soiling. However, some fat soluble substances like carotenoids that are responsible for the colors in some vegetables (e.g., carrots, tomatoes, and peppers) can cause staining. These fat soluble pigments can be a problem on plastic surfaces, requiring chlorine for complete removal. The mineral content of most vegetables is around 1%, but some vegetables high in oxalic acid (an organic acid) have a tendency to leave "stone" deposits, especially on blanchers. Oxalic acid combines with calcium and magnesium to form insoluble mineral salts. Vegetables with high oxalic acid content include celery, spinach, rhubarb, green beans, and beets (Plett 1985).

The surface to be cleaned is also a consideration in choosing a cleaner. Food plants have a complicated process machinery that speeds up production, but makes sanitation difficult. A variety of soft metal and plastic parts are included with the stainless steel and black iron. Many parts must also be disassembled for proper cleaning. Heavy duty alkaline and chlorinated cleaners are safe on stainless steel equipment, while soft metals and some plastic or rubber surfaces can only tolerate light or medium duty cleaners. Heavy duty cleaners used at excessively high temperatures accelerate the deterioration of the rubber gasketing material. Alkaline cleaners with or without chlorine must be silicated to be safely used on soft metals like aluminum. Walls, ceilings, and floors are usually cleaned with medium duty products to ensure worker safety.

A great deal of food processing equipment is cleaned-in-place. Equipment that is closed and is suitable for circulation cleaning includes pasteurizers, evaporators, transfer lines, silos, ovens, blanchers, and some cooking vessels. Heavy duty alkaline, chlorinated alkaline, and acid cleaners are used for the majority of CIP (Clean-in-Place) applications. Foam, gel, and spray cleaners are used for most other processing and filling equipment, conveyors, floors, and walls. Many surfaces are brushed and hand scrubbed after spray, gel, or foam is applied. Some equipment like rotary cutter fillers, extruders, and other portioning equipment must be disassembled and hand scrubbed or soaked to ensure cleanliness.

Films on Equipment

Visible films on "clean" equipment indicate inadequate or improper sanitation procedures. Films can be the result of poor water quality, cleaner surface incompatibility, improper choice of cleaner for a soil or water conditions, and inadequate rinsing of chemicals. These films can embed micro-organisms of many different species, which can be embedded in a protective polysaccharide layer. The removal success for such biofilms depends on proper diagnosis. There are many companies claiming that chlorine dioxide is effective in preventing biofilm formation and in killing biofilm micro-organisms. Compared to chlorine, it seems that is more efficient (factor 5–10) in killing sessile micro-organisms. (Hosni et al. 2006)

Table 18.3 is provided as an illustrative guide to troubleshooting, indicating the appearance, causes, and approaches to removing different visible films.

Standards of Cleanliness

In order to define clean, one must first define dirt or soil. Soil is simply anything that does not belong or is unwanted. The most obvious soils common to food and beverage processing facilities are raw

TABLE 18.3 Appearance, Causes, and Approaches to Removing Soils from Equipment Used in Food Processing

Appearance	Deposit	Causes	Removal
White or gray chalky	Mineral ("stone")	1. Minerals from water or food 2. Poor detergent choice 3. No acidic cleaning	Acid wash
Blue-iridescent	Protein	1. Inadequate cleaning 2. Lack of use of chlorine	Chlorinated alkaline detergent
White or gray glazed	Silicate	1. Poor rinsing of silicated cleaners 2. Inadequate cleaning with silicated cleaners	Acid wash
Blue cast	Wetting agent	Inadequate rinsing of surfactated cleaners	Thorough rinsing
Brown particulate	Fruit lignins and pectins	Inadequate removal of fruit residues	Chlorinated alkaline detergent
Greasy	Fats or oils	1. Cleaning temperature too low 2. Inadequate detergent concentration	Proper cleaning

ingredients, finished products, water deposits (hardness, iron, etc.), machinery greases and oils, and chemical residues.

Clean is an unnatural state, because it requires energy to achieve. One does not have to apply energy to soil a surface, so it is a more natural and common state. Absolute cleanliness can never be attained. When cleaning a surface one or more of the following types of cleanliness is desired: physical, chemical, or microbiological.

Physical cleanliness is defined as being free from visible soil or films. This is determined by visual examination and requires no special tools or training. One can simply look at and touch a surface and see if it is physically clean. If a surface appears clean, and it is smooth to the touch with no greasy or gritty feel and no objectionable odor, it is physically clean. This simple inspection affirms the absence of gross amounts of unwanted matter.

A surface is chemically clean if there are no microscopic residues of soil remaining and no residual detergents or sanitizing chemicals to contaminate the food product. Determination of chemical cleanliness requires tools other than the human eye. Some techniques for determining such small amounts of soil compared to nonsoiled control are listed in Table 18.4. The definition of clean depends on the sensitivity of the method.

Upon visual inspection, a surface can appear clean, but actually has a microscopic layer of soil that becomes apparent when enhanced. Since water alone cannot easily wet a soil surface, wettability is used as a means for confirming the presence of residual soil. When water is applied to a clean metal or glass surface, it will sheet completely for a period of time. Soiled surfaces become hydrophobic, so water will bead up and the sheet will "break." This is a fairly sensitive test and practical for field use on large surfaces including the interior of storage tanks and silos, but is not quantifiable in the laboratory.

TABLE 18.4 Methods of Determining Residual Soil

Technique	Characteristic Measured
Visual	Wettability of surfaces; dyes
Gravimetric	Weight of remaining soil
Optical	Reflectance or absorption of light
Tracers	Soils labeled with radioactive materials
Chemical	Removal and analyses of remaining soils
ATP	Remaining adenosine tri phosphate level

Since plastic and other synthetic surfaces are naturally hydrophobic, this technique is not appropriate for those surfaces.

Dyes with affinity for certain soils like protein or starch can also be applied to a surface to make the soil visible. This only indicates the presence of a particular soil type, while most food soils are heterogeneous mixtures. Hence, this is also not a quantifiable method.

A small clean piece of test surface can be weighed before soiling, and again after it has been soiled and cleaned to see if traces of soil remain. The gravimetric approach is only a viable alternative for small laboratory tests and cannot be used to evaluate cleanliness in a food processing facility. This method also has a high degree of error involved in it due to the small weights being measured.

Optical methods are used quite regularly in some food processing operations for detecting gross soils. Interruption of a light beam, reflectance, and absorbance are all optical means of detecting soil. Electronic bottle inspectors pass a beam of light through returnable beverage containers to check for gross soil or debris. Reflectance of light of a specific wavelength off of a surface can be used as an indication of cleanliness. This is often used in the laundry industry for fabrics, but can be applied to some hard surfaces as well if the finish does not interfere. The absorption of light by a soil is another optical means of detecting soil.

Radioactive labels can be incorporated into soils that can then be used to evaluate soil load. Because of the health risks involved with radioactive materials, the use of this is prohibited in food plants and is limited even in laboratories.

Finally, soils can be dissolved and removed from surfaces using chemicals. The removed substances can be analyzed for content to measure soil type and amount. This requires the use of expensive analytical machinery and great expertise.

Microbiological cleanliness is achieved when surfaces are *free from disease-causing* or *food-spoilage organisms*, including bacteria, yeast, and mold. This is the result of thorough cleaning and a sanitizing step. In most cases, this means that organisms that do not pose a health or spoilage threat may remain on the surface. Certain industries where aseptic packaging is used require some surfaces to be sterile or free of organisms of any type, and their spores. Sterilization requires special chemical, heat, or ultraviolet treatment beyond the traditional cleaning and sanitizing steps.

Equipment can be microbiologically clean, yet physically dirty. More often though, surfaces are physically clean, but microbiologically dirty. Viable and nonviable microbes can be considered part of the physical soil. Oftentimes, they will be dead, but still present on the surface contributing to soil load. It is much easier to achieve microbiological cleanliness if the surface in question is physically clean. Soil provides attachment sites and nutrients for organisms, and protects them from the action of detergents and sanitizing agents. It has been shown that the cleaning step removes most of the organisms from surfaces, with sanitizing chemicals killing those few that remain.

Organisms remaining on surfaces are undesirable because many have the ability to attach to surfaces and proliferate, leading to the formation of biofilms. When bacteria contact surfaces, they excrete materials that allow them to become more firmly attached. Such materials are often mixtures of proteins and sugars, and together with the organisms themselves are called biofilms. Once a foundation is established, successive layers of organisms and soils are deposited more quickly, leading eventually to visible slimy material.

Biofilms are aesthetically unpleasant because they can become slimy, foul-smelling masses, but they can be responsible for other problems as well. Trapped organisms can slough off into the food product during processing, contributing to spoilage and off flavors. Organisms in biofilms are concentrated in small areas, so their waste products build up. This often results in acidic conditions that can accelerate the corrosion of metals and the deterioration of other materials. Biofilms can also become heavy enough to block processing equipment and filters.

The goal of most food processing cleaning and sanitation programs is to achieve a physically and microbiologically clean surface. There are many techniques for enumerating organisms on surfaces, some of which are in use in processing facilities today. Most practical methods only measure the viable organisms. Surface finish and equipment accessibility play important roles in determining which technique is used.

Unfortunately, there are no universally accepted standards for the allowable levels of micro-organisms on surfaces. Many industry experts have proposed such standards, but it is an area of ongoing research. Plett has cited accepted international standards for farm bulk tanks as 50 micro-organisms/cm^2 and 2 micro-organisms/cm^2 for other dairy processing equipment. The risk factors for various levels of many organisms must be assessed for each food product, process, and the consumer. Currently, individual plant surface monitoring programs, and the microbiological contents of finished products are used to assess surface contamination. Finished product counts do not necessarily correlate to surface cleanliness. Surface counts do not indicate the degree of physical cleanliness of the surface, or the potential for the growth of any remaining organisms. Injured organisms may not show up on surface counts, giving a false indication of microbiological cleanliness.

The standard of cleanliness required varies with the equipment surface, the point in the processing, the environment, and the types of food product and their perishability. Food contact surfaces like the interior of pipelines and tanks obviously require a higher degree of cleanliness than nonfood contact surfaces like the exterior of equipment and walls. But excess soil on nonfood contact surfaces is still undesirable as it can be a source of postprocessing contamination and general unhygienic conditions in the plant. Highly perishable foods, such as meat, dairy products, and chilled prepared items require scrupulously clean surfaces and surrounding environments to minimize the chance of bacterial contamination. Canned and bottled foods are at lower risk for postprocessing contamination due to the heat treatment they receive, and the preservatives and limited oxygen content in the container. Some bottled beverages are relatively safe from contamination due to the CO_2 content and acidic nature. Others like juice-containing drinks are more susceptible.

Soil build up can compromise the efficiency of processing equipment. Membranes, filters, and screens can easily become plugged if not kept completely clean. Soiled heating surfaces can lose the ability to effectively transfer heat, and flow will be slowed in pipelines that have heavily soiled areas.

ATP (adenosine triphosphate) bioluminescence technology, used to monitor plant cleanliness and hygiene, is one of the most significant technologies to impact food processors in recent years. ATP becomes very effective by providing immediate feedback to ascertain the cleanliness of critical control areas. ATP is present in all organic material, and is the universal unit of energy used in all living cells. ATP is produced and/or broken down in metabolic processes in all living systems. Processes, such as photosynthesis in plants, muscle contraction in humans, respiration in fungi, and fermentation in yeast are all driven by ATP. Therefore, most foods and microbial cells will contain some level of naturally occurring ATP. The luminometer uses bioluminescence to detect residual ATP as an indicator of surface cleanliness. The presence of ATP on a surface indicates improper cleaning and the presence of contamination, including food residue, allergens, and/or bacteria. This implies a potential for the surface to harbor and support bacterial growth.

Swabs are moistened with a buffer that aids in the removal of any biological material (ATP) on either wet or dry surfaces, while also penetrating through any biofilm to expose underlying cells. The ATP from microbiological cells, in addition to free ATP from any food residue, is collected from the sample surface with the swab, and is then available to react with the unique liquid-stable reagent contained in the device. This reagent is derived from a naturally occurring enzyme (called luciferase) found in fireflies. When this enzyme reacts with ATP on the swab, a low level of light is produced that can be detected and quantified by a luminometer. The amount of light detected is directly proportional to the amount of ATP on the sample, thus giving a quantitative measure of the cleanliness of the surface where the sample was taken.

Summary

Cleaning in the food processing industry is not an easy task; however, it is a critical step within the food production industry and is crucial to maintain and to guarantee food safety.

Understanding the correct balance between the soil, the surface, the process conditions, and the application method will provide a satisfying hygiene result. Combined with the right chemicals and controlling methods, the correct cleaning standard can be obtained and repeated at every required cleaning interval.

References

Alfa Laval Corp. 1986. *Dairy Handbook*. Teknisk Dokumentation AB, Alfa Laval Corp., Lund, Sweden.

Burton, H. 1968. Deposits from whole milk in heat treatment plant—A review and discussion. *Journal of Dairy Research*, 35, 317–330.

Dunsmore, D.G. 1981. Bacteriological control of food equipment surfaces by cleaning systems, I. Detergent effects. *Journal of Food Protection*, 44(1), 15–20.

Harper, W.J. 1980. Sanitation in dairy food plants. In: Guthrie, R.K. (ed.). *Food Sanitation*, 2nd Ed. AVI Publishing Company Inc., Westport, CT.

Hosni, A.A., Jang, A., Coughlin, M., and Bishop, P.L. 2006. Diffusion of chlorine dioxide through aqueous and oil films. *Proceedings of the Fouling, Cleaning and Disinfection in Food Processing Conference*, Cambridge, U.K.

Plett, E.A. 1985. Cleaning of fouled surfaces. In: Lund, D.B. et al. (eds.). *Fouling and Cleaning in Food Processing*. University of Wisconsin-Madison Extension Duplicating, Madison, WI.

Salisbury, M. 1983. Cleaning of storage vessels in atmosphere. *Brewers Digest*, 58(6).

Bibliography

Belitz, H.D. and Grosch, W. 1987. *Food Chemistry*. Springer Verlag, Berlin, Germany.

Czechowski, M.H. Gasket and stainless steel surface sanitation Vl. In: *The Importance of Cleaning Surfaces*. Diversey Corporation Publication.

Fryer, P.J. and Belmar Beiny, M.T. 1991. Fouling of heat exchangers in the food industry: A chemical engineering perspective. *Trends in Food Science & Technology*, 2, 33–37. Reference Edition. Elsevier Science. (O'Brien, editor).

Jensen, J.M. 1970. Cleanability of milk filmed stainless steel by chlorinated detergent. *Journal of Dairy Science*, 52(2), 248–251.

Kane, D.R. and Middlemiss, N.E. 1985. Cleaning chemicals state of the knowledge in 1985. In: Lund, D.B. et al. (eds.). *Fouling and Cleaning in Food Processing*. University of Wisconsin-Madison Extension Duplicating, Madison, WI.

Marriott, N.G. 1989. *Principles of Food Sanitation*, 2nd Ed. Van Nostrand Reinhold, New York, pp. 71–120.

Rouillard, C. 1994. *The Chemistry of Cleaning*, Vista 6. Diversey Corporation, Plymouth, MA.

19

Electronic Assembly Cleaning Process Considerations

Introduction 283
Why Clean Electronic Assembly and Advanced Packages? 284
Electrochemical Migration Risk 284
Cleaning Challenge
Cleaning Process Design 286
Miniaturization 286
Miniaturization and Lead-Free Drive Solder Flux Requirements 287
Soldering Process Issues Create New Demands on Flux Design • Flux Types
and Their Impact on Cleaning
Science of Electronic Assembly Cleaning Agents 289
Building Blocks • Solvent Vapor Cleaning • Semi-Aqueous • Aqueous
Matching the Cleaning Agent to the Flux Residue 295
Process Cleaning Rate Theorem • Flux Residue Solubility Properties
Materials Compatibility 297
Thermodynamics versus Kinetics 299
Thermodynamic Effects • Kinetic Energy Effects • Vapor Phase Cleaning
Equipment Considerations • Semi-Aqueous Cleaning Equipment
Considerations • Aqueous Batch Spray in Air Cleaning Equipment
Considerations • Aqueous Inline Cleaning Equipment Considerations
Process Validation 312
Conclusion 315
Acknowledgments 315
References...315

Mike Bixenman
Kyzen Corporation

Introduction

Electronic device complexity brings to light an increasingly important phenomenon: designing for reliability. Product longevity—or the lack thereof—becomes more difficult when designing and manufacturing integrated circuit packages.[1] As integrated circuit manufacturing processes become more advanced, consumer demand for greater performance and system functions increases. Consistent reliability over the expected life of the product is a driving factor for designers and assemblers.

Cleaning flux residues post-soldering has been a high reliability criterion practiced by assemblers of military, aerospace, automotive, medical devices, and other value offerings. Highly dense designs reduce component spacing and standoff heights. The complexity of removing flux residue increases as standoff heights reduce. Leaving flux residue under low clearance components elevates the risk of ionic residues under low standoff components.[2] Reliability is a major concern that places increased importance on manufacturing processes that completely remove all flux residues on the surface of the assembly as well as under low clearance components. Integrated package assemblers are experiencing difficult cleaning challenges that must be overcome before they can produce reliable products that meet customers' increasing demands for faster and smaller products with higher performance.

Why Clean Electronic Assembly and Advanced Packages?

Electronic technologies have revolutionized home and consumer electronic items and have simplified complex functions in everyday life. Advances in electronics and packaging technology enable decreased size and increased features. Along with miniaturization, increased features and capabilities drive input toward functional, performance, and interface requirements.

The size of component features in 3D packaging and silicon integration reduces the space between conductors. During soldering and joining operations, flux residue wets and bridges the anode and cathode. Early circuit board failure represents a dominant concern for assemblers.

Leading-edge design failures are attributed to feature size reduction, which increases the risk of defects randomly induced by process flaws.[3] Flux residue sandwiched under high-density low standoff components creates the potential for material defects caused by the presence of ionic residue.[4] Traditional approaches to verifying reliability based on screening the output of a process are no longer effective. Rather, integrated device packagers are moving toward building-in reliability by controlling and monitoring cleaning factors that assure complete removal of flux residue.

Many industry standard cleaning processes do not remove the residue under low clearance components. Heat and voltage can cause ionic residues to migrate in the form of spherical solder patterns that eventually join and form a solder bridge across the conductors. For many of these devices, wear-out-related failure is less critical in determining product reliability over the useful lifetime of the device. The critical factors move upstream to designing a process that assures reliability.

Electrochemical Migration Risk

Electrochemical migration occurs by forming a conductive metal bridge between conductors when they are subjected to a DC voltage bias.[5] Metal conductors grow from a positively charged conductor (cathode) to a negatively charged adjacent conductor (anode), creating a short circuit between the conductors. Electrochemical migration consists of metals plating in reverse. An ionic contaminant combines with an electrolyte (water), generally forming a localized source of acid, which dissolves metal ions. Under the influence of the electrical potential, the metal ions move across the intervening space, plating out on the laminate as it goes, forming a metal filament. A more contaminated assembly has greater risk than a clean assembly.

Ion migration in an electric field is propagated by the charge balance at the interface, where the total current density entering and leaving the electrolyte causes metal ions to split and form dendrites. The spatial coupling depends on the distance of the conductors, with closer areas coupled more strongly.[4] The input/output (I/O) current decreases for smaller electrodes, while the current density increases. This phenomenon can create high current densities in high I/O devices. Highly dense assemblies and the spatial coupling are connected, and depend strongly on the geometry of the electrolyte and electrode.

Electrochemical migration is mediated by removing all flux residues, which eliminates ion migration under the influence of the electric field.[6] Partially cleaned flux residue leaves an ionic oxide film (white residue) that creates a path for metal dissolution reactions. The flux bridges between conductors, and an electrical current allows the ionic residue within the flux to disassociate and propagate.

Cleaning Challenge

Miniaturization and higher functionality in electronics packaging require the use of advanced packages and small components. This trend has translated into the use of new package types such as quad flat pack (QFP, also referred to as leadless plastic packages), increased use of chip scale packages (CSPs), as well as increased component density and tighter printed circuit board (PCB) layouts. Advanced package innovations further complicate post-assembly cleaning. It is often a challenge to clean flux residue from under these highly dense and advanced packages after reflow soldering.

Technology-based market pressures increase reliability demands as electronic assemblers continue to drive miniaturization and cost. High-performance integrated devices will be driven by multi-chip density, system in package, and chip on board, which increase the number of transistors, decrease area array pitches, and result in tighter component standoff heights (Figure 19.1).

Higher-density, smaller components, and lower clearance are changing device input requirements.[7] Design engineers must consider device operational capabilities and processing of inputs and resultant outputs. To meet performance requirements, issues such as speed, strength, response times, accuracy, and limits of operations must be considered. Temperature, humidity, shock, vibration, and electromagnetic compatibility represent important environmental factors. Interface requirements are critical to compatibility with external systems.

Chip caps, flush-mounted to the board, create a flux dam under the component during reflow.[8] The flux dam seals the underside of the component with flux residues that are difficult to remove completely. Devices placed in tightly packed arrays further increase the cleaning difficulty, as there is very limited access for the cleaning fluid to reach the contaminant. To clean under chip caps requires both improved chemical and mechanical designs.

Flux residues clean at different rates based on the flux makeup, time after reflow, reflow temperature, and the cleaning fluid design.[9] Water-soluble flux residues typically clean at a faster rate than do rosin flux residues, which typically clean at a faster rate than low solids synthetic flux residues. Flux residue becomes more difficult to clean with the passage of time after reflow. Higher reflow temperatures allow

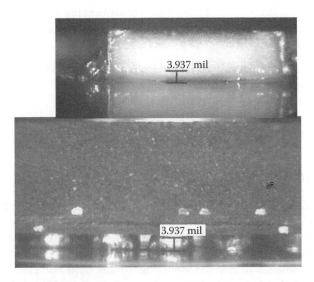

FIGURE 19.1 Component cleaning challenges.

the lower molecular weight solvent molecules to evaporate at a faster rate, leaving higher molecular weight resin molecules, which increases the difficulty of cleaning the residue.

Cleaning Process Design

Process engineers looking to clean integrated package designs typically start by identifying the cleaning equipment and cleaning material. Many use a small subset of test vehicles to qualify the process. This technique worked well when qualifying a cleaning process for conventional designs but typically fails to address numerous factors when designing leading-edge cleaning processes. The design of the cleaning process must dig into critical factors such as the integrated package designs cleaned, characterization of the soil, throughput requirements, materials compatibility, cleaning material, process parameters, cleaning equipment, spray impingement, time in the wash, temperature of the wash, controlling the chemistry, ventilation, water management, environmental constraints, and cost of cleaning.

Design for manufacturability (DFM) includes a set of techniques to modify and improve cleaning process designs in order to match the cleaning process to the soil, substrate, cleaning agent, and available cleaning methods. Designing integrated packages for ease of manufacture needs to begin with the designer's awareness of the cleaning process and its limitations. Poor solubility for the soil limits the ability to penetrate, wet, and remove the soil under low components gaps. The problem is that the driving force toward smaller components makes effective removal of flux residues more difficult.

Cleaning agents have evolved over the last few years due to the change in technology, primarily miniaturization and variations in flux residues. With miniaturization driving more complexity into the entire manufacturing process, especially with regards to cleaning, there is an emerging need to look deeper into the science of cleaning.

A deeper understanding between the uniqueness of one chemistry versus another chemistry needs to start earlier in the process. The uniqueness of cleaning agents is the ability to engineer materials using a wide array of constituents properly designed to match up with soils, cleaning equipment, materials compatibility constraints, and process requirements. One of challenges is that miniaturization is driving multiple changes in flux technology, which ultimately influences the ability to clean.

To assure complete flux removal, it is increasingly important to match the solubility parameters of the cleaning agent with the flux residue. Rather than recommending a specific cleaning agent, a five-step DFM process has been developed to match the soil, cleaning agent, part unique considerations, and cleaning equipment to the cleaning need. The process works as follows:

- Establishing the solubility parameters for independent flux soils is the *first critical step*
- Matching cleaning agents to the flux is the *second critical step*
- Materials compatibility is the *third critical step*
- Confirmation of cleaning equipment is the *fourth critical step*
- Process validation is the *fifth critical step*

Miniaturization

Moore's law holds that the number of transistors on a chip doubles approximately every 2 years.[10] Consistent with Moore's law, high-reliability electronic devices build faster processing speed and memory capacity using increasing smaller platforms. The trend toward highly dense assemblies reduces the spacing between conductors, while yielding a larger electronic field.[2] As the industry moves to higher functionality, miniaturization, and lead-free soldering, studies show that cleanliness of the assembly becomes more important. Residues under low standoff components, with gaps less than 2 mils, represent an increasingly difficult cleaning challenge.

Miniaturization and Lead-Free Drive Solder Flux Requirements

Higher functioning devices in smaller platforms are an overriding trend in the electronics industry.[11] For state-of-the-art board-level assembly, the mechanical hold size and pitch is 0.125 and 0.3 mm, respectively, while the nonmechanical hold size via diameter and pitch is 0.05 and 0.175 mm. For components, among the finest are CSP with 0.3 mm pitch and 01005 (10×5 mils) chip caps.[12] For portable devices, the leading-edge package standoff is 0.3 mm, and the pad diameter is 0.2 mm. Additionally, the cost of board assembly conversion has driven I/O cost to less than 0.2 cents, with a board assembly escape rate of 200 ppm.[13]

The move toward lead-free soldering and miniaturization represent two converging force fields that increase no-clean complexities.[11] Higher lead-free melting temperatures requires the use of fluxes with greater thermal stability. The problem is that lead-free alloys exhibit poorer wetting properties, which require higher flux capacity and strength to improve wetting and flow. Flux technology also plays an important role in reduced voiding by increasing the need for high oxidation resistance, oxygen barrier capability, high thermal stability, and low volatility.[14] Miniaturization requires the flux to be more stable at peak reflow to prevent oxidation, which requires a higher content of rosin (natural pine) or resin (modified or synthetic pine). Reducing flux volatility has a trade-off of greater amounts of circuit assembly flux residue. Halide free flux materials require higher levels of weak organic acids, which increase the level of ionic materials that can form an electrochemical cell.[14]

Soldering Process Issues Create New Demands on Flux Design

Flux technology needed to support miniaturization and lead-free circuit assemblies differs considerably from eutectic tin/lead (Sn/Pb) solder systems.[11] Factors such as higher soldering temperature, flux consistency, oxide penetration, and flux burn-off lead to graping, head-in-pillow, wetting, spattering, and tombstoning. These undesirable effects create the need for new flux designs. Many of these terms are visual descriptions developed by electronics assemblers. For example, graping is where very small solder joins resemble a tiny bunch of grapes. With a head-in-pillow, a cross-section of the component resembles a head pressed into a soft pillow. Tombstoning is a phenomenon where one end of the chip component becomes detached from the board while the other end remains bonded. Environmental factors are pushing the need to eliminate halides from flux compositions, which requires additional changes when engineering new flux designs.[14] These driving forces not only influence the material flux properties but also change the flux residue solubility parameters. As a result, cleaning agents will also need to evolve to keep pace with these changing flux packages.

Removal of flux residues trapped under low clearance components requires a cleaning agent that exhibits like solubility for the flux residue. The cleaning agent solvency behavior is proportional to the cohesive energy needed to soften, dissolve, penetrate, and remove the soil.[15] The cohesive energy of the flux residue must be overcome by the cleaning agent's tendency to rapidly dissolve the flux crystalline lattice.

Miniaturization is driving multiple changes in flux technology, which ultimately influence the solubility characteristics of remaining flux residue with the cleaning agent.[11] Small feature sizes require finer powder, with the flux employing higher thixotropic and homogeneous properties to achieve satisfactory consistency in printing and soldering.[16] Also, because of miniaturization, the assembly process is more vulnerable to bridging; therefore, the solder paste needs to be more slump resistant.

The volume of soldering materials, including fluxes and solder, reduces in proportion with decreasing pitch.[17] When the solder materials are shrunk in proportion to the pitch, the thickness of metal oxide does not shrink in proportion.[16] Consequently, the amount of oxide to be removed by unit volume of flux increases with decreasing pad dimension. To compensate for this increasing workload, the fluxing capacity per unit amount of flux needs to be increased.

While the pitch and pad size decreases, the oxygen penetration path through flux or solder paste also decreases.[16] This inevitably results in a more rapid oxidation of both flux materials and metals covered by the flux if soldered under air. Hence, a flux with greater oxidation resistance as well as a greater oxygen barrier capability is needed for finer pitch applications.

The vaporization rate of solvents used in flux compositions increases with increasing exposure area per unit volume. This can cause flux burn-off, which increases with decreasing flux quantity deposited.[17] To offset this unfavorable trend, the flux employed for finer pitch needs to be more nonvolatile, hence more resistant to flux burn-off.

Although good wetting is a desired feature at soldering, fast wetting is actually causing problems at reflow soldering.[16] Defects due to unbalanced wetting force, such as tombstoning or swimming, increase with decreasing component size, partly due to increasing sensitivity toward miss-registration. In this situation, fluxes with a slower wetting speed would allow more time for the wetting force to be balanced, and therefore promise a lower defect rate.[18]

Miniaturization brings the solder joints closer to the gold fingers, hence is more vulnerable to solder spattering.[16] Spattering can be caused by moisture pickup of the solder paste. It can also be caused by fast solder coalescence action. To minimize solder spattering, fluxes with low moisture pickup and slow wetting speed should be employed.[18]

With decreased component features, the job of the solder flux is more difficult due to more oxides and easier oxygen penetration. Desirable flux compositions will require oxidation-resistant compounds to prevent the flux from being oxidized, and oxygen barrier capability to protect parts and solder from being oxidized. The flux residue solubility parameters will also change, which creates the need to first characterize the soil when designing the cleaning process.

Lead-free is not a simple "drop-in" replacement as far as solder paste fluxes are concerned.[16] Each element and its resultant alloy has unique characteristics with regard to oxidation, surface tension, reflow, and wetting. Use of higher soldering temperature causes increased flux thermal decomposition, flux burn-off, and oxidation of fluxes and metals. Flux compositions must be designed to address these issues and other important factors such as solder balling, wetting, voiding, grain structure, solder joint appearance, shelf life, and tack life. Flux compositions will require increased thermal barrier, oxidation resistance, and activator system.[19]

Flux Types and Their Impact on Cleaning

Both lead-free and miniaturization requires flux packages to be upgraded.[14] In general, an increase in flux capacity and oxygen barrier ability on the top of existing flux systems will be needed. Flux chemistry with a soft residue will be easier to clean than that with a hard residue. This is due to the lack of crystallization formation in the soft residue, which allows the residue to be dissolved in cleaning agents that match up well with the soil structure. For hard residues, the crystal flux residue formation must be dissolved by overcoming the crystallization energy before the residue molecules can be pulled away from the main residue body.[14]

Lee identified chemical changes needed to build thermal stability, oxidation resistance, oxygen barrier, and low surface tension flux compositions (Table 19.1).[14] High thermal stability requires higher molecular weight cyclic or aromatic structures. To resist oxidation, flux compositions will require resistant chemical bonds, such as aromatic, hydrocarbon, and silicone structures. A high oxygen barrier requires highly cross-linked cyclic aromatic covalent bonds. To lower surface tension, hydrocarbon and silicon materials will be needed. Each of these changes requires high-solvency cleaning agents that dissolve the crystalline flux matrix. Temperature, time, and mechanical energy will be important factors in cleaning the flux residue.[14]

Water-soluble flux compositions require oxygenated structures high in polarity that hydrogen-bond with water. Functional groups include amides, sulfides, oxygen, nitrogen, and sulfur.[14] These highly polar residues may require water plus an additive. Aqueous reactive cleaning agents used at low concentrations in water will be needed to remove many of these residues.

TABLE 19.1 Developed by Lee (2009), Identifies Flux Features Desired in Order to Meet Miniaturization and Lead-Free Challenges

	Challenge				
Flux Feature Desired	High Temperature	High Solder Surface Tension	Miniaturization	Air	Environment
High thermal stability	X				
High resistance against burn-off	X		X		
High oxidation resistance	X		X	X	
High oxygen barrier capability	X		X	X	
Low surface tension		X			
High fluxing capacity/strength		X	X		
Slow wetting speed			X		
Low moisture pickup			X		
High hot viscosity			X		
Halogen-free					X

Source: Lee, N.C., Lead-free flux technology and influence on cleaning. *SMTAI Technical Conference*, San Diego, CA, October 2009. With permission.

Science of Electronic Assembly Cleaning Agents

Flux residues are engineered with multiple components, each possessing a different solubility parameter. In designing electronic assembly cleaning agents, best practice starts by first characterizing the flux soils. Flux residues are multicomponent compositions. During soldering, flux compositions see different temperature gradients. Multiple exposures at soldering temperature can result in a tenacious residue that is difficult to dissolve. The better matched the cleaning agent is to the flux residue(s), the broader the process window.

Building Blocks

Electronic assembly cleaning agents are designed to fit three technology groups: solvent vapor degreasing, semi-aqueous (solvent cleaning followed by water rinsing), and aqueous designs. The building blocks (Figure 19.2) used to formulate these cleaning materials are solvency to dissolve resin structures; reactive agents to buffer and saponify soils; wetting agents to lower surface tension and improve penetration under low standoff components; and nonreactive additives to improve materials compatibility and control foam propagation under high pressure.[20] Best-in-class engineered compositions are designed around performance, environmental responsibility, low volatility, nonflammability, worker safety, equipment and materials compatibility, and cost of ownership.

Solvency

Dissolution is an important driving force for cleaning today's flux residues from highly dense circuit assemblies. Solvent dissolution is a thermodynamic process, and is proportional to the cleaning agent's cohesive energy for the soil. The rate of dissolution depends on the solvent, composition of the flux residue, temperature, impingement pressure, and interfacial surface tension.

Dissolution consists of solvating flux vehicles, which commonly represent rosin, resin, and synthetic polymer structures. A wide range of solvents are used in cleaning agents, to facilitate dissolution of rosin and resin (solute) structures naturally present in many flux residues. Matching the cohesive energies of resin/rosin structures with solvent(s) formulated into the cleaning agent increases the dissolution rate. Solvents are selected on the basis of "like dissolves like," commonly referred to as the solvated state, whereby organic rosin and resin structures are dissolved by solvent molecules.[15]

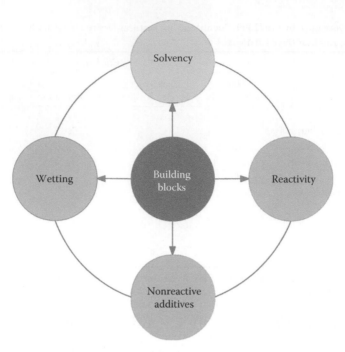

FIGURE 19.2 Electronic assembly cleaning material building blocks. (From Bixenman, M. et al., Collaborative cleaning process innovations from managing experience and learning curves. *IPC APEX Technical Conference*, IPC, Las Vegas, NV, 2009.)

The behavior of the organic materials formulated into cleaning agents is proportional to the cohesive energy of the solvating ingredients used in the formulation.[15] The cohesive energy is driven by the solvent's heat of vaporization and is quantified by its rate function for the contaminant. The rate function is influenced by kinetic forces such as the temperature of the wash bath and impingement energy. The flux residue dissolution is influenced by solvency present in the wash solution and the combination of thermodynamic and kinetic forces, which allow the wash solution to form a homogeneous solution with the soil.

Flux residues have differences based on their multicomponent compositions specifically engineered for the application need.[16] The residues can be a combination of nonionic and polar materials, which commonly exhibit different solubility properties. For polar ionic salts, water is an effective solvent for dissolving ionic salts such as chlorides, bromides, weak organic acids, and other dispersed particles that exist in equilibrium and with their saturated polar constituents present on the board and in the residue.

Many types of flux residues contain nonpolar organic structures that dissolve by breaking down the crystalline lattice.[21] Due to their chemical structures, matching the solvent molecules' cohesive energy for the soil increases the rate of dissolution. The crystalline lattice must be overcome by molecules that exhibit similar solvency behavior. Solubility theory tells us that the ideal solvents exhibit dispersive, polar, and hydrogen-bonding properties that are similar to the soils they are designed to remove.[15] Multicomponent flux residues will contain some components that will be soluble and others that will not be soluble in the solvent matrix.

Reactants

The rate of dissolution can be improved when some of the constituents in the flux residue exhibit reduction potential through hydrolysis of an ester under basic conditions to form an alcohol and the salt of carboxylic acid.[22] Reduction potential is the tendency of soils such as rosin flux to acquire electrons and thereby be reduced.[23] The process, commonly known as "saponification," is driven from a cleaning solution's metallic base oxidizing carboxylic flux residues by undergoing nucleophilic acyl substitution.[24] Rosin, resin, and polymer structures have their own intrinsic reduction potential; the more positive the potential, the greater the flux residue's affinity for electrons and tendency to be reduced.

Not all flux residue structures exhibit a reduction potential. As previously discussed, miniaturization and lead-free flux compositions require higher thermal stability, greater oxygen barrier, and protection against oxidation (charring). Some resin and polymer structures are nonpolar, and therefore require solvency (dissolution) to clean effectively. The complexity increases when the residue composition is a combination of polar and nonpolar groups that require both dissolution and reactivity. When the resin structure requires nonpolar solvent dissolution, the polar ionic constituents require reactants to remove side reaction ionic salts and complexes such as cross-linked residue.

Nonreactive Additives

Best-in-class cleaning agents perform multiple functions during the process of removing flux residues from the circuit assembly. These functions require various materials that are added to the cleaning agent formulation. These nonreactive additives are used to reduce surface tension, improve wetting, protect metallic alloys from oxidation, prevent foaming, and couple (emulsify), or, in some cases, decouple the materials used within the cleaning agent formulation with water. The nonreactive additives can add highly beneficial properties to the cleaning agent and, in many respects, form the basis for differentiating the products offered by various suppliers.

Surface Tension

The attraction of the particles in the aqueous cleaning solution creates a surface film.[25] When the tension between these particles attracts the surface layer, the liquid interface with substrate is minimized. The complexity when cleaning leading-edge circuit assemblies is increased when surface density is increased. Component miniaturization reduces the surface area within and under the part's Z-axis. In order for the cleaning solution to penetrate these highly constricted areas, the surface tension of the liquid is a critical property. High surface tension liquids, such as water, form large droplets. When these droplets do not attract the particles on the surface of the circuit assembly, the cleaning liquid exhibits a high surface tension. These large droplets repel and, as a result, fail to penetrate under the Z-axis and blind holes.

Wetting

Wetting improves when the cleaning agent maintains a film with materials on the circuit assembly. The balance between these adhesive and cohesive capillary forces helps to reduce the cleaning liquid droplet size needed to penetrate low gaps.[26] The adhesive forces between the cleaning solution and the circuit board allow the liquid to spread over the surface and drop the surface tension. In contrast, cohesive forces drive in the opposite direction when the cleaning solution repels the circuit board's surface.

To clean highly dense circuit assemblies, the contact angle at which the cleaning agent interfaces with the circuit assembly needs to be low. A small contact angle improves spreading and movement of the cleaning agent to highly restricted areas. Large contact angles with the circuit assembly minimize the cleaning solution's ability to penetrate low standoff components. This phenomenon may reduce the effectiveness of "water only, without additives" ability to remove water-soluble soils from low gaps. This shortcoming necessitates nonreactive additives in the cleaning solution to clean high-energy surfaces found under devices such as flip chips, area array components, no-lead components, and chip caps.

Protect Metallic Alloys from Oxidation

Leading-edge circuit assemblies are built with a wide range of components with varying alloys. Solder alloys used for electronic interconnections are designed with different alloys for differing applications. Common alloys include tin, lead, copper, bismuth, indium, silver, and others. In addition to solder alloys, components are constructed with a wide range of plating metals and alloys. Nonreactive materials are used in aqueous cleaning agents to inhibit or control the rate of oxidation on metal surfaces.

Reactants used in cleaning agents may oxidize (react) with some alloys and protective coatings present on the circuit assembly.[27] The interaction of the reactants with these alloys results in a change in appearance such as part discoloration. Differing alloys have specific discoloration issues.

- Tin-Lead Solder: Alkaline reactants remove the soft lead skin layer of the solder joint. The problem becomes progressively worse at an increased level of free alkalinity, higher wash bath temperature, and longer exposure time. The initial effect is a dull solder joint appearance. With elongated exposure to the reactant wash solution, the solder joint can turn blue, grey and black.
- Lead-Free Solder: Alkaline reactants exhibit a less pronounced effect on the high-lead alloys. Longer exposure results in a dull and grainy appearance.
- Aluminum (White Metals): Aluminum discoloration occurs when sensitive aluminum alloys, including copper containing 2000 series and zinc containing 7000 series, are adversely affected by the cleaning process. Alkaline reactants react strongly with most aluminum alloys, resulting in the appearance dulling with prolonged exposure. Longer exposure results in a white, spotty, bright orange, and dull grey appearance, depending on the level of free alkalinity, wash bath temperature, and exposure time.
- Copper/Brass (Yellow Metals): Copper discoloration occurs when any of the yellow metals, including copper, brass, bronze, or beryllium copper, darkens or changes color during the cleaning process as a result of metal oxidation or tarnish. The changes are more evident on highly polished surfaces than on rough surfaces. Alkaline reactants containing free amines have a greater tendency to promote tarnish of yellow metals.

The potential risk from exposing electronic components to alkaline cleaning agents can be understood by referencing Pourbaix diagrams.[28] The potential/pH diagram maps out possible stable phases for a specific alloy. Predominant ion boundaries are represented by lines. The Pourbaix diagram indicates regions of immunity, corrosion, and passivity for the selected alloy.

To inhibit attack on the range of alloys found on circuit assemblies, nonreactive materials can be formulated into the cleaning agent. The objective is to passivate the range of alloys during the cleaning process. Inhibitors form a stable oxide coating on the surface of the alloy. This protective film prevents attack from reactants used in the cleaning solution. Inhibitors are effective within a narrow range of free alkalinity in the cleaning bath. Higher levels of reactants increase free alkalinity to the point where the inhibitors can no longer form the needed level of passivation to prevent corrosion.

Decoupling/Emulsification

Aqueous cleaning agents are designed to be processed in different equipment designs. The range of ingredients engineered into the cleaning agent may be soluble or insoluble in water. The solubility can also change with the temperature of the wash bath. Some equipment types such as ultrasonic, centrifugal, bubbling, vacuum cycle nucleation, and spray-under-immersion require that the cleaning agent remain in a homogeneous mixture. Coupling agents and emulsifiers stabilize ingredients used in the cleaning agent that are not soluble in water at various temperatures and concentrations. Conversely, in spray-in-air equipment designs, decoupling may improve cleaning and defoaming.

Defoaming

Nonreactive agents are needed in some process conditions to reduce or hinder the formation of foam. High-energy cleaning machines sheer and turn over the cleaning liquid at a rapid rate. During this turbulent process, the cleaning liquid wants to entrap microscopic air pockets. These air bubbles will grow with time and energy to the point where they form a very stable foam lattice. In high-energy spray-in-air cleaning equipment, this foam lattice can cause significant processing issues such as pump cavitation, increased tank volume and high-level conditions, reduced pressure, over flows, and poor cleaning.

Solvent Vapor Cleaning

Solvent cleaning is a combination of two mechanisms.[29] The first is dissolution, where the soils are dissolved in the solvent and carried away. Condensate of the solvent in the degreaser continually replenishes the dirty solvent with fresh, clean solvent. The second mechanism is mechanical displacement of insoluble components. This can be accomplished by spray impingement, turbulence, or ultrasonic agitation.

Alternate, non-ozone-depleting, solvents have become available since the elimination of nearly all previous chlorofluorocarbons (CFCs) and hydrochlorofluorocarbons (HCFCs) as a result of the Montreal Protocol. A solvent must be chosen based on several factors, including cleaning performance, safety, regulatory restrictions, and cost (Figure 19.3). These factors can have different degrees of importance in a given situation. The best performing solvents may pose toxicological, environmental, or flammability hazards. Typically solvent mixtures operate at a stable azeotropic composition in a vapor degreaser.

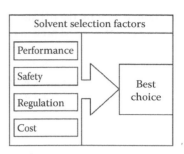

FIGURE 19.3 Solvent selection factors. (From Bixenman, M. et al., Enhanced vapor phase cleaning fluid for demanding flip chip cleaning requirements. *Georgia Tech Advanced Packaging Conference*, Georgia Tech., Atlanta, GA, March 2002. With permission.)

Vapor phase solvent blends possess good wetting characteristics and extremely low solvent viscosity and surface tension. This assures flow in all areas beneath low standoff components, thereby allowing dissolution and removal of flux residues. The solvent components within an azeotropic mixture have thermodynamic fluid properties that allow the ingredients to form a constant-boiling mixture. The composition in the boil, vapor, and rinse stages remains relatively constant over time.

Solvent vapor cleaning drives with solvency, wetting, and minor ingredients. For solvent cleaning to work, the flux residue solubility parameters must match up to the solvent composition. As multicomponent flux compositions build with higher molecular weight resins and polymers, solvent limitations arise due to the varying solubility parameters present in the flux residue. As a result, white residue remaining around the periphery of the solder bump is a common concern when cleaning with vapor-phase solvents. These residues are composed of fluxing by-products consisting mostly of unreacted soluble organic compounds and a small amount of metal salts of tin, which are highly stable and inert. These salts are encapsulated in the remaining flux vehicle. Selecting a cleaning fluid that has the strength to dissolve the nonpolar vehicle and polar salts is critical to successful removal of these residues.

Semi-Aqueous

Semi-aqueous is the process of solvent washing followed by deionized (DI) water rinsing and drying. The semi-aqueous cleaning agent's objective is to dissolve multicomponent flux residues, lower surface tension, wet the surface, and penetrate low gaps from the surface of the assembly or component. Once the wash step is completed, the parts go through a series of rinse steps to remove the solvent along with the dissolved soils (temporary soluble solder mask, nonionic, and some particulate soils) and any water-soluble (ionic) residues. The parts are then dried free of water and typically have contamination levels that are undetectable.

Semi-aqueous cleaning processes were a highly popular alternative to replace ozone-depleting solvents banned as a result of the Montreal Protocol. Semi-aqueous technology was highly beneficial by engineering cleaning agents that matched up to the flux residue soil. The building blocks for designing semi-aqueous cleaning agents consist of dissolution, reactivity, wetting, and minor ingredients.[30] Dissolution is the key driving force by combining low vapor pressure solvent molecules that were either soluble or partially soluble as the basis of the formulation. In some designs, low levels of reactants were used to increase rate. The cleaning agents were also buffered to hold high levels of acidic flux residues without dropping the pH into the acid range. Minor ingredients were used to decrease surface tension, improve wetting, and prevent oxidation of metallic alloys exposed to the cleaning process.

Semi-aqueous cleaning agents are designed to have good compatibility with most components used in electronic assembly. Some common issues are solvent entrapment, part marking degradation, seal gasket and elastomer compatibility, and solvent consumption due to evaporative and drag-out losses. Careful attention to equipment design and cleaning agent choice can minimize these problems.

Aqueous

Aqueous cleaning materials are engineered concentrated fluids that dissolve with water. Aqueous cleaning materials are nonflammable and processed in high-energy machines. Aqueous concentrated products work based on the *cleaning rate theory* that holds: The static cleaning rate (rate at which the cleaning material dissolves the flux residue at its temperature and concentration in the absence of impingement energy) plus the dynamic rate (energy and time in the cleaning machine) equals the process cleaning rate.[31]

Aqueous cleaning materials fit the following four classifications (Figure 19.4)[32]:

1. *Aqueous high reactivity*: The product contains high levels of reactants (note: there are several reactant structures commonly used by formulators) that saponify the flux residue to improve rate. The benefits of highly reactive aqueous cleaning fluids are lower operating concentration and aggressive interaction with rosin, resin, and weak organic acids. The trade-offs of highly reactive aqueous cleaning fluids are materials compatibility, short bath life, and lower effectiveness on high molecular resins and polymers used in many of the new flux designs.
2. *Aqueous mild reactivity*: The product drives with both solvent and reactivity. Mild reactivity formulation designs improve rate using higher solvency combined with reactivity. The benefits of mild reactivity aqueous cleaning fluids are improved materials compatibility, longer bath life, and greater effectiveness on resin- and polymer-based flux residues. The trade-offs are lower bath life and lower effectiveness on higher molecular weight resins and polymers used in lead-free flux compositions.
3. *Aqueous low reactivity*: The product drives with solvency combined with low reactant levels. Low reactivity formulation designs improve rate using high solvency with solubility parameters closely matched to rosin, resin, and polymer compounds used in many flux residue compositions. The benefits of low reactivity aqueous cleaning fluids are long bath life, good materials compatibility, and high effectiveness on both eutectic and Pb-free flux residues. Aqueous low reactivity provides best-in-class technology due to high solvating power and excellent materials compatibility.
4. *Aqueous neutral*: The product drives with solvency combined with low levels of reactants. Aqueous neutral formulation designs are similar to semi-aqueous cleaning agents in that they are designed to clean with solvency and rinsed with DI water. Similar to vapor degreasing solvents, the solubility parameters must match up with the flux residue soil. When the solvency package does not match up, the cleaning agent will exhibit a boundary condition for the soil. In cases where the soil does match up, higher concentration levels, temperature, and mechanical energy will be required.

PWB Aqueous Cleaning Material Design Options				
	Solvency	Reactants	Wetting	Nonreactive additives
Aqueous strong reactivity	Low	High	Low	Low
Aqueous mild reactivity	Medium	Medium	Low	Medium
Aqueous low reactivity	High	Low	Low	Medium
Aqueous neutral	High	Low	Low	Medium

FIGURE 19.4 Aqueous cleaning material design options.

Matching the Cleaning Agent to the Flux Residue

Flux residue variation impacts the electronic assembly cleaning process. Understanding the effects of the solder flux residues is critical to the cleaning process. The thermal phases during the soldering process influence the flux residue solubility parameters. The number of heat excursions and time above alloy liquidus can cause flux burn-off, polymerization, and hardening of the flux residue. These hardened flux residues, when partially cleaned, form white residue, which represents one of the most complex problems in electronic assembly. White residue left after the cleaning process is commonly the result of extracting and removing only some soluble flux residue ingredients, while leaving behind an insoluble ionic white powder.

Fluxes used in solder paste formulations comprise resins, activators, solvents, and rheological additives.[18] Some rosin compounds are chemically modified to impart additional features such as higher tackiness, better thermal stability, and greater fluxing activity. Resin/rosin structures are organic materials such as rosin, synthetic resins, and polymers with a medium to high molecular weight. During the reflow process, heat drives rosin/resin to undergo isomeric transformations and thermal dimerization. During the soldering process, activators and solvents in the flux are reacted and volatilized. Post-soldering, the rosin and higher molecular weight additives form a clear translucent solid residue.[18]

With the wide acceptance of no-clean soldering processes, boards may see multiple reflow cycles (top side, bottom side, through-hole, and selective soldering) since the residue was not designed to be cleaned. This was made possible largely due to advances in no-clean flux technology. Each consecutive reflow exposure hardens the rosin/resin translucent film by driving out solvent molecules. As a result, the flux residue forms a hard shell that is increasingly difficult to dissolve and clean.

Lead-free soldering and miniaturization complicates this process, due to increased melting temperature and component density.[14] The surface tension of tin-silver-copper (SAC) alloys is approximately 20% higher than eutectic tin-lead, which results in poorer wetting. This wetting deficiency needs to be compensated with improved flux compositions that lower surface tension and increase thermal stability. Additionally, miniaturization aggravates the burn-off factor of flux due to the need for lower volatility and greater oxidation resistance. Reduced volatiles mean a greater amount of flux residue, with a higher molecular weight.[14]

New flux compositions may contain more oxidation resistant chemical bonds that change cleaning properties. To prevent tombstoning, a longer soak period is needed for lead-free soldering, which may cause greater cleaning difficulty. White residue formation of these various flux types is more prone to formation in high-temperature soldering conditions. As a result, lead-free and miniaturization creates a condition where high thermal heat is generated, making the flux more difficult to clean and prone to the formation of white residue.[33]

Process Cleaning Rate Theorem

The process cleaning rate theorem holds that the static rate (solubility parameter for the flux residue) plus the dynamic rate (thermal and mechanical energy) equals the process cleaning rate.[31] The static cleaning rate represents the cohesive properties of cleaning agent for the flux residue. Flux soils with similar solubility parameters will be miscible and will dissolve in cleaning agents whose solubility parameters are not too different from their own.

The basic principle is that "like dissolves like." Solvents that do not match up to the soil do not usually dissolve the soil.[15] Temperature is a key driver, as flux residue softens when heat is applied. Mechanical energy works when the cleaning agent dissolves the residue. Poor cleaning agent performance can rarely be overcome with stronger mechanical forces.

Flux Residue Solubility Properties

Flux residues dissolve in cleaning agents whose solubility parameters are not too different from the flux residue solubility parameter. Flux residue solubility parameters are directly related to the cohesive energy parameters, which represents the energy required to dissolve the flux residue into the cleaning agent.[15] The cleaning agent's ability to soften and dissolve the flux residue arises from dispersive forces, permanent dipole–permanent dipole forces, and hydrogen-bonding forces. Dispersive cohesive interactions occur from the cleaning agent's free energy to dissolve the nonpolar rosin, resin, and polymer flux residue structures. The permanent dipole–permanent dipole interactions occur from the flux residues molecular interactions with reactants and water present in the cleaning agent. Hydrogen-bonding cohesive energy occurs from the cleaning agent's use of oxygenated solvents, reactants, and water to attract and share electrons.[15]

It is important to know whether the flux residue is a single-phase or a multiple-phase mixture. Most flux compositions contain multiple phases, which require cleaning agents engineered with building blocks that dissolve the various components present in the flux residue. Since flux compositions contain multiple components, a composite solubility parameter for residue soil can be established. By selecting a wide range of solvent families to characterize the flux residue, the investigator gains needed insight into the material properties needed to rapidly dissolve and remove the soil.

By first characterizing the flux soil, an interaction zone can be developed for dissolving or dispersing the soil. Desirable cleaning agents positioned within the interaction zone rapidly dissolve the flux residue in question. Matching the solubility parameters of the cleaning agent with the flux residue interaction zone assures excellent cleaning performance.

In the search for the best cleaning agent, it may be determined that less optimal cleaning agents may clean the part using aggressive process conditions, but may not be an ideal fit for cleaning the specific flux residue. To address this concern, the ideal cleaning agent selection needs to be defined using solubility research by applying low levels of cohesive and kinetic energy factors to establish the cleaning agent that holds the ideal solubility match.

To determine the solubility parameters for a given flux residue (water-soluble flux, rosin flux, no-clean flux, lead-free no-clean flux), a simple series of tests are performed.[11] For tests performed in these experiments, samples of the various flux residues are heated to peak reflow temperatures, and held at that temperature until low boiling solvents are removed (approximately 5 min) and cooled to room temperatures. An alternative method of preparing samples is to reflow circuit boards with the solder paste in question, and then cut up pieces where flux residues are present and use these small samples for testing.

Matching the Cleaning Agent to the Flux Residue Solubility Parameters

Choosing the best-fit cleaning agent for removing a specific flux residue requires characterization of both cleaning agents and the flux residue. Solubility parameter characterization can be used for single and multiple component compositions for both the soil composite and the cleaning agent blend. The solubility characteristics of multiple component compositions can be compared to known values, just as similar values of solvents and polymers are compared to determine if the cleaning agent can dissolve the flux residue.

To characterize the solubility parameters of the cleaning agent for the soil, similar tests are run on the soils at their various concentrations, temperatures, and time levels. The goal is to identify differences at the lowest applied levels for concentration, temperature, and time. When testing selected cleaning agent candidates using elevated levels for concentration, temperature, and time, solubility differences can be clearly identified. This methodology helps the investigation more closely match the solubility characteristics of the cleaning agent to the soil. The testing also identifies optimal process parameters for wash time, temperature, and concentration.

Figure 19.5 is a snapshot of data generated from a water-soluble cleaning study. More than 50,000 data points were generated from this study. The objective was to match up the solubility parameters of

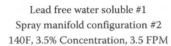

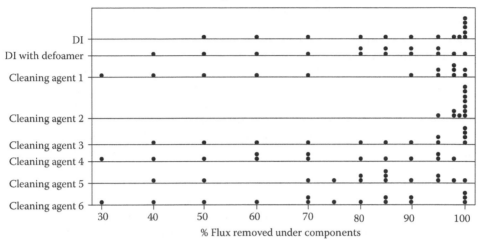

Each symbol represents up to four observations.

FIGURE 19.5 Data snapshot of a DOE designed to match the best fit cleaning agent with flux residue.

various water-soluble flux residues with cleaning agents and mechanical impingement levels. The test boards were populated with 1210 and 1815 chip cap resistors. Following cleaning, the components were removed and graded for the level of flux residue left under the components. Minitab software was used to analyze the data.

When determining the best-fit properties of multicomponent compositions, it is important to elevate the factors and levels in the designed experiment to see differences. In this particular example, notice that the cleaning agent concentration was 3.5%, with a wash temperature of 140°F and a belt speed of 3.5 ft/min. This equates to approximately 1.25 min of wash time. At these levels, the best-fit cleaning agent properties are easier to identify. Cleaning Agent #2 was the best-fit cleaning agent. From this testing, we can infer that Cleaning Agent #2 closely matches the solubility parameters of the Lead-Free Water-Soluble #1 flux residue.

Selecting a cleaning agent that matches up to the flux residue opens the process window and lowers cost of ownership. Cleaning agents that work at lower concentration levels, lower temperatures, higher processing speed, and lower impinging force carry numerous benefits. The less time exposure there is to the electronic assembly hardware, the better the materials compatibility window. The benefits of characterizing the flux residue and matching to the best-fit cleaning agent save time and money over the commonly used method of trial and error.

Materials Compatibility

When designing the cleaning process, materials compatibility is often overlooked. Much of the focus is targeted at the selection of the cleaning equipment and cleaning agent. Engineers responsible for the cleaning process spend considerable time and effort to assure that the cleaning agent removes flux residues on production hardware. Of all the possible issues that could go wrong, materials compatibility is high on the list.

Engineers responsible for the designing new cleaning processes commonly source and select the cleaning equipment first. Once the cleaning equipment choice is made, they turn their focus on selecting a cleaning agent. Test board selection is typically chosen from industry kits that support surface insulation resistance (SIR) comb patterns and ion chromatography analyses. Many of these test boards are not representative of the product hardware geometries and process conditions that the cleaning process is designed to remove. In a number of cleaning process qualifications, it is only after implementation of the cleaning process on the manufacturing floor that materials compatibility issues come to the surface.

There are several materials compatibility issues that engineers responsible for the cleaning process must consider. Factors of concern are board laminate, surface finishes, components, metal alloys, adhesive bond strength, part markings, plastics, the mix and configuration of materials in the assembly, and the impact of entrapped contaminants. Other factors must also be considered, such as the chemical characteristics of the cleaning material, cleaning temperature, exposure time to the cleaning process, including rework cycles and impingement energy. The possible interactions of all these factors are especially challenging today with the much wider availability of board and component lead finishes, as well as joining metals. The difficulty of locating data on each and every possible combination of factors associated with this issue creates a high level of complexity.

The selected cleaning process must be validated prior to production of the new assembly. Designers, engineers, quality control personnel, shop floor personnel, safety/environmental advisors, and other key personnel should be involved in the cleaning process evaluation. The cleaning process should be acceptable to process owners; otherwise, the new process is less likely to be successful.

When cleaning high-density integrated circuit assemblies, the design engineer needs to ensure that every opportunity is taken to enhance the ability to clean. Reducing the component size and spacing increases cleaning difficulty on both component terminations and under the components. This may mean that components should be oriented in a specific direction to improve the cleaning process window.

Longer wash time is needed when cleaning flux residue under low standoff components. Highly reactive cleaning materials create several compatibility concerns in the form of solder joint attack, anodized aluminum attack, dry film solder mask removal, part marking removal, component attack, polymer/adhesive attack, and a range of other issues. Highly reactive cleaning materials saponify rosin, which can create a foam condition as the wash bath loads.

Cleaning Agent Selection: The following points are worthy of consideration:

- How do the cleaning agent's solubility parameters match up the flux soils represented in your process?
- What is the cleaning concentration needed to remove the flux soils represented in your process?
- What is the wash temperature needed to remove the flux soils represented in your process?
- What is the concentration level of reactants used in the cleaning agent?
- What is the flux loading potential?
- Does the cleaning agent foam when flux loaded?
- What are the materials compatibility properties of
  a. Metallic alloys
  b. Board laminate
  c. Plastics
  d. Elastomers
  e. Coatings
  f. Components
  g. Part markings, labels, and inks
- What is the history of use?
- What do other users have to say about the product?
- What is the cleaning supplier's support level and reputation?
- Is the supplier globally located, and will the supplier provide global support?
- Is the product compliant with Restriction of the Use of Certain Hazardous Substances (RoHS)?

- Is the product hazardous?
- What is the odor in the workplace?
- What are the evaporative and exhaust losses?
- How does the cleaning agent change over the life of the bath?
- What is the cleaning agent's surface tension?
- How is the cleaning agent monitored and controlled?
- What could go wrong?

Thermodynamics versus Kinetics

As previously defined, the process cleaning rate is a function of the cleaning agent solvency for the flux residue and internal energy in terms of temperature, pressure, and volume. The solubility parameter is an important quantity for predicting solubility relations.[15] The solubility parameters of the cleaning agent for the flux residue are important in determining the solubility of the system. The solubility of the cleaning agent for the flux residue leads to a zero change in non-combinatorial free energy and the positive entropy change. Cleaning agents whose solubility parameters match up with the flux residue will be thermodynamically better than a boundary cleaning agent.

Thermodynamic Effects

The first law of thermodynamics states that internal energy is equal to the amount of energy added by heating the part and cleaning agent minus the amount lost as a result of the work done from mechanical impingement.[34] Most flux residues soften when heat is applied. In most cases, this softening effect changes the cohesive energy of the cleaning agent for the soil. It essentially defines the cleaning agent's ability to do the work.

The second law of thermodynamics states that the entropy of the cleaning process and its surroundings must experience an overall performance increase.[35] Matching the best-fit cleaning agent at the optimal temperature, volume, and pressure creates a process for doing work. When the system is in equilibrium, the differences in temperature, pressure, and chemical potential tend to even out in the cleaning system. The cleaning process theoretically is in balance with its environment, and as long as the process window is maintained, the system should deliver consistent results.

The third law of thermodynamics is a statistical law of nature regarding entropy and the impossibility of reaching absolute zero temperature.[36] If the thermal system effects move outside the designed process window, the ability to do the desired work decreases. This brings to light the characteristics of the soil and the properties of the cleaning agent for removing that soil. Knowing these effects helps to determine available entropy within the system and the limitations when going out of those limits.

Kinetic Energy Effects

Mechanical systems provide energy in motion by accelerating flow, velocity, and pressure effects to move the cleaning agent to the soil.[37] To remove flux residue under components with low standoff gaps, kinetic energy must be targeted and delivered to source of contamination. The cleaning agent represents the potential energy within the process. Kinetic energy, in the form of heat, fluid flow, pressure, and directional forces, moves the cleaning agent's potential energy to the area of contamination. To remove all residues, there must be channels of flow under the part.

Time is a critical factor. The time in the cleaning agent section is dependent on part geometry and on the solubility characteristics of the flux residue. Conventional circuit assemblies cleaning in a continuous inline cleaning process typically require 1–3 min. of wash time. Leading-edge circuit assemblies with low clearance components require upwards of five times longer in the wash section to clean. This is a critical factor that must be known before deciding on cleaning equipment selection.

Vapor Phase Cleaning Equipment Considerations

Today's vapor degreasing solvents are more costly than their ozone-depleting predecessors. This has driven the development of cleaning systems with low evaporative and drag-out losses. Design improvements include subzero chillers and extended freeboard height. These modifications have greatly reduced solvent emissions.

Solvent cleaning is a combination of two mechanisms. The first is dissolution, where the soils are dissolved in the solvent and carried away. Condensate of the solvent in the degreaser continually replenishes the dirty solvent with fresh, clean solvent. The second mechanism is mechanical displacement of insoluble components. This can be accomplished by spray impingement, turbulence, or ultrasonic agitation.

Vapor Degreasing Batch

Two basic equipment configurations are used to clean boards.[29] A batch configuration allows for a higher solvent exposure time by allowing boards to be processed in a parallel fashion. Canisters of boards can be loaded into cascading solvent tanks, and moved to subsequently cleaner and cleaner tanks (Figure 19.6). Generally, at least three cascading tanks are used, consisting of clean, rinse 1, and rinse 2. This configuration provides more throughput in a smaller amount of space and allows a system to accommodate a wide range of board sizes. Impingement normal to board surfaces is not possible in this configuration. However, pumps can be used to generate turbulence in the solvent tanks. In the batch system, chemical dissolution time can be maximized, but mechanical displacement is limited.

Vapor Degreasing Continuous

An inline configuration, on the other hand, consists of spray impingement of every board. Boards travel through the cleaner on a wire mesh belt with spray impingement normal to the plane of the boards (Figure 19.7). Since boards travel through the machine in a serial manner, the overall system for cleaning and rinsing can be very long. This transport design is mechanically more complex than the batch system and limits changes in board width. An inline configuration provides a high degree of mechanical energy, but limits the time of solvent exposure, thus limiting the time available for dissolution.

Semi-Aqueous Cleaning Equipment Considerations

Semi-aqueous is the process of solvent washing/water rinsing. When cleaning electronic assemblies, the objective is to solvate the soils from the surface of the assembly or component. Once the wash step is completed, the parts go through a series of rinse steps to remove wash chemistry, spot mask, and ionic residues. The parts are dried free of water and typically have contamination levels that are undetectable.

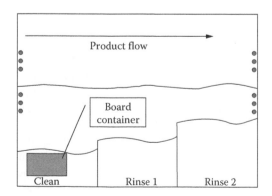

FIGURE 19.6 Example of batch cleaner.

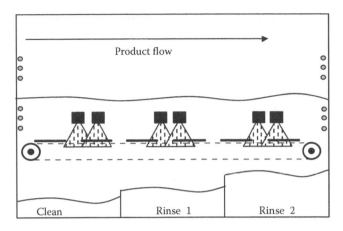

FIGURE 19.7 Example of an inline cleaner.

Inline Semi-Aqueous Equipment

Spray-under-immersion cleaners position spray nozzles located below the liquid level in the wash tank (Figure 19.8). The mechanical energy moves the chemistry across the surface of the board, allowing the chemistry to soften and dissolve the contaminant, removing it from the part. Because the mechanical energy is lower than that found in a spray-in-air system, the solvency of the chemistry and the exposure time become more important. Sizing the length of the wash tank is done by multiplying required chemical exposure time by required belt speed, and adding the length required to lower the parts into and raise the parts out of the solvent. There should be an air knife at the end of the wash section to remove residual chemistry. This should not dry the part, but remove the bulk of the chemistry to minimize the drag-out.

Initial rinse performed at stage two is commonly referred to as chemical isolation. The goal of this section is to remove and isolate the cleaning chemistry from the remaining rinse sections. Sizing the chemical isolation section is very important. The chemical isolation section should be at least 1.5 times longer than the largest part on each side of the spray nozzles. This is done to prevent bridging and contaminating the wash or rinse sections. Bridging occurs when the part acts as a bridge to allow the chemical isolation water to be carried into the other section. Tap water is sufficient for chemical isolation sections, or, if a cascading rinse is used, the final cascade will work well for chemical isolation, as shown above. The air knife in this section is important to reduce the contamination levels in the rinse section.

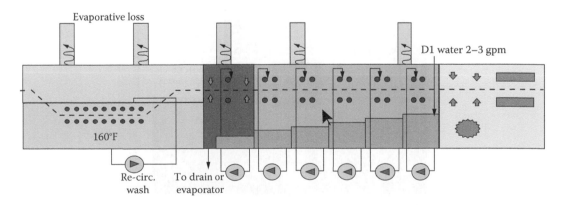

FIGURE 19.8 Semi-aqueous inline cleaning equipment.

The rinse section is very important. This is the final stage of cleaning, ensuring that the parts are clean before drying. Two or more stages in the rinse section, with good quality DI water, works best. The dryer section is the last section of the cleaner. The dryer section should be sized to ensure that the parts will be completely dry when they come out of the machine. The faster the belt speed, the larger the dryer section required. If there are any hidden spaces or blind holes, extra drying may be required.

Semi-Aqueous Centrifugal Systems

Centrifugal batch semi-aqueous systems fixtures and rotates the assemblies about its center of gravity, causing two accelerations to be experienced: coriolis and centrifugal (Figure 19.9). By spinning the circuit board and introducing a cleaning solution to the circuit board, the accelerations become forces and obey the formula $F = Mw^2r$ (force = mass × weight squared × radius). The magnitude of the energy is directly proportional to the square of the angular velocity (RPM).

However, for centrifugal cleaning, the value of this energy is not its magnitude, but its direction. Other cleaning processes add energy to the cleaning process by spraying the cleaning solution at an oblique or perpendicular angle to the surface of the printed circuit board. The cleaning solution strikes the board, some of which rebounds underneath a component, alternately striking the board and the bottom side of the component. At every impact, however, energy is consumed, making penetration under closely spaced components very difficult. Poor penetration of the cleaning solution under the component hinders effective cleaning.

Semi-aqueous, aqueous, alcohol blends, alcohols, esters, terpene, and hydrocarbon blends can all be used in the centrifugal cleaner. A spin-under-spray rinse of DI water follows, flushing solvent residue from the product. Drying is accomplished through centrifugal force extracting water from the product; while dry, heated air excludes vapor from the chamber.

Benefits of the semi-aqueous centrifugal batch cleaner include less time required for cleaning, low fluid consumption, efficient size, and solvent cleaning agent flexibility. Disadvantages of a batch process include the requirement of operator loading/unloading, the inability to handle the volume of parts an inline machine can, and the need to run the product in batches instead of continuously.

Semi-Aqueous Batch Spray under Immersion/Ultrasonic Systems

Spray-under-immersion batch cleaners position spray nozzles located below the liquid level in the wash tank. The mechanical energy moves the chemistry across the surface of the board, allowing the chemistry to soften and dissolve the contaminant, removing it from the part. Because the mechanical energy is typically lower than a spray-in-air system, the solvency of the chemistry and the exposure time become more important.

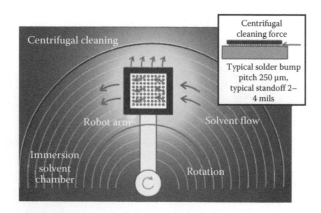

FIGURE 19.9   Centrifugal cleaning.

Ultrasonic batch cleaners transform low-frequency AC current into high-frequency sound waves via an electro-acoustic transducer. The transducer creates sinusoidal waves, which in turn cause cavitation—the formation and violent collapse of minute vacuum bubbles in the solution. These implosions thoroughly scrub every surface with which the solution makes contact, yet are not harsh on delicate items.

In the cavitation process, micron-size bubbles form and grow due to alternating positive and negative pressure waves in a solution. The bubbles subjected to these alternating pressure waves continue to grow until they reach resonant size. Just prior to the bubble implosion, there is a tremendous amount of energy stored inside the bubble itself. Temperature inside a cavitation bubble can be extremely high, with pressures up to 500 atm. The implosion event, when it occurs near a hard surface, changes the bubble into a jet about one-tenth the bubble size, which travels toward the hard surface at speeds up to 250 mph. With the combination of pressure, temperature, and velocity, the jet frees contaminants from their bonds with the substrate.

Semi-Aqueous Spray in Air Batch Systems

Semi-aqueous batch machines (dishwashers) use medium-pressure spray to remove contaminants from the parts. The machines typically have top and bottom rotating spray bars, and some have side mounted spray nozzles. The impingement of the chemistry loosens contaminants and removes them from the parts. The chemistry also helps to soften and dissolve contaminants through solvency. Typical spray pressures range from about 30–60 psi. These machines are usually inerted with nitrogen or carbon dioxide.

Aqueous Batch Spray in Air Cleaning Equipment Considerations

Batch cleaning equipment designs provide a small footprint, low cleaning fluid consumption, and low cost of ownership (Figure 19.10). Batch cleaning machines use flow, time, temperature, and advanced cleaning fluids as critical drivers for delivering a clean part.[38] Increased density, low standoff components, and lead-free flux residues place increased importance on the cleaning fluid design. There is a need for improved cleaning fluids to remove lead-free flux residues from populated circuit assemblies in batch cleaning machines.

While modern defluxing chemistries require contact between the chemical and the target, there is much more to the process than that. There are fourteen fundamental elements across four specific design criteria to a best-in-class defluxing machine.[9] Mike Konrad of Aqueous Technologies defined the essential factors for batch processing as follows:

FIGURE 19.10 Batch cleaning machine design.

Wash Cycle Heat

- Contact
- Spray design
- Segregation of wash solution
- Chemical dosing

Rinse Cycle

- Contact
- Spray design
- Ionic contamination detection

Dry

- Cubic feet per minute (CFM)
- Convection heater power
- Radiant heater power

General Equipment Guidelines

- Chemical compatibility
- Operator safety
- Environmental safety
- Process control

Wash Cycle (Heat)

All best-in-class aqueous chemistries require heat. If operated unheated, disastrous results can occur, including poor cleanliness and extreme foaming. Most cleaning chemicals produce optimum results when operated at temperatures between 50°C and 70°C. The defluxing machine must be capable of heating and maintaining the wash solution to these temperatures.

Wash Cycle (Contact)

No defluxing chemistry can ever remove flux if it cannot come into contact with the assembly. Modern surface mount assemblies feature complex geometries. Both large and small components may be mounted in close proximity to each other, allowing one component to shadow another.

In the industry's most popular format (batch), shadowing is of particular concern as assemblies are stacked much like dishes in a dishwasher. While batch defluxing machines utilize upper and lower rotating spray bars that produce thousands of possible angles of attack, there remains a potential for shadowing. This remaining potential may be mitigated with the implementation of an oscillating device that moves the assemblies in a forward/rearward motion simultaneous to the spray-arm rotation. The rack oscillating device increases contact by reducing the possibility of shadowing.

Wash Cycle (Spray Design)

There are two competing theories when it comes to fluid diffusion. All cleaning begins with contact—contact between the cleaning fluid and the cleaning target. Even though a defluxing machine may be equipped with rotating spray arms and even a rack oscillating device, at the core is the fluid delivery device. Some equipment designs utilize spray nozzles, while others do not. The purpose of a spray nozzle is to "bend" the fluid to a shape that best fits the target, much like placing your thumb over the end of a hose to increase the fluid's diffusion pattern. This action forces fluid through a smaller hole, increasing its velocity and therefore its impact pressure, while reducing the water drop size and associated surface tension.

While some defluxing equipment utilizes nozzle-based fluid distribution, others do not. These nozzle-less machines simply utilize a hole with a specified diameter to produce a coherent stream of fluid with no diffusion. The advantage of a coherent spray is that the flow can travel further before losing velocity. This is due to the fact that coherent fluid flows produce larger fluid drop sizes and are therefore capable of greater travel before losing velocity.

The debate between nozzle-based and coherent-based fluid delivery designs is based on indisputable factors. The more water is "bent," the faster it loses its impact pressure. On the other hand, nozzles produce smaller droplet sizes, aiding in under-component penetration. As we will discuss later, under-component penetration (impingement) is among the most critical elements in a successful defluxing process. Nozzles, by widening the fluid's trajectory, ensure full (and even overlapping) contact with target assemblies.

Coherent fluid distribution maintains fluid velocity for a longer distance, but produces the largest fluid droplet size, impeding its ability to penetrate under low standoff components. In addition, coherent spray patterns do not overlap. Thorough assembly coverage is only possible if the fluid hitting the assembly ricochets in a manner to allow thorough coverage. It should be noted that when the fluid changes direction (like with ricochets), it loses the majority of its velocity, and rapidly becomes ineffective.

Wash Cycle (Segregation of Wash Solution)

Most modern defluxing chemicals are prepared and shipped as a concentrate. They are mixed with water (normally DI water) to form a wash solution. Common in-use percentages are 10%–20% concentrated defluxing chemistry and 90%–80% DI water. Most modern defluxing chemistries provide a relatively wide process window. Commonly, ±5% concentration or dilution still produces acceptable cleanliness results. Due to many corporations' increased environmental sensitivities and budgets, many defluxing equipment manufacturers have incorporated wash solution recyclers in their equipment. With a wash solution recycler, the same chemical/water mixture may be used dozens of times, over the course of days or even weeks. Because of this, it is vital that the defluxing machine have incorporated safeguards within its basic design to prevent the dilution of wash solution with rinse water.

Design elements such as segregated spray and drain/transfer pumps reduce the chance of chemical dilution. Anti-drag-out features such as programmable rest (drainage) times, and self-purging wash pumps also contribute to the reduction of chemical dilution. A highly effective drying system will also prevent chemical dilution by eliminating any residual rinse water from mixing with the upcoming cycle's wash solution.

Wash Cycle (Chemical Dosing)

As previously stated, to reduce the cost to their customers, most chemical manufacturers provide their defluxing chemical in concentrated forms. Most of the failures in a defluxing process witnessed by this author have been caused by inaccurate chemical mixing, mostly by equipment operators. A well-designed automatic chemical dosing technology combined with periodic monitoring (via titration or refractometer) will provide consistent and accurate chemical concentrations without the need for operator intervention. This reduces operator errors, and ensures that the process stays within the required guidelines.

Rinse Cycle (Contact)

Like the wash cycle, contact between water (the rinsing agent) and the assemblies is required. Because most batch defluxing systems utilize the same chamber for all cycles (wash, rinse, and dry), the spray technology used in the wash cycle will be used in the rinse cycle. All required design attributes associated with the wash section (contact and spray design) are identical.

Rinse Cycle (Spray Design)

While both contact and spray designs are identical between wash and rinse cycles, it is vitally important to make the following statement. *The most critical aspect of a successful defluxing process is the rinse cycle.* While most attention is bestowed upon the wash cycle, cleanliness results would be catastrophically worse if the rinse cycle were not performed properly. While conventional aqueous defluxing chemicals perform substantially better than their obsolete solvent counterparts, they cannot be allowed to remain on an assembly. Most modern defluxing chemistries maintain a pH level in excess of 11. While anti-corrosion (brightening) agents prevent dulling of the solder joints during the wash cycle, the defluxing chemical must be thoroughly removed.

After the wash cycle, the assemblies are covered in wash solution, both above and below the components. A thorough rinsing process must be initiated to remove all traces of wash solution from the assembly. Because the wash chemical contains surface tension reducing components that can reduce the surface tension of the wash solution from 72 dynes (water) to 25 dynes, under-component penetration is much more easily achieved in the wash cycle than in the rinse cycle. This is when the small water droplet attributes of spray nozzles come into play. The only way to effectively chase out 25 dyne fluid with 72 dyne fluid is to manipulate the water droplet size mechanically, with the use of precision cut spray nozzles and a very large pump (to provide significant pressure and velocity).

Rinse Cycle (Ionic Detection)

A successful defluxing process relies on the successful removal of flux into the wash solution during the wash cycle and the successful removal of the wash solution during the rinse cycle. Fortunately, all aqueous defluxing chemistries have one thing in common. They contain highly ionic properties. The incorporation of an ionic residue detection device (resistivity sensor) into the rinse plumbing is highly effective at detecting ionic contamination in the normally nonionic DI rinse water. A defluxing machine equipped with this technology can automatically add or subtract rinse cycles until the rinse effluent's ionic properties reach a limit preset by the user. Use of this technology ensures the complete removal of wash solution (and the flux it contains) consistently, batch after batch.

Dry Cycle (CFM)

Dry assemblies are essential to a successful defluxing process. Most batch format defluxing systems utilize a mechanical blower to provide air exchange within the process chamber. The larger the blower (CFM), the greater frequency of complete air exchange within the process chamber. For rapid and thorough drying, the objective is to exchange the moisture-saturated air with hot and dry moisture-receptive air. Depending on the specific location of the defluxing equipment, a particle filter may be required to remove unwanted particles from the rapidly moving air.

Dry Cycle (Convection Heat Power)

As mentioned previously, hot, moisture-saturated air needs to be replaced with hot dry air. This requires the incorporation of convection heaters to heat the incoming air before it enters the process chamber. The degree of power (wattage) should be proportionate to the CFM of the blower.

Dry Cycle (Radiant Heat Power)

While the convention heater is working to produce hot incoming heat, a radiant heater will allow the assemblies to absorb heat and themselves become mini-heaters. As the assemblies absorb heat, water trapped below components (and between layers) begins to evaporate. A successful drying process will produce assemblies that measure a lower post-defluxing weight than pre-defluxing.

General Equipment Guidelines (Chemical Compatibility)

Obviously, the equipment must be compatible with the defluxing chemical. There are two levels of compatibility: material and process.

Materials compatibility requires that all wetted surfaces of the equipment be compatible with the defluxing chemical. All seals (pumps, doors, covers, etc.) must be compatible. Materials such as rubber, Buna, Viton, and other similar materials are not generally compatible with many defluxing chemicals. Materials such as Teflon, EPDM, and EPR are widely compatible.

The defluxing machine must also meet the process requirements of the chemical. If a chemical requires heat, so will the equipment. If the chemical requires mixing before use, the equipment must be equipped with a mixer. Additional considerations such as ventilation, chemical re-use capabilities, dosing requirements, foam control, and other factors need to be considered when choosing a defluxing machine.

General Equipment Guidelines (Operator Safety)

Operator safety is paramount. Fortunately, modern aqueous defluxing chemicals, while maintaining their ability to remove all flux types, are nonflammable. The use of nonflammable chemicals has greatly increased the overall safety of cleaning equipment. Additional desired safety features include hands-free chemical dosing, over-heat protection devices, and keyed maintenance functions.

General Equipment Guidelines (Environmental Safety)

Today's defluxing chemicals and equipment are widely considered environmentally responsible. Many defluxing machines utilize evaporators to eliminate any discharge of effluent (wash or rinse solution) into the drain. While most municipalities allow the discharge of effluent from modern defluxing systems, zero-discharge configurations are becoming widely preferred, as they eliminate the concern of unknown future environmental regulations.

General Equipment Guidelines (Process Control)

The big question is "Who controls your process?" Best-in-class equipment provides a level of process control that ensures a predictable and consistent result. The operator interface should be clear and intuitive. Closed-loop process feedback eliminates operator panic (Did I press start? Is the water turned on? Is there chemical in the machine?). Password protected sections of the interface prevent both unintentional and unauthorized process changes. Statistical Process Control (SPC) data logging allows cleanliness analysis and historical review of process trends.

Aqueous Inline Cleaning Equipment Considerations

Aqueous inline cleaning machines are designed to wash, rinse, and dry the part, and typically support fast processing (Figure 19.11). The cleaning machine designs deliver the cleaning fluid to the part, and create force using pressure and volume. Fluid management is critical in maintaining an economic cleaning process. Individual module containment, specifically with the wash chemistry, is essential. Fluid delivery is critical for penetrating and rapidly breaking the flux dam under low standoff components. Air management is critical to reducing chemical odors in the workspace while minimizing the amount of wash fumes exhausted from the machine. Fluid storage is critical for long wash bath life. Fluid control is critical in maintaining the proper wash bath concentration within the cleaning process tolerance.

Cleaning equipment design issues in any of these areas can and will upset the cleaning process over time. Issues such as high wash consumption, steam out of the machine, foaming in the wash and/or rinse, exhaust losses, and poor cleaning all result from an imbalance caused by one or more of these factors. Process issues may not show up when the machine is initially charged with cleaning chemistry and started up, but slowly creep in over time. Lack of process optimization results in higher defect rates, which typically render white residue formation and unacceptable levels of ionic residues on the surface and under component gaps. Bixenman et al.[10] identified a series of best practice considerations for optimizing and controlling inline cleaning systems.

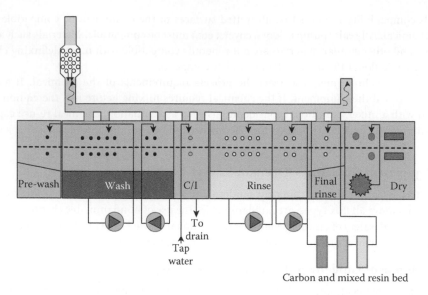

FIGURE 19.11 Aqueous inline cleaning system.

Fluid Delivery

To improve cleaning under low standoff components, research data indicates that fluid flow, pressure at the board surface, directional forces, and time in the wash improve the process cleaning rate. The wash section of the cleaning machine is highly important. Research data findings indicate that flux not adequately removed in the wash will *not* be removed in the rinse sections. Cleaning data studies show that high levels of fluid across the board surface decrease needed cleaning time. Directional forces that provide a 360° impingement pattern during the wash exposure decrease time in the wash. Maintaining pressure with flow also decreases the amount of time required in the wash section.

Wash impingement effects can be generated using various nozzle and pump technologies. To improve cleaning efficacy within continuous systems, boards are initially sprayed in the pre-wash section using fan jets. The pre-wash zone brings the circuit card up to process temperature, which starts the flux softening process. In the wash section, nozzle jets provide uniform wash coverage. Board geometry, density, and component types are impinged upon using a combination of nozzle technologies that provide various levels of fluid flow, pressure at the board surface, and directional forces. Printed circuit boards with increased density and component shadowing require a longer wash time to allow wash fluid to penetrate blind gaps.

To remove all flux residues under gaps less than 2 mils, time in the wash and wash temperature are critical parameters. The wetting effects of flux during the reflow soldering process cause the flux to penetrate under small component gaps and create a flux dam (Figure 19.12). To break the flux dam, the cleaning fluid and impingement energy must first dissolve the residue to create an opening for the wash fluid to flow under the component. Hard flux residues take a longer time to dissolve than do soft flux residues, which increases wash complexity. The static cleaning rate (dissolution in the absence of impingement energy) of the wash chemistry is driven by the cleaning materials compatibility with flux soil, rate of dissolving the flux soil, concentration, part fixturing, and wash temperature effects. The cleaning material static cleaning rate may vary on different flux residues. To address these complexities, best-in-class cleaning material designs are formulated to work on most flux residue types, but the rate varies for both hard and soft flux residues, with the key variable representing time in the wash stage.

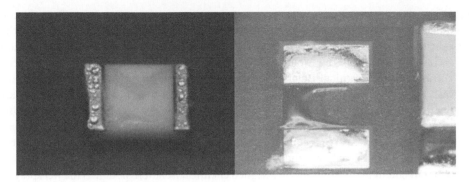

FIGURE 19.12   Flux dam illustration.

Fluid Storage

The size of the wash holding tank is often overlooked in the cleaning equipment design. The wash tank surface area (width × length) and tank volume (surface area × depth) are an important cleaning machine design criterion. High fluid flow nozzles increase the level of wash turns per minute of operation. If the wash tank capacity is too small, rapid tank turnover can cause air bubbles to migrate deeper into the wash tank. When this condition exists, foam build is greater than foam break. Additionally, rosin/resin-based flux soils saponify with the reactive agents in the cleaning material, which can couple or emulsify the anti-foam minor ingredients. The combination of these two factors allows air to eventually reach the pump intake, resulting in a highly stable foam condition. Foaming leads to pump cavitation or micro air pockets, which reduces the machine's operating spray pressures, reduces the effectiveness of the wash process, and leads to costly mechanical pump seal failure.

Wash Bath Life

The volume of the wash tank is also important in maintaining a long wash bath life. The affinity of the wash fluid to hold another substance in solution is bound by the volume of the available fluid and critical soil loading limitation. Critical soil loading represents the contamination level at which cleaning is no longer acceptable. The dynamics at work when running an inline cleaning process find losses through evacuation (ventilation effects) and drag-out. As soil is introduced, the wash holding tank will start to accumulate contamination. Larger wash tank volume provides a wider processing window for holding the flux contaminants introduced to the cleaner. As water and wash chemistry are replenished, the ideal condition occurs when the cleaning material additions maintain a wash tank volume below the critical soil loading limitation.

Fluid Management

One of the risk factors from high fluid flow, high impingement pressure, and directional force nozzles is spray deflection into the chemical isolation section. Shorter wash sections have less distance from the spray manifold to the chemical isolation section. Deflecting wash fluid into the chemical isolation section can cause high wash chemistry consumption. To address this issue, larger wash sections provide distance from the final spray manifold to the chemical isolation section entrance. Wash spray deflected into the chemical isolation zone must be captured and returned to the wash tank. Chemical isolation innovations that dramatically reduce wash chemistry consumption and good fluid management designs are the keys to reducing chemical consumption and operating cost.

The chemical isolation module provides the ultimate fluid management by separating the wash and rinse sections. The conveyor belt, circuit cards, and fixtures are wetted with the cleaning fluid. As these wetted components leave the wash section, they are carried into chemical isolation section. One of the objectives of the chemical isolation section is to remove and return the wash solution from the conveyor belt, circuit boards, and fixtures back to the wash tank. Recent equipment innovations strip off the wash fluid from parts as well as capture defected wash spray that enters the chemical isolation section. This important economic function reduces chemical usage and saves operation cost.

At the entrance of the chemical isolation section, an air knife above and below the conveyor belt removes wash fluid mechanically from the conveyor belt, circuit boards, and fixtures. A bulkhead is positioned within the chemical isolation section after the air knife but in front of the chemical isolation spray rinse manifolds. The deflected spray from the bulkhead isolates is carried into the chemical isolation section. The front isolated area is equipped to collect and drain wash chemistry removed from both the air knives and deflection back into the wash tank.

The next management tool used in a well-designed chemical isolation module is the wet spray section. This wet spray ensures the dilution of chemistry that remains in and around tightly-spaced components and underneath low standoff devices missed by the isolation air knives. This wet section manifold is powered from the rinse module pump, which provides high quality third-use water for wet isolation. The slipstream from the rinse improves rinse quality and prolongs carbon and ion exchange purification media. This cascade function allows the rinse tank to be replenished and regenerated by the gravity cascading final rinse.

The last important section of the chemical isolation is another set of air knives above and below the conveyor belt. These air knives mechanically strip the wet section spray from the circuit board and conveyor belt before they enter the rinse section. This prevents any diluted chemistry from reaching the recirculation rinse, and is especially important when running a closed-loop rinse. Even if the rinse tank is not closed looped, chemistry carryover can lead to rinse section foaming and excessive consumption. When close-looping the rinse, the chemical isolation spray section is a must to ensure that DI filter columns are not degraded due to chemistry contamination.

Air Management: Ventilation

Airflow management prevents steam from exiting the entrance or exit ends of the machine. Cleaning machines that run wash chemistries require a separate exhaust plenum for managing the wash section, and a separate exhaust plenum for managing the dryer sections. The air must be balanced to provide a slight negative draw at both the entrance and exit ends of the machine. Excessive air draw in the wash section can cause wash fluid fumes to be evacuated to the roof. Excessive air draw from the dryer vent can draw wash fluid fumes into the rinse section, which can cause rinse foaming. Properly balanced, the system does not fill the room with wash odors and manages the airflow so that minimal wash fluid is evacuated. Mist arrestors are commonly used to condense the wash fumes, which allow much of the cleaning solution to be returned to the wash tank.

Fluid Control (Controlling the Wash Chemistry)

To achieve a tight quality range, wash tank chemistry control is critical. A properly balanced wash tank provides cleaning consistency over time and increases the life of the chemistry. Best-fit cleaning fluids work well at lower cleaning concentrations and readily condense, which reduces ventilation losses. The benefit for users is much lower cleaning material consumption, which equates to lower cost of ownership. For example, when running a 10% wash concentration, only 5%–6% cleaning fluid additions is needed to maintain the 10% concentration over time.

Typically, an inline cleaner will consume from 3 to 10 gal of wash fluid per hour of pump time operation. Since new-generation cleaning chemistry losses are minimized by the cleaning material design, very little cleaning chemistry is needed to maintain wash concentration. The problem is that, without a consistent addition of cleaning material along with water additions, excessive amounts of water can be added over time, diluting the wash bath concentration. Without the consistent addition of wash chemistry with makeup water additions, the wash tank will eventually lose strength, and cleaning performance will drop off. To prevent this condition, a chemical proportioning device should be used to maintain the wash tank concentration to process specifications. Proper additions increase cleaning consistency over time, and increase wash bath life.

Automated Monitoring and Control

More and more high-reliability products are being cleaned that require full-time data logging of the monitoring and control of the wash bath concentration. These capabilities are being driven by governmental agencies and original equipment manufacturers (OEMs) that require product traceability all along the manufacturing process. Process control systems with these capabilities can be easily integrated with the machine wash tank. Proportional injections of chemistry can be accomplished based on feedback from automated monitoring methods. This automation reduces the requirement for manual concentration monitoring, freeing process engineers for other important duties. The process control system will maintain the wash bath concentration, keeping the cleaning process within the determined tolerance with minimal effort and with captured data.

Process Monitoring

With current and future electronic assemblies becoming densely populated with miniaturized components, qualifying a cleaning process has become a long and costly procedure. Highly dense assemblies populated with low standoff components not only create difficult areas for fluid penetration during the cleaning process but also when cleanliness tests are conducted. The days of validating a cleaning process solely utilizing an ionograph are over. This is due to the lack of sufficient agitation and averaging of surface area when looking at ionic contamination levels. The vast majority of electronic manufactures are now sending assemblies to analytical labs for localized ion chromatography and SIR testing when validating cleanliness levels of a new product or a new cleaning process. While this method provides the manufacturer a higher level of accuracy when checking assemblies for ionic contamination, this form of validation is too costly to be used as a standard quality assurance test. Another reliable method used is a destructive test in which components are removed to visually check for residues. This again can be costly; products being cleaned are usually expensive in nature, so again this is not a cost-effective method for frequent quality assurance testing.

This has lead to the implementation of a new control system for monitoring and tracking cleaning data. The new control system is a Windows-based computer with powerful yet simple user interface software that utilizes digital and analog I/O control.[32] This system provides complete control that tracks all process set points for the entire cleaning process. The system can be configured to data log process parameters per board for future process traceability, or can be set as a time-based function. This logged information provides application and quality engineers with quantifiable data. The collected data can be reviewed to ensure that the cleaning process is operating within the upper and lower control limits that were originally developed from SIR and ion chromatography cleanliness results. This provides manufactures with a cost-effective means for frequent quality assurance checks and with reviewable board-based data that can also be used for process troubleshooting.

Drying

Dryer innovations remove all moisture from dense substrates by using high-velocity, high-temperature regenerative blower technology using air knives to mechanically strip the water from the board. This high-temperature air helps maintain a constant board temperature from the final rinse to the convection zones.[5] Two independently temperature-controlled convection zones are then used to evaporate the remaining water from the board. The use of heated convection drying provides manufacturers with test-ready assemblies at high throughput rates and a lower cost of operation when compared to most standard air knife dryers. The dryer design is free from the typical belt-driven blowers that are used in most inline cleaners. The technology utilizes direct drive blowers, which provide manufacturers with a more reliable process along with less down time and maintenance.

Cleaning Equipment Selection: A number of factors need to be considered when selecting cleaning equipment.

- Will the cleaning machine and cleaning agent clean your most demanding assemblies?
- Does the ventilation design isolate the wash section from the rinse and dryer?
- Will the heating elements maintain desired temperature settings?
- What is the response time to bring the machine up to production readiness?
- What is length of the wash tank?
- Does steam dissipate from the front, end, or windows?
- Is the equipment assessable?
- How effectively does the machine isolate the wash section?
- Does the wash fluid removed in the chemical isolation section drain back to the wash tank?
- Does the cleaning machine have a sampling port?
- How many times per minute does the wash tank turn over?
- How is the cleaner injected and controlled?
- Is the machine controlled by a programmable logic controller (PLC) or is it computer controlled?
- Are the seals and gasket materials compatible with the cleaning agent?
- Will the machine dry your assemblies?
- What is the throughput rate needed to clean your most challenging assemblies?
- What is the noise level?
- What is the cost of the equipment?
- What is the cost/part to clean?

Process Validation

Process engineers tasked with designing the cleaning process often start by selecting cleaning equipment. Once the cleaning equipment decision is made, they often seek out cleaning agents designed for the specific chosen cleaning equipment. The initial feasibility assessment typically includes material and process compatibility testing, test vehicles, and formal test plan. Most follow the protocols called out in the J-STD-001 in the IPC Joint Industry Standards for validating cleaning agents.

Process engineers commonly prefer designed test kits. The benefits of these kits include board and component combinations to run a standardized test. A common mistake is that the test vehicles are not representative of the production assemblies. The better approach is to first evaluate the chemistry set to assure it will clean without getting into the specifics of the cleaning equipment and particular test vehicles. Segregate the flux soils and cleaning agent options from hardware.

Work with a reputable cleaning agent supplier to first establish the flux soil solubility parameters. As previously stated, the best-fit cleaning agent will have solubility parameters that closely match that of the flux soil. A second common oversight comes from the fact that many cleaning agents will adequately clean the flux residue. Which cleaning agent is the best match for the specific process in question?

The best-fit cleaning agent choice reduces kinetic and mechanical impingement forces. Few process engineers and cleaning agent suppliers understand how to properly characterize the solubility parameters for the flux residue and determine the cleaning agent that is the best fit. Performing this work upfront typically saves time addressing process issues down the road.

Once the best-fit cleaning agent is determined, materials compatibility testing should be initiated to assess interactions on hardware sets. As integrated circuits become smaller, time, temperature, cleaning agent concentration, and impingement energy typically increase. Identifying the process gaps early in the design for manufacturing will help in establishing the process window and eliminating process issues.

Following the characterization of the flux soil and cleaning agent that fits within the interaction zone, the process engineer now has a good understanding of the time, temperature, and cleaning agent

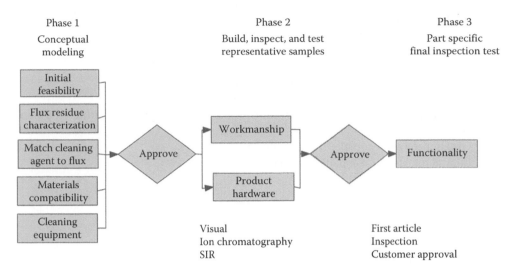

FIGURE 19.13 Modified OEM Aerospace cleaning process methodology. (From Valladares, H., Process validation. *IPC Cleaning and Alternatives Handbook*, IPC, U.K., 2010. With permission.)

concentration needed to dissolve the flux soil. Better definition of these factors provides an understanding of the cleaning equipment needed to process the hardware set. At this point, representative test vehicles should be run at cleaning equipment or cleaning supplier's application labs. If the process meets the cleaning requirement, move forward with J-STD-001 process validation. If not, stop and reassess the chemical material sets and cleaning equipment.

After working with customers to design and implement new cleaning processes, I have come to the conclusion that this methodology is flawed. A better approach is to use Lean Sigma methodology when designing new cleaning processes. Start by defining the problem. For example, the problem may be that miniaturization and lead-free soldering require more active flux residues that must be removed to assure product reliability. In the problem definition stage, engineers must ask what could go wrong. Using Lean Sigma methodology, the process of designing a new cleaning process has a much better probability of meeting the initial design criterion. Figure 19.13 is a modified Honeywell[38] cleaning process qualification methodology on product hardware.

1. Prior to committing resources to qualify new cleaning materials, an initial feasibility should include the following:
   a. Occupational, safety, and environmental impacts
   b. Customer's requirements and expectations
2. Characterize the flux residue
   a. Review process flow
   b. Review thermal profiles and document peak reflow temperature and time above liquidus
   c. Review bake-out procedures before cleaning
   d. Review the number of soldering exposures before cleaning
   e. Process test boards with flux residues that represent the worst-case scenario
   f. Work with your cleaning agent supplier to develop the Hansen Solubility Parameter (HSP) for all flux residues that will be cleaned by the new cleaning process
   g. Analyze the data with the cleaning agent vendor to understand the characteristics of the flux residues used in your process and the cleaning material properties needed to clean the flux residues
3. Match the cleaning agent to the soil
   a. Using HSP methodology, conceptually model the best fit cleaning agent.

       i.   Determine process levels
          1.   Wash concentration
          2.   Wash time
          3.   Wash temperature
   b.   Work with your cleaning agent vendor to run validation testing
   c.   Analyze the data to assure removal in the desired cleaning equipment

4. Once the desired cleaning agent is known, run materials compatibility testing on production hardware
   a.   List all materials compatibility possibilities
   b.   Common materials compatibility concerns
       i.   Metallic alloys
       ii.   Board finishes
       iii.   Components
       iv.   Entrapment
       v.   Labels
       vi.   Part markings
       vii.   Adhesives
       viii.   Coatings
   c.   Work with your cleaning agent supplier to run materials compatibility testing on the production hardware set
       i.   Analyze the data to identify potential problem areas
   d.   Ask the cleaning agent supplier for an equipment materials compatibility document
       i.   Seals
       ii.   Gaskets
       iii.   Curtains
       iv.   Piping
       v.   Plastics
       vi.   Motors
       vii.   Ventilation
       viii.   Water recycling filtration

5. Evaluate cleaning equipment options
   a.   Throughput rate
   b.   Low clearances and ability to penetrate and remove residue under components
   c.   Ability to contain cleaning agent
   d.   Cost of ownership
   e.   Manufacturers' testing, quality, service, and support
   f.   Equipment compatibility and capability
   g.   Waste generation and disposal
   h.   Process capability and robustness
   i.   Process control requirements
   j.   Cleanliness monitoring—type of contaminants, cleanliness levels, and measurement
   k.   Measurement system capability evaluation
   l.   Cost impact

6. Process validation
   a.   Select a test board representative of process hardware
   b.   Design an experiment to test factors and levels
   c.   Confirm visual cleaning effectiveness
   d.   Run materials compatibility screening
   e.   Test vehicles should meet the requirements defined in J-STD-001

Conclusion

Integrated circuit design failures are attributed to feature size reduction, which increases the risk of defects randomly induced by process flaws.[2] White residue sandwiched under highly dense, low standoff components creates the potential for material defects caused by the presence of ionic residue. Traditional approaches to verifying reliability based on screening the output of a process are no longer effective. Rather, integrated circuit packagers are moving toward building in reliability by controlling and monitoring cleaning factors that assure complete removal of flux residue.

Higher density, smaller components, and lower standoffs are changing the definition of integrated package cleanliness. With the reduction in component size and low standoff clearances, the ability to extract and see measurable residues that correlate to product quality is much more suspect. Cleaning processes take on a whole new cleaning definition of removing residue that can be seen visually and residue that is commonly out of sight, entrapped under components.

Cleaning process optimization requires a balance of chemical and mechanical effects. Best-fit cleaning materials match up to the flux residue solubility parameters. By first characterizing the flux soil, an interaction zone can be developed for dissolving or dispersing the soil. The challenge is to find the solubility sphere for the flux residue. Matching the solubility parameters of the cleaning agent with the interaction zone for the flux residue assures excellent cleaning performance.

The five-step process of first establishing the flux solubility parameters, finding the best-fit cleaning agent design, evaluating materials compatibility on the hardware set, confirming performance in the cleaning machine, and validating the process uncovers issues early in the design. Selecting a cleaning agent that has the best fit for dissolving the flux residue in question assures the desirable properties for opening the process window.

Acknowledgments

Cited content in this chapter includes content from technical papers jointly authored with Dr. Ning Chen Lee of Indium Corporation, Steve Stach of Austin American Technology, Mike Konrad of Aqueous Technologies, David Ihms of Delphi Electronics, Wayne Sozansky of Delphi Electronics, John Neiderman of Speedline Technologies, Dirk Ellis of Kyzen Corporation, Fernando Rueda of Kyzen Corporation, Serge Tuerlings of Kyzen Corporation, JoAnn Quitmeyer of Kyzen Corporation, Ram Wissel of Kyzen Corporation, Art Thompson of Kyzen Corporation, Phil Zhang of Kyzen Corporation, and Chris Shi of Kyzen Corporation.

References

1. Santarini, M. (March 2008). Consumer ICs: Designing for reliability. March 6, 2008, Retrieved from www.edn.com
2. Hugh, S. (October 2008). Qualifying lead-free medical electronics. *IPC/SMTA Higher Performance Electronics Assembly Cleaning Symposium*, Rosemont, IL.
3. Perng, S. (October 2008). Spot cleaning assessment under components following inline aqueous cleaning. *IPC/SMTA High Performance Cleaning Symposium*, Rosemont, IL.
4. Byle, F. and Eichstadt (June 2005). Enhancing flip chip reliability. Advanced packaging. www.apmag.com
5. Moosa, M.S., Poole, K.F., and Grams, M.L. (1996). EFSIM: An integrated circuit early failure simulator. *Quality and Reliability Engineering International* 12(1996), 229–334.
6. Moore's Law (2009). Retrieved from http://en.wikipedia.org/wiki/Moore%27s_law
7. Bixenman, M., Ellis, D., and Neiderman, J. (2009). Collaborative cleaning process innovations from managing experience and learning curves. *IPC APEX Technical Conference*. IPC, Las Vegas, NV.

8. Osterholf, A., Ellis, B., Naisbitt, G., and Pauls, D. (2008). Component cleanliness. IPC Technet Archives. Retrieved from http://listserv.ipc.org/scripts/wa.exe?A2=ind0002&L=TECHNET&P=R72868

9. Christoph, J. and Elswirth, M. (March 2002). Theory of electrochemical pattern formation. *American Institute of Physics* 12(1), 215–230.

10. Bixenman, M. and Stach, S. (September 2006). Optimized static and dynamic driving forces for removing flux residue under flush mounted chip caps. *SMTAI Technical Conference*, Rosemont, IL.

11. Bixenman, M. and Konrad, M. (October 2009). Cleaning for reliability post QFN rework. *SMTAI Technical Conference*. San Diego, CA.

12. Bixenman, M., Lee, N.C., and Stach, S. (October 2009). Ionic cleanliness testing research of printed wiring boards for purposes of process control. *SMTAI Technical Conference*, San Diego, CA.

13. IPC. (2006). *IPC International Technology Roadmap for Electronic Interconnections*. IPC Association for Connecting Electronic Assemblies, Chicago, IL.

14. iNEMI. (2006) iNEMI Technology Roadmap. iNEMI, Herndon, VA.

15. Lee, N.C. (October 2009). Lead-free flux technology and influence on cleaning. *SMTAI Technical Conference*, San Diego, CA.

16. Lee, N.C. and Bixenman, M. (2000). Lead-free: How flux technology will differ for lead-free alloys & its impact on cleaning. *NEPCON Technical Conference*, Anaheim, CA.

17. Bixenman, M., Rueda, F., and Tuerlings, S. (November 2008). Innovative printed circuit board cleaning fluid designed to remove process soils from populated assemblies, stencils, and rework in batch and inline cleaning equipment. *Global SMT Romania Technical Conference*, Timisoara, Romania.

18. Hansen, C.M. (2007). *Hansen Solubility Parameters: A User's Handbook*. CRC Press, New York.

19. Lee, N.C. (May 2009). Achieving high reliability lead-free soldering—Materials consideration. ECTC Short Course. San Diego, CA.

20. Bixenman, M. (September 2009). Science of aqueous cleaning. SMTA Capital Vendor Day. John Hopkins University, Baltimore, MD.

21. Saponification. (2009). Retrieved from http://en.wikipedia.org/wiki/Saponification

22. Reduction Potential. (2009). Retrieved from http://en.wikipedia.org/wiki/Redox_potential

23. Nucleophilic Acyl Substitution. (2009). Retrieved from http://en.wikipedia.org/wiki/Nucleophilic_acyl_substitution

24. Surface Tension. (2009). Retrieved from http://en.wikipedia.org/wiki/Surface_Tension

25. Wetting. (2009). Retrieved from http://en.wikipedia.org/wiki/Wetting

26. Quitmeyer, J. (September 2009). Tips in solving aqueous cleaning problems. *Metal Finishing* 107(9), 18–24.

27. Pourbaix Diagram. (2009). Retrieved from http://en.wikipedia.org/wiki/Pourbaix_diagram

28. Bixenman, M., Ihms, D., and Sozansky, W. (March 2002). Enhanced vapor phase cleaning fluid for demanding flip chip cleaning requirements. *Georgia Tech Advanced Packaging Conference*. Georgia Tech. Atlanta, GA.

29. Bixenman, M., Thompson, A., and Wissel, R. (1999). *Semi-Aqueous White Paper*. Kyzen Corporation, Nashville, TN.

30. Stach, S. and Bixenman, M. (September 2004). *Optimizing Cleaning Energy in Batch and Inline Spray Systems. SMTAI Technical Forum*. SMTA, Rosemont, IL.

31. Lee, N.C. (2002). *Reflow Soldering Processes and Troubleshooting SMT, BGA, CSP and Flip Chip Technologies*. Newness, Woburn, MA.

32. Bixenman, M., Zhang, P., and Shi, C. (May 2008). White residue on printed wiring boards post soldering/cleaning. *SMTA China Conference*. Shanghai, China.

33. First law of Thermodynamics. (2009). Retrieved from http://en.wikipedia.org/wiki/First_law_of_thermodynamics

34. Lee, N.C. (December 2007). Combining superior anti-oxidation and superior print—Is it really possible? EPP Europe.

35. Second Law of Thermodynamics. (2009). Retrieved from http://en.wikipedia.org/wiki/Second_law_thermodynamics

36. Third Law of Thermodynamics. (2009). Retrieved from http://en.wikipedia.org/wiki/Third_law_of_thermodynamics

37. Bixenman, M. (2007). Engineered cleaning fluids designed for batch processing. *IPC APEX Technical Conference*. Los Angeles, CA.

38. Valladares, H. (2010). Process validation. *IPC Cleaning and Alternatives Handbook*. IPC, U.K.

20

Precision Cleaning in the Electronics Industry: Surfactant-Free Aqueous Chemistries

Surfactants..319
Historical Perspective • Structure of Surfactants • Classification of
Surfactants
Critical Cleaning in the Electronics Industry..322
Cleaning Requirements and Qualification • Main Equipment-Process
Limitations
Surfactant-Free Aqueous Cleaning Products...325
Attributes of Surfactant-Based Products versus Surfactant-Free
Products • Process Integration: Controlling the Chemical Wash
Fluids • Encountered Residues and Associated Risks
Overall Cost of Ownership of the Cleaning Process..............................329
Cleaning Agent Technology • Vapor Recovery • Bath Monitoring
Conclusions: The Most Important Characteristics of
Surfactant-Free Aqueous Cleaning Agents 330
Acknowledgment...330
References...331

Harald Wack
ZESTRON

Surfactants

Historical Perspective

Surfactant is an abbreviation for surface active agent, which means active at a surface. This means a surfactant is characterized by its tendency to adsorb contaminants at surfaces. Surprisingly, the preparation of a surfactant is potentially the oldest documentation of a defined chemical synthesis, as the oldest man-made surfactant that was used as a "consumer product" is soap. Documents exist from the early Sumerians, who inhabited the region between Euphrates and Tigris rivers, which clearly describe the production of soap in 2500 BC. In these documents, there are detailed instructions on which quantity the two starting materials, oil and wood ash, have to be mixed prior to heating. They used this soap for the washing of their woolen clothing. Ever since these former times there are various documents in the Egyptian, German, and Arab civilizations talking about the preparation and usage of soap as a detergent for body and clothing and even sometimes for medicinal purposes.

This situation remained unchanged until the late 1800s—so people only had the choice to use washing products such as bar soap, soft soap, and powdered soap. In the further development of detergents, chemists very soon encountered some insuperable problems caused by the structural properties of soap: its alkaline reaction and its sensitivity to water hardness. So there was the need to look for some other surfactant, which did not have these disadvantages.

In 1917, one German chemist from BASF succeeded in the alkylation and sulfonation of naphthalene, the first attempt to synthesize a soap substitute. These first soap substitutes did not possess all the positive aspects of classic soaps, and so there was a very fruitful process of developing new artificial surfactants during the next decades. In the 1950s, soap had almost been completely replaced as a surfactant in detergents. And the market of synthetic surfactants really exploded, resulting in serious ecological problems, as the aspect of biodegradability was not sufficiently considered. Although today strict regulations are in effect for biological degradation, the question of the ecological compatibility of surfactants continues to be one main focal issue. This resulted in a great variety of surfactants, such as typical commodities and very specific selective designed ones. In the following section, we will outline a short overview of the surfactant structure and classification, which will of course not be able to cover all aspects.

Structure of Surfactants

Surfactants are substances with molecular structures consisting of a hydrophilic (water attracting) and a hydrophobic (water repelling) part (Figure 20.1). The hydrophobic part is normally a hydrocarbon fragment, and the hydrophilic part consists of ionic or highly polar groups.

Due to this characteristic structure, these compounds have some special properties, namely their tendency to adsorb at interfaces, which means that their concentration at the interfaces is higher than in the inner regions of the solution. This behavior is attributable to their amphiphilic structure (the Greek word amphi, meaning both, covers the aspect that we have hydrophobic and hydrophilic areas in one molecule). Another fundamental property of surfactants is that the monomers in solution tend to form aggregates called micelles (Figure 20.2). Micelle formation can be viewed as an alternative mechanism to adsorption at the interfaces for removing the hydrophobic groups from contact with water. This results in some of the most important reasons surfactants are used: lowering of interfacial tension, such as wetting and foaming. Micelles are generated at low surfactant concentrations. However, there is a threshold level; the concentration at which micelles start to form is called the critical micelle concentration or CMC.

Classification of Surfactants

Surfactants are primarily applied in aqueous solutions and so normally the classification by the hydrophilic group is appropriate. Classification by the hydrophilic group, one differentiates

- Anionic surfactants
- Cationic surfactants
- Non-ionic surfactants
- Amphoteric surfactants

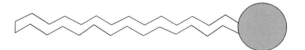

Hydrophobic group
"Fat loving end"

Hydrophilic group
"Water loving head"

FIGURE 20.1   General surfactant molecule structure.

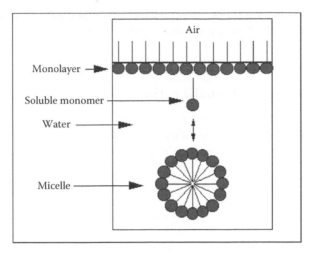

FIGURE 20.2   Step-by-step micelle formation.

Anionic surfactants are molecules in which the hydrophobic group is connected with one or more hydrophilic groups that dissociate in water into a negatively charged ion (anion) and a positively charged ion (cation) and where the negative-charged ion is a carrier of the surface-active properties.

Some of the most important types of anionic surfactants are

- Alkyl sulfates
- Alkyl ether sulfates
- Alkylbenze sulfonates
- Alkane sulfonates
- Soaps
- Alkyl phosphates

Cationic surfactants also dissociate under aqueous conditions but here the cation is the carrier of the surface-active properties.

Examples of cationic surfactants are

- Tetra-alkyl ammonium chloride
- Hydroxyethyl-methyl-ammonium-bisalkylester

Non-ionic surfactants are surface-active compounds that do not dissociate in aqueous solutions. The solubility and the surface-active properties here are generated by the presence of a hydrophobic group and one or more polar groups like polyglycol ether or polyol groups.

- Fatty alcohol polyethylene/polypropyleneglycol ethers
- Fatty acid alkanol amides
- Alkyl glycosides
- EO/PO adducts
- Amine oxides

Amphoteric surfactants contain, in aqueous solutions, both negative and positive charges in the same molecule. Depending on the conditions of the cleaning medium (pH value), the molecules are negatively or positively charged. In alkaline pH medium, for example, they are negatively charged, and vice versa.

- Alkyl betaines
- Alkyl sulfo-betaines

Commercially available surfactants are not uniform substances, but are mixtures of homologous substances, for example, the length of the hydrocarbon chains encompasses a certain range.

Critical Cleaning in the Electronics Industry

Cleaning Requirements and Qualification

The use of cleaning agents in the electronics industry has been commonplace for decades. The main purpose of adding a cleaning step to the manufacturing process is to increase the reliability of the final assembly. In most cases within the electronics industry, users remove flux residues, which are seen as unwanted residues. Fluxes are organic acids (OA) and are used to help activate metals during the soldering process. Unfortunately, if not subsequently removed, their later presence leads to a variety of quality-related failures [1].

In the North American market, the majority of electronic assemblies are built for military, aerospace, medical, and other higher-end market segments. It is essential that products are produced according to the highest cleanliness levels. A large number of users rely on using OA fluxes and use DI water as a cleaning agent. Alternatively, companies who use RMA (rosin moderately activated), no-clean, or synthetic fluxes require a chemistry-assisted process to fully remove the residues. In general, it appears that a chemistry-assisted process shows a much improved cleaning result.

With chemically assisted processes on the rise, users are well advised to carefully qualify their cleaning process.

Main Equipment-Process Limitations

Prior to qualifying a new cleaning process, one has to be made aware of numerous process limitations. Some of the most significant limitations are the ability of DI water to effectively remove OA fluxes, or the limited chemical solvency of some products (especially for no-clean residues) as well as temperature limits for processes using plastic construction materials.

Chemical Isolation

Recent market studies suggest that water is beginning to reach its cleaning limitation. A number of reasons supporting this trend can be cited which are mentioned later on in this chapter.

The usage of chemistry in the long run seems therefore to be overall most beneficial. Benefits include but are not limited to better cleaning through lower ionic contamination which in turn provides much higher product reliability. Recent studies have also demonstrated better bonding and coating after the introduction of chemistry-assisted cleaning. To offset the added cost, users can operate at lower temperatures and, with wider process windows, not only when one cleans to remove OA but also when removing RMA and no-clean fluxes. That will become a necessity in the North American market as contract manufacturers are moving to lower volume, higher mix, and significantly higher reliability products. At the end, the introduction of a chemistry-assisted cleaning process will increase your cleaning-process window and permit the de-fluxing of all production boards with a single cleaning process.

Despite all of the valid arguments encouraging the use of a chemically assisted process, most equipment currently using only DI water are not properly designed to use chemistry without continuous dilution, i.e., losses. This means that they lack a chemical isolation section. The latter is an essential part to not only conserve chemistry but to also minimize foaming, for example. DI-water machines take advantage of cascading DI water tanks from back to front. Employing a chemical product in the wash tank would lead to continuous dilution of the recommended application concentration by DI water cascading into it. Companies that are strategically planning their capital purchases are therefore well advised to incorporate the mechanical option to run aqueous chemistries.

Solvency

"Like dissolves like." This simple but adequate statement describes the principle of solubilization. While inorganic materials such as salts are easily solubilized in water due to their common polarity, organic contamination such as flux residues can only be solubilized if the user employs matching organic cleaning agent. The physical chemistry phenomena responsible for this observation include polarity, Van der Waals forces, dipole moments and others. For solvency to occur, no chemical reaction is required. Various chemical blends are currently available on the market, and for a user to find the most appropriate for his or her need, one has to conduct cleaning experiments. Each and every residue reacts respectively to a cleaning agent, i.e., it cleans fully, partially, or not at all. Chemical solvency is related to temperature, which is relayed in any basic physical chemistry text book. Let's look at IPA, for example: being a small organic molecule with a slight polarity, it can hardly be considered a good solvent. This is also reflected by a low kauri-butanol value, a measuring technique that assigns numbers to solvency [2].

IPA is widely used as extraction fluid for Ion Chromatography (IC) and ionic contamination measurements as well as for manual cleaning processes after rework. Most users will concur that IPA spreads the residues around but does not fully remove it from the assembly. The same is true for extraction purposes. Given the current no-clean usage in the industry, it has already become clear that IPA/Water is not capable of fully solubilizing these modern flux residues anymore [3].

Solubility increases with temperature and modified Resistivity of Solvent Extraction (ROSE) tests are used to increase the solubility of IPA/Water [4]. In this case, the solvency is not sufficient and future alternatives are needed to overcome this analytical limitation.

In addition to the limitation of IPA in its respective applications, another widely used cleaning agent is reaching its solvency limits. As briefly mentioned before, DI water is being challenged by modern water soluble fluxes used in combination with increased use of lead free alloys. This results, for example, in more burnt fluxes that are much harder to remove as they create water-insoluble contamination. DI water alone has a very limited ability to solubilize non-ionic residues on the board's surface.

Second, the cleaning of water-soluble fluxes (especially under components) has also become more difficult. Water with its high surface tension of over 70 dynes/cm cannot effectively penetrate low-standing components. And as the standoff heights are decreasing and component densities are increasing, companies will have to improve their existing cleaning process. Chemistry-assisted cleaning can reduce the surface tension to 30 dynes/cm and below. Interestingly, the industry so far has mostly reverted to adjusting the cleaning process to its respective limits. This entails, for example, an increase in operating temperature to above 150°F, increase of the spray pressures, or the lowering of the belt speed to prolong the exposure time. With pure water soluble fluxes in a eutectic tin–lead environment such measures can provide sufficient cleaning results. Given the introduction of lead free, however, the solubility of residues in DI water becomes the limiting aspect. If non-ionic contamination is produced, water alone cannot chemically dissolve such contamination. It is essential at this point to point out that solvency will also increase with the use of non-surfactant-based products.

Another often overlooked consequence is that higher pressures might allow the water to penetrate low-standing components by forcing water underneath or into the capillary spaces. Unfortunately, the cleaning equipment is not capable of removing the water later on in the drying sections. It is of utmost importance to verify a dry and water and flux free environment under components after cleaning, to limit the formation of electrochemical migration or leakage currents. Cleaning agents other than water tend to be more easily rinsed and dried as the lower surface tension allows for quick removal. Numerous surfactant-based products contain inorganic ingredients. Users should be considerate of this fact as it might impact the nature of residues that are left under components after cleaning. Cleaning agents that do not contain any solids as part of their formulation will clearly limit the amount of introduced contamination. In general, any product technology should be compared in the virgin state and the contaminated state by users. Once the cleaning agent has been in use for some time, the respective product will have been loaded with a variety of contaminants. The latter contributes to the issue of remaining residues under components. Please also check virgin cleaning agents in regards to their solid contents (typically inorganic salts).

With cleaning agents used in an automated Printed Circuit Board (PCB) or stencil-cleaning process, solvency plays an equally important factor. The user has to be able to rely on his or her equipment to reliably remove the remaining flux residues. It is therefore of utmost importance that cleaning agents do not deplete quickly. In other words their solvency should be as constant as possible. For automated processes, solvency is supported by mechanical energy. Numerous studies have been published on the topic. The main conclusion is that both solvency and mechanical energy equally contribute to the overall cleaning result. Mechanical energy can help overcome chemical limitation on solvency through higher impact pressures, increased numbers of spray nozzles, longer exposure times, or improved pre-soaking stages. Numerous variations are of course possible, and users are well advised to first ascertain the pure chemical ability of the cleaning agent to solubilize the residue in question, prior to soliciting mechanical support. Such a planning sequence will allow for a successful cleaning-process implementation in the long run. Depending on the cleaning agent (surfactant or non-surfactant), users might experience a difference in solvency [5].

Temperature

As mentioned previously, temperature plays an important role in any cleaning process, especially within electronics precision cleaning. Think about some of the effects temperature can have on the product:

- Material compatibility with components
- Material compatibility with the equipment
- Evaporative losses caused by excessive operating temperatures
- Effect on the temperature on the flux residues

Most current equipment is built with polypropylene and or stainless steel. The advantage of steel of course is a higher potential operating temperature. For example, numerous DI water users revert to an increase in temperature in order to remain with DI water instead of using a chemically assisted process. Keep in mind that for large in-line equipment energy consumption can be significant. Polypropylene materials cannot exceed 160°F, which at present seems to be the agreed on process limit for de-fluxing applications. Keep in mind that this temperature setting was established 20 years ago, when only surfactant-based products were available. At that time, more mechanical energy was needed, as most of these traditional surfactants did not contain any organic constituents. Recently, a study suggested that higher temperatures above 160°F for a chemically assisted process can provide added benefits [6].

Stencil-cleaning processes on the other hand rely on room temperature. This is mainly due to the ease of raw solder paste removal and the fact that higher temperatures can lead to material compatibility issues with stencils and adhesives used.

Environmental

Most of the environmental limitations observed in the precision-electronics industry have been around specified VOC (volatile organic compound), BOD (biological oxygen demand), COD (chemical oxygen demand), and pH levels, but should also include political directives such as REACH and RoHs, which were implemented recently due to environmental concerns in EU. For an average customer, it is important to know what effects his rinse discharge has, including potentially contaminated cleaning agents. The latter could be pure solvent-based products or water based. As the industry trend is toward water based, spray in air process for most PCB de-fluxing applications, the dragged out or evaporated chemistries have to be approved by local environmental organizations for appropriate disposal.

It is also important to know that, in most cases, the cleaning chemistry itself is not considered the hazard but rather the contamination that builds up during use. For example, lead, tin, and other metals are partially dissolved during the exposure to the cleaning cycle and therefore invariable end up in the cleaning bath. This process is also heavily dependent on the pH level of the product in use. The lower the pH value, the more beneficial it will be for any user to cope with environmental regulations. Standardized analytical methods exist to gauge their respective limits without the need for external labs. However, it is advised that users should always check the requirements for each manufacturing location.

Users also have an opportunity to work on process improvements. Evaporative losses can be minimized through an appropriate cleaning agent (which allows for lower operating temperatures) as well as vapor recovery systems. With the help of the equipment manufacturers, viable solutions can be suggested to reduce the amount of drag over from the wash to the rinse section. In addition, users should select cleaning products which are, preferably, approved by all global and local environmental bodies. With the industry becoming one big global platform, companies are able to transfer entire production lines from one continent to another. This requires that throughout the transition the qualified processes (especially the cleaning processes) do not prompt re-qualifications with local environmental organizations. This is of particular interest in cases where cleaning agents (originally qualified for use in one area) do not meet respective limits/regulations of the future location [7].

Surfactant-Free Aqueous Cleaning Products

Attributes of Surfactant-Based Products versus Surfactant-Free Products

Effects on Temperature and Concentration

As discussed above, one main characteristic feature of surfactants is their tendency to accumulate at interfaces associated with a lowering of the interfacial tension. That is one main effect for which surfactants are used: lowering of the surface tension of the water-based cleaning agent. A good surfactant is able to lower the surface tension of water from approx. 72–73 mN/m down to values from <30 mN/m. This enables the cleaning agent to fully wet the surface of the area being cleaned so that the contamination can be readily loosened and removed. Furthermore, due to the hydrophobic reservoir in the formed micelles, the surfactant solution is also able to remove even greasy or oily dirt and keep them emulsified so that they do not settle back onto the surface being cleaned.

The CMC of the aqueous solution of some surfactants is one very important variable, as after reaching the CMC, the concentration of the monomers that are responsible for most of the properties of the surfactants will no longer increase. Typical CMCs are in a range of 10^{-2} mol for ionic surfactants and down to 10^{-5} mol for non-ionic surfactants. After reaching the CMC, there is no further lowering of the surface tension but there is an increase of the solubilization and capacity of the cleaning agent. The CMC of a surfactant is not very temperature sensitive, but the solubility of the surfactant is very temperature sensitive (Krafft point),* and especially in the case of the non-ionic surfactants most of them are insoluble below a certain temperature. This means that one can observe a phase separation, which allows for completely different behaviors and performances.

If you are formulating a surfactant-free cleaning agent, you also have to lower the surface tension of your aqueous system. This can be achieved by the usage of some solvents that also show a very low surface tension. Mixing them with water can affect the hydrogen bonding structure of water and lower the surface tension down to values that are in the area of surfactant-based systems or even lower. Also, surfactant-free systems are typically not as sensitive to the solubility of added ingredients. Furthermore, high temperatures present the potential of "surfactant separation." This phenomenon means that individual raw materials can precipitate due to the higher temperatures, which in turn can affect the cleaning consistency. Another benefit of surfactant-free systems is the fact that they do not

---

* The temperature at which the aqueous solubility of an ionic surfactant equals the CMC is referred to as the Krafft point. Below the Krafft point, micelles will not form and the surfactant solution will have no solubilization potential.

form micelles, which means the concentration dependence becomes different. The influence of concentration is normally a more or less linear one and not so sensitive to low concentration, as there is not the danger to reach regions below the CMC.

Exposure Times

Effective cleaning agents allow the reduction of exposure times. This can in turn lead to higher throughput and reduce overall process costs. The shorter the exposure times, the lower the chances of negative impacts. This is already becoming more and more important as material compatibility requirements are the prime focus of almost every modern cleaning process. A suitable surfactant-free cleaning agent gives you the options to realize the shortest exposure time possible.

Builders and Conditioners

When formulating a cleaning agent, one sometimes has to add ingredients which are salts (e.g., builders, corrosion inhibitors) and which also affect the formation of the micelles and the solubility of your surfactant. This may also lead to some problems in terms of stability and product performance. In the case of a surfactant-free product, the compatibility with some salt-like components is mostly dominated by the solvents used. This becomes more important as the sheer number and respective sensitivity of the materials in use generates significant challenges. Furthermore, it enables you to strike new paths by formulating, for example, pH-neutral products.

pH Level

Ionic surfactants are particularly pH sensitive, as the dissociation under aqueous conditions is affected. pH level can accelerate or decelerate micelle formation. So let us take an example of an anionic surfactant, like soap: under acidic conditions, the free acid, which is not surface active, results in a loss of cleaning effectiveness. Also, it is not possible to use cationic surfactants or a combination together with anionic systems at alkaline conditions. Only the non-ionic systems are relatively insensitive to the pH of the cleaning agent. In the case of formulating surfactant free, there are almost no limitations in terms of the pH range as you can adjust the base of your cleaning agent to the necessary pH, i.e., neutral for the stencil and screen cleaning applications and alkaline for the PCB cleaning. Here, the limitations of the pH value are mainly driven by the worker's safety and environmental aspects.

Material Compatibility Considerations

One important point with regard to material compatibility is the very specific behavior of some non-ionic surfactants. They can show a phenomenon which is described as stress corrosion cracking (SCC). SCC is the formation of brittle cracks in a normally sound material through the simultaneous action of a tensile stress and a corrosive environment. That effect is very often induced by the existence of non-ionic surfactants and can result in the destruction of some sensitive plastics. That's really a problem as most products are formulated with a base because it's very often the only chance to formulate low-foaming products. In the case of formulating surfactant-free products, all the raw materials can easily be selected to maximize the likelihood of material compatibility.

Residue

Finally, there is one very important point which is a side effect of the general behavior of surfactants. As mentioned above, a surfactant is characterized by its behavior to adsorb at interfaces, which include the interface of cleaning agent and parts that have to be cleaned. Especially in the case of ionic surfactants, this results in remaining residues on the surface of the cleaned parts, even after rinsing the parts very well. This has to be considered if there are some further processing steps after cleaning where this may cause incompatibilities. Very often there are processing steps (i.e., bonding or the application of some adhesives which are very sensitive to some surfactant residues) resulting in some failures or lower yields.

Process Integration: Controlling the Chemical Wash Fluids

Process integration is a powerful term in an industry that includes numerous well-respected equipment manufacturers and renowned cleaning-agent suppliers. The reference list for studies completed is extensive and begins to pinpoint the actual difficulty of "process integration." It is true that each process is unique and requires special support. The assembly alone provides endless levels of differences, be that geometrical, component related, materials related, or simply based on the specific process used for assembly. If all those were to be equal, you will still be hard-pressed to find a company that used the identical equipment for printing, soldering etc. It becomes quickly apparent that the magnitude of variables leaves little room for general recommendations.

Undoubtedly, users should know that in order to achieve a stable cleaning process, a thorough analysis of their needs is required, which means time, patience, and a lot of designed cleaning trials. Second, users should be made aware that each cleaning process will require the aid of cleaning product and cleaning equipment, i.e., chemical and mechanical energy. No process exists that relies solely on either one. As much research has been conducted on the importance of flow versus pressure, the author will not elaborate but will instead refer the reader to the reference section. It is clear at this point that the more users you ask, the more you realize that variations from both extreme opinions (high pressure, high flow vs. low pressure, high flow) are actually mostly used. The role of the cleaning agent has not yet been mentioned but with the extent of variation for each process, its role also becomes critical.

Users are best served with products that work not for single processes, but are capable of covering most of the known cleaning challenges. As this chapter mostly focuses on surfactant-free aqueous solutions, the author felt the need to discuss the importance of controlling the wash fluids once in use. The importance of the latter is oftentimes understated and requires a detailed analysis, which serves readers as a guideline and starting point.

Titration

Historically, surfactant-based products did not contain organic solvents, meaning they were based on inorganic constituents and a number of pH alkaline raw materials. This meant that process control (i.e., am I working at the right concentration?) was only viable through the use of a pH measurement. The latter is a tedious procedure and holds a large margin for error, based on users and procedural understanding. The trend to more modern cleaning-agent formulations indicates that the pH value measurement is becoming a less important process-control variable.

Refractive Index

With the introduction of modern cleaning agents in the late 1990s (their improved cleaning performance and associated cost savings), the industry has been converging more and more toward refractive index as a process-control measurement. Now, users are able to combine pH measurements with refractive index measurements and assess the organic and alkaline level of ingredients in their product. Companies even started to introduce automated concentration monitoring systems to the market, based on the refractive index measurement. But there are also the limitations to the refractive index measurement [8].

The refractive index, n, of a medium is defined as the ratio of the phase velocity, c, of a wave phenomenon such as light or sound in a reference medium to the phase velocity, v_p, in the medium itself. The refractive index of a medium is a measure of how much the speed of light (or other waves such as sound waves) is reduced inside the medium. For example, a typical soda-lime glass has a refractive index of 1.5, which means that inside the glass, light travels at $1/1.5 = 0.67$ times as fast as the speed of light in a vacuum. Two common properties of glass and other transparent materials are directly related to their refractive index. First, light rays change direction when they cross the interface from air to the material, an effect that is used in lenses. Second, light partially reflects from surfaces that have a refractive index different from that of their surroundings.

Since the refractive index is a fundamental physical property of a substance, it is often used to identify a particular substance, to confirm its purity or to measure its concentration. Refractive index is used to measure solids (glasses and gemstones), liquids, and gases. When talking about a solution of sugar, the refractive index can be used to determine the sugar content ("Brix" measurement).

But let's ask ourselves what happens if the sugar solution becomes contaminated with flux residues? In a current study, it has been discovered, that all cleaning agents used in the electronics industry are affected by the contamination (i.e., flux) they remove. And I mean *all* of them! In extreme cases, the difference between the perceived concentration and the actual concentration deviates by over 15%! For example, the operator is reading a 12% concentration solution, which seems fairly close to his specified bath concentration of 14%. The operator now adds a concentrated cleaning solution to make up the difference. He needs to be certain that his cleaning process is in full compliance, as most North American companies produce highly reliable products. He really cannot afford any inaccuracy at this crucial point of the production process. The boards are either about to be shipped to his or her customer or will be undergoing a conformal coating step. Now, let's remember that the refractive index is prone to deviation. Given that the operator was not using a freshly prepared cleaning agent in his cleaning equipment, there is a high possibility that this 12% concentration is more likely to be a 5% one. This means that the cleaning agent might not be in a position to clean at its full strength, possibly resulting in board failure due to contaminant-induced electrochemical migration. But there is no way of knowing, unless the operator sends the contaminated sample to the cleaning-agent manufacturer and allows them to perform more elaborate analytical tests, such as GC (gas chromatograph) measurements. Owning a GC is much too costly for regular bath maintenance purposes. Therefore, alternative methods are indeed required.

Furthermore, the contamination can also artificially decrease the operator's reading. In other words, if you are actually using a 20% concentration, but the contaminated solution makes you believe that you are at 14%, you are running at significantly higher concentrations than necessary and increasing your chemistry cost. Again, any contamination added to the cleaning agent has a different effect on the refractive index and cumulatively it will affect the accuracy of actual bath concentration.

Other Monitoring Techniques

Apart from pH measurements, and refractive index values, other methods can be used in combination with the former. For most processes, users are able to define one or two lead control methods that (given a stable and repeatable cleaning process) behave consistently over time. Conductivity is an often cited method that allows users to measure the build up of contamination in the wash bath as well as the quality of the water being used in a process. Ionic contamination, be that cationic or anionic, is constantly introduced into the cleaning process. This could happen through the flux formulation or the materials of construction; and even some cleaning agents have ionic raw materials as part of their formula. Non-conductive residues can be traced through methods such as FTIR (Fourier transform infrared spectroscopy), NMR (nuclear magnetic resonance), UV spectroscopy (ultraviolet-visible spectroscopy) just to name a few. One low-cost alternative is a method called solid-content measurement. It is designed to quantitatively measure the amount of residue collected in a bath sample, after the volatile liquids (including water) are evaporated.

While one can always find solutions to any technical challenge, the importance to the author in this case is to pinpoint process-control issues that can arise after the initial process qualification. Oftentimes, decisions about implementing processes are made without being fully informed about common problems the user will encounter after start up. Discussing all viable and appropriate monitoring methods is crucial and should be an integrated part of any cleaning-process specification. The possible variations are simply too large in number and the risk too high to discover process-control issues after production ramp up is in full force. This advice to users has been previously mentioned, but is reiterated once again to stress the importance: Select your partners and then conduct a very thorough DOE (Design of Experiment) to address each aspect of the cleaning process prior to implementation.

Encountered Residues and Associated Risks

Most residues encountered in the electronic precision cleaning market are fluxes; however, others do exist as well. Among the flux residues one can differentiate between RMA, water-soluble, no-clean, and synthetic flux materials. They all have in common the property of exhibiting hygroscopic behavior, which is one of the precursors to induce climatic failure mechanisms [1].

Residues do occur at different stages of the manufacturing process; and residues become attached to different surfaces. Assemblies, stencils, wave solder pallets, and processing equipment all do attract contamination, and therefore require regular cleaning. Cross contamination is quickly introduced and all measures should be taken to limit its ability to occur. One frequently encountered example is cross contamination from previously condensed flux residues in reflow ovens. Condensation traps and heat exchangers do require regular maintenance schedules as they carry a high potential for cross contamination of assemblies.

Surfactants typically contain solid contents and builders (i.e., ionic materials), and other potential future contamination. It is imperative to encourage process design and practices so as to limit introduction of possible contamination. With surfactant-based products, users have to pay particular attention to the chemical isolation and rinse section of the equipment. The objective is to fully rinse all potential contaminants off the board prior to the drying stages. Once the cleaning agent is trapped under components, it becomes challenging to remove. Therefore, it is oftentimes more advantageous to select cleaning processes that introduce very little ionic contamination.

Cross contamination, however, can also occur due to improper process separation, for example, if a de-fluxing process is being used for other cleaning processes concurrently. This could include but is not limited to wave solder pallets/frames, stencil, and other process utensils. It is noteworthy to point out that 1 solder pallet is equal to 1000 PCB relative to the flux residues they introduce. This puts not only a strain on the stability of the cleaning agent in use, but it disproportionally increases the risk of climatic failure. On a side note, the proper process separation also contributes to fewer material compatibility issues with the substrates cleaned.

Overall Cost of Ownership of the Cleaning Process

Many companies at present are on a mission to survive, reverting to payroll cuts and line item reductions in an effort to meet these unprecedented challenges.

Not surprisingly, cleaning processes are also under investigation, and purchasing or engineering departments are tasked to verify whether the current solution used is the best one available.

Superseding all thoughts on cleaning process savings mentioned above, internal task groups have to ensure that with any potential change the cleanliness requirements are still being met or exceeded.

Then, it is the responsibility of these internal groups to compare apples to apples. In other words, the customer has to take into account all the elements of overall cost associated with cleaning and rinsing agent usage as well as associated hidden costs for disposal, power usage, compatibility, equipment maintenance. Recent studies have demonstrated that for any in-line equipment cost breakdown, the chemistry drag-out and evaporative losses are deemed the most significant. The top three cost drivers are described below in order of priority:

Cleaning Agent Technology

Comparing the price per gallon is a mistake that many of us still make. Many "cheaper by the drum" products have severe drawbacks and at the end of the year are more costly overall. For example, a product that can be used at half the concentration will provide 50% annual savings. This is the difference between running at 7.5% and 15%. Another commonly overlooked cost driver is "work around" that companies devise to make a product work. These redundant and costly additions are not needed as modern products can easily eliminate them. For example, a recent customer was manually pre-cleaning all assemblies to ensure the product in the equipment was actually cleaning well enough. Other examples include wastewater collection and disposal, as when the product in use is hazardous to the environment and cannot be drained into the regular wastewater system. The latest product technologies can be operated as low as 5%–10% and even include pH-neutral versions that show unprecedented material compatibility and disposal characteristics. The choice of product in use has been shown to be the biggest potential cost driver for conveyorized cleaning processes, as operating concentration directly affects the "drag-out" losses. Take the time and investigate your viable options thoroughly.

Vapor Recovery

This element is what is called a peripheral, yet noticeable cost driver. Evaporative losses double with every 18°F (10°C) increase in operating temperature. Products designed to recover cleaning agents can reduce as much as 85% of the evaporative losses. With exhaust rates of 800–1500 cfm, such savings cannot be overlooked. Active and passive cooling systems allow for the evaporated cleaning agent to be returned to the wash tank to be reused for future cleaning cycles. Products with lower operating temperatures and vapor pressures might help in minimizing evaporative losses.

Bath Monitoring

Proper bath control has been simply overlooked as a potential area for cost savings so far. The reason is simple: there was no better way available to determine concentration than refractive index, which as described earlier, has significant drawbacks.

Thankfully, new product introductions have overcome this issue to allow operators to be able to personally verify that their actual bath concentration is very close to the specified value.

Cleaning is a tremendously important "value added part" of the electronic manufacturing process. This does not imply though that one should simply accept the status quo. To the contrary, please use the time now to talk to your cleaning service provider of choice and work on process savings together. Yes there are hidden costs that can be eliminated, but please do ensure that during this exercise that NO compromises are done in terms of cleanability. Remember, board failures usually price at 100 times the cost of the board.

Conclusions: The Most Important Characteristics of Surfactant-Free Aqueous Cleaning Agents

This chapter illustrated the advantages and disadvantages of a surfactant-free and a surfactant-based concept of formulating cleaning products for electronic applications. If one looks at the different aspects discussed, one can see that there are distinct advantages to choosing the most appropriate cleaning agent. For example, formulating surfactant-free products gives many degrees of freedom to rely on many more raw materials—thereby increasing the process window for users. Surfactants have earned their place in precision cleaning over the years and they serve specific niches of the market. The key will be to select the right product for the right application.

Acknowledgment

The author extends his gratitude to the numerous contributors including but not limited to Dr. J. Becht and Dr. H. Schweigart for their joint effort and collaboration.

References

1. H. Schweigart, 2011. Contamination-induced failure of electronic assemblies. In: *Handbook for Critical Cleaning: Applications, Processes, and Controls.* Taylor & Francis: Boca Raton, FL.
2. B. Kanegsberg, 2011, Cleaning Agents: Overview. In: *Handbook for Critical Cleaning: Cleaning Agents and Systems.* Taylor & Francis: Boca Raton, FL.
3. H. Wack, S. Ahmad, and J. Becht. Research of ionic cleanliness testing alternatives to IPA/water. *Circuits Assembly Magazine*, September 2009.
4. Foresite et al., Apex Presentation 2009 (Las Vegas) on modified ROSE test.
5. S. Stach and M. Bixenman, April 2008. Cleaning process integration of cleaning material with cleaning equipment. Global SMT& Packaging.
6. S. Stack and H. Wack, Thermal Residue Fingerprinting. Apex 2009 (Las Vegas).
7. B. Kanegsberg, 2001. Chapter 6.1. In: *Handbook for Critical Cleaning*, 1st edn. CRC Press: Boca Raton, FL.
8. H. Wack, *Circuits Assembly Magazine*, January 2009.

21

Contamination-Induced Failure of Electronic Assemblies

Contamination-Induced Climatic Failure...333
Background • General Categories of Failure Mechanisms in
Electronics • Critical Applications
Humidity and Pollution Effects on Electronic Assemblies 336
Preconditions • The Nature of These Phenomena • Electrochemical
Migration • Corrosion-Induced Leakage Current and Cutoffs • Other
Mechanisms • Cases
Conclusions...343
Acknowledgment... 344
Reference 344
Bibliography 344

Helmut Schweigart
ZESTRON

The increasing usage of electronic assemblies having exposure to environmental conditions is causing stresses that are not only threatening the individual circuit but the reliability of the entire product. Failure mechanisms such as electro-chemical migration, corrosion-induced creepage currents and interruptions, destabilization of surface resistance, and interference with the signal integrity of high frequency (HF) circuits will be discussed in this article. Knowledge of these mechanisms and the influence of contamination will help us to better understand the importance of circuit cleaning and coating, and will assist in optimizing costs through targeted reduction of expenditures.

Contamination-Induced Climatic Failure

Background

The use of electronic circuitry in consumer and industrial applications is increasing greatly. Electronics are also being introduced into other areas of technology since the introduction of field sensors. The increased use of electronics in different applications has caused reliability concerns for circuits operating in varying ambient conditions. For example, automobiles use identical components for the European market as they do for vehicles in tropical regions and the functional reliability of microelectronic components must be maintained under both conditions. This development has been associated with and supported by a higher performance of the components with increased component packing density. This has invariably resulted in closer conductor spacing and the use of components with an increasing number of pins. Miniaturization of component dimensions and complete electronic assemblies, the trend toward no-clean processes, competition, and manufacturing demands are making it extremely difficult to achieve the necessary quality and reliability of the electronic circuits. The potential operating stresses to which electronic modules are exposed must not result in component damage or, even worse, system failure.

The climatic reliability of modules is determined largely by the cleanliness of the circuits' surfaces and the effectiveness to which they are screened against the influences of moisture. Contamination promotes moisture absorption, and with it, electrochemical migration and corrosion-induced leakage currents. Because of their conductivity, most production-induced contaminations reduce the surface resistance of the assembly:

- Flux residues change the capacitance of the trough-connection contact areas, which primarily affects the signal integrity of high-density integrated and high-frequency circuits.
- Dust on nonconformally coated miniaturized circuits, in conjunction with moisture, can result in the formation of so-called dust dendrites, i.e., tree-like short-circuit bridges.
- Contaminants can cause long-term delaminations, i.e., the peeling of conformal coatings. Both hygroscopic effects and hydrolytic processes impair the adhesion of coatings and the protection they typically provide.

General Categories of Failure Mechanisms in Electronics

During their production and use in the field, electronic equipment and assemblies are exposed to the influence of moisture and environmental conditions (air, weather, location of the assembly, storage, and cleaning). Air, humidity, and water lead to an electrical conductive connection of adjacent metal surfaces, which may have different electrical potentials and may thus cause disorder of the electrical insulation by developing additional transconducting paths. Other influences like fluctuations of temperature or strain caused by harmful substances, vibration, and mechanical stress lead to changes in the electrical conduction properties and destruction of the conductor and insulation materials.

Malfunction caused by water or moisture often disappears in a dry environment. Also, changes of the electrical conductivity caused by temperature disappear when used at moderate ambient temperatures. However, material which was destroyed by thermal or mechanical strain remains in its destroyed condition even after those stressors are removed.

A disorder caused by moisture may disappear in favorable cases; however, in view of functional safety considerations of electronic assemblies, such damages must be avoided. The most important failure mechanisms in this context are electrochemical migration, corrosion-induced leakage currents and interruptions, destabilization of surface resistance, and interference with the signal integrity of HF circuits.

An electronic assembly is only suitable if an acceptable performance is guaranteed for a specified time. The majority of assemblies is installed in final equipment without any insulation and they operate throughout their lifetime without failure. However, assemblies are increasingly used under more severe environmental conditions. In such cases, acceptable operation of an assembly is only guaranteed by a protective coating.

Critical Applications

Reliability of an electronic product and customer satisfaction are of utmost importance for success in the market. Product liability requires manufacturers to take the necessary steps in order to ensure safe and reliable operation of electronic equipment under critical end-use conditions. This mainly applies to assemblies which are used in space technology and aviation, defense technology, medicine, and automotive applications.

The functional reliability of electronic modules determines the overall reliability of the products in which they are incorporated. They are exposed to climatic influences of increasing severity, particularly with regard to moisture and harmful gases. Also, miniaturization and the use of high-ohmic components render these assemblies more sensitive to environmentally induced disruptions. These enlarged demands contrast with reduced product cycle times of five to seven years, in particularly in the automotive industry, and current warranty demands of up to 15 years. In addition to avoiding costly, image-damaging product recalls, there is the urgent need to focus on the prevention of expensive operating standstills, particularly in communication technology. Large-scale serial production, complex supply chains and the ever-growing trend toward company mergers throughout the world make it imperative to reveal assembly weakness as early as possible to avoid logistical problems.

The effect has been to extend climatic reliability and environmental testing beyond the traditional realms of military and aviation applications to consumer-oriented markets.

For example, automobiles use identical components for the European market and for vehicles in tropical regions; and the functional reliability of microelectronic components must persist under both conditions. The application of electronic assemblies in cars is increasing rapidly. A high-end car has up to 100 electronic control modules. According to the VDI (Verein Deutscher Ingenieure), in 2010, 35% of the total production costs for car will be for electronics. The global market for automotive electronics is expected to grow about 6% per year and reach $230 billion by 2015. The failure of electronic assemblies is 40%, and constitutes the main reason for car breakdown according to the world's third largest automobile club ADAC (Allgemeiner Deutscher Automobil Club = U.S. equivalent is AAA) (see Figure 21.1).

The reasons for climatic failure of a hybrid were investigated in detail by the Modern Electronic Packaging Circuits Engineering Corporation (see Figure 21.2). About 8% were directly caused by contamination. However, a major proportion of the failure of so-called electrical connection and active

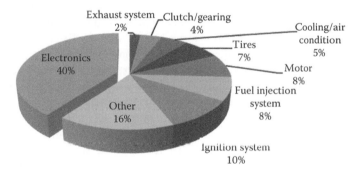

FIGURE 21.1 Causes of car breakdowns (ADAC).

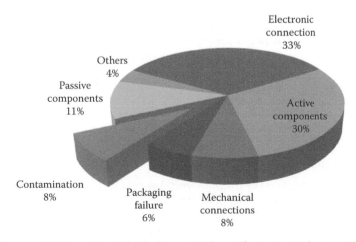

FIGURE 21.2 Causes of failure of a hybrid. (From Modern Electronic Packaging Circuits Engineering Corporation. With permission.)

components failures are induced by contamination, for example, corrosion and short circuits. Thus, about a third of the failures are directly or indirectly caused by contamination, mostly in connection with humidity.

Humidity and Pollution Effects on Electronic Assemblies

Preconditions

The main humidity and pollution effects on electronic components are electrochemical migration, transconductance currents, corrosion-induced cutoffs, and signal-integrity disruption of high-frequency systems. In order for a climatic reliability failure of the electronic circuitry to occur, four main factors need to be present. These include, but are not limited to

- Differences in the electrical potential
- Sufficient atmospheric humidity
- Suitable metal alloys that can electrochemically migrate or corrode
- Remaining contamination on the surface

The Nature of These Phenomena

Electrochemical migration, one of the most common failure mechanisms, can be initiated by as little as a few monolayers of moisture from humidity. This can start the corrosion process at critical humidity levels of 60%–70% RH at ambient temperature, depending on the polarity of the substrate and its respective surface energy. Hygroscopic pollutants are known to lower the critical humidity level to as little as 30% RH. This implies that, for example, under most common climatic conditions (i.e., standard office climate) even the functionality of common household electronics (i.e., computers) is endangered.

Historically, electronic assemblies were mostly used in controlled environments, in which the risk of electrochemical migration was limited. However, now that electronics are utilized in the harshest environments and in critical applications, the risk of migration is indeed increasing.

Furthermore, the exposure of electronic circuits to atmospheric pollutants, such as dust, inorganic particles, as well as harmful gases like sulfur and nitrous oxides (i.e., SO_x, NO_x) dramatically accelerate this type of failure mechanism. The pollutants then act as condensation seeds and promote the adsorption of moisture.

Besides the above-mentioned critical moisture film, a metallic alloy with a tendency to migrate is also required. This is evidently related to the composition of the alloy. For the simplest cases, the tendency of a metal to migrate in pure, condensed water can be assessed with a so-called Pourbaix diagram, which correlates the pH value against the electrical potential. An example is shown in Figure 21.3. It should be pointed out that while such Pourbaix diagrams are representative for pure water, they enable an estimate to real life situations.

Electrochemical Migration

Electrochemical migration processes consist essentially of three steps as shown graphically in Figure 21.4:

Anodic Metal Dissolution

Anodic metal dissolution is a process by which the metal atoms (M) are transformed into their metal ion state (M^{n+}) and the corresponding number of electrons at the anode (ne^-), i.e., signal line, and begin to migrate toward the cathode, i.e., mostly ground (see Figure 21.4).

The driving force for this transformation of metal-to-metal ion is the electrolysis ($2H_2O + 2e^- \rightarrow 2OH^- + H_2$) of the moisture film on the surface, caused by the operating voltage during normal board operation (Figure 21.5). It is known that the electrolysis process occurs at voltages above 1.5 V. With most

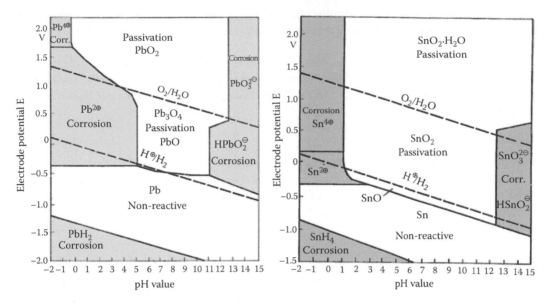

FIGURE 21.3 Pourbaix diagrams for tin and lead.

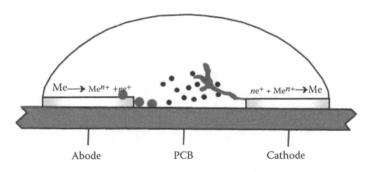

FIGURE 21.4 Mechanism of dendrite growth.

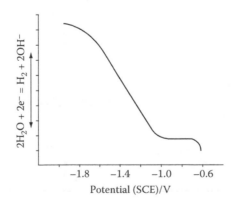

FIGURE 21.5 Electrical dissociation of water.

circuitry operating at upward of 3.3 V, this phenomenon is therefore prone to occur. Contaminants, especially flux activators, are stabilizing the moisture film due to their hygroscopy and thus are promoting the mechanism.

As hydrogen (H_2) leaves the moisture film as gas into the atmosphere and only the hydroxide ions (OH^-) remain, the pH value rises in just a second. At alkaline pH values above 12, the anodic metal dissolution starts immediately, as shown in the Pourbaix diagram of lead and tin (see Figure 21.3), for example.

Diffusion of Metal Ions

Metal diffusion is characterized by the migration of dissolved metal ions (M^+) from the anode to the cathode. The metal-ion concentration gradient between the anode and the cathode is determined by the operational current density, the size of the active anodic area, and the mobility of the metal ions. The conductivity of the electrolyte depends on the amount of dissolved metal ions and hydroxide ions.

Deposition of Metal Ions

Dendrite formation is driven by the deposition of the dissolved metal ions at the cathode ($M^{n+} + ne^- \rightarrow M$). The deposition occurs preferentially at places of high electrical field strength. If the concentration of metal ions is low, the dendrite grows in tree-like bridge structures. Should the metal ion concentration be higher, one will observe the growth of more band-like pattern as depicted in Figure 21.6.

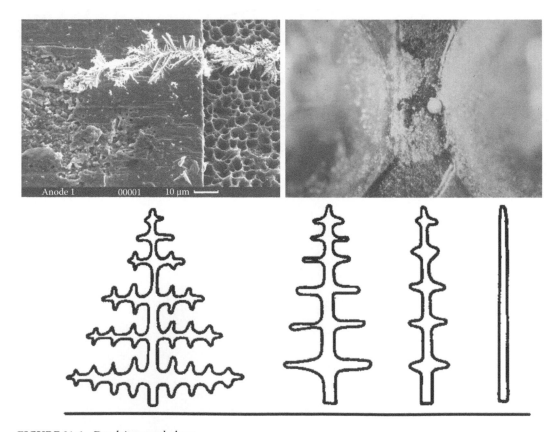

FIGURE 21.6 Dendrite morphology.

Corrosion-Induced Leakage Current and Cutoffs

Corrosion-induced leakage currents arise primarily with hybrid circuits or on copper contacts, both are promoted by sulfidic atmospheres with more than 60% relative humidity. Corrosive atmospheres prevail in industrial plants and in the vicinity of the engine in motor vehicles.

The corrosion attacks the copper lines at nonconforming pores and cracks in the solder mask (Figure 21.7). This mechanism of corrosion is the formation of unwanted conductive paths by the precipitation of saline corrosion products with very high water content and hygroscope potential. The presence of an electrical potential is not required. The formation of the conductive path is powered by the diffusion of the corrosion salts. Therefore, even distances of inches are easily bridged over time.

Sulfur-vulcanized seals used for hermetically sealed housings of electronic assemblies are also a source of these phenomena. The mechanism is as follows. Temperature changes produce pressure differences between the inside of the housing and the atmosphere in seemingly hermetically encapsulated circuits. Over extended periods of time, this allows penetration of humidity into the hermetically sealed housing by pressure compensation. The sulfur is outgassing from the seals producing a corrosive atmosphere inside the encapsulation. A concentration of only a few ppm of sulfidizing compounds at a relative internal moisture of more than 70% is sufficient to initiate a sulfidizing attack of thick-film pastes of hybrid circuits (Figure 21.8).

In addition to corrosion-initiated leakage currents, the intrinsic conductivity and electro-diffusion effects of most contaminants lower the surface insulation resistance (SIR). This mechanism conducted by hygroscopically induced moisture absorption can intensify due to a number of factors such as dissociated hydronium ions, i.e., from activators, to result in malfunctions and further complete assembly failures.

In extreme cases, the board material overheats along the leakage paths; there is the potential for smoldering and even fires to occur. Similarly, activator residues can change the impedance of connecting surfaces and through-holes, causing statistical fluctuations and virtual pad geometry enlargements. This effect is described in detail in the section "Signal-Integrity Disruption of High-Frequency Systems."

The conductor and connecting contact spacing have to be continuously decreased to achieve higher performance densities. Therefore, a design-induced lowering of the insulating effect of the substrate material must be accepted along with the concomitant increasing susceptibility to contamination-induced leakage currents. The intensified hygroscopy of the flux residues due to no-clean processes promote increased resistance reduction. This contamination-induced reduction of the SIR is the result

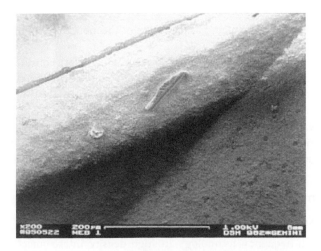

FIGURE 21.7 Bridge of copper sulfate, originating from a crack in the solder mask.

FIGURE 21.8 Conductive path in a hermetically sealed hybrid circuit due to sulfur corrosion.

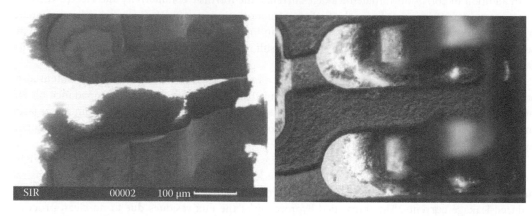

FIGURE 21.9 Confirming the presence of leakage current paths by charge contrast representation with a raster electron microscope.

of hygroscopic moisture accumulation and acidification of the moisture film due to activator residues (Figure 21.9).

Electronic modules usually do not have the necessary product safety considerations against leakage currents. Various contaminants can be deposited on the circuits during manufacture and operation of the electronic modules. These contaminants, in conjunction with moisture, will more than likely result in failures. Residues of mould-releasing agents on resistors and other components as well as glass hydrates formed as a result of the incorrect component support, can lower the surface resistance due to their intrinsic conductivity or cause adhesion problems when protective coatings are applied. Similar problems can also be caused by conducting oils leaking out of non-conformally coated coils.

Other Mechanisms

Signal-Integrity Disruption of High-Frequency Systems

The application of high-frequency technology in high-density interconnection (HDI) assemblies, together with a wider use of lead-free solders, have served to initiate a closer scrutiny of the flux removal cleaning process. Because adequate climatic operating conditions cannot always be assumed, system signal integrity is vulnerable to failure through the parasitic capacitances of hygroscopic activators as well as of resin and rosin residues. Further, such contamination, particularly with new lead-free solder formulations, is no longer detectable by ion-equivalent measurement alone. This is because the levels detected are far below $1.5\,\mu g/cm^2$ due to the paste formulations. That results in an assurance of long-term reliability of assemblies per ISO 9001 certified manufacture. This not only requires suitable cleaning methods, but also demands an appropriate testing technology for effective monitoring of all assembly line processes.

The use of HDI assemblies, particularly in motor vehicles for operating-data acquisition, in the logistics area and in modern building machines, is increasing rapidly. Hence, more of these systems are exposed to widely differing climatic influences, including moisture and harmful gases, which effectively threaten their functional reliability and that of the products and devices in which they are housed. Moreover, the sensitivity of these circuits to environmental interference is exacerbated by the use of high-ohmic components. High-frequency circuits between 30 MHz and 5 GHz, a requirement in communications electronics are highly susceptible to environmental interference because of their frequency.

Thus, to maintain signal integrity, the systems not only require an adequate ohmic-insulation resistance, they also must have a stable complex impedance. For this reason, capacitive surface effects must be taken into account in the circuit design.

Increasingly, corrosion-induced assembly malfunctions (e.g., electrochemical migration and leakage currents) are the sources of diminished component reliability and service life. In high-frequency designs, the parasitic capacitances can distort the "ramp-up" of the signal, thereby disrupting its integrity to the point of causing equipment malfunction. Because guaranteed reliable operation of a product for extended periods now is imperative, an increasing importance is placed on ensuring its quality. For high-frequency assemblies; this standard primarily is determined by circuit surface cleanness (see Figure 21.10).

Malfunctions in the interconnected assemblies of motor vehicles, for example, may result in responses that are difficult to interpret. Speed sensors for automobile wheels are monitored not only by the ABS system but also by the engine management and other assemblies. A malfunction of a monitored component often causes other assemblies to generate misleading readings.

With frequencies higher that 1 GHz, the circuit designer must calculate even the low (but limited) resistance of conductive lines. If residues enlarge pad areas, the electrical layout may be changed, and due to a support capacitor, might lead to malfunctions by causing a time delay at the watchdog of a controller, further resulting in a function-status error at that location. Additionally, SIR might be diminished locally and cause a similar effect by crossing leakage currents. Finally, as well as the static effects

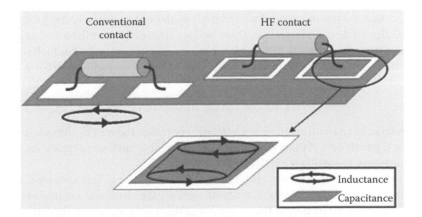

FIGURE 21.10 Signal distortion through contaminant-induced capacitances in HF circuits.

described, dynamic effects also can be present: Parasitic capacitors will distort the ramp slope: edge-triggered active components might not recognize the signal if the ramp slope is too flat; and the signal integrity of highly integrated, high-speed or high-frequency circuits primarily are affected.

Reductions in SIR and the capacitive potential of the type that can result from build-up of activator residues can be shown qualitatively under a scanning electron microscope (SEM). Visualization is possible via a test that responds selectively to carbon acid-based activators of fluxes by a corresponding color reaction. The test not only detects the activator residue from fluxes but also makes their distribution visible.

Impedance spectroscopy promises to be a direct technique to measure electronic performance parameters. For example, the ohmic-resistor quota under chip capacitors can be determined by this method. The technique has the potential to characterize the aging behavior of assemblies and components over time when they are placed in a climatic chamber under operating conditions during the measurement.

The intensified use of high-frequency technology, HDI assemblies, and lead-free solders are giving rise to new aspects in flux removal. As a result, any decision concerning cleaning or "no-clean" manufacture must be discussed intensively with respect to quality requirements. In spite of the diversity of efforts to circumvent cleaning as a critical step via new joining techniques, it has become quite clear that cleaning is inseparably associated with electronics manufacturing. Accordingly, the creation of qualified cleaning processes that meet ISO 9001 guidelines also requires provision for optimal testing and monitoring procedures. Cost-optimized solutions that guarantee the highest possible long-term reliability of assemblies can be realized only through close cooperation between the manufacturers, designers, and suppliers of cleaning processes.

Reliability of Lead Free/Silver-Based Solders

Silver compared to tin (Sn) and lead (Pb) forms hydroxides that are readily water soluble. So these silver hydroxides readily diffuse in moisture films, which are in turn formed at a relative humidity as low as 50%–60%. The latter value is known as the upper moisture range of non-air-conditioned building interiors.

Climatic reliability studies carried out by Tabai Espec in Japan revealed the formation of short-lived silver dendrites with a brief life span of 10–15 min due to electrochemical migration. Tests performed on un-cleaned comb structures (at 85°C/85% RH) with a permanently applied voltage of 50 VDC showed on the contrary, that silver-free eutectic tin–lead solders did not follow the same, temporary failure pattern as silver containing solders.

With only low concentrations of silver hydroxides present, "delicate" dendrites, characteristic of their minimal current carrying capacity, are formed. These dendrites lower the surface resistance primarily in the final stage of their growth, before being burned off by the ensuing short circuit. The original surface resistance is then reestablished.

Due to the fact that the rate of renewed silver supply to the surface of the solder joint is slower than the attack by the electrochemical migration, these bridges are mostly short lived and only infrequently transform into constant short circuits. This behavior in turn, results in inexplicable malfunction patterns in the circuitry that can neither be predicted nor discovered by discontinuous resistance measurement.

Cases

There have been failures and malfunctions of electronic assemblies induced by climatic impact and contamination since the early days of using electronic assemblies. Therefore, some typical and pathbreaking examples of automotive and military applications are described.

During the 1950s, problems with the central door locking system of cars occurred. Due to hygroscopic contamination, air humidity moisture condensed on the electronic control units and caused short circuits. As a result, the owners of these cars found them unlocked in the morning at the parking lot, although they had locked the car when they left it in the evening. As a final result of that experience, the pathbreaking BMW dewing test was developed; that test established the link between cleanliness and climatic reliability of electronic assemblies for automotive applications.

On partial conformally coated electronic control units for window lifters, electrochemical migration was observed due to condensation of moisture. The resulting uncontrolled opening and closure of car windows was a high risk of injury for the passengers. This problem was solved after a change in the geometry of the car door that avoided the condensation of air moisture on the electronic assembly. Therefore, the importance of the application situation was recognized. Currently, the product application situation is always taken into account when climatic tests are designed.

Contamination-induced condensation also caused short circuits on antenna control units of car radios. The electronic assemblies were destroyed by smoldering due to the high current density applied. Smoldering also generated corrosive bromide-containing gas, which damaged the wiring harness. The resulting costs for the elimination of all destructions were extremely high. Cleaning and subsequent conformal coating of the antenna control units solved this reliability problem.

For an application in a military all-terrain vehicle, an electronic assembly was hermetically encapsulated. On the back side of the housing was a big copper connector. Because the encapsulation was conducted under normal environmental conditions, air moisture was embedded. During the operation in the field the temperatures changed (e.g., night-to-day cycles). The moisture condensed at the copper connector, the most thermally inert material (i.e., remained coldest). In combination with vibrations, the result was a rupture of the connector due to corrosion fatigue. This climatic reliability issue was resolved with cleaning and subsequent conformal coating of the electronic assembly. Cleaning before coating ensures a good adhesion of the coating and a reliable protection against climatic impacts. In this case, the flooding of the encapsulation with inert dry nitrogen gas was not possible in the field. This case demonstrated the importance of cleaning and coating even, or especially in, seemingly climatic protected applications.

As these examples prove, there are various failures and malfunctions of electronic assemblies induced by climatic impact and contaminations. Thus, the product safety standards for aviation and aerospace applications are high. In general, all assemblies are cleaned and conformally coated as protection against extreme climatic impact during operation.

Conclusions

The climatic reliability of electronic circuitry is becoming ever more critical due to their increased requirements for long-term functionality (medical, safety systems, etc.) and the harsher environments they have to operate in (aerospace, automobile, etc.). In addition, with the ever-increasing use of HF in high-density interconnection assemblies, increasing package component densities, decreasing component stand offs, and the introduction of lead free solders, the issue of climatic reliability and contamination effects is becoming an important consideration.

The electrochemical migration and induced leakage current failure mechanisms are outlined. The interesting observation of short-term dendrite formation with silver-containing alloys was described. The susceptibility of HF assemblies to failure due to surface contamination was outlined.

All the above must be seen in the context of the importance of the cleaning process in the whole electronic assembly production process and the positive effects cleaning can have on the whole product quality and long-term climatic reliability. Cleaning is becoming increasingly more important, if not vital, to mitigate the above effects especially in critical high-value applications.

Flux residue removal after soldering is only one process step when using cleaning solutions to solve or avoid climatic problems. Cleaning is still proving to be essential in production despite efforts to introduce new connecting techniques designed to avoid the process. Cleaning not only removes contaminants, but also activates surfaces in such a manner that bonding and coating are enhanced, as is the case with Chip on Board (COB) technology. This can be achieved by optimal co-ordination of the cleaning media with the corresponding cleaning machines. The adhesion of conformal coatings can also be influenced by cleaning. It is not only a matter of surface cleanliness, but also of surface priming to achieve better coating adhesion. The combination of cleaning with proof of purity supports module manufacture as a whole by providing cost-optimized problem solutions. Cleaning also contributes significantly to process safety and ultimately to the operating reliability of the manufactured modules.

Circuit manufacturers are increasingly seeking the support of specialists who focus on solutions for specific cleaning problems. Cleaning processes are being increasingly implemented with close cooperation between the supplier of the cleaning media and the cleaning-line suppliers. These efforts are always geared to the individual requirements of users. A close partnership between the user, the supplier of the cleaning agents, and the manufacturer of the cleaning lines is gaining increasing importance due to the complexity of the associated problems. Cleaning-process selection should above all be guided by technical requirements. In this particular case, the primary consideration would be the adhesion of coating substances and the stabilization of surface resistance. The second marginal peculiarity is concerned with the available space at the production floor and the clarification of ambient conditions, for example, application of inline or batch equipment. This forms the basis for a cost assessment to predetermine the optimal cleaning process.

Acknowledgment

I am pleased to acknowledge my thanks for contributions from Harald Wack and Alexandra Rost.

Reference

Modern Electronic Packaging Circuits Engineering Corporation.

Bibliography

Ellis, B.N. *Cleaning and Contamination of Electronic Components and Assemblies.* Ayr, U.K.: Electrochemical Publications Ltd., 1986.

Engelmaier, W. Reliability of lead-free (LF) solder joints revisited. *Global SMT & Packaging*, 3(7), 26–27, November 3, 2003.

Krumbein, S.J. Metallic electromigration phenomena. Chapter 5, in: Christou, A. (ed.) *Electro Migration and Electronic Device Degradation*. New York: Wiley Interscience, 1994, pp. 139–166.

Muehlbauer, A. and Schweigart, H. Lead-free and no-clean: A contradiction of terms? *SMT*, Pennwell, October 2, 2002a.

Muehlbauer, A. and Schweigart, H. Preventing contamination—Caused assembly failure. *SMT*, Pennwell, March 2, 2002b.

Muehlbauer, A. and Schweigart, H. In-field assembly failures—Expensive and avoidable? *SMT*, Pennwell, January 03, 2003.

Schmitt-Thomas, K.G. and Schmidt, C. The influences on the quality of electronic assemblies. *Soldering & Surface Mount Technology* (18), 4–7, 1994.

Schweigart, H. Humidity and pollution effects on electronic equipment, CEEES Publication, 1999.

Schweigart, H. The reliability of silver based solder. *EPP Europe*. March 4, 2003.

Tautscher, C.J. *Contamination Effects on Electronic Products*. New York: Dekker, 1991.

22

Surface Cleaning: Particle Removal

Introduction 345
Adhesion Forces
Particle Removal 347
Hydrodynamic Removal • Ultrasonic Removal • Megasonic
Removal • Brush Cleaning • Particle-Removal Mechanism •
Chemistry-Enhanced Cleaning • Removal of Nanoparticles
References..356

Ahmed A. Busnaina
Northeastern University

Introduction

Surface contamination by small particles and other contaminants is a major problem in many industries such as semiconductor, storage, imaging, aerospace, pharmaceutical, automotive, food, medical, etc. Contaminant particles range in size from several hundred microns to less than 20 nm. In today's semiconductor manufacturing, the minimum size that needs to be removed keeps decreasing, and currently it is below 20 nm. Surface contamination can result from particle deposition in the manufacturing environment as well as particle generation by the manufacturing process or process tool. Improving the clean environment further or isolating products can only solve part of the contamination problem. There is always a need for effective and economical technique for surface cleaning of a variety of substrates.

Adhesion of small particles to substrates presents a serious problem to many industries. Particulate surface contamination is one of the reasons for yield problems in these industries. The adhesion forces of these particles are greatly affected by many of the processes that the substrate may go through such as cutting, polishing, etching, rinsing, and drying that follow the particle deposition. The adhesion forces to be considered in the process include van der Waals forces (vdW), electrostatic forces, and chemical bonds. Chemical bonds are usually orders of magnitude larger than vdW bonds.

As the size of the circuit line width approaches one-tenth of a micron in the semiconductor industry, the situation will become much more serious with respect to very small particles. There is a need for efficient and reliable particle-removal techniques capable of removing very small particles without causing surface damage. Many studies have been conducted using various methods to detach particles from surfaces [18–45]. In this chapter, the most common and widely used particle-removal (surface cleaning) techniques will be reviewed.

Adhesion Forces

Adhesion forces are the forces responsible for adhering a particle to a surface. It is important to know the adhesion forces for particles we need to remove in order to ensure that the removal force we apply is sufficient for particle removal. Adhesion forces depend on the particles and substrate material and the medium they are in (water, air, etc.). The adhesion force is also a function of size; it is proportional to the radius of the particles. The reason smaller particles are more difficult to remove than large particle is not because the adhesion force is larger. On the contrary, it's smaller. But since the removal forces we apply depend on the area (R^2) of the particles (such as removal using hydrodynamic, megasonic, etc.) and particle mass (R^3) (such as removal using centrifugal, gravity, vibration, etc.), the force we can apply to a particle decreases much faster than the adhesion force.

Adhesion forces have been classified by Krupp [1] into three classes:

Class I: includes vdW and electrostatic forces that act in the periphery of the adhesive area as well as in the contact area [2–5].

Class II: includes various types of chemical bonds as well as intermediate bonds (hydrogen bonds). Chemical bonds are usually an order of magnitude stronger than vdW bonds.

Class III: includes sintering effects such as diffusion, condensation and diffusive mixing. These forces are usually known as interfacial reactions.

Class I forces, i.e., vdW and electrostatic forces are the major contributors to particle adhesion. Dry uncharged surfaces in contact with dry uncharged particles will experience vdW and electrostatic double layer forces as the only adhesion forces. Charged particles and surfaces will introduce an additional force (electrostatic image force). Wet surfaces, on the other hand, can shield these forces and thus reduce them significantly. The vdW forces can be reduced by a factor of two and the electrostatic forces can be more or less eliminated. Visser [6–8] has published several papers on particle adhesion. A detailed review of the adhesion forces and particle–surface interaction is presented in other references [9–15]. Class II forces (chemical bonds and intermediate bonds) can also occur on silicon substrates following certain conditions of treating the substrates chemically followed by rinse and dry processes. Adhesion forces resulting from chemical bonds are usually orders of magnitude larger than vdW adhesion forces.

The vdW force (an intermolecular adhesion force acting between molecules that arises due to the polarizability of the molecules) is the dominant adhesion force for small particles (less than 50 μm). It arises from the short-period movement of the electrons in the atoms or molecules giving rise to momentary areas of charge concentrations called dipoles [14]. The vdW force is given as

$$F_0 = \frac{AR}{6z_0^2} \tag{22.1}$$

where

A is the Hamaker constant
R is the radius of the spherical particle
z is the separation distance between the particle and the substrate

The average separation distance (z) between the two surfaces is taken as 4A (for smooth surfaces). For the ideal case in which both the spherical particle and surface are not deformed, the vdW is proportional to the radius of sphere as shown in Equation 22.1. However, when deformation occurs, the magnitude of the adhesion force will also depend on the contact area between the particle and the surface. When a sphere and a flat substrate come into contact with each other, the attractive force deforms the interface and a circular adhesion area is formed. The total adhesion force consists of two additive components, the force acting between the adherents before deformation, and the force acting on the contact area due to the deformation, $F_{(vdW\_deform)}$ [12,15]:

$$F_A = \frac{AR}{6z_0^2}\left(1 + \frac{a^2}{Rz_0}\right) \tag{22.2}$$

Krishnan and Busnaina showed that the adhesion induced deformation is plastic in the case of polystyrene latex (PSL) particles on silicon and it may take up to 70 h to reach equilibrium [12,15]. Adhesion induced deformation will occur when the particle is softer than the substrate, or vise versa.

Electrostatic force constitutes the main force of attraction for particles larger than 50 μm in diameter. For dry particle–substrate system the electrostatic force becomes important. The presence of electrostatic charge can drastically alter the total adhesion force. Zimon [13] reported that the force of adhesion can be increased by a factor of two when the net number of unit charges per particle on 40–60 μm particles increases from 700 to 2500. Most particles carry some electric charge, and some may be highly charged. Particles at low humidity were found to retain their charge and are held to surfaces by an attractive electrostatic force [14]. A charged particle experiences an electrostatic force in the vicinity of a charged surface or other charged particle. The charge on a particle can be negative or positive depending on whether the particle has an excess or deficiency of electrons. If a charged particle carrying a charge Q comes in contact with an uncharged plate, the charged particle induces an equal and opposite charge on the surface, this is known as the columbic, or the electrostatic image force [2–4]. This electrostatic force will deteriorate with time due to the dissipation of the charge.

When moisture is present in the air medium, condensation can take place between the particle and substrate. The capillary condensation gives rise to a capillary force [2–4]. The capillary force is made up of two components; surface tension at the perimeter of the meniscus and the capillary due to the difference in pressure between the liquid and vapor phases. The existence of tension in a liquid–gas interface causes a difference in hydrodynamic pressure across the interface if the interface is curved. The capillary force depends on several parameters such as particle size, the surface tension of the condensed fluid, the wettability (contact angle) of the substrate surface [16]. Capillary forces are proportional to particle size and the adhesion of large particles is found to increases with the relative humidity of the air. Zimon [13] found that in air relative humidity near 100%, the majority of particles are held with force larger than the van der Walls adhesion forces. Kordecki and Orr [17] observed that capillary condensation begins to appear at relative humidity above 50%. Luzhnov [18] reported that adhesion due to capillary force occurs when the relative humidity exceeds 70%. The same effect was also reported by Zimon [13], who concluded that at relative humidity of 50% and particularly at humidity below 50%, capillary forces have no effect on the adhesion force. But all agreed that at relative humidity between 50% and 65%, the capillary force starts to have an effect on the total adhesion force. Zimon went on to conclude that between 70% and 100% relative humidity, the capillary force dominates the other adhesion forces and should be the only adhesion force considered.

Busnaina and Elsawy [19] showed that the effect of relative humidity on the adhesion and removal for the 22 μm PSL particles on silicon substrates was very significant. The removal of PSL particles was very low at high and low relative humidity. The lowest adhesion force (highest removal efficiency, 49%) occurs at 45% relative humidity as shown in Figure 22.1.

Particle Removal

Hydrodynamic Removal

Hydrodynamic removal is the removal of particle by using the inertia of a moving fluid on a particle. This can be done using a jet, overflow rinse in a tank, and using many other applications. The hydrodynamic force on a particle is applied through the drag and the lift force the fluid applies on anybody in its path as shown in Figure 22.2. The hydrodynamic forces depend on the cross-sectional area of the particle and fluid velocity and density. The force is directly proportional to these parameters.

Drag on a sphere in a uniform flow can be obtained by utilizing available experimental data [2–4]. For low Reynolds number flows, Oseen's approximation can be used. Submicron particles exist in the boundary layer and usually in that part known as the viscous sublayer. In order to calculate the correct drag on the sphere, the sphere is broken into small discrete cylinders. The velocity across the cylinder

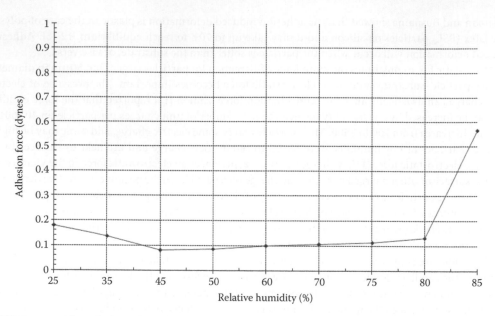

FIGURE 22.1   Effect of relative humidity on the adhesion force (PSL particles/silicon).

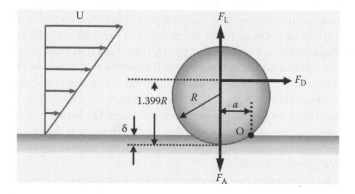

FIGURE 22.2   Lift force F_L, drag force F_D, and adhesion force F_A acting on a particle in a shear flow.

is derived using boundary-layer analysis. For parts of the sphere that lie within the viscous sublayer, standard *law of the wall* is used. Using this velocity and expressions for the drag coefficient, a local drag force is found. These local drag forces are then summed over the sphere producing a total drag force. The hydrodynamic lift force is calculated in the same manner [2–4,20].

Visser [21], in addition to his later work in theoretical aspects of adhesion, conducted experiments in 1970 concerning particle removal. The apparatus consisted of two concentric cylinders, the outer one was fixed; the inner one was capable of rotating at a maximum of 5000 rpm. The adhering system involved 0.21 μm carbon black particles deposited on cellulose film on the inner cylinder. Visser assumed the criterion of 50% removal as a measure of the adhesion force [21].

Musselman and Yarbrough [22] used a model of viscous drag from a high-velocity spray to predict the drag force on particles at different spray nozzle pressures. They explained the difficulties in hydrodynamic drag removal due to "particle hideouts" in the boundary layer. Although free stream velocities may be substantial, the local fluid velocity at the particle is small due to its proximity to the wall. Both the turbulent and laminar boundary layers cause this problem. Drag force on the particle was calculated by a summation of the local drags at different heights on the sphere. Musselman and Yarbrough predicted the drag versus particle size at different nozzle pressures.

Kurz et al. [5] used a rotating disk (silicon wafer) to generate hydrodynamic force to remove 1 μm or larger particles. They used PSL spheres on bare silicon in deionized water as the medium. Removal rates above 90% were reported for particles larger than 2.0 μm. Taylor et al. [2–4] measured the removal force (using hydrodynamic drag and lift forces) of submicron particles on silicon substrates and correlated it with the theoretical adhesion force. The results indicate that the theoretical adhesion force (using the Hamaker equation) was in agreement with the experimental measurements. Most of the hydrodynamic removal was effective at removing micron size particles or larger. The efficiency of submicron particle removal has been shown to be small [2–4].

There are three hydrodynamic factors acting on the adhering sphere, a Saffman [20] lift force, Stokes drag, and turbulent bursts. Saffman lift results from the gradient in the shear flow. Drag originates due to a pressure difference across the sphere. In the case of a very slow flow around a sphere, Stokes drag provides adequate formulation. Turbulent bursts are present in turbulent flows and act to move fluid very rapidly from one section of the flow. They are influenced by vortex patterns among other things. The bursting activity is not yet wholly understood but its location cannot be predicted. Cleaver and Yates [23] stated that the lift force was due to the bursting activity.

Ultrasonic Removal

Ultrasonic cleaning, where transducers frequency operates between 25 and 200 kHz, is widely used in many aerospace, automotive, electronic hardware, medical industries. In reality, any frequency higher than 17 kHz is considered ultrasonic. However, because the dominant cleaning mechanism is different at low frequency (25–200 kHz) as compared to frequencies higher than 360 kHz, the lower frequency cleaning is known in the industry as ultrasonic cleaning while the high frequency is known as megasonic cleaning. The boundary between the two has not been clearly defined yet but it lies somewhere between 200 and 360 kHz.

Ultrasonic cleaning has been an accepted cleaning method for decades and still proves to be valuable today. A typical ultrasonic cleaning system consists of piezoelectric transducers attached to the bottom of a tank. These transducers typically vibrate at a single frequency. The vibration energy generated by these transducers is transmitted to the cleaning solution, creating longitudinal pressure waves. The key to ultrasonic cleaning is the physical implosion of gas or vapor bubbles in a cleaning solution. The major drawback to ultrasonic cleaning is the damage to a surface caused by cavitation. Cavitation forms when the tensile strength of a liquid is exceeded due to rapid alternation between positive and negative pressure of the sound wave propagating through the liquid. The cavitation bubbles are formed when the pressure is in the negative area. There are two types of cavities formed: transient and stable. The transient cavities occur in low frequency and change in size and implode where stable cavities change very little in size and do not implode [36–38]. The transient cavities undergo a series of expansion and contraction within the fluid until it reaches a critical size. At the critical size, the bubble implodes creating a high-velocity jet and increasing the local temperature to well above thousands of degrees Kelvin. It has been calculated that the velocity of the jet and the local temperature during the implosion can reach up to 130 m/s and 3000 K, respectively. These implosions remove particulate or film contaminants on the surfaces to be cleaned. These implosions can also cause damage on the surface [32–34]. Many studies on cavitation impact have been done and found that cavitation damage can be minimized by changing the liquid property such as temperature and gas content as well as the transducer frequency sweep [39]. Cavitation is the dominant cleaning mechanism in ultrasonic cleaning but it is not the only one. Acoustic streaming, which is the dominant cleaning mechanism in megasonic cleaning, is as equally important to particle removal in ultrasonic cleaning. Figure 22.3 shows the effect of ultrasonic cleaning time and temperature on the removal of submicron PSL particles using 68 kHz frequency and 400 W of input power. The figure shows that the optimum

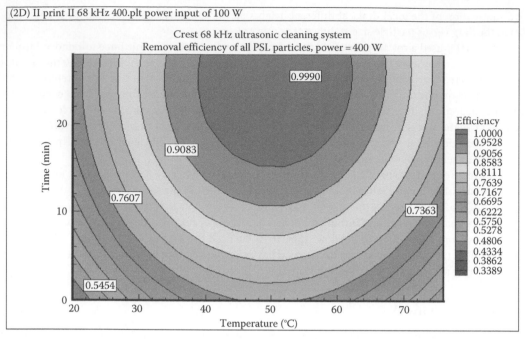

FIGURE 22.3   Effect of cleaning time and temperature on ultrasonic cleaning using 68 kHz frequency and 400 W of input power.

temperature occurs at an intermediate value in the considered range used and that the cleaning time is optimum after 15 min of cleaning.

Megasonic Removal

Megasonic cleaning, where transducer frequency typically operates between 360 and 1200 kHz, is widely used in many semiconductor (wafer cleaning) and hard-disk industries. Cavitation implosion does not occur in megasonic cleaning and therefore the dominant cleaning mechanism is the acoustic streaming. The acoustic streaming removes particles by exerting a hydrodynamic removal force (drag and lift) but at much higher velocity near the surface as compared to typical hydrodynamic cleaning using a jet or spin rinse.

Recently, megasonic cleaning at high frequencies near 1 MHz has gained attention as an efficient, though poorly understood, Si surface cleaning technique. Olaf [24] made early observations of sonic cleaning of glass surfaces in the range from 15 kHz to 2.5 MHz. Rosenberg [25] used ultrasonic cleaning for removing contaminant films and concluded that the removal was due to cavitation. McQueen [26,27] recognized the importance of acoustic streaming in decreasing the boundary layer thickness, based upon his studies of removing small particles from surfaces. Megasonic-cleaning applications were first described in detail by RCA engineers [28,29]. Kashkoush et al. [30–38] studied ultrasonic and megasonic particle removal, focusing on the effects of acoustic streaming. Removal percentage increased with power. Their results also indicated different removal efficiencies for PSL, silica (SiO_2), and silicon nitride (Si_3N_4) particles. Greatly enhanced particle-removal efficiency on Si from megasonics in SC1 and SC2 solutions was reported by Syverson et al. [40]. Again, removal increased with increasing power, up to a maximum tested value of 150 W. Wang and Bell [41] performed experiments using megasonics for cleaning after RIE planarization. Of the parameters they tested, power had the greatest influence on the results. Cleaning improved with increased power up to the maximum tested value of 300 W [41], another result consistent with what was observed by Kashkoush et al. [30–38].

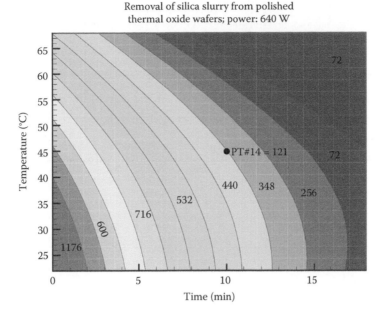

FIGURE 22.4 Effect of cleaning time and temperature on megasonic cleaning using 760 kHz frequency and 640 W of input power.

Megasonic power exerts a greater influence on particle-removal efficiency than solution temperature, both in water and in SC1 solution. Removal efficiency increases with increasing power up to an intermediate point above which it decreases slightly. In deionized water, removal efficiency decreases slightly at temperatures above 50°C, whereas in SC1 solution it is generally highest at temperatures above 50°C. Though SC1 removes particles more efficiently than DI water, particularly at lower megasonic powers, it is still possible to achieve 100% removal in DI water under the proper conditions. SC1 solutions which are significantly more dilute in NH_4OH content than the standard 5:1:1 recipe, work well in the presence of megasonic energy. Particle-removal efficiency decreases when the ammonia content is decreased slightly from the 5:1:1 ratio, but increases again as ammonia content is further decreased. The efficiency then remains high even for R as low as 0.01. Figure 22.4 shows the effect of cleaning time and temperature on megasonic cleaning of polished thermal oxide wafers using silica slurry using 760 kHz frequency and 640 W of input power. Figures 22.5 through 22.7 show the effect of cleaning time, input power, and cleaning temperature on the removal of submicron silica slurry particles using megasonic cleaning at 760 kHz. Figure 22.5 shows the effect of cleaning time using 41.5°C and 345 W of input power. The figure shows that the optimum cleaning time is 20 min after which the cleaning efficiency goes down. This is due to particle redeposition on the substrate.

Figure 22.6 shows the effect of input power using 41.5°C and 20 min of cleaning time. The figure shows that the optimum power is about 500 W. This is due to the fact that high power generates many more bubbles that interfere with the acoustic streaming thereby decreasing the removal efficiency. Figure 22.7 shows the effect of the cleaning temperature using 20 min of cleaning and 345 W of input power. The figure shows that the optimum cleaning temperature is higher than 40°C.

Brush Cleaning

Brush cleaning works by using a soft brush that englufs a particle and removes by applying a torque through the brush rotation. The pressure helps the brush engulf the particle and the rotation applies the torque that will overcomes the adhesion moment and remove the particles.

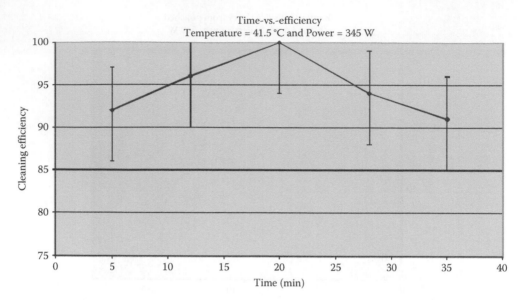

FIGURE 22.5 Effect of cleaning time on megasonic cleaning using 760 kHz frequency, 1.5°C, and 345 W of input power.

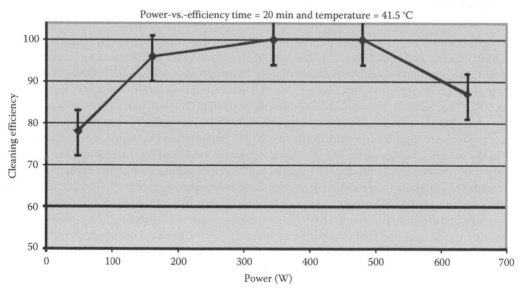

FIGURE 22.6 Effect of input power on megasonic cleaning using 760 kHz frequency, 41.5°C, and 20 min of cleaning time.

Brush cleaning is widely used in the industry especially following chemical–mechanical polishing processes of silicon or metal substrates. There are few scientific published studies on the effectiveness of brush cleaning and its effectiveness in removing small particles adhered by vdW forces or chemically bonded. There has also been some recent work on cleaning oxide silicon wafers using PVA brush with DI water, basic chemistry, or surfactants [43]. Brush cleaning can be effective if applied properly by optimizing the water flow, rotational speed, and brush pressure. Using chemistry during brush cleaning can enhance particle removal [42]. Research shows that the brush pressure is one of the most important parameters in removing particle [42]. The pressure helps the brush engulf the particle to be removed.

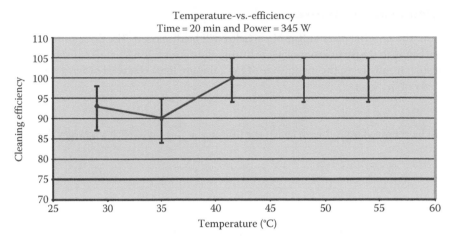

FIGURE 22.7 Effect of cleaning temperature on megasonic cleaning using 760 kHz frequency, 20 min of cleaning, and 345 W of input power.

The rotation of the brush applies the torque that will overcomes the adhesion moment and remove the particles.

Complete removal using the brush scrubber with DI water is achieved in cleaning thermal oxide silicon wafers dipped in STI silica slurry by Busnaina et al. [43]. They showed that intermediate brush pressure, speed, and time gave the best overall particle-removal efficiency. High pressure and long cleaning time will cause scratches (more defects) in the substrate. Figure 22.8 shows the effect of cleaning time and brush speed on brush cleaning using a PVA brush at 40 psi pressure between the brush and the substrate. The figure shows that the optimum cleaning time is longer than 30 s at 40 psi brush pressure [43].

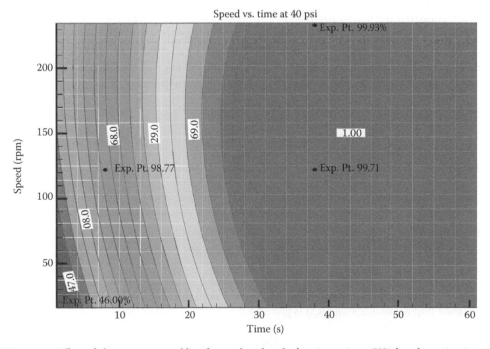

FIGURE 22.8 Effect of cleaning time and brush speed on brush cleaning using a PVA brush at 40 psi pressure between the brush and the substrate.

Particle-Removal Mechanism

In this section, the removal mechanism of particles will be discussed. The three different mechanisms that may contribute to the particle removal are lifting, sliding, or rolling. Let us consider a deformed PSL particles on silicon substrates, the magnitude of the adhesion force is several orders larger than removal forces [12,15,45–47].

Lifting

Particles will be removed from the surface if the lift force acting on particles is larger than the adhesion force.

$$F_L \geq F_A \tag{22.3}$$

Sliding

Particles will also be removed by sliding if the drag force [45], lift force, and adhesion force satisfy the following equation:

$$F_D \geq \kappa(F_A - F_L) \tag{22.4}$$

where κ is the coefficient of friction. The ratio of drag force over adhesion force, RS, is defined to judge whether detachment by instantaneous sliding occurs or not (if we assume that the lift force is very small and can be neglected compared to adhesion force).

$$RS = \frac{F_D}{F_A} \tag{22.5}$$

If $RS > \kappa$, particles will be removed by sliding.

Rolling

Hubbe [48] evaluated the torque balance on a spherical particle in contact with the surface. Sharma et al. [49] further included a factor of 1.399 in the Equation 21.6 since the drag force and the hydrodynamic torque on a particle near the wall could be substituted by an effective force at a distance of 1.399 R from the surface. When large deformation ($a/R > 0.1$) and the lift force are considered, the torque balance equation about point O can be described as the follows (as shown in Figure 22.2):

$$(1.399R - \delta)F_D = (F_A - F_L)a \tag{22.6}$$

where
  a is the contact radius
  $\delta = R - (R^2 - a^2)^{0.5}$ is the relative approach between the particle and the substrate

The particle will be removed instantaneously when the removal forces are applied, if the removal force overcomes adhesion force F_A. The ratio of the hydrodynamic rolling moment to the adhesion resisting moment, RM, is given by (neglecting a very small lift force):

$$RM = \frac{F_D(1.399R - \delta)}{F_A a} \tag{22.7}$$

When $RM > 1$, particles are removed by the drag force instantaneously.
  The relationship between RS and RM is given by

$$\frac{RM}{RS} = 1.399 \frac{R}{a} \tag{22.8}$$

Chemistry-Enhanced Cleaning

Chemicals such detergents, surfactants, etchers, etc. are often used to enhance physical cleaning. The chemical can be used to increase the wetting, wash and dissolve organic contaminants, or change the charge on particles to make them more repulsive to facilitate the removal process. In this chapter, only chemicals used to remove particles will be discussed.

Basic chemistry is often used in particle removal from silicon and metal substrate. It is usually accompanied by megasonic cleaning or overflow rinse. Basic chemistry has been used to remove silica, alumina, PSL, and silicon nitride particles [30–40]. Basic chemistry is used mainly to increase the repulsive charge (zeta potential) between the particle and the substrate. The zeta potential of the particle and the substrate has to be known in order to use the solution with a proper solution pH that provides the maximum repulsion. A number of investigators have studied the effects of zeta potential on deposition of particles onto surfaces [50–57]. Marshall and Kitchener [44] examined the deposition of carbon black particles from dilute aqueous suspensions onto glass. They observed that deposition was greatest when the zeta potentials of the particles and the substrate were of opposite sign. Hull and Kitchener [50] studied deposition of PSL particles onto a rotating disk. They found that when the particles and the substrate have opposite charges, deposition followed the expected diffusion-limited behavior. However, when the particles and substrate had like charges (repulsive interaction) there was considerably less deposition. Ali [55] measured the zeta potentials of a number of different particle types in semiconductor-processing liquids with an emphasis on applications to ultrafiltration of semiconductor chemicals. Albaugh and Reath [56] correlated particle counts in a process bath with surface counts following deposition from the bath onto hydrophilic wafers. They demonstrated the strong influences of pH and ionic strength on deposition.

The dependence of zeta potential on pH plays a significant role in surface cleaning. Zeta potential decreases as pH increases; it is typically positive at low pH, and negative at high pH. The point at which the zeta potential of a solid surface is zero is referred to as its isoelectric point or point of zero charge (pzc). The pzc of different solids depends upon the H^+ and OH^- ion concentrations in the solution, and therefore occurs at different pH values (pH $= -\log[II^+]$). At high pH, the particle can release H^+ ions into solution, resulting in a negative charge for the particle. At the pH of water, a silicon surface with a native oxide has a negative zeta potential (the pzc for a hydrophilic silicon substrate is approximately 2.6) [55]. Thus, negatively charged particles will be repulsed from a hydrophilic substrate at this pH, and even more strongly at higher pH. In deionized water, silica, and PSL particles are both negatively charged, whereas silicon nitride (Si_3N_4) particles typically carry a positive charge [57]. Thus, silica and PSL will be repelled from a wafer surface in water while silicon nitride will be attracted. The reduction of zeta potential at high pH contributes to the success of the SC1 solution (pH $= 11$) as a particle-removal chemistry. Figure 22.9 shows the zeta potential (particle charge) of colloidal silica particles as a function of pH [37–38].

Ionic strength of the liquid also affects electrical double-layer interaction. When ionic strength is high, the Debye length decreases and the strength and range of double layer interactions are significantly reduced [57]. Thus, where repulsion between particles and surfaces is expected, an increase in ionic strength will increase deposition. Similarly, low ionic strength give rise to a thicker double layer and more repulsion between the particle and the substrate.

Removal of Nanoparticles

The removal of nanoparticles is becoming increasingly challenging as the minimum line width continues to decrease in semiconductor manufacturing. In a recent study, the removal of nanoparticles from flat substrates using acoustic streaming was investigated [58]. Bare silicon wafers and masks with a 4 nm silicon-cap layer are cleaned. The silicon-cap films are used in extreme ultraviolet (EUV) masks to protect Mo–Si reflective multi-layers. The removal of 63 nm PSL particles from these substrates is

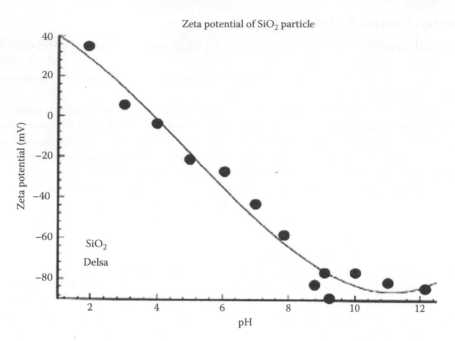

FIGURE 22.9   Zeta potential (particle charge) of silica particles as a function of pH.

conducted using single-wafer megasonic cleaning. The results show higher than 99% removal of PSL nanoparticles from bare silicon wafers and wafers with a 4 nm silicon-cap layer. Particle removal from the 4 nm Si-cap substrate is slightly more difficult as compared to bare silicon wafer due to a lower double layer repulsive force of the Si-cap surface. The moment ratio analysis exhibits good agreement with the experimental results. The experiments also show that dilute SC1 results in higher moment ratio than DI water due to the larger repulsive charge on the particle at a higher pH. Dilute SC1 provides faster removal of particles which is also verified by the analytical analysis. The simulation shows that once a nanoparticle is detached from the substrate it undergoes oscillatory motion and multiple re-depositions and removal takes place. The particle oscillatory motion and re-deposition is the main reason for the long time required for nanoparticle removal as compared to micron and submicron size particles. The modeling and experimental results show that 63 nm and larger PSL particles can be completely removed using acoustic streaming from bare silicon and 4 nm Si-cap layer wafers. According to the particle-removal experiments, complete removal of the 63 nm PSL particles is achieved after 4 min in dilute SC1 and after 6 min in DI water for both substrates. The long time due to particle oscillatory motion and re-deposition is not observed in the removal of submicron or larger size particles [58].

References

1. H. Krupp, *Adv. Colloid Interface Sci.*, 1, 111–140, 1967.
2. J. Taylor, MS thesis, Clarkson University, Potsdam, NY, December 1990.
3. J. Taylor, A. A. Busnaina, F. W. Kern, and R. Kunesh, In: *Proceedings, IES 36th*. New Orleans, LA, April 23–27, 1990, pp. 422–426.
4. A. A. Busnaina, J. Taylor, and I. Kashkoush, *J. Adhes. Sci. Technol.*, 7, 5, 441, 1993.
5. M. Kurz, A. A. Busnaina, and F. W. Kern, In: *Proceedings, IES 35th*. Anaheim, CA, May 1–5, 1989, pp. 340–347.
6. J. Visser, *Adv. Colloid Interface Sci.*, 2, 331–363, 1972.

7. J. Visser, *Adv. Colloid Interface Sci.*, 15, 157–169, 1981.

8. J. Visser, *Adv. Colloid Interface Sci.* In: *Surface and Colloid Science*, ed., E. Matijevic, Vol. 8, John Wiley & Sons, New York, 1976.

9. H. C. Hamaker, *Physica*, 4, 1937.

10. D. Tabor, *J. Colloid Interface Sci.*, 58, 1977.

11. L. P. DeMejo, D. S. Rimai, and R. C. Bowen, *J. Adhes. Sci. Technol.*, 2, 331–337, 1988.

12. D. S. Rimai and A. A. Busnaina, *J. Particulate Sci. Technol.*, 13, 249, 1995.

13. A. D. Zimon, *Adhesion of Dust and Powder*, Plenum Press, New York, 1969.

14. W. C. Hinds, *Aerosol Technology*, John Wiley & Sons, New York, 1982.

15. S. Krishnan, A. A. Busnaina, D. S. Rimai, and D. P. DeMejo, *J. Adhes. Sci. Technol.*, 8(11), 1357–1370, 1994.

16. C. N. Davies, *Aerosol Science*, Academic Press, New York, 1966.

17. M. C. Kordecki and C. Orr Jr., *Arch. Environ. Health*, 1, 7, 1960.

18. Yu. M. Luzhnov, *Research in Surface Forces*, Consultants Bureau, New York, 1971.

19. A. A. Busnaina and T. M. Elsawy, In: *The Adhesion Society Proceedings, 21st Annual Meeting*, Savannah, GA, February 22–25, 1998.

20. P.G. Saffman, *J. Fluid Mech.*, 22, 385–400, 1965.

21. J. Visser, *J. Colloid Interface Sci.*, 34, 1970.

22. R. P. Musselman and T. W. Yarbrough, *J. Environ. Sci.*, 51–56, 1987.

23. J. Cleaver and B. Yates, *J. Colloid Interface Sci.*, 44, 1973.

24. J. Olaf, *Acustica*, 7(5), 253, 1957.

25. L. D. Rosenberg, *Ultrasonic News*, Winter, 1960, p. 16.

26. D. H. McQueen, *Ultrasonics*, 24, 273, September 1986.

27. D. H. McQueen, *Ultrasonics*, 28, 422, November 1990.

28. S. Schwartzman, A. Mayer, and W. Kern, *RCA Review*, 46, 81, March 1985.

29. A. Mayer and S. Schwartzman, *J. Electronic Mater.*, 8, 855, 1979.

30. I. Kashkoush, A. Busnaina, F. Kern, and R. Kunesh, In: *Particles on Surfaces 3: Detection, Adhesion, and Removal*, ed., K. L. Mittal, Plenum Press, New York, 1991, pp. 217–237.

31. I. Kashkoush and A. Busnaina, In: *Proceedings, IES 38th Annual Meeting*, San Diego, CA, May 6–10, 1991, pp. 861–867.

32. I. Kashkoush and A. Busnaina, *Particulate Sci. Technol.*, 11, 11, 1993.

33. A. Busnaina and I. Kashkoush, *Chem. Eng. Commu.*, 125, 47, 1993.

34. I. Kashkoush and A. Busnaina, In: *Proceedings, IES 40th Annual Meeting*, Chicago, IL, 1993, p. 356.

35. I. Kashkoush, PhD thesis, Clarkson University, Potsdam, New York, 1993.

36. G. Gale, A. Busnaina, and I. Kashkoush, In: *Proceedings, Precision Cleaning '94*, Rosemont, IL, May 17–19, 1994, pp. 232–253.

37. A. A. Busnaina, I. I. Kashkoush, and G. W. Gale, *J. Electrochem. Soc.*, 142(8), 2812–2817, 1995.

38. G. W. Gale and A. A. Busnaina, *J. Particulate Sci. Technol.*, 1995.

39. Y. Hirota, Y. Homma, and K. Sugii, *Appl. Surf. Sci.*, 60, 619, 1992.

40. W. Syverson, M. Fleming, and P. Schubring, In: *Second International Symposium on Cleaning Technology in Semiconductor Manufacturing. Electrochemical Society Proceedings PV92-10*, Pennington, NJ, 1992, p. 10.

41. P. Wang and D. Bell, In: *Third International Symposium on Cleaning Technology in Semiconductor Device Manufacturing. Electrochemical Society Proceedings PV94-7*, Pennington, NJ, 1994, p. 132.

42. S. R. Roy, I. Ali, G. Shinn, N. Furusawa, R. Shah, S. Peterman, K. Witt, and S. Eastman. *J. Electrochem. Soc.*, 142(1), 1995, 216–226.

43. A. A. Busnaina, N. Moumen, and J. Piboontum, In: *Proceedings of the VLSI Multilevel Interconnection Conference (VMIC)*, Santa Clara, CA, February 8–12, 1999.

44. J. K. Marshall and J. A. Kitchener, *J. Colloid Interface Sci.*, 22, 342, 1966.

45. F. Zhang and A. Busnaina, *J. Electrochem. Soc.*, 146(7), 2665–2669, 1999.

46. F. Zhang and A. A. Busnaina, In: *The Adhesion Society Proceedings, 21st Annual Meeting*, Panama City, FL, February 21–24, 1999.
47. F. Zhang and A. Busnaina, *Electrochem. Solid-State Lett.*, 1(4), 1998.
48. M. A. Hubbe. *Colloid Surf.*, 12, 1984.
49. M. M. Sharma, H. Chamoun, D. Sarma, and R. Schechter, *J. Colloid Interface Sci.*, 149, 1992.
50. M. Hull and J. A. Kitchener, *Trans. Faraday Soc.*, 65, 3093, 1969.
51. G. E. Clint, J. H. Clint, J. M. Corkill, and T. Walker, *J. Colloid Interface Sci.*, 44, 121, 1973.
52. E. Ruckenstein and D. Prieve, *J. Chem. Soc. Faraday II*, 69, 1522, 1973.
53. D. Prieve and E. Ruckenstein, *J. Colloid Interface Sci.*, 60, 337, 1977.
54. W. Brouwer and R. Zsom, *Colloids Surf.*, 24, 195, 1987.
55. I. Ali, Ph.D. Thesis, University of Arizona, Tucson, AZ, 1990.
56. K. B. Albaugh and M. Reath, In: *Proceedings, Microcontamination 1991*, San Jose, CA, October 16–18, 1991, p. 603.
57. D. J. Riley and R. G. Carbonell. *J. Colloid Interface Sci.*, 158, 259, 1993.
58. K. Bakhtari, R. O. Guldiken, P. Makaram, A. A. Busnaina, and J. G. Park, Experimental and numerical investigation of nanoparticle removal using acoustic streaming and the effect of time. *J. Electrochem. Soc.*, May 2006.

Cleaning Processes for Semiconductor Wafer Manufacturing (Aluminum Interconnect)

Introduction 359
Basic Operations in Wafer Fabrication 360
Photoresist Chemistry • Photolithography and Masking
Process • Radiation-Sensitive Polymers • Comparison of Positive and
Negative Resists • Negative-Acting Photoresists • Positive-Acting
Photoresists • Photoresist Performance Factors
General Wafer Cleaning Techniques...368
FEOL Cleaning Processes • Cleaning Process Optimization • Temperature
Effect • Ultrasonic and Megasonic Effect
FEOL Cleaning Processes..369
Sulfuric-Peroxide Chemistry • Quaternary Ammonium Hydroxides—
Choline-Surfactant Chemistry • TMAH Chemistry • Ozone–Water
Mixtures
BEOL Cleaning Processes 372
Chemistry of Positive Photoresist Strippers 373
NMP-Based Strippers • Non-NMP-Based Organic Strippers
Chemistry of Negative Photoresist Strippers.............................373
Chemistry of Post–Plasma Etch Polymer Removers.................373
HA/Amine Chemistry • HF/Glycol Chemistry
Challenges of Future Technology...375
Copper Interconnects • Low-k Dielectric Material
References...377

Shawn Sahbari
*Applied Chemical
Laboratories Inc.*

Mahmood Toofan
*Semiconductor
Analytical Services*

John Chu
Consultant

Introduction

In semiconductor device manufacturing, silicon wafers are processed to fabricate very-large-scale-integrated (VLSI) or ultra-large-scale-integrated (ULSI) circuits. Since early stages of the semiconductor wafer processing in the 1960s, significant improvements and advancements have been made in chip manufacturing. However, the chemistry of wafer cleaning material and basic cleaning operations have remained fundamentally unchanged. During recent years, the geometry of the microcircuits, the diameter of the silicon wafers, and the processing equipment and methods have been significantly improved and updated. In the early stages, simple immersion tanks of cleaning solutions were employed with manual agitation. Today, more advanced cleaning solutions are applied on sophisticated wet benches and spray tools with automated chemical delivery systems and robotic arm movements for the displacement of wafers.

In the past three decades, wafer fabrication technology has made significant advancements in terms of the density of microcircuits, the reduction of feature size, and increase in wafer diameters. Diameters of the wafers have increased from 2–3 to 8–12 in. On the other hand, as diameters of the wafers have increased, the geometry of microcircuits and interconnects have reduced from 6–8 μ (10^{-6} m) to 0.2–0.3 μ and beyond. These dramatic changes in wafer processing technology require more precise cleaning solutions with ultrahigh purity and advanced cleaning formulations.

The cleanliness of the wafer surfaces and the purity of cleaning chemicals used in wafer fabrication processes are essential requirements to yield improvement in microelectronic device manufacturing. In order to meet the required specifications of sub-half micron substrate geometry in wafer processing, the surface cleaning chemistry must meet stringent quality of clean room packaging, filtration, and ultrahigh purity of sub-ppb (part per billion) level ionic contamination. Trace ionic impurities, such as sodium or potassium cations and chloride anions and particulates, are especially detrimental if present on wafer surfaces during the thermal processing.

Basic Operations in Wafer Fabrication

Wafer fabrication is a series of processes used to create semiconductor devices on a silicon wafer surface. The polished silicon wafers with blank surfaces undergo hundreds of process steps and end up producing hundreds and thousands of chips with multiple and diverse functions. The designs of the devices and circuits are based on different transistor structures. Among the major structure designs, bipolar and MOS (metal oxide semiconductor) transistors are the most widely manufactured and used with numerous variations. Furthermore, there are several choices of processes and materials available to create each individual layer of any particular device structure.

Regardless of process diversity and a variety of hundreds of process options, only four major operations are performed during the fabrication process. These major operations are layering, patterning, doping, and heat treatment.

Layering is the operation used to add thin layers of materials to the wafer's surface. The layers are added to the surface in multiple major techniques: growing a silicon oxide or silicon nitride layer on the wafer using a thermal process, and chemical vapor deposition (CVD). Rapid thermal operation (RTO) or rapid thermal process (RTP) technology is a natural choice for the growth of oxides used in MOS devices. Other techniques such as evaporation, physical vapor deposition (PVD), spin-on deposition, and sputtering are also used to add layers on the wafers.

Patterning is a series of steps that results in the removal of selected portions of the added surface layers (Figure 23.1a and b). After removal, a pattern of the layer is left on the wafer surface. The material left or removed may be in the form of a hole in the layer or just a remaining island of the material.

The patterning process is named photomasking, photolithography, or microlithography.

Photolithography is a multistep pattern transfer process similar to stenciling or photography. In photolithography, the required pattern is first formed in photomasks and transferred into the surface layers of the wafer through the photomasking steps. The polymeric materials used in photolithography to transfer patterns to the wafer are called photoresist. Figure 23.2 shows the 10-step process of pattern transfer to the wafer surface using photomasking process for a negative-acting photoresist.

The foregoing was a brief description of some of the basic operations in wafer fabrication. Since the focus of this chapter is the review of cleaning technology in semiconductor wafer fabrication, the emphasis of the following sections will be on the chemistry of photoresist, its cleaning solutions, and cleaning processes in wafer operations, especially advanced cleaning methods for film removal such as photoresist strippers, post plasma etch polymer removers.

For additional information on fabrication processes, the readers are encouraged to refer to *Microchip Fabrication, a Practical Guide to Semiconductor Processing,* by Peter Van Zant [1].

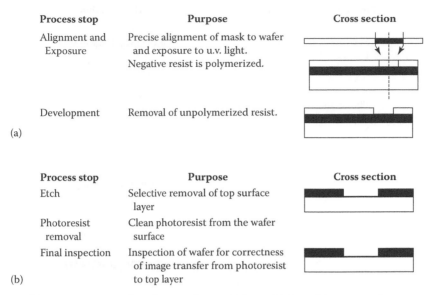

Process stop	Purpose	Cross section
Alignment and Exposure	Precise alignment of mask to wafer and exposure to u.v. light. Negative resist is polymerized.	
Development	Removal of unpolymerized resist.	

(a)

Process stop	Purpose	Cross section
Etch	Selective removal of top surface layer	
Photoresist removal	Clean photoresist from the wafer surface	
Final inspection	Inspection of wafer for correctness of image transfer from photoresist to top layer	

(b)

FIGURE 23.1 (a) First pattern transfer Dfrom mask to resist layer. (b) Second image transfer.

Photoresist Chemistry

Photoresists have been used in the printing industry for over a century. In the 1920s, photoresists found a wide range of applications in the printed circuit board industry. The semiconductor industry adopted this technology to wafer fabrication in the 1950s. Photoresists specifically designed for semiconductor use were first developed by the Eastman Kodak Company. In the late 1950s, they introduced Kodak Photo Resist (KPR), Kodak Metal Etch Resist (KMER), and Kodak Thin Film Resist (KTFR)—*negative photoresists*. At around the same time, Shipley Company introduced a line of *positive-acting photoresists*. Since that time, some other companies also have entered into the market with photoresists designed to keep pace with increasing demand in the industry for printing narrower lines in fabrication of fine geometry integrated circuits. Today, different manufacturers offer a wide range of products designed to match a variety of applications.

Photoresists are used in the masking process for patterning the wafers in the process of *photolithography*. Other terms used in the industry for these steps are called *photomasking, masking,* or *microlithography*.

Photolithography and Masking Process

Photolithography is one of the most critical operations in semiconductor manufacturing processes. It is a patterning process that sets two-dimensional horizontal patterns on the various parts of the circuit designing on the wafer. The photoresist material performs the function of transforming a two-dimensional circuit design into a three-dimensional electric circuit. The photoresist materials used in photolithography are generally formulated from polymeric materials with photosensitive additives. Most photoresist materials consist of four basic ingredients, each one having a different function. Table 23.1 shows the basic components of photoresists.

Radiation-Sensitive Polymers

The photosensitive ingredients of the photoresist material are special polymers. Polymers are macromolecules containing carbon, hydrogen, and oxygen atoms that are formed by repeated patterns of their monomers or simple molecules. Most plastics are a form of polymers. Photoreactive polymers

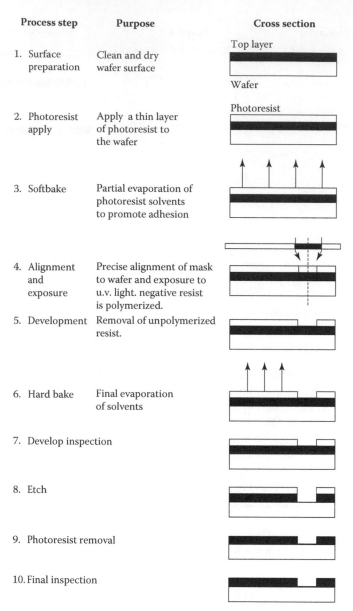

Process step	Purpose	Cross section
1. Surface preparation	Clean and dry wafer surface	Top layer / Wafer
2. Photoresist apply	Apply a thin layer of photoresist to the wafer	Photoresist
3. Softbake	Partial evaporation of photoresist solvents to promote adhesion	
4. Alignment and exposure	Precise alignment of mask to wafer and exposure to u.v. light. negative resist is polymerized.	
5. Development	Removal of unpolymerized resist.	
6. Hard bake	Final evaporation of solvents	
7. Develop inspection		
8. Etch		
9. Photoresist removal		
10. Final inspection		

FIGURE 23.2 Pattern transfer process.

TABLE 23.1 Photoresist Components and Their Functions

Component	Function
Polymer	Changes structure due to reaction with radiation energy (polymerization or photo-solubility)
Solvent	Used as thinner to allow application of thin film layer of the spun material
Sensitizers	Control modification of chemical reaction when exposed to light
Additives	For special purposes

are radiation sensitive and react with some type of light energy, ultraviolet or laser. Those photoresists that contain these types of polymers are called *optical resists*. Other resists respond to x-ray radiation or e-beams which are *i-line* or *j-line* resists.

Comparison of Positive and Negative Resists

Up to the mid-1970s, negative resist was dominant in the masking process. The advent of VLSI circuits and image sizes in the 2–5 µm range strained the resolution capability of negative resists. Positive resists had been around for over 20 years, but their poorer adhesion properties were a drawback and their superior resolution capability and pinhole protection were not needed.

By the 1980s, positive resist became the resist of choice. The transition was not easy. To switch a fabrication line from negative to positive resist requires changing the polarity of the masks or reticles from clear field to dark field. Unfortunately, it is not a simple matter of reversing the fields in the mask-making process. The dimensions have to be adjusted to accommodate the different characteristics of the positive resist. The determination of the correct mask of reticle dimensions is a lengthy procedure.

Positive resists have a higher *aspect ratio* compared to negative resists. In other words, positive resists have a better *resolution capability* and can resolve smaller geometry such as wire lines and via openings.

Another problem with negative resists is oxygenation. This is a reaction of the resist to oxygen in the atmosphere, and can result in a thinning of the resist film by as much as 20%. Positive resists do not have this property. Cost is always an important consideration. Negative resists sell for about one-third of the cost of positive resists.

Developing characteristics differ between the two types of resists. Negative resists develop in readily available solvents and possess wider developer process latitude. Positive resists require carefully prepared developer solutions and temperature control of the process.

The next-to-last step in the masking process is a photoresist removal, which can take place in chemical solutions or in plasma systems. Generally, the removal of positive resists is easier and takes place in chemicals that are more environmentally sound. While positive photoresists are the resists of choice for fabrication areas processing state-of-the-art circuits, there are many lines still producing devices and circuits with image sizes greater than 5 µm. A great many of these lines use negative resists. Table 23.2 shows a comparison of properties of the two resists.

Negative-Acting Photoresists

Negative photoresists are normally based on polyisoprene-type polymers. Polyisoprene polymers naturally occur in rubber material. The Hunt Corporation developed the first synthetic polyisoprene polymer structure (Figure 23.3).

TABLE 23.2 Comparison of Negative and Positive Resists

Parameter	Negative	Positive
Aspect ratio (resolution)		Higher
Adhesion	Better	
Exposure speed	Faster	
Pinhole count		Lower
Step coverage		Better
Cost		Higher
Developers	Organic solvents	Aqueous
Strippers		
Oxide steps	Acid	Acid
Metal steps	Chlorinated solvent compounds	Simple solvents

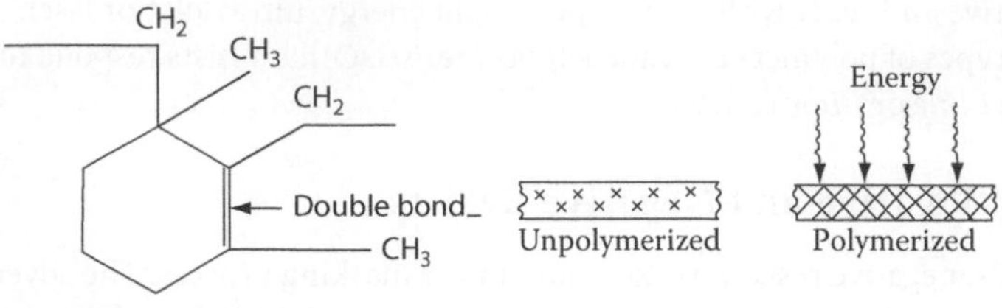

CH₂
CH₃
CH₂

←— Double bond_
CH₃

Unpolymerized

Energy

Polymerized

FIGURE 23.3　Chemistry of negative photoresist (isoprene monomer).

Before exposure to the light, the negative-resist polymers exist in their unpolymerized condition (under which the polymers are not chemically linked to each other). When the photoresist is exposed to proper light or energy, the polymers become cross-linked, or in chemical term, polymerized. This process may also be achieved when the photoresist materials are exposed to heat and/or visible light. To prevent this deterioration, the photoresist material is normally packaged in amber glass bottles or dark color, brown or black, plastic packaging. During the application process, in order to prevent accidental exposure, photomasking and resist processing areas use yellow color filters or yellow lighting.

Depending on the photoresists' response to the type of energy or radiation, photoresists are normally referred to by their general category, such as ultraviolet, deep UV, x-ray, i-line, etc.

Positive-Acting Photoresists

Positive-acting photoresists are based on the phenol-formaldehyde polymer, also called phenol-formaldehyde novolak resin (Figure 23.4).

The novolak resin within the unexposed photoresist is relatively insoluble. After exposure to the proper radiation energy, the photoresist converts to a more soluble state. This reaction is called photosolubilization. Table 23.3 contains a list of commonly used photoresist polymers used for the photolithography process in the semiconductor industry.

Solvents

The largest ingredient by volume in a photoresist composition is the solvent. It is the solvent that converts the solid resist material to a liquid and allows the liquid photoresist to be applied to the wafer surface as a thin layer by spinning. Photoresist is analogous to paint, which is composed of the coloring pigment and polymer dissolved in an appropriate solvent. It is the solvent that allows the application of the paint onto a surface in a thin layer. For negative photoresist, the solvent is an aromatic hydrocarbon such as xylene.

In positive resist, a variety of solvents are used, depending on the type of polymer. The most commonly used solvents are ethoxyethyl acetate (EEA), 2-methoxy propyl acetate (propylene glycol monomethyl ether acetate, PGMEA), and ethyl lactate (ELS).

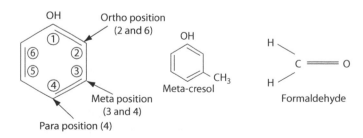

FIGURE 23.4　Phenol-formaldehyde novolak resin structure.

TABLE 23.3 Commonly Used Photoresists

Resist	Polymer	Polarity	Sensitivity (C/cm²)	Exposure Source
Positive	Novolak (M-cresoformaldehyde)	+	$3–5 \times 10^{-5}$	UV
Negative	Polyisoprene	−		UV
PMMA	Poly-(methyl methacrylate)	+	5×10^{-5}	E-beam
PMIPK	Poly-(methyl isopropenyl ketone)	+	1×10^{-5}	E-beam/deep UV
PBS	Poly-(butene-1-sulfone)	+	2×10^{-6}	E-beam
TFECA	Poly-(triflouro-ethyl chloroacrylate)	+	8×10^{-7}	E-beam
COP	Copolymer-(cyano ethyl	−	5×10^{-7}	E-beam/x-ray
PCA	Acrylate-(amido ethyl acrylate)			
PMPS	Pentene-1-sulfone	+	2×10^{-7}	E-beam

Source: Van Zant, P., *Microchip Fabrication, a Practical Guide to Semiconductor Processing*, 2nd edn., McGraw-Hill, Inc., New York, 1990. With permission.

Sensitizers

Chemical sensitizers are added to the resists to cause or control certain reactions of the polymer. In negative resists, the untreated polymer responds to a certain range of the ultraviolet spectrum. Sensitizers are added to either broaden the response range or narrow it to a specific wavelength. In negative resists, a compound called bisaryldiazide is added to the polymer to provide the light sensitivity. In positive resists, the sensitizer is o-naphthaquinonediazide.

Additives

Various additives are mixed with resists to achieve particular results. Some negative resists have dyes intended to absorb and control light rays in the resist film.

Photoresist Performance Factors

The selection of a photoresist starts with the dimensions required on the wafer surface. The resist must first have the capability of producing those dimensions. Beyond that, it must also function as an etch barrier during the etching step, a function that requires a certain thickness for mechanical strength. In the role of etch barrier, it must be free of pinholes, which also requires a certain thickness. In addition, it must adhere to the top wafer surface or the etched pattern will be distorted, just as a paint stencil will give a sloppy image if it is not taped tight to the surface. These, along with process latitude and step coverage capabilities, are resist performance factors. In the selection of a resist, the process engineer often must make trade-off decisions between the various performance factors.

Resolution

The smallest opening of space that can be produced in a photoresist layer is generally referred to as its *resolution capability*. The smaller the line produced, the better the resolution capability. Generally, smaller line openings are produced with thinner resist film thickness. However, a resist layer must be thick enough to function an etch barrier and to be pinhole-free. The selection of a resist thickness is a trade-off between these two goals. At present, the more advanced photoresists offered by Shipley Co. is the 0.193 μ resist with a resolution sensitivity of 193 nm. Shipley is also offering a more advanced line of positive resists in the UV range with an exposure wavelength as low as 154 nm.

The capability of a particular resist relative to resolution and thickness is measured by its *aspect ratio* (Figure 23.5). The aspect ratio is calculated as the ratio of the resist thickness to the image opening. Positive resists have a higher aspect ratio compared to negative resists, which means that for a given image-size opening, the resist layer can be thicker. The ability of positive resist to resolve a smaller opening is due to the smaller size of the polymer. It is similar to the requirement of using a smaller brush to paint a thinner line.

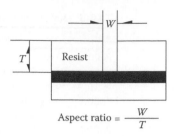

Aspect ratio = $\dfrac{W}{T}$

FIGURE 23.5 Aspect ratio.

Shipley is offering a line of advanced photoresists that can generate a via opening with an aspect ratio of up to 10:1 in copper interconnect technology. Figure 23.6 shows an advanced device structure with high aspect ratio via opening.

Adhesion Capability

In its role as an etch barrier, a photoresist layer must adhere well to the surface layer to faithfully transfer the resist opening into the layer. Lack of adhesion results in distorted images. Resists differ in their ability to adhere to the various surfaces used in chip fabrication within the photomasking process; there are a number of steps that are specifically included to promote the natural adhesion of the resist to the wafer surface. Negative resists generally have a higher adhesion capability than positive resists.

Exposure Speed and Sensitivity

The primary action of a photoresist is the change in structure in response to an exposing light or radiation. An important process factor is the speed at which that reaction takes place. The faster the speed, the faster the wafers can be processed through the masking area. Negative resists typically require 5–15 s of exposure time while positive resists take 3–4 times longer.

The sensitivity of a resist relates to the amount of energy required to cause the polymerization of photosolubilization to occur. Further, sensitivity relates to the energy associated with specific wavelength for the exposing source. Understanding this property requires a familiarization with the properties of the electromagnetic spectrum (Figure 23.7). Within nature, we identify a number of different types of energy: light, short and long radio waves, x-rays, etc. In reality, they are all electromagnetic energy (or radiation) and are differentiated from each other by their wavelengths, with the shorter wavelength radiations having higher energies.

— 100 nm

FIGURE 23.6 Advanced aspect ratio.

Name	Gamma rays	X-rays	Ultra violet (UV)		Infrared (IR)	Short radio waves	Broadcast radio waves
Wavelength (cm)	10^{-11}	10^{-8}	10^{-6}		10^{-3}	10^2	10^4

Visible

10^{-4}

FIGURE 23.7 Electromagnetic spectrum.

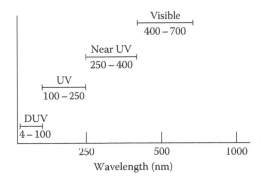

Visible
400 – 700

Near UV
250 – 400

UV
100 – 250

DUV
4 – 100

250   500   1000

Wavelength (nm)

FIGURE 23.8 Ultraviolet and visible spectrum.

Common positive and negative photoresists respond to energies in the UV and deep ultraviolet (DUV) portion of the spectrum (Figure 23.8). Some are designed to respond to particular wavelengths (peaks) within those ranges. When we speak of resist sensitivity, we refer to the specific wavelengths the resist reacts to. This property is also called the *spectral response characteristic* of the resist. Figure 23.9 is the spectral response characteristic of a typical production resist. The peaks in the spectrum are regions (wavelengths) that carry higher amounts of energy.

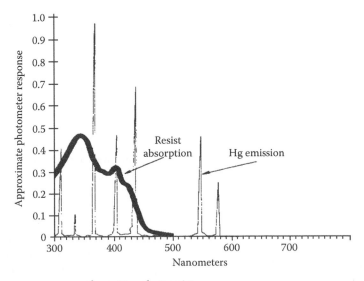

FIGURE 23.9 Exposure response of a positive photoresist.

General Wafer Cleaning Techniques

Impurities on the surface of the silicon wafers come from various sources at different stages of the manufacturing process. These impurities must be removed following each process step in order to keep the substrates clean for next process. Depending on the type of surface contamination or impurities, different cleaning solutions and techniques need to be applied [1].

Wafer cleaning solutions, depending on the process and their functions, are classified as

1. Cleaning solutions of bare silicon and oxidized wafers involving pre-metal processes or so-called front-end-of-line (FEOL) cleaning processes.
2. Cleaning solutions of post-metal processes used in different stages of metallization, or so-called back-end-of-line (BEOL) cleaning processes [2].

Surface conditioning process or "FEOL" is a pre-metal process and normally uses acids and oxidants to precondition and clean the wafer surfaces.

Post-metal processes or "BEOL," which includes photoresist stripping, post-plasma etch residue removing, and post-CMP polishing slurry removal, use sophisticated solvent formulations that are not the traditional cleaners. These two processes can be distinguished as residual or contamination cleaning versus bulk material removing, such as resist stripping.

The solvent formulations for bulk material removal or BEOL processes are either semi-aqueous solvent formulations or aqueous solutions of more advanced cleaning solutions that will include corrosion inhibitors and wetting agents for the protection of fine geometry in metal layers, and high aspect ratio via openings.

Depending on whether a semiconductor device is DRAM (dynamic random access memory) or Logic, a wafer can undergo 25–30 different steps, including ash processing followed by cleaning step using one of the two chemistries.

FEOL Cleaning Processes

The addition of hydrogen peroxide to sulfuric acid (piranha) is perhaps the most common wafer-cleaning solution used in semiconductor processing plants. This mixture is commonly known as sulfuric peroxide mixture (SPM) or piranha, and is used in pre-diffusion cleans and is also used as a photoresist stripper. When hydrogen peroxide is mixed with sulfuric acid, the exothermic resultant reaches up to 130°C providing the required cleaning efficacy.

Other cleaning chemistries used in FEOL processes include RCA clean, choline chemistry, and quaternary ammonium hydroxide or tetramethyl ammonium hydroxide (TMAH)-based chemistry in either spray or immersion equipment. These will be discussed in more detail shortly.

Cleaning Process Optimization

Depending on the chemistry of the cleaning solution involved and the type of wafers, the process conditions such as process temperature, process time, and the equipment used may vary from one fabrication site to another. Process engineers optimize their cleaning processes to achieve high yields and low defects.

Temperature Effect

An important factor in cleaning wafers is the bath temperature of the processing material. In an FEOL cleaning process such as RCA clean chemistry, the chemical is normally heated to an optimum temperature (typically 75°C–85°C) to achieve the best results. Photoresist developers and edge bead removers (EBR) in BEOL process are normally applied at ambient (23°C) temperature. These materials only

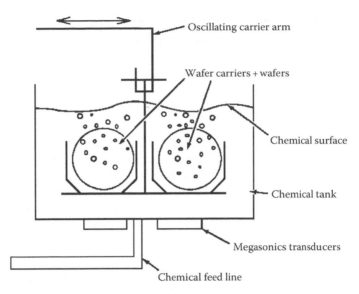

FIGURE 23.10 Megasonic tank configuration. (After J. Ruzyllo and R. Novac.)

dissolve the uncured or soft-baked photoresist which is not cross-linked or polymerized. In resist stripping process, however, the cross-linked and polymerized resist may require heated stripping solutions and a longer time for complete removal of the hardened photoresist. If plasma-treated photoresist residues (post-plasma polymers) are not oxygen plasma ashed, they are even harder to clean and will need more aggressive solutions at higher temperatures. In any case, whether a FEOL or BEOL process is in mind, the temperature of the chemistry in immersion tanks or spray tools is optimized and preset to safe operational conditions.

Ultrasonic and Megasonic Effect

For certain cleaning applications such as metal lift-off processes or stripping of an ion implanted and deep UV-cured photoresist, without a plasma ashing, ultrasonic or megasonic agitation may be necessary for a complete dissolution. In FEOL cleaning using RCA clean process, megasonic energy has shown significant improvement in cleanliness of the wafers and particle removal efficiency of the solution. Figure 23.10 shows the configuration of a megasonic wafer cleaning tank.

FEOL Cleaning Processes

Sulfuric-Peroxide Chemistry

Early cleaning processes of silicon wafers involved using concentrated inorganic acids such as boiling nitric acid; aqua regia; concentrated hydrofluoric acid; and mixtures of phosphoric, acetic, and sulfuric acids. Mixtures of sulfuric acid and hydrogen peroxide, or so-called piranha solutions, are still being used in FEOL wafer cleaning applications. In terms of sulfuric chemistry and acid-to-peroxide mix ratios, process engineers use their own selections. Instead of having premixed, stabilized mixture, one can prepare two chemicals on-site, as is needed in different ratios.

Oxidizing agents such as mixtures of sulfuric acid and chromic acid were also used as a general purpose glass cleaner or silicon wafer surface cleaner. This type of cleaners, however, caused ecological toxic pollution and waste disposal problems.

Sulfuric Acid and Ammonium Persulfate Chemistry

Hydrogen peroxide (H_2O_2) is essentially unstable and readily disassociates to water and oxygen at elevated temperatures.

$$H_2O_2 \rightarrow H_2O + \frac{1}{2}O_2$$

An alternative oxidizing agent used in wafer cleaning is ammonium persulfate in sulfuric acid mixtures. Ammonium persulfate (AP) is added to sulfuric acid baths in a concentration 40–80 g/L. Since AP is less reactive at room temperature than H_2O_2, it is therefore safer to store with a longer shelf life. Being less reactive, AP ensures a more stable and steady release of oxygen to the cleaning bath and more consistent and stable bath life to the chemical. Sulfuric acid–AP mixture is used in FEOL cleaning process, for general wafer cleaning, and also in resist stripping processes of non-metalized wafers.

RCA Chemistry

The first systematically developed cleaning process for unprocessed or oxidized silicon wafers was developed at RCA and published in 1970 and was called RCA Clean. The RCA cleaning process involved a two-step process of peroxide treatment with a high pH alkaline solution (normally ammonium hydroxide mixed with hydrogen peroxide) as RCA1 or SC-1, followed by a treatment with a mixture of hydrochloric acid and hydrogen peroxide as SC-2 [3].

The solutions are made using ultrafiltered deionized (DI) water, electronic grade ammonium hydroxide (29% wt/wt% as NH_3), electronic grade hydrochloric acid (37 wt/wt%), and high purity unstabilized hydrogen peroxide.

In the first treatment step, the wafers are exposed to a hot mixture of water-diluted hydrogen peroxide and ammonium hydroxide. This procedure was designed to remove organic surface films by oxidative breakdown and dissolution to expose the silicon or oxide surface for concurrent or subsequent decontamination reactions. In this treatment, metal impurities such as copper and zinc are dissolved and removed by a complexing agent of ammonia, for example, in forms of [Cu $(NH_3)4]^{2+}$ amino complex.

The second treatment is designed to remove alkali ions, cations such as Al^{3+}, Fe^{3+}, and Mg^{2+} that form water-insoluble hydroxides in SC-1 ammonia solution.

The volume ratios for the RCA standard clean 1, SC-1 clean used in first treatment step are $H_2O:H_2O_2$ (30%):NH_4OH (29% as NH_3) as 5:1:1, and the volume ratios for the RCA standard clean 2, SC-2 clean used in second treatment step are $H_2O:H_2O_2:HCl$ as 6:1:1.

The processing temperature should be kept at 75°C–80°C to sufficiently activate the mixture without causing thermal decomposition due to higher temperatures.

The original RCA cleaning processes were based on a simple immersion technique. Several different improved techniques have been introduced over the years. More advanced automated wet bench immersion systems for large-scale production are now available and being offered in the industry by equipment manufacturers such as FSI International.

Quaternary Ammonium Hydroxides—Choline-Surfactant Chemistry

Among the other alternative alkaline cleaning solutions that have been studied on silicon wafers, choline (2-hydroxyethyl trimethyl ammonium hydroxide) [5], a strong base with a chemical formula [N $(CH_3)_3$ $CH_2CH_2OH]OH$, and $pK_b = 5.06$, which is metal ion free, has drawn much attention [5–7]. The immersion of HF-etched silicon wafers in a choline solution followed by water rinsing has shown very clean results.

Muraoka et al. [5] have developed several techniques to clean silicon wafers using choline. They have reported that dilute aqueous solutions of choline with selected nonionic surfactants can remove heavy metals from the silicone wafers' surface and prevent replating of these metals from solution on the wafer. Harri and Hockett [6] compared wet cleaning involving choline with those of other solutions and concluded better electrical properties on those with choline process.

TMAH Chemistry

Another strong base that is relatively stable at ambient temperature and is also metal ion free is tetramethyl ammonium hydroxide (TMAH), $N(CH_3)_4OH$. TMAH is widely used as a positive photoresist developer in a relatively low concentration (2.5 wt%) in aqueous solution with surfactants. TMAH-based photoresist developers have replaced the traditional alkaline developers because of their low ionic impurities and high polymer dissolution capability.

In recent years, TMAH base formulations in organic solvents have been used for positive photoresist stripping applications [8–10]. Even though the TMAH-based strippers are now commercially available and are being used in the industry, they have shown some drawbacks and do not provide a robust resist stripping process. A major disadvantage of TMAH-based strippers is the product's high pH value that makes it corrosive to sensitive metals, especially Al and Al alloys.

Aluminum reacts with alkaline solution in aqueous media, which results in etching of the aluminum lines, especially in the submicron geometry.

$$2Al + 2OH^- + 2H_2O \rightarrow 2AlO_2^- + 3H_2(g)$$

Another disadvantage of TMAH strippers is the instability of the quaternary ammonium hydroxide at elevated temperatures. An independent laboratory study of a commercially available TMAH/NMP base stripper using proton NMR spectroscopy and gas chromatography/mass spectrometry (GC/MS) has indicated the following: At high temperatures, TMAH disassociates to trimethylamine and methyl alcohol in aqueous media [10].

$$N(CH_3)_4OH^- \rightarrow N(CH_3)_3(g) + CH_3OH(g) \text{ at } t > 80°C$$

The bath life of a typical TMAH/NMP stripper is approximately 2–4 h at 85°C.

At ambient temperatures, however, the TMAH base developers are widely used in the industry for positive photolithography process.

Ozone–Water Mixtures

Another oxidizing agent that has historically been used for waste water treatment, drinking water sterilization, and in swimming pools is ozone in water. In recent years, ozone has been introduced into the microelectronic industry in both wafer cleaning (FEOL) applications and in photoresist residue removal (BEOL) processes.

Ozone has basically the same role in oxidizing and cleaning organic residues as H_2O_2 has in the RCA clean. Ozone and H_2O_2 decompose virtually in the same way:

$$H_2O_2 \rightarrow O^- + H_2O$$

$$O_3 \rightarrow O^- + O_2$$

The biggest advantage of ozone over RCA clean is that ozone leaves no harmful decomposition residues or by-products. It is partially soluble in water, especially at lower temperatures. Generally, ozone is about 10 times more soluble in water than oxygen. The lower the water temperature, the higher the ozone solubility. The half-life of ozone in high purity DI water is about 20 min.

In recent years, ozone chemistry has been receiving considerable attention because it has a potential to be used in both FEOL and BEOL and eliminate or reduce the usage of organic solvents. In a study to compare the effectiveness of ozone chemistry and modified RCA clean, FSI International used its centrifugal spray processors to clean silicon wafers. The wafers then were examined for their metal ionic contamination and change in particles before and after cleaning process. Residual metals following the cleaning processes were measured using both secondary ion mass spectroscopy (SIMS) and total reflection x-ray fluorescence (TXRF) methods. Results of those studies are presented in Tables 23.4 and 23.5 [4].

BEOL Cleaning Processes

Bulk material removing processes normally takes place following three major process steps in wafer fabrication:

1. Photolithography process (photoresist stripping and EBR)
2. Metal and oxide etched by plasma or reactive ion (polymer removing)
3. Chemical mechanical polishing, CMP (slurry removing process)

The major photoresist residues due to photolithography process are normally cleaned using a photoresist stripping process involving organic solvents.

The EBR process utilizes the following solvents:

1. PGMEA based
2. Ethyl lactate based
3. Organic solvents as ketones and esters (acetone, MEK, MIBK, and NBAc)
4. Environmentally preferred alternatives (VOC exempt)

After plasma-etch of metal and oxide, the tough sidewall polymers are cleaned by advanced formulations of post-etch polymer removers, or by ashing with oxygen followed by wet clean.

TABLE 23.4   SIMS Analysis of RCA Clean vs. Ozone Clean (Atoms/cm^2)

	Relative Elemental Abundance						
	F	Na	K	Cu	Mg	Al	Ca
Clean starting wafers							
Modified RCA clean	140	225	65	<20	22	1575	113
Ozone clean	200	105	25	<20	10	1064	98
Contaminated starting wafers							
Modified RCA clean	145	330	160	<20	N/A	N/A	N/A
Ozone clean	250	125	30	<20	N/A	N/A	N/A

TABLE 23.5   TXRF Analysis of RCA Clean vs. Ozone Clean (Atoms/cm^2)

	Mn	Fe	Zn	Br	Cr	Co	Cu
Clean starting wafers							
Modified RCA clean	<0.4	0.3	0.2	1.4	<0.6	<0.4	<0.2
Ozone clean	<0.4	<0.3	<0.2	<0.3	<0.6	<0.4	<0.2
Contaminated starting wafers							
Modified RCA clean	<0.4	<0.3	<0.2	1.8	<0.6	<0.4	<0.2
Ozone clean	0.5	<0.3	<0.2	<0.3	<0.6	<0.4	<0.2

After CMP, the slurry particles are rinsed away with water-based, diluted solvents. In many instances, scrubbing with brush and/or megasonic agitation is needed to dislodge the much heavier slurry particles.

Chemistry of Positive Photoresist Strippers

NMP-Based Strippers

Among other aprotic solvents, *N*-methyl pyrrolidone (NMP) have attracted particular attention in photoresist stripping formulations. In some applications, pure NMP is used for stripping soft-baked resists. For relatively hard and cross-linked resists, a more aggressive alkaline mixture with high pH values is needed to mix with NMP. Organic amines have shown to possess the desired characteristics when mixed with the aprotic solvents.

During the past two decades, the NMP/amine-based strippers have dominated the positive resist stripping market due to their low toxicity and resist cleaning efficiency. NMP/amine solutions are not only effective in cleaning hard-baked cross-linked resists at elevated temperature, they are also 100% water soluble and biodegradable, which makes these formulations particularly popular than the more toxic phenolic or chlorinated solvents.

For more advanced applications, different wetting agents or non-ionic surfactants such as polyalkylene glycol (ethylene glycol or propylene glycol) are added to the stripper formulations [11–15].

Non-NMP-Based Organic Strippers

Other aprotic solvents such as tetra hydrothiophrene 1,1-dioxide (Sulfolane), dimethyl sulfoxide (DMSO), dimethyl formamide (DMF), and dimethyl acetamide (DMAC) are also used in different stripping formulations [11,12].

Several other stripping solutions using dibasic esters, alcohols, ketones, glycol ether, or other organic solvents have also been reported and are being used. Aqueous-based (water/surfactant) strippers using dibasic esters as active ingredient with neutral pH values have also shown favorable results in stripping soft-baked and hard-baked bulk resists [13,14].

Chemistry of Negative Photoresist Strippers

Negative photoresists are polymerized rubber and are normally soluble in aromatic hydrocarbons or phenolic solvents. The most commonly used solvent/thinner for negative photoresist is xylene mixture. In the early 1960s, the first organic stripper containing chlorinated aromatic hydrocarbon as solvents and an alkylbenzene sulfonic acid as surfactant was introduced to the market by a company called Industri Chem. This formulation utilized phenol to create a water-rinseable solution, as such the first organic resist stripper J-100 was born.

Stripping photoresist by J-100 requires a heating bath in the range of 90°C–120°C and followed by a series of post-strip rinse solutions and a DI water rinse and spin dry. Since the introduction of J-100, a number of suppliers have developed similar products, some designed for direct water rinse. Other manufacturers have offered similar products containing dodecylbenzene sulfonic acid, phenol, and chlorinated benzene solvents [15–17]. These products were successfully being used as workhorse strippers for two decades. However, in the 1970s, the environmental concerns over the toxic ingredients in these formulas, led to the development of non-phenolic and non-chlorinated solvents with less waste disposal difficulties.

Chemistry of Post–Plasma Etch Polymer Removers

Tough sidewall polymers are created during the plasma etch process. The cross-linked and hardened polymers need to be cleaned by advanced chemical formulations. These advanced formulations are normally referred to as post-plasma etch polymer removers. In general, after plasma etch, the wafers are

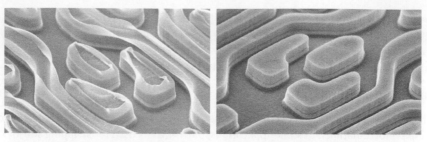

FIGURE 23.11   SEM images of sidewall polymers on metal lines, before and after clean.

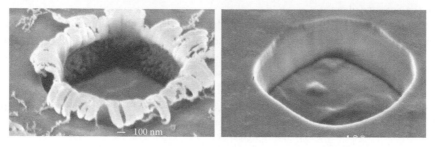

FIGURE 23.12   SEM images of via polymers, before and after clean.

normally ashed in oxygen plasma ashers, in which majority of the photoresist material is oxidized and removed from the surface of the wafer. However, the organometallic polymer formed on the sidewall of the metal layer (Figure 23.11) or inside the via polymer openings (Figure 23.12) does not react and remains. If the polymer is not removed properly, the residue will cause failure in connections and subsequently the device will fail the electrical test.

A number of post-etch polymer removers have been introduced to the market by chemical manufacturers and are being used since the early 1990s. EKC Technologies first introduced hydroxyl amine (HA) chemistry for post-etch polymer removal applications. Following that, Ashland Chemical Co. and a number of other chemical manufacturers followed the lead and offered various different formulations. Among many products available commercially, the following basic formulations have been used more frequently:

Hydroxyl amine chemistry
HF/glycol chemistry
Other organic alternative solvents

HA/Amine Chemistry

Hydroxylamine, NH_2OH free base, commercially available as 50% by weight in water, is a strong reducing agent and a weak base with pH about 8.0–8.5. Commercially used for polymer removal application, EKC-265 manufactured by EKC Technology and ACT-935 manufactured by Ashland Chemical are hydroxyl amine and organic amine (diglycolamine or monoethanolamine) mixtures. Although these products are widely used in the industry for post-plasma cleaning applications, the products have a number of disadvantages and drawbacks. Hydroxylamine/amine mixtures are not good resist strippers and do not strip the photoresist if it is not ashed. The products have short shelf life and are not stable at elevated temperatures. In order to increase their efficiency and improve corrosiveness, the manufacturers add chelating agents such as *catechol* up to 5 wt%. Despite problems associated with HA chemistries, most DRAM manufacturers have been using HA/amine-type products for metal and via polymer cleaning applications. The metal stacks on those wafers are typically Al-Cu/Ti/TiN and the dielectric layers

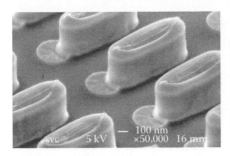
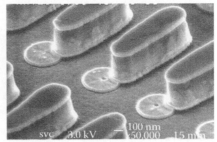

FIGURE 23.13 Exposed tungsten (W) via with sidewall polymer on Al metal, before and after clean.

on vias are silicon dioxide. HA/amine products with some process modifications can be adapted to those cleaning processes with little process difficulties [18–21].

In new technologies of copper metallization and with increasing applications of different low-k dielectric material, the usage of HA chemistry has become very limited and, therefore, more advanced formulations have become necessary for those processes. HA/amine mixtures are corrosive to copper and dissolve copper layers to form a water-soluble complex of Cu $(NR_3)_4$. Copper metallization and cleaning will be covered in Chapter 24. Other sensitive metal alloys are also being used as tungsten (W) in via interconnects as tungsten plugs, which are also susceptible for amine corrosion. Figure 23.13 presents a via structure with exposed W openings, before and after cleaning process with a non-corrosive polymer remover.

HF/Glycol Chemistry

Other formulations used in cleaning post-plasma polymers are hydrofluoric acid (HF) and ethylene glycol (EG) mixtures. Low concentrations of HF in EG are not corrosive to sensitive alloys and have strong residual cleaning capability and can be used as alternative to HA chemistry. However, HF attacks silicon oxide layer on the wafer and hence is not suitable for via cleanings. For metal cleaning with controlled temperature and process latitude, HF/EG mixtures can be used successfully with limited oxide loss. Ashland Chemicals Co. manufactures different products of HF/EG mixtures for various cleaning or oxide etching applications. Another similar product called NOE (natured oxide etchant), which is buffered HF with ammonium bifluoride in a polyglycol mixture, is commercially available from ACSI (Advanced Chemical Systems International), another specialty chemical company.

Challenges of Future Technology

The real challenge for the future of wafer cleaning technology is the integration of copper interconnects and new low-k dielectric material in wafer manufacturing. As we enter the third millennium, major advancements and dramatic changes are taking place in manufacturing and applications of electronic products. Computer manufacturers are motivated to come up with more powerful systems having more complex power transistors in smaller sizes.

Copper Interconnects

In chip manufacturing, a dramatic shift from aluminum (Al) to copper (Cu) interconnects is taking place. IBM Corporation was the first to produce products with 100% copper wires with substantial shipment of Power PC 750 microprocessor starting late 1998. Other major memory and DRAM manufacturers are in various stages of making this transition from Al to Cu processes. Semiconductor equipment manufacturers such as Applied Materials and Novellus and R&D institutions such as SEMATECH, a

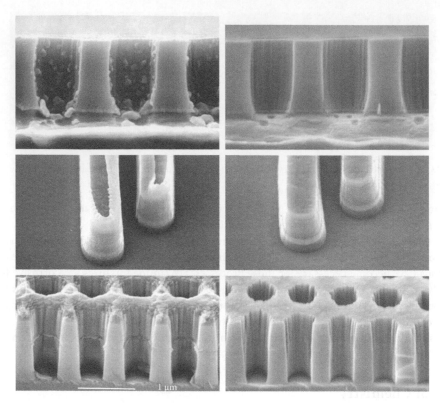

FIGURE 23.14   SEM images of sidewall polymers on via and metal lines, before and after polymer removal process. (Courtesy of Silicon Valley Chemlabs, SVC, Inc.)

consortium of major semiconductor manufacturers in Austin, Texas, are following the lead to implement copper as a replacement for aluminum.

Chip performance is the motivation for this transition since copper has significantly better conductivity (60%–70%) than aluminum. There is also continuous improvement in transistor performance of about 20%–30% per technology generation. As performance improves, the geometry of interconnects (including wires) is shrinking about 30% per generation, providing a constant cost per circuit reduction, and higher density patterning. All of these improvements force the wiring together of transistors with thinner metal lines (wires) which, because of their reduced size, are less conductive. The transition to Al to Cu helps to solve this problem.

The integration of copper technology requires completely different line of equipment and material, including cleaning chemicals. Low-k dielectric material is a major part of this emerging technology transition. Chemical mechanical planarization (CMP) slurries and polishing material and post-CMP cleaners are also part of this transition.

Low-k Dielectric Material

In recent years, several different low-k dielectric materials have been developed and introduced to the market by the chemical manufacturers. These spin-on films include HOSP, FLARE, HSQ, FOX, and SiLK. Allied Signal has offered HOSP and FLARE; Dow, Dow Corning, JSR offer HSQ and MSQ, and SiLK films, respectively. These products, having lower dielectric constant k compared to silicon dioxide, replace the traditional spin-on glass or other thermally grown oxide materials.

These, together with other newly invented CVD carbon-doped oxide low-k material, present challenges to the cleaning chemicals. The traditional solvent and amine chemistry may interact with these materials and change their low-k property. They could have been left behind on the wafers, causing contamination and via poisoning. Fortunately, new chemistry continues to be developed to offer engineers solutions for emerging technologies. Copper metallization and low-k films will be covered in Chapter 24.

References

1. P. Van Zant, *Microchip Fabrication, A Practical Guide to Semiconductor Processing*, 2nd edn., McGraw-Hill, Inc., New York, 1990.
2. A. E. Braun, Photoresist stripping, *Semicond. Int.*, Oct. 1999.
3. W. Kern, J. Ruzyllo, and R. E. Novak, Semiconductor cleaning technology/1989, In: *Proceedings*, Vol. 90–99, The Electrochemical Society, Pennington, NJ, pp. 5–15, 67–78.
4. W. Kern, J. Ruzyllo, and R. E. Novak, Cleaning technology in semiconductor device manufacturing, In: *Proceedings*, Vol. 92–12, The Electrochemical Society, Pennington, NJ, pp. 11–27, 1990.
5. H. Muraoka, H. Hiratsuka, and T. Usami, Abstracts 238. p. 570, *The Electrochemical Society Extended Abstracts*, Vols. 81–82, Denver, CO, October 1981.
6. A. Harri and R. H. Hockett, *Semicond. Int.*, 8, 74 (1989).
7. G. Gould and E. Irene, *J. Electrochem. Soc.*, 135, 1535 (1988).
8. N. Haq et al., U.S. Patent # 4,744,834.
9. H. Steppan et al., U.S. Patent # 4,776,892.
10. Silicon Valley Chemlabs, Instability of TMAH bases strippers. Technical Note, Silicon Valley Chemlabs, Sunnyvale, CA.
11. I. Ward et al., U.S. Patent # 5,554,312.
12. J. Sizensky et al., U.S. patent 4,617,251.
13. J. Sahbari et al., U.S. Patent # 5,741,368.
14. J. Sahbari et al., U.S. Patent # 5,909,744.
15. G. Schwartzkopf et al., U.S. Patent # 5,308,745.
16. W. Corbey et al., U.S. Patent # 3,673,099.
17. E. Thomas et al., U.S. Patent # 4,791043.
18. W. Lee et al., U.S. Patent # 4,824,763.
19. W. Lee et al., U.S. Patent # 5,279,771.
20. W. Lee et al., U.S. Patent # 5,902,780.
21. W. Lee et al., U.S. Patent # 5,911,835.

24

Advanced Cleaning Processes for Electronic Device Fabrication (Copper Interconnect and Particle Cleaning)

Chemical Mechanical Planarization 379
CMP Primer • Post-CMP Clean
Back End of Line Cleaning...382
Copper Interconnects • BEOL Cleaning for Copper Low-k
Wafer Backside and Bevel Cleaning..385
Edge Exclusion • Chemical Cleaning
Particle Cleaning: An Introduction...387
Cleaning Stationary Particles..388
Cleaning Methods for Removing Strongly Bonded Stationary
Particles • Weakly Bonded Particles (Ionic Bonds or Hydrogen Bonds)
Cleaning Mobile Particles..391
Simple Blow-Off Using an Inert Gas • Laser-Induced
Shockwaves • Supercritical CO$_2$ Process • Cleaning by Electrolyzed Water:
Comparison with SC1 and SC2
Cleaning and Rinsing...395
Conclusion 396
References...396

Shawn Sahbari
*Applied Chemical
Laboratories Inc.*

Mahmood Toofan
*Semiconductor
Analytical Services*

Chemical Mechanical Planarization

CMP Primer

Large-scale integration (LSI) transitioned to very-large-scale integration (VLSI) and more recent technology nodes moved to ultra-large-scale integration (ULSI). As a result, over 3 million individual transistors are fabricated into a modern microprocessor roughly the size of a postage stamp. The various dielectric and metal films that make up the individual layers of the device need to be planarized (smoothed) through chemical mechanical planarization or polishing (CMP). To achieve miniaturization requirements of this magnitude, it is mandatory that thin film planarization steps be integrated into the process flow.

Chemical mechanical planarization requires specialty slurries, polishing pads, and cleaning chemicals matched to the particular required level of polish. Typically, silicon oxide and tungsten slurries are employed for aluminum interconnect applications. In 1997, IBM announced the first commercial integrated circuit with multiple copper layers. Since that announcement, TI, Motorola, AMD, Intel, and a host of other industry leaders have transitioned from aluminum to copper metallization. Copper interconnects provide higher device speed and power; but they were previously not considered due to electromigration problems.

Because copper is a soft alloy, the only form of patterning that has been commercialized is a process called dual damascene. Essentially, in a damascene process, the oxide or dielectric film is patterned into trenches that are eventually filled via an electroplating process that forms the wiring and copper circuitry. Dual damascene will be discussed in the next section. In this process, there is no etching of metal films to pattern and thus the need for CMP to remove excess copper becomes mandatory.

As the damascene technique replaces subtractive aluminum technology, the need for CMP and particularly CMP of bulk copper post electrochemical deposition continues to rise. In addition to the complexities of the various thin films required to integrate copper metallization, and the required CMP consumables, the area of post-CMP clean continues to challenge engineering groups toward improving and optimizing the process.

Post-CMP Clean

Companies that employed CMP as a technique for aluminum device fabrication achieved the desired cleaning through dilute inorganic chemistries such as ammonium hydroxide (NH_4OH). However, because copper tends to corrode readily, these traditional chemicals cannot be used in the damascene process.

Contaminants from the polishing process must be removed during the post-CMP cleaning process. Cleaning is accomplished by any of a number of methods including immersion, megasonic, or the traditional double-sided scrubber (DSS) with PVA (poly vinyl alcohol) sponge contact brushes as in Figure 24.1. These contaminants include particles from the slurry, particles from the material being polished, chemical contamination from the slurry, and cross-contamination from the inlaid metal.

Copper itself is a contaminant that diffuses quickly into silicon and silicon dioxide. Therefore, it must be removed from all wafer surfaces (front, back, and bevel edge) in order to prevent adverse effects on device performance or reliability. Acceptable surface copper levels vary but are usually between 1E10 and 1E11 atoms/cm^2.

During the copper CMP process, the copper layer is oxidized and forms copper oxides. Specific oxides include CuO, Cu_2O, and copper hydroxides ($Cu(OH)_2$), depending upon the slurry pH, the electrochemical potential, and the additives. In a basic or neutral pH cleaning environment, these copper oxides and hydroxides do not dissolve and may be easily transferred to the PVA brushes. If the brushes

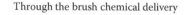

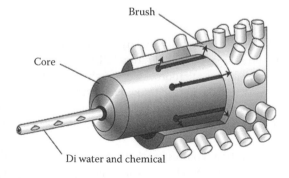

FIGURE 24.1   Contact brush for removal of contaminants from the polishing process.

become contaminated, and the pores loaded by the copper oxides, copper contaminants may be transferred onto subsequently processed wafers. This brush-loading effect would then cause severe copper cross-contamination.

Brush loading may also result from the alumina abrasive particles used in the polishing slurry, if the proper cleaning chemistry is not used. For example, in a neutral or inorganic acid cleaning environment, the electrostatic attraction between the alumina particle and the silicon dioxide surface makes it difficult to remove the abrasive particle. Because of this electrostatic attractive force, the alumina particles may also adhere to the PVA brush. The outcome of brush loading, whether it be from copper oxides or alumina particles, is the same, cross-contamination to downstream wafers. Therefore, the goal for a successful PCMP clean solution is to prevent both types of brush loading, while removing the slurry from the surface of the wafer and preserving the quality of the copper layer (i.e., inhibiting copper corrosion).[1]

Particle reattachment and brush loading are also problems when operating at an acidic pH. Below pH 8, alumina particles have a positive charge (zeta potential) while the oxide and brush surfaces are negative. The attraction causes dislodged particles to reattach to the wafer or brush surface.

It has been discovered that organic acids alter the surface of alumina particles, and consequently, their zeta potential. Without altering the alumina particles, reattachment would occur below pH 8. The altered alumina particles however, have negative zeta potential values above pH 4. This allows for the use of acidic cleaning solutions because particle reattachment does not occur in the acidic pH range.

It is possible to reduce metal contamination depending on the pH and chelating ability of the organic acid. Etching is the only way to remove metal contamination buried within the oxide film from the CMP process. Consequently, fluoride is added to the solution if optimal metal contamination levels are required. Time of flight scanning ion mass spectroscopy (TOF-SIMS) and total x-ray fluorescence (TXRF) are accepted measurement techniques in the evaluation of Cu-contamination levels on wafer surfaces.

Single wafer megasonic tools have also been reported in literature as viable PCMP cleaning equipment. Table 24.1 summarizes the surface metal levels PCMP clean of copper wafers between megasonic and brush cleaning systems (Figure 24.2).

Chemical manufacturers, capital equipment manufacturers, and device fabrication companies continue to invest in improving post CMP clean processes. In the decade following the introduction of

TABLE 24.1 Metal Contamination Levels Determined by TXRF (10^{10} Atoms/cm²)

	K	Ca	Fe	Zn	Cu
Megasonic	<DL	1.0	0.5	<DL	<DL
Scrubber	0.5	1.0	0.5	<DL	<DL

Note: DL, detection limit.

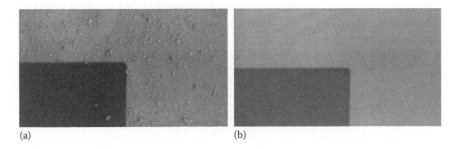

(a)   (b)

FIGURE 24.2 (a) Pre- and (b) post-CMP cleans of copper wafer showing slurry particles and other surface contamination. (From Krulic, G., Post CMP clean technology update, *ISMI Wafer Cleaning Workshop*, May 2002, Austin, TX.)

copper as an interconnect metal, more patents were filed in the area of CMP and CMP clean than in any other area of device fabrication, and performance-based industry cleaning has been in the forefront of this development. However, there are numerous other stakeholders in this arena. With CMP consuming millions of gallons of water per day, environmental groups are striving for improvements to help with conservation efforts. Copper and waste water treatment is costly and management is always looking for cost improvements over existing methods. Thin film engineers are continuously struggling to keep up with the deluge of new chemicals, slurries, polishing pads, and other critical consumables used in the CMP process. Adopting new materials requires that films withstand the chemical and mechanical burdens of the process. Surface roughness, refractive index, and other optical properties, thin film adhesion, film hardness, dielectric constant, and a host of other requirements will continue to provide opportunities in this exciting area of device cleaning.

Back End of Line Cleaning

Copper Interconnects

The rapid pace of technology development makes it challenging to identify an inflection point in device development. However, after decades of miniaturization and advances in integrated circuit design, the single most disruptive transition can be classified as the move to copper from aluminum metallization.

Advanced fabrication involves sub–65 nm (65×10^{-9} m) patterning of feature sizes. The International Technology Roadmap for Semicondcutors (ITRS) outlines the various device parameter targets. Table 24.2 illustrates that with every node there are multiple integration considerations:

1. Feature sizes are rapidly shrinking toward the single-digit nanometer scale or are approaching the atomic level.
2. The interlayer dielectric films have to continue to provide a lower dielectric constant compared to the previous generation.
3. The number of metal layers on logic BiCMOS devices is exceeding 10 layers. (*Note:* BiCMOS is an integrated circuit that integrates CMOS technology with bipolar junction transistors).

In addition to the above mentioned challenges, the simultaneous integration of electrolytic copper has exponentially increased the difficulty in wafer level cleaning.

In order to be able to build a structure with eight or more metal layers, it is imperative that each layer be perfectly flat and planarized before subsequent films are deposited. To achieve this, a process called chemical mechanical planarization (CMP) is employed. Copper CMP and post-CMP cleaning challenges will be discussed under the CMP section.

BEOL Cleaning for Copper Low-k

In standard aluminum processes, the insulating dielectric layer is etched to create via holes that make the contact between the top and bottom metal layers. Aluminum films are also etched separately to form the wiring making up the circuitry.

TABLE 24.2   Device Parameter Targets

Production Year	2002	2004	2007
Technology Node MPU ½ Pictch (nm)	130	90	65
Interlevel metal insulator effective dielectric constant (κ)	3.0–3.6	2.6–3.1	2.3–2.7
Number of metal levels	8	9	10

In copper interconnect, there is no metal etch process. Circuitry is formed by utilizing a dielectric etch process where the via holes and wiring trenches are created simultaneously through the dual damascene process. A traditional via-first damascene process incorporates a via etch in the low-k dielectric stack as illustrated in Figure 24.3.

The subsequent trench is etched into the stack to create the channel where the copper electroplating will form the wiring for the circuitry as indicated in Figure 24.4.

As can be deduced, there are numerous films and there are a variety of mechanical, optical, chemical, and electrical issues that must be addressed with each material.

Bulk low-k films can be spin-on or chemical vapor deposition (CVD) films. Successful integration in design schemes employing dielectric constants between 2.4 and 3.7 with copper metallization have been published in literature. Carbon doped oxides (CDO) deposited in a dry process have fewer integration challenges compared with spin-on low-k films.

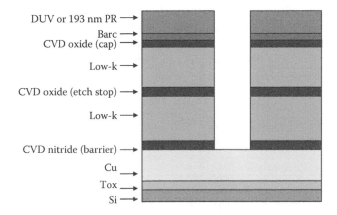

FIGURE 24.3 Traditional via-first damascene process.

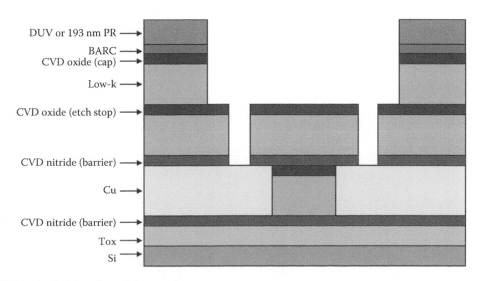

FIGURE 24.4 Etching of trench into stack.

A dual damascene process sequence introduces complex cleaning requirements in such steps as photoresist strip, bottom antireflective coating (BARC) removal, and post-damascene etch polymer removal.

Some of these requirements result in challenges that need to be addressed. One is the need to achieve residue-free cleaning steps. A second is to implement chemicals and processes that are compatible with copper and with the variety of films incorporated into the dielectric stack. There must be no degradation of the costly low-k value. With 65 nm technology nodes, there is very little critical dimension (CD) tolerance, and thus cleaning processes cannot damage or change the CD within the vias or trenches.

The shear number of films in the material stack creates uncommon post-etch cleaning requirements. The insulating layers employ new etch chemistries designed to remove multiple layers of dielectric, barrier, etch stop, and cap layers while maintaining vertical, smooth side walls necessary for a subsequent void-free copper fill. During the etch process, the sidewalls are passivated with a layer of fluorocarbon polymer induced by $C_xH_yF_z$ gases to protect the low-k and to reduce edge roughness. These modified plasma processes designed specifically for damascene process flows will cross-link the surface of the photoresist to create a "crust" that is unlike the "bulk" resist underneath. The organic and inorganic materials that need to be completely removed after etch are illustrated in Figure 24.5.

Highly carbonized, cross-linked crust on the surface of the resist is impervious to standard resist strip chemicals. Standard solvents and cleaners may remove the bulk photoresist; however, they are not effective on the crust and the etch residues are basically left intact. Figure 24.6 is a pre-clean SEM showing the post via etch condition of the film with the stack configuration.

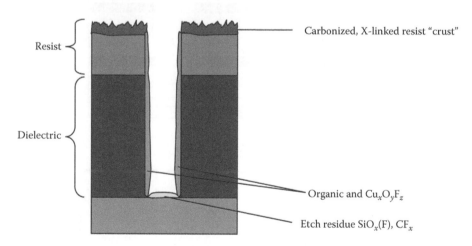

FIGURE 24.5 Organic and inorganic residue that must be removed after etch.

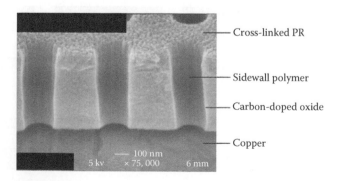

FIGURE 24.6 Residue, post-via-etch.

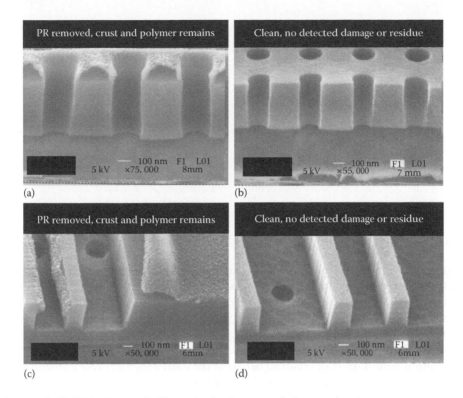

FIGURE 24.7 (a–d) SEM micrographs illustrating inadequate and adequate cleaning.

In Figure 24.7a through d, SEM micrographs illustrate inadequate and adequate cleaning of these small features.

Depending on the tool sets and material make up, copper-damascene technology will create opportunities for specialty chemical manufacturers to provide application-specific solutions for copper device fabrication.

Traditional solvents and cleaners are ineffective in this arena and have reached their maturity. The successful chemistries are proprietary formulations that have been developed over years of research and applications testing. The formulations are closely guarded secrets by manufacturers and device fabricators alike.[3]

Wafer Backside and Bevel Cleaning

With the commercialization of copper technology, device fabricators have found that copper contamination is a major problem. One of the contributors to copper contamination and cross-contamination across tools is copper formed at the periphery and bevel areas on the wafer as well as the wafer back side. These regions were previously not highly critical in total wafer cleaning.

Edge Exclusion

Figure 24.8 illustrates the edge of the silicon wafer where the barrier layer and seed copper layers converge. Ta, or TaN barrier layers cover the entire front surface of the wafer with a copper seed layer through physical vapor deposition (PVD). A shadow mask is employed to create an approximate 2 mm edge exclusion. If not properly addressed, the resulting copper islands at the transition area can flake and cause downstream contamination.

FIGURE 24.8   Convergence of barrier layer and seed copper layer at edge of wafer.

The flakes can create copper contamination across tool sets and modules, can cause scratches and other physical damage during the CMP process, and can cause particulate defects at any time during the process. As a result, a backside, bevel, and edge copper clean is necessary.

The SEM image in Figure 24.9 illustrates a clean copper edge exclusion zone followed by the adjacent seed copper.

Figure 24.10 provides the sequence of copper processing in the absence of a bevel and edge clean process. The resulting residual seed and residual barrier at the wafer edge will cause a downstream defect and process problems that will reduce yield and increase cycle times.

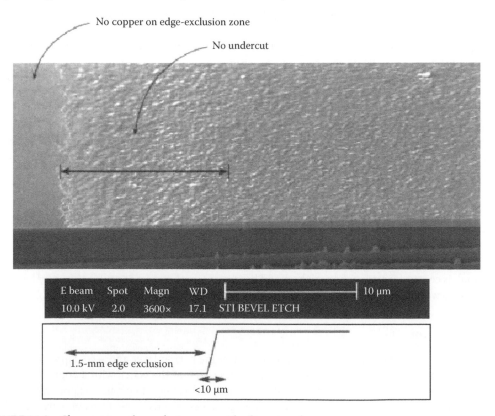

FIGURE 24.9   Clean copper edge exclusion zone with adjacent seed copper.

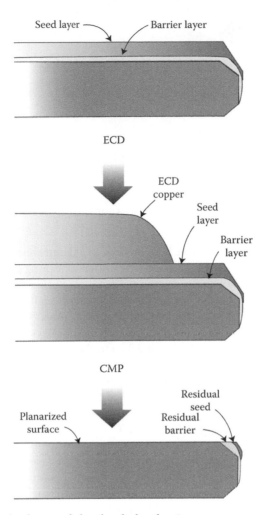

FIGURE 24.10 Residues in the absence of a bevel and edge cleaning process.

Chemical Cleaning

Total x-ray fluorescence (TXRF) measurement of control, ECD copper, and various wafers cleaned with different chemistries illustrate that dilute hydrogen peroxide/HF, and sulfuric acid-hydrogen peroxide mixtures are able to clean the bevel, edge, and backside of the wafer to achieve acceptable levels of copper. The choice of chemistry depends on the barrier films and other integration and process considerations.

The adoption of dilute chemistry common to the fab provides an easy-to-implement solution for backside and wafer bevel cleaning. These commodity materials are available, well understood, and currently employed in all major fabs, and the handling and disposal does not pose any difficulty for process or equipment engineers.[4]

Particle Cleaning: An Introduction

As chip manufacturing process technologies continue to advance at a rapid pace, the critical cleaning of particles at the surface of electronic devices is ever more necessary. The functionality and performance, lower power consumption, higher yield, and ultimately lower cost translating into greater market share are some of the drivers impelling the control of defects due to particles. Minute

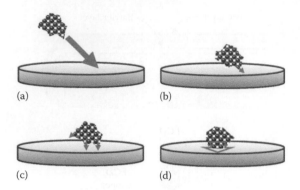

FIGURE 24.11   Presentation of particle deformation on the surface to minimize its potential energy.

contaminating particles, which could have been ignored in older process technologies, pose a significant challenge for manufacturers.[5-7] Unfortunately, traditional cleaning procedures may not be able to remove particles from extremely small geometries.[8] Cleaning solutions, used in manufacturing steps of these chips, must be free of contamination. As an example, a small amount of metal ions, such as iron in a cleaning solution, can be adsorbed on the surface of a wafer causing problems for following steps in lithography.[9]

In cleaning processes, most of particles are attached to the surface by the van der Waals force. The van der Waals force is a summation over all atoms of dispersive interactions between instantaneous dipoles and those induced in neighboring atoms. Therefore, the sensitivity of the van der Waals potential depends on the interparticle distance. The attached particles to the substrate surface via van der Waals forces may create more contact between attached particles to the substrate surface by deformation, to minimize their potential energy (Figure 24.11).

One of the most challenging issues in electronic device fabrication is the selection of an appropriate cleaning technique that not only influences the current state of the cleaning process but also affects the outcome of the following fabrication steps. Usually, the major contaminants on the surface are either particles, or traces of organic or inorganic compounds. In case of particles, they could be organic or inorganic, or a combination of both. They could also be charged or neutral particles, and could either be stationary or mobile.

The removal of particles from the surface depends on the nature of the surface and on the physical and chemical characteristic of the adherent particles. It is not uncommon for the surface to be hydrophobic or hydrophilic. In a hydrophobic surface, since there are no attraction forces between the surface and the charged particles, particle removal is rather easy. This statement can also be applied to hydrophilic surfaces. Due to non-physical or chemical attraction forces between uncharged particles and the surface, uncharged particles could also be easily removed from hydrophilic surfaces. The cleaning process thus becomes challenging when both particles and surfaces have attractive forces between each other. In such situations, the selection of a cleaning technique becomes very critical.

Cleaning Stationary Particles

Stationary particles are those particles that are somehow attached (bonded) to the substrate surface. These particles are divided into two different groups:

Strongly bonded particles (covalent bonds)
Weakly bonded particles (ionic or H-bonds)

The cleaning procedures for the strongly bonded particles are different from those for the weakly bonded particles.

Cleaning Methods for Removing Strongly Bonded Stationary Particles

Strongly bonded particles are more difficult to remove from a surface. The common methods for removing these particles may involve the use of an appropriate chemical cleaning solution. Particles with covalent bonding can be divided into two important groups. One group includes particles whose molecular structures or physical natures are different from the molecular structure of the surface (S1). The other group contains particles with molecular structures or physical natures that are similar to the surface molecular structure (S2).

To clean S1 particles from the surface, one should apply an appropriate chemical solution that can dissolve the particles without harming the surface, such as an organic solvent or inorganic solution to dissolve these organic or inorganic particles respectfully. Therefore, advance knowledge of the chemical structure of the particles and of the molecular structure of the substrate is very helpful. As an example, in Figure 24.12, one can remove a residue of organic–inorganic silicon polymer after a long period of exposure of optical waveguides to a dry etch process, by simply dipping the contaminated surface in a dilute HF solution for less than 20 s.

In instances where there are no suitable solutions for removing these types of particles by the dissolution process, a cleaning solution that can etch the surface may be employed where the undercutting process removes these particles from the surface. In the undercut process, the solution affects the substrate surface rather than the particles. As a result, in this wet etch process, the clean solution gradually undercuts the particles and removes them from the substrate surface (Figure 24.13).

For cleaning the (S2) particles, due to the existence of similar molecular and chemical structures between both the substrate surface and the bonded particles, it is difficult to remove the particles by chemical procedures because the solution will have the same effects on both the bonded particles and

(a)   (b)

(c)

FIGURE 24.12 Effects of cleaning solution on the optical waveguide: (a) formation of polymeric contamination on the waveguide before cleaning, (b) effects of cleaning solution on the contamination after the cleaning process, and (c) magnification of a trace of hard polymer on the waveguide.

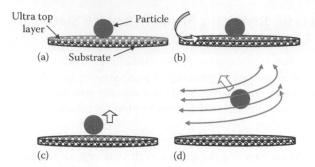

FIGURE 24.13 (a) Particle contamination on the substrate surface, (b) effect of etch clean solution on the top layer of the substrate surface, (c) particle lift off by electro-repulsive forces, and (d) movement of particle far from the substrate surface by diffusion or convection.

on the substrate surface. In this situation, it is possible to employ a physical contact technique such as the buffing or the mechanical polishing method. Usually, such particles can be detected on the substrate before subsequent processes during standard metrology and defect analysis.

Weakly Bonded Particles (Ionic Bonds or Hydrogen Bonds)

Electrostatic or ionic bonds, as well as hydrogen bonding (H-bonding), are weaker bonds, as compared with covalent bonds. The best methods to remove these types of particles are to have a wet clean process using a solution with an appropriate pH value. Cleaning these surfaces by exposing them to high concentrations of hydronium ions (H_3O^+) or hydroxide ions (OH^-) is one of the most successful cleaning methods currently used. When excess ions, such as hydroxide or hydronium ions reach the substrate surface, a layer of hydroxide or hydronium ions (depending on the chosen pH) will immediately cover the substrate and particle surfaces. The result of this effect creates a similar charge on both particle surfaces as well as the substrate surface, and immediately a repulsion force will form between the substrate and particles.

As illustrated in Figure 24.14, the strong electro-repulsion force between the particles and the surface will repel the particles away from the surface. The particles can then be removed by an appropriate rinsing method.

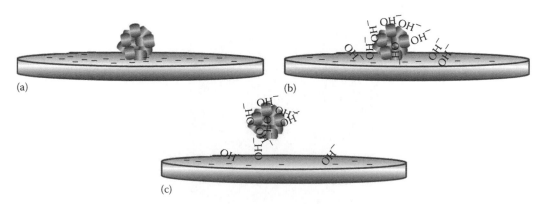

FIGURE 24.14 Illustration of the separation of the particle from the substrate via the mechanism of electro-repulsion forces (immediate formation of similar charges on the surface of both substrate and particles). (a) Particle on the substrate before exposure to the cleaning solution. (b) Instantaneous formation of similar changes on the surface of both substrate and particle after the surface has been exposed to the clean solution. (c) Separation of particle from the surface due to electro-repulsion forces.

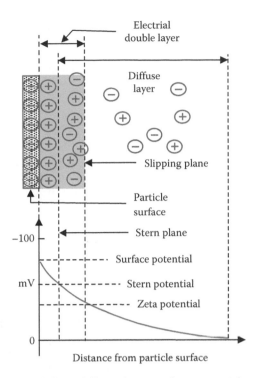

FIGURE 24.15 Presentation of double layer, diffusion layers, and zeta potential.

In general, it has been found that a solution with higher pH has better cleaning efficiency on silicon and silica when compared to acidic solutions.

This method is also called the zeta potential cleaning process. By definition, the zeta potential is an electrical potential that exists across the interface of solid surface and solution (Figure 24.15). When a cleaning solution is in contact with the substrate surface, it forms a double layer in the interface of the solid and liquid surface. The double layer is made up of ions of a one-type charge that are fixed to the surface of the solid (positive or negative), and an equal mobile number of ions on the liquid surface but with an opposite charge (compared to the solid surface charge), accumulating on the liquid surface. This layer is termed the electrochemical double layer. So, by movement of ions in liquid or any change in the nature of the cleaning solutions, displacement of all mobile ions in the double layer can be changed.

One reason for the low cleaning efficiency is due to the sudden dilution of the cleaning solution at the beginning of the rinsing period by DI-water, when most of the particles are still floating near the substrate's surface. Wherever the concentration of the effective ions, such as hydroxide or hydronium ions, is significantly decreased, the surface of the substrate becomes more susceptible to contamination. To solve this problem, new cleaning solutions are formulated in such a way as to protect the substrate surface against contamination during the first stage of the rinsing process. Figures 24.16 and 24.17 show the cleaning results of these solutions on various substrates as illustrated with a commercially available cleaning agent (SurfPurge Series), from Semiconductor Analytical Services, Inc. (www.sas-page.com).

Cleaning Mobile Particles

There is also another particle type that is neutral called the mobile particle. Usually, these particles are not attracted to the surface, and they stay on the surface due to the force of gravity or by carrying a very

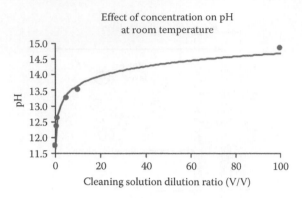

FIGURE 24.16   Achieving desirable high pH rinse with a commercial cleaning solution during DI water rinsing.

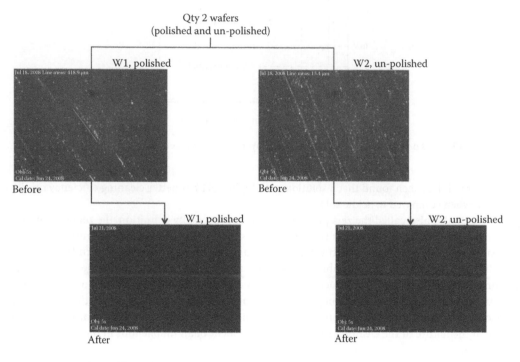

FIGURE 24.17   Effects of cleaning solution on polished and unpolished glass wafers before and after the cleaning process.

weakly induced electrostatic charge (by the surface). The most common cleaning methods for removing these particles include

  Simple blow-off using an inert gas
  Laser-induced shockwaves
  Supercritical CO_2 cleaning
  Electrolyzed water techniques

Simple Blow-Off Using an Inert Gas

If used in a clean environment, a particle-free inert gas, such as argon gas, is effective in removing uncharged particles. In this cleaning process, the size of the particle plays an important role. Usually, smaller particles (submicron particles) are harder to remove from the surface when compared to larger particles. Cleaning submicron particles from a surface with particle-free inert gas is not very efficient due to the lack of sufficient force between the inert gas atoms and the particles of small dimensions. The advantage of argon gas over other inert gases is its high density, its high atomic weight, and its commercial availability. Because argon gas has a higher density than air, once it is applied to a surface it quickly covers the substrate's surface, preventing any particle of a lower density than argon gas to reach the surface.[10]

Another way of removing uncharged particles from a substrate is to use an appropriate cleaning solution. Any cleaning solution that can be managed safely and that is easily rinseable can be used for this purpose. In wet cleaning processes involving submicron particles, it is important to choose a cleaning solution with a high specific gravity compared to the density of water to allow particles to become suspended in solution, and subsequently, particles moved away from the surface via diffusion (Figure 24.18).

Laser-Induced Shockwaves

Laser-induced shockwaves provides a dry cleaning method that can be used for removing nonbonded particles from the substrate surface. In this method, the laser-induced shockwave can remove free particles without any direct interaction between the laser beam and the surface.[11,12] Particles can be removed from the surface if the force of the shockwave is larger than the adhesion forces between the particle and the substrate Figure 24.19.

In this situation, particles will begin to detach from the surface. The velocity of the shockwave depends on the input power of the laser beam and the gap distance, where the required power of the applied laser beam depends on the nature of the particles. For an effective removal of metal particles, more energy or a shorter wavelength is needed. In this method, the percent removal of particles from the surface depends on the distance between the focused point of the laser and the surface of the substrate.

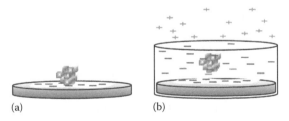

(a)   (b)

FIGURE 24.18 Effects of cleaning chemistry with a specific gravity (higher) on the cleaning process before (a) and after (b).

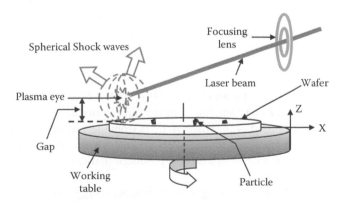

FIGURE 24.19 A schematic diagram of laser-induced shockwaves. The particles are removed if the shockwave force is greater than the adhesive forces between the particles and the substrate.

Certainly, a lower gap distance provides better cleaning.[13] Unfortunately, a low gap distance can cause damage to dense electronic devices. In this technique, one can use a nitrogen or argon gas atmosphere (preferably argon gas) to avoid any undesirable surface oxidation during the laser shockwave. The common wavelength in this method is approximately 1064 nm with the laser energy density at a focus point of about 10^{11} W/cm^2.

Supercritical CO$_2$ Process

Supercritical carbon dioxide (SC-CO$_2$) technology has been recognized as a viable cleaning method.[14] Technologists and researchers are interested in this method due to its high efficiency in removing trace organic contamination. In this technique, all cleaning is done in supercritical fluids at temperatures in the range of 32°C–49°C and pressures in the range of 1070–3500 psi. One advantage of SC-CO$_2$ is that the amount of waste generated by the cleaning process is minimum. Supercritical fluids by definition are fluids with a temperature and pressure equal or greater than the critical temperature and pressure. SC-CO$_2$ is able to spread out over the surface of parts much faster than any liquid due its lower surface tension. SC-CO$_2$ has the great capability of dissolving oil and most organic contaminations from the surface, even when organic contaminants are trapped in very small areas. Additional details about this technology are available in Poliakoff's review.[15]

Cleaning by Electrolyzed Water: Comparison with SC1 and SC2

Cleaning by electrolyzed water is another cleaning procedure that has recently been introduced into the semiconductor industry for removing contamination from the surface of photomasks. The electrolyzed water system works via a mechanism similar to that of suppressors in an ion chromatography system.[16,17] In this technique, deionized water (DI water or DIW) is passed through two separated chambers in the presence of two ion exchange membranes and two electrodes. Water is ionized to hydroxide ions (OH$^-$) and hydronium ions (H$_3$O$^+$) by the following reaction:

$$4H_2O + 4e^- \rightarrow 4OH^- + 2H_2 \text{ (gas)}$$

$$6H_2O \rightarrow 4H_3O^+ + O_2 \text{ (gas)} + 4e^-$$

$$\overline{\phantom{6H_2O \rightarrow 4H_3O^+ + O_2 \text{ (gas)} + 4e^-}}$$

$$2H_2O \rightarrow 2H_2 \text{(gas)} + O_2 \text{(gas)}$$

This electrolyzed water unit (Figure 24.20) generates hydrogen gas from the cathode chamber and oxygen from the anode chamber plus ultra pure water in both chambers. By following the above equation, it is clear that the water from the hydrogen-generated chamber has a higher pH compared with the oxygen-generated chamber due to the formation of hydroxide ions, and the water from the oxygen-generated chamber has a lower pH compared with the hydrogen-generated chamber due to the formation of hydronium ions.

The acidity or pH of the anodic chamber can be controlled simply by the amount of current that is passed through the electrodes and by the amount of additive (NH$_4$OH or HCl) in the electrolyzed water. It has been demonstrated that if a small amount of ammonium hydroxide is added to the cathode water, the new electrolyte can remove most of the silica particles from the surface of the silicon wafers after a CMP process.[18,19] On the other hand, by adding a small amount of hydrochloric acid to the anode chamber or the hydronium chamber, a new electrolyte is produced that can remove all copper and sulfate residue from a silicon surface.[20]

Therefore, the electrolyzed water cleaning process has been suggested as an alternative replacement process for the traditional cleaning processes—SC1 (Standard Cleaning 1) and SC2 (Standard Cleaning 2). The chemical composition of SC1 and SC2 and their applications are shown in Table 24.3.

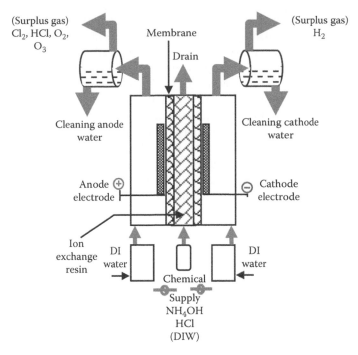

FIGURE 24.20 Diagram of an electrolyzed water system.

TABLE 24.3 SC1 and SC2 Applications and Formulation

Object to Be Cleaned	Chemical Mixture Ratio	Popular Name
Particles	APM ($NH_4OH/H_2O_2/H_2O$) (mixing ratio 1:1:5)	RCA (SC1)
Metallic	HPM ($HCl/H_2O_2/H_2O$) (mixing ratio 1:1:5)	RCA (SC2)

Although both SC1 and electrolyzed water processes are equally effective, there is some risk in using SC1 in place of electrolyzed water on a photomask as SC1 could cause some soft defects after DUV lithography.[21] These defects are formed from chemical residue on the reticle through photochemical reaction with eximer light. Therefore, this problem becomes more acute when the lithography process approaches shorter wavelengths such as 193 or 157 nm.

Cleaning and Rinsing

Today, most cleaning procedures rely on the wet techniques. Factors favoring wet processes include lower cost of operation, batch processing capability, and global availability of high purity chemicals.

There are two major cleaning processes used in a wet process. One is the immersion or the dipping process, and the other is the spray process. In general, an immersion-cleaning operation offers a lower operational cost compared with a spray process. In the immersion-cleaning process, a cleaning tank with a certain amount of cleaning chemical is used for cleaning parts. In this process, all contaminations from the parts are left behind in the cleaning solution and remain in contact with the parts. Any highly contaminated part can contaminate all the other parts in this process or subsequent parts yet to be cleaned.

In the spray process, the parts can always be in contact with the fresh cleaning solution and the contamination will wash away from surface without the possibility of cross-contamination or redeposition on other areas. Therefore, the spray process offers a major advantage over the immersion-cleaning processes. Megasonic devices usually carry out the agitation in the immersion-cleaning process. Sonication can cause mechanical damage to high-density features in sensitive devices. Spray processes can offer adequate agitation without the compromise of significant damaging to the sensitive features if the spray pressure is controlled in advance. Rinsing the parts with DI water in the spray process is more efficient when compared to immersion water rinsing. In the immersion process one needs a high volume of DI water to rinse the parts while in the spray process, the volume of rinsing DI water is much less.

While there are pros and cons with spray and immersion rinse systems, spray techniques have been used widely for semiconductor cleaning applications for decades.

Rinse considerations include temperature, electrostatic discharge (ESD), corrosion control, and other factors. In both procedures, spray and immersion, it is preferable to choose the temperature of the DI water close to that used for the cleaning process in order to eliminate thermal shock that causes micro cracks on the devices due to different fabricated layers with different thermal expansion coefficients. Some electronic components, such as disk drive heads, are extremely prone to ESD and thus require proper grounding to prevent damage and yield loss. In recent years, tool manufacturers have introduced carbon dioxide (CO_2) sparge into the rinse cycle. This is to slightly acidify the deionized water (DIW) so as to offset the alkaline cleaning agents. This helps to minimize water-induced corrosion.

Conclusion

As it is clear, contamination plays a major role affecting the fabrication yields of electronic devices. Therefore, elimination of contamination from every stage of fabrication is a serious necessity and a task. During the cleaning procedure, paying attention to the nature of the contamination, the utilized cleaning tools, and the assumed cleaning procedures are quite important. Understanding possible molecular structures of particles or trace contamination as well as their bonding forces toward a substrate surface will help to choose the best cleaning procedure, tools, and cleaning chemicals. The selection of a cleaning methodology such as immersion versus the spraying process or megasonic agitation versus ultrasonic agitation, or recirculation agitation versus inert gas bubbling agitation requires detailed analysis and decision making. The selection of an efficient cleaning procedure such as concentration of the cleaning solution, the pH of the cleaning solution, and the temperature of the cleaning process all are important contributing factors to higher yields. In addition, another factor in the cleaning process is the selection of the rinsing procedure, the choice of the correct temperature, and the rinsing time. An incomplete removal of all the cleaning solution from the surface of the device during a DI water rinse process can potentially create defects and device failure downstream.

References

1. S. Sahbari, Post CMP clean technical conference CMP user's group AVS December 2002, Santa Clara, CA.
2. G. Krulic, Post CMP clean technology update, *ISMI Wafer Cleaning Workshop*, May 2002, Austin, TX.
3. S. Sahbari, *Cleaning Chemistries for the New Millennium*, European Semiconductor, April 2002, Angel Business Publications, London, U.K.
4. C. R. Simpson and T. Ritzdorf, Building Copperoppolis II, *Curtis Dundas Micro Magazine*, October 2000.
5. M. Toofan, A. L. Baker, and H. K. Yun, Thermal optical switch apparatus and methods with enhanced thermal isolation, U.S. Patent 689, 5157, May 17, 2005.
6. W. Xiaolin, M. Toofan, and D. Yi, Double sides optoelectronic integrated circuit, U.S. Patent application 200,402, 45538, December 9, 2004.
7. M. Skinner, B. Chaudhuri, C. Yoo, M. Toofan, G. Chang, and A. Bhowmik, Waveguides with optical monitoring, U.S. Patent 6,934,455, April 21, 2005.

8. W. Kern and D. A. Puotinen, Cleaning solutions based on hydrogen peroxide for use in silicon semi-conductor technology, RCA Laboratories, Princeton, and RCA Solid-State Division, Somerville, NJ, *RCA Review* 31, 187, 1970.

9. J. Atsumi, S. Ohtsuka, S. Munehira, K. Kajiyama, Semiconductor cleaning technology, 1989, in: Ruzyllo, J., Novak, R. E., Eds., *The Electrochemical Society Proceeding Series*, 1990, The Electrochemical Society, Pennington, NJ, PV90-9, p. 59.

10. M. Toofan and F. O. Eschback, Pellicle-reticle methods of reduction haze or wrinkle formation, U.S. patent 10919548, October 16, 2004.

11. L. M. Lee, K. G. Watkins, and W. M. Steen, Angular laser cleaning for effective removal of particles from Si surface, *Applied Physics A* 71(6), 671, 2000.

12. A. A. Busnaina, J. G. Park, J. M. Lee, and S. Y. You, Laser shock cleaning micro and nano-scale particles, in: *Proceedings of ASMC*, 41, Munich, Germany, 2003.

13. J. M. Lee and K. G. Watkins, Removal of small particles on silicon wafer by laser-induced airborne plasma shock waves, *Journal of Applied Physics* 89(11), 6496, 2001.

14. Y. Kikuchi, S. Fukuda, K. Masuda, and N. Kawakami, in: *Proceedings of SPIE's Microlitography Symposium*, 101, 2002, p. 4688.

15. S. Poliakoff, Introduction to supercritical fluids, 2001. Retrieved from the World Wide Web at: www.nottingham.ac.uk/supercritical/scintro.html

16. J. R. Stillian, V. M. Barreto, K. A. Friedman, S. B. Rabin, and M. Toofan, U.S. Patent 5,248, 426, September 28, 1993.

17. J. R. Stillian, V. M. Barreto, K. A. Friedman, S. B. Rabin, and M. Toofan, U.S. Patent 5,352 360, October 4, 1994.

18. K. Yamanaka, T. Imaoka, T. Futatsuki, S. Yamasaki, and H. Aoki, Cleaning technologies using electrolytic ionized water and analysis technology of fine structures for next generation device manufacturing, in: *Proceedings of Semiconductor Pure Water and Chemicals Conference*, Santa Clara, CA, February 1995, pp. 1–22.

19. H. Aoki, T. Nakajima, K. Kikuta, and Y. Hayashi, *Proceedings of Symposium on VLSI Technology*, Tokyo, Japan, May 1994, pp. 79–80.

20. H. Aoki, S. Yamasaki, Y. Shiramizu, N. Aoto, T. Imaoka, T. Futatsuki, Y. Yamashita, and K. Yamanaka, in: *Proceedings of International Conference on Solid State Devices and Materials*, Osaka, Japan, August 1995, pp. 252–254.

21. B. J. Grenon, C. R. Peters, K. Bhattacharyya, and W. Volk, Formation and detection of sub-pellicle defects by exposure to DUV system illumination, in: *Proceedings of 19th Annual Symposium on Photomask Technology*, SPIE 3873, 1999, pp. 162–176.

25

The Cleaning of Paintings

Richard C. Wolbers
University of Delaware

Chris Stavroudis
Stavroudis Art Conservation

Aqueous Cleaning 401
Cleaning Paintings 404
Surface Cleaning • Varnish Removal • Overpaint or Retouch Removal

The cleaning of paintings is a specialized task performed by painting conservators. Conservators are trained in art history, studio art, and chemistry. While each conservator may have different strengths and weaknesses in those three broad areas of study, all three are brought to the surface of a painting before a treatment begins.

The most powerful tool used by the conservator when cleaning a painting is her or his eyes. Because no two paintings are alike, each cleaning requires a delicacy of hand, a sharpness of mind, acuity of eye, and years of practice. Even though in broad classes, there are only a limited number of ways to create a painting (pigments in drying oil, pigments in wax, pigments in egg yolk, pigments in acrylic dispersion) on a limited number of substrates (fabric, wood, metal, plaster), the nuances of each are manifold. Add to that the variables of light exposure, changes in humidity and temperature, and the ravages of previous treatments, and it is not surprising that no two treatments are the same.

In the general context of conservation, "cleaning" a painting is a broad and all-encompassing term that refers to the removal of any non-original material from the original paint surface created by the artist as well as the removal of a discolored layer of varnish either applied by the artist or one applied subsequent to the painting's completion. Cleaning is most often undertaken to bring the appearance of the painting to a state visually appropriate and comparable to its appearance and effect when first created. This is often not fully possible due to irreversible changes to the paint itself and the nature of the interaction between soils, varnishes, and overpaints with the original paint surface.

One fundamental problem in cleaning a painting is that the material to be removed is often very similar in material and chemistry to the materials of the artwork itself. "Cleaning" is often finding a means of dissolving or suspending one material without affecting the other by exploiting very subtle differences in chemistry, age, or physical arrangement within the materials.

Another fundamental problem is what might be referred to as a kind of "selective" restraint required by the conservator in cleaning painted surfaces. As with other forms of cleaning, a central concern is efficacy of course, but here conservation cleaning is idiosyncratic and varies from other critical cleaning situations. Soil on a surface may be considered "patina" or a valued aspect of the aging process and may appeal to one's aesthetic sensibilities. The complete removal of soiling materials, for instance, may also not be a desirable goal in an art-historical and artifactual sense. Soils, as naturally accruing materials, can signal or demarcate "age" on a painting's surface and can therefore carry other types of information as well. The "use" of an object for instance—the variation in accumulation and removal (e.g., in wear patterns that may exist in the soiling material) can be valued as artifactual evidence of its history, and may help place the painting or painted object in a particular context or set of conditions, and may increase its value historically and economically. Equally problematic sometimes is the complete removal of varnish materials that have degraded (oxidized, discolored) with age. "Age" can be signaled just as effectively by the yellowed nature of a surface varnish. If it is completely removed, the brightness, freshness of the original paint—however carefully uncovered—may appear inappropriate visually and aesthetically in some cases. A central desire in terms of cleaning varnishes that have degraded substantially, therefore, is the ability to reduce the varnish layer but not necessarily remove it completely.

Additional restraints in designing cleaning systems for painted surfaces are what might be termed "ethical" concerns. Consistent with the American Institute of Conservation's (AIC) guidelines for standards of practice, processes, and materials are mandated to be "reversible." In this context, the tenant of "reversibility" simply means that cleaning, as a conservation or restoration process, must leave essentially no residuals behind on a cleaned fine art surface, and further, the process of cleaning itself should not damage or alter irreversibly the paint surface or the normal aging and fundamental appearance of the painting.

As paintings, and artwork in general, are composite objects often built from many types of organic materials, the method of application of a cleaning system is restricted to techniques that will not cause unwanted side effects. While immersion, sprays, and working at elevated temperatures are common in industrial precision cleaning, these processes would damage the most common substrates of a painting. While numerous substrates have been used historically and currently, the most common painting substrates are fiber-based textiles and wooden panels, both of which react unfavorably to copious amounts of liquid.

Because each painted surface is different, each surface and often many different areas must be tested to evaluate the efficacy of a cleaning system. These tests are discretely done in as small an area as can be managed, often with magnification or under a stereobinocular microscope.

The accumulation of contaminants is built up in a layering process, and the conservator often tries to work down through each layer, or even a portion of a layer, in developing the cleaning regimen. Ideally, the conservator will have the option to remove each layer and decide if the cleaning is sufficient or should proceed further. It is during this decision-making process that the conservator considers the historical, aesthetic, and ethical impact of that cleaning in both positive and negative terms. Often, these subtleties are discussed with other conservators, curators, or art historians and/or conservation scientists before the decision is made to proceed with a cleaning.

The options for applying and removing cleaning systems are fairly restrictive. The surface must be visible during the cleaning process so that the removal of surface grime, varnish, or overpaint can be monitored. Because a painting's surface is never homogenous and each color may have very different solubilities, the cleaning must be carefully watched for both efficacy of the cleaning process and, more importantly, monitored for damage to the painting's surface.

The methods of bringing a cleaning formulation to the surface of a painting are quite primitive, but in skilled hands, they can be exquisitely effective. Most commonly, cotton swabs are used to apply, remove, and, when working with systems that contain nonvolatile components, clear the cleaning system from the paint surface. The swab allows different degrees of mechanical contact with the surface as it can be rubbed with different amounts of pressure or, more commonly, rolled with very little applied pressure, controlling the physical interaction with the surface. The cotton itself offers many benefits: (1) it is white so the material being removed can be observed easily, particularly as the swab is being rolled, (2) cotton has a very large surface area and a negative zeta potential that can physically attract and hold dirt and swollen, but not dissolved, polymers, (3) the degree of wetness of the swab can be controlled so as to apply a large amount of material on the surface or a relatively small amount that is largely pulled back into the swab itself as pressure is reduced, and, (4) finally, as we normally roll our own swabs as we work, we can control the size and thickness of the cotton at the end of the swab.

Other application methods are often used to control the application and removal of cleaning systems. These include the use of the artist's brushes of various sizes, as well as specialty sponges, makeup applicators, and various synthetic swabs.

Aqueous Cleaning

There has been a renewed interest in the past few decades in the development of more appropriate and more "engineered" aqueous cleaning methods. This largely stems from a variety of concerns (less toxicity, higher specificity for certain soils, more control over processes, greater safety for the paint materials, etc.). Very few proprietary products have been manufactured or directed to this area of application however. (In fact, very few commercial materials or tools are made specifically for conservators as we constitute a very small market.) Generally, when aqueous cleaning materials are used, they are prepared on an as needed, case-by-case basis, in small batches. Shelf life is generally not an issue, nor are storage and transportation. These solutions are generally consumed "in studio" so that marketing, product aesthetics (to some extent), labeling, and other normal consumer type issues are not usually concerns with these materials.

Because these are generally prepared and used "by hand," the handling, formulation, and health and safety issues *are* important concerns in preparing aqueous cleaning materials from stocks of acids, bases, buffers, surfactants, etc. Disposal of waterborne materials and wastes is usually via domestic means in small, limited amounts, or as solid disposal after the water has evaporated from the cotton swabs.

What have emerged as critical factors in designing aqueous cleaning systems for fine art surfaces are

- Water purity (distilled, deionized, etc.)
- pH (acid, bases, buffers, etc.)
- Ionic species (specific ions, ionic strength)
- Surfactants (type and amount)
- Chelating agents (type and amount)
- Other "adjuvant" materials (enzymes, "redox" materials, etc.)
- Viscosity (polymer type and amount)

Water quality should be of the highest possible (e.g., pure water obtained through distillation, deionization, or reverse osmosis). Conductivities as low as reasonably achievable (5–10 μS) are usually associated with the low-cost lab or studio scale purification systems that have found their way into the profession thus far, and this is suitable for formulating these materials.

Setting and controlling the cleaning solution's pH is of crucial importance. When working with oil-based paints, extremes of pH need to be avoided. Oil films tend to exhibit pronounced swelling at pHs close to the pK_as of the free fatty acids present in the films (palmitic, stearic acids pK_as= 4.5) setting the safe limit for working at a low pH. As the pH is increased, saponification of the triglyceride materials that make up the film-forming components of oil paint becomes a danger in the pH 8.5–9.5 range. Consequently, buffers with pK_as within the range of about 5.5–8.5 are used to stabilize the pH of aqueous cleaning preparations within this "safe" range. Both inorganic (carbonate (pK_1), phosphate (pK_2)), and organic (tris, triethanolamine, borate (pK_1), citrate (pK_3), morpholino ethane sulfonic acid, and acetate) buffers have found use in these preparations. Generally, the amounts of buffer are commensurate with concentrations that yield ionic strengths below 10–20 times the isotonic point of the paint films. In practice, this amounts to buffer levels in the range of 25–50 mM. The presence of a sufficient buffer will prevent the precipitation of deprotonated oxidation products as more and more acidic moieties are dissolved into the cleaning solution. In contrast, an unbuffered system's pH will decrease as acidic fractions are dissolved into the cleaning solution, risking the precipitation of acid groups with higher pK_as.

The control of ionic strength in aqueous cleaning preparations is also important. A paint film will contain ionic species (pigments, binders, and other paint film and preparatory constituents) that will contribute to the overall ionic strength of these surfaces when wetted. Isotonicity, when the ionic strength of the cleaning solution equals the ionic strength at the surface of the paint film, is usually in the range of 50–300 μS for normal oil-bound artist's paint. This number may vary quite a bit because of variation in the preparation of the artist's paints and because of the intrinsic variability (and sometimes poor quality) of the various pigments and additives used. This number also changes the type of paint. Acrylic dispersion paints, for instance, introduced into the artist's repertoires by the late 1950s tend to exhibit much higher isotonic points (50–500 µS), largely due to the incorporation of a variety of ionic materials, e.g., surfactants, thickeners, and other ionic species fundamental to those paint film types. Hypertonicity is generally considered to be higher than about 10–20 times the isotonic condition of these surfaces, and is generally avoided because of the substantial risk of swelling and softening the artist's paints, with the possible concomitant trapping of cleaning solution components within the paint film. The absorption of cleaning materials in these softened films complicates the clearing of the cleaning solution's components from the paint surface.

Surfactants incorporated into cleaning systems have generally been chosen with relatively low HLBs (12–20). Often, these have been nonionic structures like Triton X-100™, Triton XL-80N™, Synperonic N™ (in Europe), Brij™ 35 and 700, Tween™ 20, 30, etc., because of their ease in formulation. As nonionics, they are more easily compatible with other common ionic materials like buffers, chelators, etc. The cloud points of these nonionic surfactants tend to be well above room temperature to avoid phase separation and adsorption onto oil paint surfaces. These nonionics are also generally used in dilute ionic strength preparations and, generally, in the presence of monovalent ions like Na^+, K^+, NH_4^+, and triethanolate ion to avoid any substantial cloud point depression. Higher HLB surfactants have also found use (Sodium Lauryl Sulfate [SLS], Vulpex™ notably), but these tend to be rather strong solubilizers of hydrophobic paint film constituents, including the oil binding materials themselves, and also are the strongest w/o type emulsion formers. (Prolonged contact with SLS will "blanch" or create a cloudy cast to the paint by emulsifying the water phase into the paint.) A point that must be kept in mind is that the artist's oil paints cannot be considered as *hard surface* materials in the traditional sense as they exhibit pronounced swelling and dissolution effects at elevated ionic strengths and pHs, with strong or high HLB surfactants, or with water-miscible solvent additions to the aqueous preparation.

In addition to considering surfactants based on their HLB values, and whether or not they are anionic, cationic, or nonionic, conservators try to use surfactants with low critical micelle concentrations (cmc). A low cmc allows less surfactant to be used in a cleaning system and yet still possess useful detergency. A low aggregation number also helps keep the surfactant concentration low. Less surfactant in the solution means that there is less that must be rinsed from the paint surface.

The specific chemical nature of the hydrophobic moiety of the surfactant is also exploited in the formulation of aqueous cleaning systems for paintings conservation. The surfactant Maypon® 4C (Inolex Chemical Company) is a potassium salt of a coco fatty acid and polypeptides derived from hydrolyzed collagen. As a surfactant derived from proteinacious precursors, it has an affinity for animal-based glues. If grime is held on the surface of a painting in part by animal glue residue (perhaps from a glue lining in the past), including Maypon in the cleaning solution will make a marked difference in the efficacy of the cleaning. Other affinity surfactants include deoxycholic acid (bile acid) and abietic and alueritic acids (natural resin acids) that collectively have been referred to in the conservation field as "resin soaps." These acids, when deprotonated, work as surfactants and have been adopted for their ability to interact with various natural resin substrates. The hydrophobic portion of the surfactant molecule possesses similar solubility parameters to those of the natural resins traditionally used as varnishes on painted surfaces. The deoxycholate ion, as a salt of the bile acid, tends to bind specifically free fatty acids but also shows an affinity for the natural resins—dammar and mastic. Free fatty acids or their metal salts often comprise the weathering materials at oil paint surfaces. Soaps based on the two resin acids have been used specifically to solubilize aged di- and tri-terpenoid materials (dammar and mastic in the case of the abietate ion; and shellac, the aleuratae ion) largely based on the stereospecificity of these resins.

Chelating materials have found widespread use in aqueous preparations for cleaning fine art surfaces. They function in these preparations both as aids in dissociating relatively low pK_{sp} salts present on these surfaces in the form of soils, or deterioration products from paints themselves (pigments), and as anti-biologicals. By far the most universally used chelator is citric acid. But EDTA, HEDTA, DPTA, and NTA have all found use, and their adoption simply follows from the specific formation constants required and the solubility products of the materials to be solubilized. As surface grime ages on the paint surface, its removal with surfactants alone diminishes. At the same time, chelating agents added to the cleaning system show much greater activity toward the aged dirt layer. This is presumed to be caused by the ubiquitous divalent metal ions Ca^{2+} and Mg^{2+} essentially cementing the grime to the surface of the paint or varnish layer. The presence of a chelating agent sequesters the ion allowing the grime to be removed.

Some limited use has been made of enzymes as aqueous adjunct materials to facilitate the breakdown and the solubilization of selected biopolymers associated with both art-making and restoration materials. To date, these have included aqueous cleaning preparations that have included enzymes to specifically hydrolyze or breakdown the primary ester linkages in drying oils (lipases), the glycosidic linkages in starches (α amylase, pullulanase), and the peptide bonds in proteinaceous materials (proteases) to facilitate their removal from painting supports or surfaces. Enzyme usage in this context is extremely hampered because of the lack of availability of substrate materials to the solution-bound enzymes. Because substrates are generally solid materials (adhesives, coatings), enzyme kinetics are slowed because of their relatively insoluble nature compared to the enzymes acting on them. Diffusion of enzymes and the concomitant solubility of enzymatic products also play a major role in the rates of these reactions. Consequently, enzyme usage has been at the protein solubility limit (10 mg/mL) in aqueous systems, and at the highest specific activities (>1000 μ/mg) possible, to be the least bit effective in these preparations.

Mild oxidants (or for that matter, reductants) are usually not used in these preparations because of fear of altering the paint film constituents and their aging, reactivity, etc. However recently both carbimide peroxide (5%) and the per-oxy compounds, ammonium and sodium persulfate, have been used to increase the water solubility of hydrocarbon materials accumulated on these surfaces. Their use generally must follow, of course, with compatibility with other aqueous preparation ingredients.

Anti-biologicals are normally not added to these solutions because of the small scale and rapid turnover of these types of preparations in normal studio use. But propyl and ethyl parabens (food grade) have been used as well as aromatic alcohols, such as phenoxyethanol and benzyl alcohol. Generally, anti-biologicals are chosen for their nonreactivity to paint film materials. Formaldehyde-generating materials like the hydantoins are avoided in these types of preparations, as are azides and halogenated materials, and other potentially reactive species.

Often aqueous preparations include viscosity-modifying materials to restrict the aqueous solutions from penetrating into paint layers by absorption and to manage surface application to prevent the spreading of solutions beyond the immediate application area or down into cracks. These polymers undoubtedly function additionally as suspension aids for soils when coupled to surfactant use. To date, polyacrylic acids (PAAs, Noveon/Lubrizol® Carbopols®); acrylate cross-polymers (Noveon/Lubrizol's Ultrez® 10, Pemulen® TR-1, TR-2); xanthan gums (Vanderbilt Chemical's Vanzan® NF-C); a few of the more common cellulosic ethers (methyl cellulose (MC), e.g., Dow A-4C; and hydroxy propyl methyl cellulose (HPMC)) all have been used in various preparations. The "ideal" viscosity range is about 4000–6000 cps, and all of the cited polymers can raise the viscosity of aqueous solutions to this point within the pH range of 5.5–8.5, which, as noted above, is taken as the most stable pH condition for cleaning oil films. Polymers that exhibit relatively low yield points (can be stirred into, wiped physically from surfaces) are also advantageous. Most of the polymers that have been adapted into conservation have relatively high flocculation temperatures, but as nearly all cleaning applications are accomplished at room temperature, this has rarely been a problem.

Perhaps, as important as formulating the cleaning solution, is the clearing of any nonvolatile components from the surface of the painting. As with any critical cleaning, the residue of the cleaning agent itself and the dispersed material being removed cannot be left behind, and the process of clearing the surface cannot be allowed to affect the surface any more than the cleaning system itself. Again, the control of pH and ionic strength is necessary, and, because immersion and spray are not options, clearing is by necessity by sequential dilution and removal, again most commonly with cotton swabs.

Cleaning Paintings

For clarity's sake "cleaning" painting surfaces will be divided into the three (still) broadly defined but more representative activities or endeavors that employ very different materials and methods: (1) surface cleaning or the removal of accumulated soils, (2) varnish or clear coating removal from a painted surface, and (3) overpaint or retouching removal—paint applied by restorers that cover or obscure original artist paints. A discussion follows on each area of concern as well as the current "state-of-the-art" materials and methods to help illustrate "best practices" at the moment.

Surface Cleaning

Removing soiling material adhered to the paint surface or "surface cleaning" has been practiced since antiquity. Historically, materials as widely ranging as bread, spoiled wine, a potato or garlic clove sliced in half and rubbed on the surface, simple soap preparations, and saliva have been used to reduce or remove accumulated soils on paint surfaces.

Saliva on cotton swabs is still often used by painting conservators and can be very effective. (This is normally called "spit cleaning" or by the less vulgar but less precise "cleaning with a mild enzymatic solution.") Various commercial household cleaners have also been used to assist in the removal of surface grime as have erasers and other mechanical methods.

With the advancement of the characterization of the nature of grime and its changes upon aging by researchers within and outside of the conservation profession, cleaning systems used by conservators are becoming more sophisticated. The conservator can now build cleaning systems with the aqueous components discussed above. As a result, surface cleaning has become a much more precise, controlled, and safer process. The conservator can vary the intrinsic properties of an aqueous cleaning system (the pH and ionic strength) and has the option of employing one of a number of cheating agents; surfactants of various types to exploit both their HLB characteristics and affinity-type interactions with the surface grime, and various thickeners.

While all conservators have some background in chemistry, most are stronger in art history, connoisseurship, and studio art. To assist the conservator with the chemistry of mixing the various components of an aqueous cleaning system, the Modular Cleaning Program (MCP) was developed to provide assistance in the formulation and understanding of the many parameters that can be varied when trying to determine the most effective cleaning system. The MCP is a database of physical properties of all the materials used to formulate cleaning systems in the conservation of paintings. The database is programmed to formulate cleaning systems using the physical properties of the materials and standard thermodynamic equations. In this way, the conservator can request that the computer calculate a new test cleaning solution based on changing one or more of the solution parameters (pH, ionic strength, presence or absence, and type of chelating agent, presence or absence of one of many surfactants, and the presence or absence of one of a number of thickening agents). The program then calculates mixing instructions that allows the conservator to formulate the solution that they have specified in the software.

A typical formulation for removing an aged layer of surface grime from a varnished painting is shown in Table 25.1.

This solution would typically be applied with a small white-bristle nylon artist's brush. The gel would be lightly agitated with the brush allowing the color change in the gel to be observed in contrast with the white brush bristles. When appropriate (after seconds to a minute or so), the gel would be wiped away with a dry cotton swab using a counter-rolling motion to pick the gel up onto the swab and move it away from the paint surface. Immediately, the area would be rolled with a rinse to clear the gel and

TABLE 25.1 Sample Aqueous Oil Painting Cleaning Preparation

Ingredient	Total (%)	Molarity (M)	Function
Part A			
Water	91		Diluent
Citric acid	0.5	0.025	Chelator
Triethanolamine	5.0	0.034	Base; buffer
Ammonium chloride	0.5	0.093	Ionic strength
Part B			
0.5 mL Triton XL 80N	0.5	0.0113	Surfactant
Part C			
Methocel A-4C	2.0		Thickener
Part D			
Tegosept	0.5		Preservative

Note: Typically, the ingredients of Part A are added together and the pH adjusted to 8.5; this mixture is then warmed on a stirring hot plate to 80°C and the Part D material is added, and dissolved. The mixture is then cooled back to room temperature, then the Part B material added, followed by the Part C material with constant stirring, until a smooth clear gel is obtained (approx. viscosity 4000 cps; conductivity 4000 μS). Additional water can be added to adjust the overall conductivity of the composition.

dissolution products. The rinse solution is always selected before the cleaning is begun. In this case, a very dilute mixture of acetic acid and ammonium hydroxide adjusted to a pH of 8.5 (matching that of the cleaning gel) and with a conductivity of somewhere between 50 and 500 μS would be used. The rinse solution would be rolled over the surface with cotton swabs, wetting the surface and removing residues from the gel by dilution and suction back into the swab. After the area is dry, a final rolling is often done with a mild organic solvent to further reduce residue.

Varnish Removal

The removal of discolored varnish layer(s) is also imprecisely referred to by the generic term "cleaning" in many cases (in the United Kingdom and the United States for instance), while in others, the terms varnish removal or reduction or "de-varnishing" has been the norm (France, Italy, etc.) to describe this type of activity. The stereotypical cleaning of a painting is the removal of a discolored varnish from the surface of an oil painting using organic solvents.

Historically, varnishes are natural resins dissolved in a solvent and applied to the surface of the painting some months after the painting has been finished. Initially, the varnish is readily soluble in mild solvents so that it can be safely applied by brush to the dried (but not completely cured) paint surface. The varnish is typically applied to saturate the paints and further the illusion of looking into a different scene or space. (The traditional frame serves as a window into the scene or person depicted.)

Over time, the varnish oxidizes and yellows due in large part to photo-induced reactions in the natural resins. In time, the varnish becomes so yellow (a deep orange-brown) that it obscures details, alters colors, and changes value and color relationships in the painting.

At some point, the decision may be made that the varnish should be removed and the painting revarnished with a colorless transparent material. This decision is not taken lightly as cleaning is one of the only processes undertaken in art conservation that is not reversible. Once removed, the varnish cannot be unremoved. Whether a painting is to be cleaned or not is often a collaborative decision made with the input of conservators, curators and art historians, owners, and conservation scientists.

Historically, the technology for removing discolored varnishes is tied to advances in chemistry. As the technology of distillation matured and, later, advances were made in organic chemistry, new solvents were incorporated into materials that could be used in the dissolution of aged varnishes. However, very oxidized varnish layers or varnishes mixed with drying oils would not respond to the solvents available, so often early restorers would resort to cleaning with highly alkaline mixtures that could saponify oils in the varnish and render acidic fractions in the resin soluble by deprotonating the free acids. Unfortunately, caustic solutions could damage the paint surface as well and aggressive cleanings of this sort often damaged the delicate modeling and glazing on the painting's surface.

As organic chemistry advanced in the last century, more solvents became available to aid the restorer in removing aged and discolored varnish layers. The modern conservator has quite a range of solvents and solvent mixtures to choose from to assist in removing or reducing a varnish. As the health hazards of working with solvents has been recognized, a number of solvents are no longer used in the conservation studio. (These would include benzene, methylene chloride, carbon tetrachloride, Cellosolve, Cellosolve acetate, methanol, and dimethyl formamide, to name but a few.) Considerable work is going into systems that allow mixtures of less harmful solvents to replace solvents that pose greater health and safety risks to the conservator.

A typical varnish reduction/removal might be performed with a mixture of a nonaromatic petroleum distillate and isopropyl alcohol. As the varnish ages and oxidizes, a more and more polar cleaning solution is required to dissolve the varnish layer. By trial and error, the conservator increases the proportion of alcohol until the varnish can be dissolved. This solvent mixture is then tested to make sure that the solvent mixture will not dissolve the original paint below the varnish layer. This is a gross oversimplification of the process. Often, an aged varnish layer can require very high polarity solvents to dissolve or swell it but those same solvents dramatically increase the risk of damaging the original paint surface. If a varnish cannot be removed safely, that is without disruption of the original paint surface as well as original varnish glazes that might have been applied by the artist, the varnish will not be removed.

One of the great advantages to using organic solvents is that after the solvents evaporate, there is no concern about residues or residual effects. While some solvents are retained in paint films for surprisingly long periods of time (months), eventually they leave the paint film entirely. However, as paint films swell with solvent during varnish removal, soluble fractions present in the paint film can leach out. This can leave the paint film thinner, more dense, and embrittled as the extracted lower molecular weight fractions serve as plasticizers to the paint film.

Many aged varnishes cannot be removed with mixtures as simple and relatively mild as mineral spirits and isopropanol. Often, the varnish will contain additives that complicate the removal. A drying oil added to the varnish will form cross-links within the resin matrix preventing the true dissolution of the varnish layer. The traditional natural resins (dammar and mastic) may have been mixed with harder resins that require solvents with much higher polarities (e.g., copal) to yield a physically tougher varnish layer.

One can imagine that a natural resin, as it ages and oxidizes, acquires carboxylic acid groups that are very polar. As long as these polar domains are few, increases in the polarity of the solvent can remove these fractions along with the less oxidized material that constitutes the bulk of the varnish layer. However, as the fraction of oxidized material increases, it may not be possible to find a solvent or solvent mixture that can safely remove the most oxidized fractions of the degraded varnish layer. However, solvents will often dissolve the less oxidized fraction of the varnish layer. Over time, as the cleaning cycle continues—a painting is cleaned, revarnished, and, after a number of years, again cleaned and revarnished—the most oxidized fractions of the varnish layer are left behind as the soluble varnish fractions are leached out. This process can leave a dark brown residue, most commonly in the valleys of brush strokes in the paint.

This leached, oxidized, and highly polar varnish fraction often seems intractable. Usually, a solvent strong enough to remove this residue from the low areas in the brush stroke will "skin" the higher parts of the stroke (i.e., remove the top surface of the paint layer). Even though this material cannot be safely

TABLE 25.2 Sample Aqueous Preparation for Removing Oxidized Natural Resin Varnish

Material	Total%	Molarity	Function
Water	92%		Diluent
Bicine (free acid)	0.8%	0.05 M	pH buffer
Citric acid	1.0%	0.05 M	Chelating agent
Sodium hydroxide (10%)	8.3%		Sets pH to 8.5
Brij 700	0.1% (10 × cmc)	0.0002 M	Nonionic surfactant
Sodium deoxycholate	2% (10 × cmc)	0.05 M	Affinity surfactant
Hydroxypropylmethylcellulose	2%		Thickening agent
Benzyl alcohol	as needed 0–5%		Co-solvent

removed by increasing the polarity of the cleaning solvents, it often can be quite successfully dissolved with an aqueous cleaning system.

By raising the pH of the water (but staying below a pH of 8.5–9 to avoid saponification of the paint's oil binder), the acid groups can be deprotonated and brought into solution. Chelating agents and surfactants can assist in this process. Often, an affinity surfactant (resin soap) can be used to bring the less polar resin fractions into solution. If the aqueous polarity needs to be reduced a bit more, a small amount of benzyl alcohol or *n*-butanol can be added to help coax the varnish residue into aqueous solution.

Table 25.2 shows a typical aqueous solution that can be used to remove a highly oxidized natural resin varnish. This can be diluted to reduce the solution's ionic strength.

As with the surface cleaning preparation above, this gel must be cleared from the surface to minimize any residue left on the painting. As above, a dilute pH 8.5 and ionic strength-adjusted solution of acetic acid and ammonium hydroxide would be appropriate as a clearance agent.

When a safe, suitable solvent or solvent mixture cannot be found, gelled solvents employing a lower polarity solvent mixture can sometimes be effective. A "weaker" gelled solvent mixture can be applied to a paint surface for a longer duration, commonly between 30 s and 5 min. The gel restricts the solvent from flooding the surface and may reduce the amount of leaching of the paint film below the varnish layer. Very often, a gelled solvent can work down through a layer of varnish slowly and with great precision.

Most commonly, solvents are gelled with Lubrizol's Carbopol line of polyacrylic acid resins. To make a solvent gel, the polyacrylic acid groups must be paired with an organic amine of a polarity appropriate for the solvents to be gelled. A small amount of water is needed to form the adduct between the amine and the carboxylic acid group.

There are three organic amines that are commonly used with Carbopol solvent gels. Ethomeen® C/25 (polyoxyethylene(15) cocoamine) is used for solvent solutions with a high polarity and can even gel water. Ethomeen C/12 (diethanol cocoamine) will gel moderately nonpolar solvents and Armeen® 2C (dicocoamine) is used to gel the lowest polarity solvents.

A short segment of the Carbopol backbone studded with the adducted Armeen 2C molecules is illustrated in Figure 25.1. The long hydrocarbon arms surround and shield the low polarity solvent from the relatively polar polyacrylic acid. Each amine–carboxylic acid union (it is not a chemical bond) requires the presence of two water molecules.

The formulation of a typical low polarity solvent gel is shown in Table 25.3.

Carbopol based solvent gels are more than the sum of their components. The presence of the polyacrylic acid probably confers some sequestering capacity to the gel, the amine probably confers some mild detergency, and the structure of the gel itself may offer different microdomains that broaden the effective solubility parameter of the gel. Whether these explain the difference between a Carbopol®-based solvent gel and the same solvent mixture suspended in an inert suspension agent or if there is another explanation, Carbopol®-based solvent gels can work on problems that cannot be solved with liquid solvents.

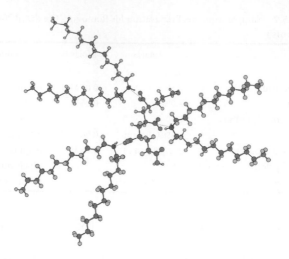

FIGURE 25.1   A short segment of a Carbopol®/Armeen® 2C solvent gel.

TABLE 25.3   Typical Formulation for a Low-Polarity Solvent Gel

Material	Total Amount	Function
Carbopol® e.g., 934	2 g	Gelling agent
Armeen® 2C	10 g	Amine
Xylenes	97 mL	Low polarity solvent
Benzyl alcohol	3 mL	Increases polarity
Distilled/deionized water	0.9 mL	Forms gel

To prepare the solvent gel, mix the solvents together (xylenes and benzyl alcohol) and set aside. Mix Carbopol® and Armeen, adding a little solvent mixture to form a paste. Mix the paste until it is smooth and uniform. This step should be done quickly as moisture from the air can react with the Carbopol®/amine mixture that will ruin the gel. When the paste is smooth, add the rest of the solvent mixture. While stirring, add the water and stir or shake the mixture until the gel forms. Note that the gel will take 3–5 min to form and transitions suddenly from a slightly thickened liquid to a heavy gel.

Solvent gels are applied to the paint surface with white-bristled nylon brushes. The gel is gently agitated and the dissolution of the varnish layer is observed in contrast with the white bristles. Finally, the solvent gel is wiped off of the surface with a slightly dampened or dry cotton swab, and then the area is rinsed with an appropriate solvent mixture. As with complex aqueous solutions, solvent gels contain nonvolatile components that must be thoroughly cleared from the painting's surface. Most commonly, a solvent or solvent blend with slightly lower polarity than the solvent gel is used to clear the gel components from the surface.

Overpaint or Retouch Removal

The removal of material added deliberately by a later hand, often in a previous restoration, is referred to by specific references to what is being removed, e.g., overpaint removal or the removal of overfills.

Again, the conservator must use extreme care in removing older restorations and retouching. First and foremost, overpaint should not be removed if it was applied by the artist themselves later in the history of the painting. In fact, there must be very clear evidence that an old retouch was not added by the artist before it would be removed. In addition, historic retouching is often preserved as part of the history of the painting; say a change in hairstyle of a sitter in a portrait to reflect changing tastes in fashion.

However, other types of overpaint added later to the painting are commonly removed in a cleaning. The most common overpaint was applied as part of a restoration campaign. While modern conservation strives to make all interventions reversible and we use retouch paints that will remain readily soluble as they age, previous restorers were often not as obliging. Old overpaints have often yellowed with age and no longer match the colors they matched when originally applied. Sometimes, overpaint was applied over a yellowed varnish and matched the discolored paint. When the varnish is removed in a cleaning, the old retouch will no longer match the cleaned surface. Previous restorers (and sometimes it seems like gross amateurs) will repaint an area much larger than the damage being compensated for. On occasion, one can find that an entire sky was repainted to conceal a small tear. In such cases, when safely possible, the overpaint is removed in the course of a conservation intervention.

An interesting special case of overpaint that is historic but is often removed by conservators (in consultation with curators or owners) are additions added to conceal material judged to be offensive to later eyes. The most common example of this sort of overpaint is the judiciously applied fig leaf to cover male and female genitalia.

The processes used to remove overpaint are often the same as those used to remove discolored varnishes. Although, often, much more polar solvents and solvent gels may be tested. However, aged oil paint on top of aged oil paint can be impossible to remove from the original with solvents. Sometimes, mechanical means (working carefully with a scalpel under a microscope) can be employed, and, sometimes, the overpaint is left behind and retouched to make it less noticeable.

It must be emphasized that cleaning a painting or other type of artwork should only be undertaken by a trained conservator. Conservators often compare themselves to surgeons. We hope that you would no more consider cleaning your own painting than you would remove your own appendix. While there are obvious differences between the conservation profession and the medical profession (besides salaries), conservators often lament that while doctors can bury their mistakes, they have to hang theirs on the wall.

26

Road Map for Cleaning Product Selection for Pollution Prevention

Overview 411
Introduction...411
Goals 412
Alternative Product Selection...412
Performance-Based Selection • Health and Safety Screening
Selection • Comprehensive Health and Safety via P2OASys • On-Site Pilot
Plant and Scale-Up Feasibility Studies
Metal Cleaning Summary...426
Conclusions and Lessons...426
Acknowledgments..426
References...427

Jason Marshall
*Massachusetts Toxic Use
Reduction Institute*

Overview

Technical assistance in pollution prevention activities has had marginal success over the years with a national average adoption rate of around 10%.[1] The Toxics Use Reduction Institute's (TURI's) Laboratory (Lab) has traditionally had slightly higher success, with about a third of companies adopting the changes that recommend solvent substitution in cleaning application.[2] In an attempt to improve the adoption rate of the work conducted at the Lab, a more comprehensive on-site follow-up assistance program was implemented in 2006. The process utilized is described in this chapter with several case studies illustrating the successful practices.

Introduction

The passage of the Toxics Use Reduction Act (TURA) in 1989 by the Massachusetts legislature marked the creation of the TURI at the University of Massachusetts Lowell. Fully operational since late 1993, the TURI Laboratory (Lab) is the research and testing facility, designed to evaluate the performance of cleaning formulations and related equipment. That same year, the U.S. Environmental Protection Agency's Toxic Release Inventory (TRI) revealed that many of the state's industries such as metalworking and electronics use several hazardous chemicals recognized for their excellent solvating powers.

Goals

The objective of the laboratory is to assist in the development and promotion of safer alternatives to these hazardous materials, primarily organic and chlorinated solvents used to clean metal surfaces (Figure 26.1), without causing economic hardship or a loss in cleaning performance. The figure reveals that hazardous organic and chlorinated solvents make up nearly 30% of the chemicals replaced at the Lab. Interestingly, a quarter of the products replaced have been aqueous cleaners that were not working as well as the companies would like.

Specifically the Lab looks to reduce the use of solvents that contribute to ozone depletion, global warming, and volatile organic compound (VOC) emissions. In addition, decreases in exposures to flammable, carcinogenic, and other toxic substances are attempted.[3]

Termed environmental indicators, these factors are important not only to workers directly involved in surface cleaning but also to the communities in which the processes are conducted and, ultimately, the consumer. The Lab screens the potential alternatives using five indicators[4] and then using the Institute's Pollution Prevention Options Assessment System (P2OASys),[5] both are described later in the chapter. The chemical damage associated with any given cleaning solvent can be reviewed by using the state-of-the-art health and safety related databases available online or from technical libraries.

Alternative Product Selection

Performance-Based Selection

Prior to beginning a project, a brief questionnaire (Figure 26.2) should be filled out so that testing can be tailored to meet the individual firm's needs. Designed with templates from both industry and government, this format also can assist plant personnel involved with cleaning to become familiar with all aspects of the process.

Prior to the search for safer and greener chemical cleaners, solvent cleaning was often performed via single-species vapor degreasing. In addition to immersion, vapor degreasers were employed because of

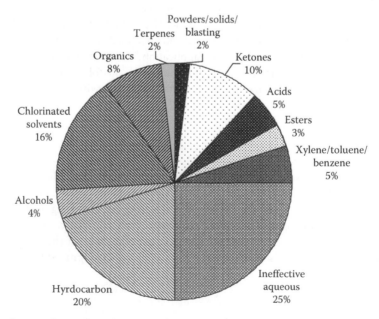

FIGURE 26.1   Cleaning chemicals replaced at TURI Lab 1994–2009.

Describe the part/product to be cleaned

Materials of construction: Aluminum   Brass Stainless Steel   Other Metals   Plastic   Glass   Other:
Please specify type:
List percentages cleaned (if more than one substrate):
(for example, 60% of parts are aluminum; 40% are 304 stainless steel)
Surface (circle two):   Rough or Smooth   Hard or Soft
Approx. size (dimensions in inches):
Geometry: Simple (e.g., flat)   OR   Complex (contains inaccessible areas)
Gram weight: Min.   Max.
What is this part/product used for?

Describe the current cleaning process

Contaminants to reduce or eliminate (circle all that apply):   Oil   Grease   Wax   Flux   Dirt   Salts
Combination (describe):   Other:
Are samples of contaminants available? No   Yes (if available, attach MSDS)

Manufacturing step immediately before/after cleaning:

Number of parts cleaned **per week** (or shift, etc.):   **per batch**:

Equipment in use (circle all that apply):
Vapor degreaser Agitation/air sparging unit   Immersion/soak/dip tank   Ultrasonics   Pressure spray washer
(approx. psi _____) Other (Specify vendor, if possible):
Cleaning Chemical(s):   (attach MSDS)
Concentration: %   Time: min. Temp.:   °F   Water source, if applicable: DI /Tap
Rinse Cycle, if any: Time: min. Temp.:   °F   Water source: DI/Tap
Drying Cycle, if any: Time: min. Temp.:   °F
Any problems with present cleaning system?
After cleaning, parts are (circle one):   Used Immediately   OR   Stored
   If stored, How: How long:
What is the purpose of cleaning (i.e., desired product specifications)?
Methods employed for evaluating cleanliness: None   Visual   Microscopic   UV   Other performance test,
if any (please describe):
Comments or Areas of Concern:

FIGURE 26.2   Sample cleaning process questionnaire.

traditional solvents' superior ability to dissolve organic matter. In vapor phase cleaning, the final "rinsing" of the surface was accomplished by the condensation of solvent vapors.

Unlike these chlorinated organic solvents, environmentally friendlier aqueous (i.e., water-based) detergents may not depend on their penetrability for their cleaning efficiencies, especially for the removal of petroleum-based surface debris. They rely instead on a number of chemical processes such as solubilization, wetting, emulsification, deflocculation, sequestration, and saponification.[6] To complete these tasks, aqueous cleaners can be complex mixtures of surfactants, emulsifiers, and other additives in an alkaline (pH > 7.0) formula.

The traditional outline of a cleaning test plan is shown is Figure 26.3. Due to their decrease in chemical energy relative to chlorinated organic solvents, several aspects of an aqueous cleaning system must be optimized prior to switching to the new cleaning formulation. These included but are not limited to agitation methods, cleaner concentration, temperature, and cycle time as reflected in Phase II of Figure 26.3.

As this figure illustrates, metal surfaces initially cleaned are flat coupons (Phase I) matched to client parts' material(s) of construction, followed by the eventual cleaning of actual parts (Phase IV). Phase IV is concerned with the significance of part shape (screw configurations and blind holes et al) on metal cleaning.

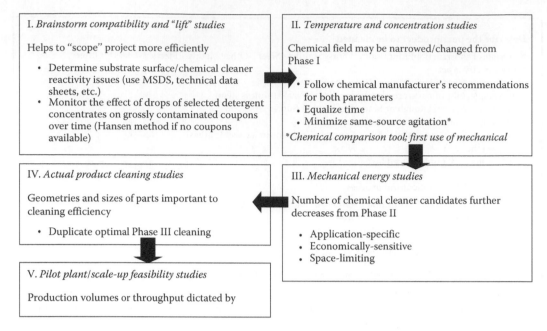

FIGURE 26.3   Phases of an aqueous-based surface cleaning test.

Note that in this model, mechanical energies (Phase III) are critically examined only after the proper cleaners, concentrations, and temperatures (Phase II) are ascertained. Other testing protocols exist that may be as effective; it is essential that the experimental design be logical and consistent.

The initial Phase I of the testing process has been able to be mostly eliminated due to the development of the CleanerSolutions database (www.cleanersolutions.org). From the testing performed at the Lab, data has been accumulated on the performances of industrial cleaning products. In order to use this information effectively, the CleanerSolutions database was created. This web-based interface is field searchable by surface contaminants, surface substrates, cleaning equipment, solvents replaced, and vendor product data. The resulting list of alternatives helps companies to select products that are most likely to work for their specific situation. The most basic search method on the site requires the user to know the general classification of the contaminant that needs to be cleaned. In addition, substrate and equipment being used can be included in the search as shown in Figure 26.4.[7]

By starting at the traditional Phase II of the testing process, the online database allows for faster testing of alternatives. Testing would still follow Phases II–V involving experimentation with as many as six chemical cleaners, over a period of 2–8 weeks.[8]

Health and Safety Screening Selection

During the selection process of possible alternatives using the CleanerSolutions database, performance should remain at the highest level of importance. However, almost as relevant as performance is whether the products to be selected are safer than the current cleaning solvent. To that end, when choosing an alternative it is important that the risk is not shifted from the worker to the environment or from the environment to the worker. Selection of a product should be safer for one or the other; ideally it would be best if the product is safer for both as compared to the current cleaning product.

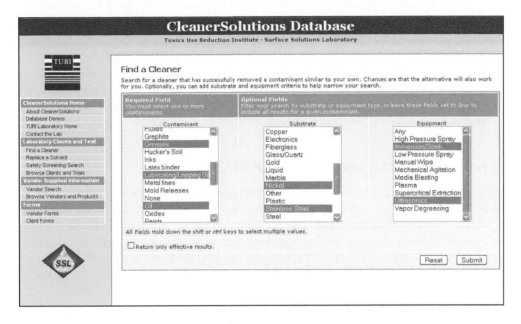

FIGURE 26.4 CleanerSolutions Find-a-Cleaner search options.

To help make the chemical safety selection process easier, a screening of products for health and safety issues using a small subset of EHS criteria. Five such environmental indicators for surface cleaning could include the following:

- Volatile organic compounds (VOCs)
- Global warming potential (GWPs)
- Ozone depletion potential (ODPs)
- Hazardous material information system/ National fire protection association (HMIS/NFPA)
- pH

Each indicator is assigned a value up to 10 points. Individual indicator points are then combined (with equal weighting) to give a safety score out of 50 points. A higher score implies a potentially safer product. An example safety screening score for a terpene based product is shown in Figure 26.5. It is important to note that the indicators used are not the only indicators that can be used for evaluating risk hazards associated with these products. For further details on how the safety scoring

Safety score details

Indicator	Value	Points
VOC :	780	0
GWP :	0	10
ODP :	0	10
HMIS H :	0	
HMIS F :	2	8
HMIS R :	0	
pH :	8	9
Total: 37		

FIGURE 26.5 Sample safety screening score.

system works, information is available at the CleanerSolutions Web site. However, these indicators do cover a broad range of worker and environmental hazards and provide the user with a good starting point.[7]

With the selection of past effective products that are potentially safer than the existing cleaning process, Phases II–IV should be completed. Once the list of potential cleaning alternatives has been narrowed down to a few likely candidates, a more thorough health and safety evaluation can be completed.

Comprehensive Health and Safety via P2OASys

P2OASys is one tool that has been developed for systematic assessment that helps companies determine whether the toxics use reduction options they are considering may have unforeseen negative environmental, worker, or public health impacts. P2OASys allows companies to assess the potential environmental, worker, and public health impacts of alternative technologies aimed at reducing toxics use. The goal is more comprehensive and systematic thinking about the potential hazards posed by current and alternative processes identified during the planning process. The tool will assist companies in two ways:

- To systematically examine the potential environmental and worker impacts of TUR options in a comprehensive manner, examining the total impacts of process changes, rather than simply those of chemical changes
- To compare TUR options with the company's current process based on quantitative and qualitative factors.[5]

Embedded formulas in P2OASys provide a numerical hazard score for the company's current process and identified options, which can then be combined with other information sources and professional expertise to make decisions on possible implementation.

Companies will input both quantitative and qualitative data on the chemical toxicity, ecological effects, physical properties, and changes in work organization as a result of the proposed option. A large portion of the data needed for P2OASys is available through vendors, existing databases, or through the technical assistance providers.

With the performance confirmed on supplied parts and the EHS profiles compared, the project would then shift to the facility to better assess the likelihood of adoption of the new cleaning process.

On-Site Pilot Plant and Scale-Up Feasibility Studies

Having identified a handful of laboratory successful cleaning formulations, Phase V of the process involves on-site testing and assistance in the form of equipment loaners, formulation samples, and personnel training. Additionally, work should be done with connecting the company with the appropriate vendors. Setting up and running batch cleaning testing with parts that come directly off the product line will allow for real time evaluation of the proposed cleaning process. The on-site process is intended to include the employees who are responsible for cleaning for the company. Input into the process allows for a greater opportunity of successfully adopting the laboratory recommended alternative cleaning process.

Prior to the interactive, hands-on assistance, the adoption of safer alternatives has been estimated to reach around a third of the companies looking for a new cleaning process. With an expanded effort, acceptance of safer and effective products into the companies operating plans has been shown to reach nearly three quarters of those conducting on-site process specific testing of bench scale proven products and equipment.

CASE STUDIES

To help better understand the process described above for finding effective and safe cleaning formulations, three case studies have been provided. The work involved in these projects was conducted during the years 2008 and 2009.

STEEL CLEANING

The company makes tools for other companies that make screwdrivers. The company makes the bits or tools for other machinists to make the screwdrivers. At this facility, raw steel is cut to form small pointed cylinders, which are then cold iron shot blasted and then sent to a machine to start the tooling process and are shown in Figure 26.6. Cleaning was focused on four soils, two of which were aerosols lubricants, a third was a cutting fluid and the final contaminant was a wax like oil. The company wanted to replace an ineffective solvent and an aqueous product.

PHASES I AND II: PRODUCT SELECTION AND TEMPERATURE AND CONCENTRATION STUDIES

The initial was to evaluate selected products (typically selection includes between 8 and 10 products) on first supplied soil, Safety Draw fluid, using immersion cleaning. These products were selected from the online cleaning results database, www.cleanersolutions.org, based on supplied information from client. Each product was diluted to 5% using DI water in 600 mL glass beakers. Solutions were heated to 130°F on a hot plate. Safety Screening scores were incorporated into the selection process by comparing potential alternatives to the safety screening score of the current solvent. Products with better safer screening scores and a history of effective testing for similar situations were chosen when possible for laboratory evaluations.

PHASE III: MECHANICAL ENERGY STUDIES

No product was very successful using immersion cleaning to remove the Safety Draw fluid (contains chlorinated paraffin and mineral oil). Three of the products had marginal success, removing over a third of the soil. The products that were partially successful were evaluated using ultrasonic cleaning. Additional products also were selected from the Lab's inventory.

FIGURE 26.6   Steel parts.

The addition of ultrasonic energy improved cleaning results for the four previously tested products. The additional four products were found to be effective as well. Of the eight products, seven removed in excess of 80% of the Safety Draw within 5 min of cleaning.

This cleaning process was repeated for the other supplied soils in an effort to find cleaners that would work on all of the contaminants. Six of the seven products removed more than 90% of the Tru-Edge cutting fluid in 5 min of cleaning.

The third soil, ChemSearch Aerolex Plus Moly dry film lubricant (alcohol mixture with molybdenum disulfide and aromatic petroleum distillates), resulted in limited success. With the inconsistent results using ultrasonic cleaning, a batch of non-aqueous based products was evaluated using manual wiping. Of the six new products evaluated, four of them worked well. The other two removed some of the lubricant but were noticeably slower. Additional evaluation of these solvents may be conducted on the other three soils in the future if necessary.

The six aqueous-based products were again selected for ultrasonic cleaning of the fourth supplied contaminant, the Dow Corning G-n metal assembly spray lubricant (contains hydrocarbon aerosol propellant). Several of the products were found to remove most of the lubricant within the short cleaning time. Overall, cleaning of the supplied contaminants yielded Mirco 90, Daraclean 282 GF, Polyspray Jet 790 XS, and Armakleen M Aero as the top four products with averages above 90%. These products were taken to be evaluated on supplied parts.

PHASE IV: ACTUAL PRODUCT CLEANING STUDIES

A set of three soiled steel parts (with metal grit/fines) were immersed into each product and cleaned for 5 min and another set for 10 min using ultrasonic agitation. Following cleaning, coupons were observed for cleanliness. At room temperature, parts were cleaner after 10 min of ultrasonic cleaning. Daraclean 282 GF was the most effective cleaner. Similar results were obtained when the products were heated to 130°F.

During the testing, the company requested the inclusion of a product that was available to them through their current supply chain network. This product was evaluated at 3% on the four contaminants and then on the supplied dirty parts. The parts coated with the metal shot grit after 5 min of ultrasonic cleaning were comparable to the parts previously cleaned using Polyspray Jet 790 XS and Micro 90. This would place the product between second and fourth best after 5 min of cleaning and was found to be one of the better cleaning products when used for 10 min in ultrasonic cleaning.

Parts cleaned in the laboratory setting were then brought to the facility to see what the company thought of cleanliness. In addition to the cleaned parts, a loaner Crest 19L, 40 Hz ultrasonic unit was provided for the company to borrow to facilitate on site evaluation of the alternative products. Samples of the Poly Spray Jet 790 XS and the Daraclean 282 GF were made available to the company.

PHASE V: PILOT PLANT/SCALE-UP FEASIBILITY STUDIES

The parts were presented to the company in the order of cleaning at the Lab. The parts were acceptably clean with the Daraclean 282 GF being the best. They were very happy with how the parts looked especially since there was little to no rust on all of the parts. The company was impressed at how the well the ultrasonic cleaning did with removing the iron shot. The Lab facilitated getting a gallon sample of the 282 for the company to try onsite.

During the on-site evaluation, a review of the top performing cleaning products was conducted to determine the potential impacts to the worker and/or the environment. Upon review of the collected data, one of the company's existing cleaning products was among the top products when using the five-indicator screening process. When comparing the wider EH&S profiles after testing was completed, the current products were the top two products. The only problem with the two products was therefore not the health and safety concerns, but with the performance. The driving force for the company switching was to find a more effective product. To best accommodate a safe and effective product, the U.S. Polychem Polyspray Jet 790XS and the BCS Green Spray 400 would be the best options for the company to evaluate on-site. A summary of the screening and comprehensive EHS comparison can be found in Table 26.1.

ADOPTION

The company has changed almost all their cleaning lines to the BCS Green Soak at 3% (an alkaline aqueous based product with a mix of alkali and wetting agents). They have bought two used spray wash units and are using them at two points in their process line. The first is used to clean blanks after they are processed with cold iron shot to ready them for the rest of the manufacturing line. In addition, the company started using the new cleaner in an ultrasonic cleaning tank to remove buffing compound where they found the product worked better and faster then their previous cleaner. Up to this time, they were using a biobased solvent, Spartan Biorenewables Industrial Cleaner and Degreaser at full strength at room temperature in an immersion bath.

The exception was for about 20% of the blank line after iron shot processing that goes to a copper sulfate plating process. This process was being affected by the new cleaning process. Questions

TABLE 26.1 Steel EHS Comparison of Current and Potential Products

Products		Current Product		Potential Alternative	
EH&S Data	Category	Value	Pts	Value	Pts
Safety screening (conducted during Phase I)	VOC	25	9	30	9
	GWP	0	10	0	10
	ODP	0	10	0	10
	HMIS H	2	6	1	9
	HMIS F	2		0	
	HMIS R	0		0	
	pH	NA	10	10.3	7
Safety Screening Score (best = 50)		**45**		**45**	
Full EHS evaluation (conducted during Phase III or IV)	Acute human effects	6		6	
	Chronic human effects	2		2	
	Physical hazards	2		2	
	Aquatic hazard	ND		ND	
	Persistence/ bioaccumulation	ND		ND	
	Atmospheric hazard	2		2	
	Disposal hazard	2		2	
	Chemical hazard	6		7	
	Energy/resource use	2		6	
	Product hazard	2		2	
	Exposure potential	6		6	
EHS Score (best = 0)		**3.33**		**3.89**	

Note: NA, not applicable; ND, no data.

were brought up about whether temperature or water introduction was causing plating problems. This line was reverted back to being cleaned in a small dip tank (3–5 gal) using mineral spirits.

LESSONS LEARNED

For this steel cleaning company, their main driving force was to find a more effective cleaning product and to increase process times. The overall safety of the cleaning process was maintained by conducting a comparison of the current product with potential replacement chemicals. Every effort was made by the company to find one cleaning product that would work for multiple contaminants used in their facility. In the final stages of the process, the company determined that nearly 80% of their production line could be cleaned adequately with an alkaline aqueous cleaner at a low concentration. The low concentration and moderate temperature of the new process helped the company to use fewer materials and reduce the amount of energy needed for cleaning.

ALUMINUM CLEANING

An aluminum anodizing job shop was having trouble with VOC emissions and concerns for its worker safety. They were moving their operation to another building across the street and wanted to address these concerns before the move. They needed assistance in the replacement of methyl ethyl ketone (MEK) in the company's lacquer removal process. Typically, the demasking process utilized two and half tons on an annual basis. About 200 parts were being cleaned in a week's time. Cleaning was at room temperature and lasted 12 h. Figure 26.7 shows a typical part with the applied lacquer. Ideally a product that would remove the lacquer from the 5052 aluminum parts using wipe and ultrasonic cleaning was needed. Workers were using premoistened swabs for the wipe application and multiple 5 gal buckets were used in the immersion operation.

PHASES I AND II: PRODUCT SELECTION AND TEMPERATURE AND CONCENTRATION STUDIES

Using the information provided about the current cleaning process from a cleaning process questionnaire, 15 products were selected from the online database based on past testing results matching client-supplied information. Seven of these products were diluted to 5% using DI water in 600 mL beakers. The other eight products were used at full strength as recommended by the vendor. All eight products were heated to 130°F on a hot plate.

FIGURE 26.7   Aluminum parts coated with lacquer.

Several of the full-strength products showed some signs of removing the Microshield Red Stop-Off lacquer (high-solids, air-dry lacquer) from the aluminum coupons within 5 min of cleaning. The products that changed to a pink/red color will be retested using a longer cleaning time, with a higher temperature utilizing ultrasonic energy.

PHASE III: MECHANICAL ENERGY STUDIES

The six products that showed signs of dissolving the lacquer from the immersion cleaning were heated to 130°F in 300 mL beakers suspended in a 40 kHz Branson 2510 ultrasonic unit and degassed for 5 min. The top two performing products were then evaluated using room temperature immersion cleaning for an extended period of time (1–2 h). After 60 min, both products showed good removal of the coating.

PHASE IV: ACTUAL PRODUCT CLEANING STUDIES

Having successfully identified two potential alternative cleaning products, the next phase of the laboratory work was to pilot these products to determine the optimum cleaning conditions on supplied parts. Aluminum Alternatives 1 and 2 were heated to 130°F in a 40 kHz Branson 2510 ultrasonic unit and degassed for 5 min. One supplied dirty part coated with the lacquer was immersed in each solution and cleaned at 10 min intervals until completely cleaned. Observations were made and estimated removal was recorded at each interval. Once clean, parts were rinsed in tap water for 15 s at 120°F and dried using compressed at room temperature for 30 s.

Cleaning of the lacquer was successful in about 20 min of ultrasonic cleaning. Cleaned parts were returned for client inspection. The next experiment was to pilot the same two products using immersion cleaning only at room temperature.

Both products did not completely remove the coating after soaking for 8 h. However, the Alternative 2 only had about 5% of the coating remaining, mostly in the external threading. The Alternative 1 was only moderately successful, removing about 75% of the coating.

Three supplied dirty aluminum coupons coated with the Red Stop Off were wiped with cotton tip swabs until the outer portion of the coupons were free of lacquer. The intent of this cleaning was to simulate removal of a masking process and only the lacquer outside the etched lines was to be removed. Figure 26.8 shows the supplied masked aluminum coupons before cleaning and after cleaning.

The Alternative 2 was more effective at spot cleaning, requiring less time and fewer swabs to remove the coating from the parts.

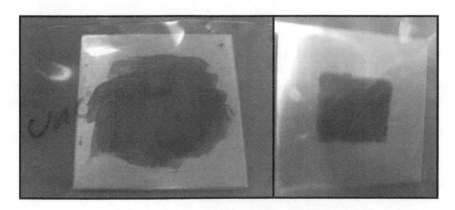

FIGURE 26.8 Supplied masked aluminum coupons (un-cleaned and cleaned).

PHASE V: PILOT PLANT/SCALE-UP FEASIBILITY STUDIES

With successful testing in the laboratory setting on the supplied parts, the company was contacted to arrange for on-site testing of the two alternatives with ultrasonic cleaning. During the on-site testing, the company wanted to test parts that were handled differently then the Lab-cleaned parts. These painted parts were more typical of what the company was looking to have cleaned, removing the red lacquer from areas of the painted parts. The company wanted to see how the ultrasonic energy and solutions acted on the paint.

Periodically the parts were removed from the cleaning to see the progress of the lacquer removal. The company staff seemed pleased with both products. Inspections were also looking for any damage to the integrity of the ink coating on the part, pitting, change in appearance, or obvious removal of the paint. The Alternative 1, which the workers liked the least in terms of smell, started to take off the black paint. This concerned the company workers. There was some debate on if the removal was harming the coating or not. Some workers said the parts felt tacky or slick and some said they looked iridescent, meaning the paint coating was being compromised.

During the on-site testing, the company brought workers from the wipe application process to participate in the piloting so they could ask questions and observe. In addition, the president of the company sat in on the testing as did some of their own lab and EH&S staff. Comments from the workers noted that they did not like the smell of the products and expressed concerns that cleaning may take more effort to remove the lacquer over all using in the wipe applications. Taking the workers concerns into considerations, the president noted that the company had to make a switch and understood that a new cleaner would be less flammable and have a better EH&S profile for the workers.

At the end of the first on-site pilot cleaning, the company requested to keep the two small ultrasonic units for a month and keep the remainder of the gallon of each product so they could continue more testing in house.

During the on-site evaluation, a review of the top performing cleaning products was done to determine the potential impacts to the worker and/or the environment using both the screening tool and the comprehensive evaluation.

The safety screening values of the two alternative cleaning products were shown to be a potential improvement over the current solvent, MEK. Most important to the company was the fact that both the alternatives had a lower flammability rating for HMIS than the MEK. When looking at the expanded EH&S assessment done in P2OASys after product testing was completed, the Alternative 2 had the best profile of the three products evaluated. Table 26.2 lists the results from both EH&S evaluations for all three products.

ADOPTION

Through follow-up conversations between the TURI Lab and the company, the company had made the decision to go with the Alternative 2 based on low odor and superior performance. Also the EH&S comparison performed by the Lab showed that the Shopmaster RC (ester-based solvent) will reduce the health factor of the HMIS from a 3 for MEK to a 2 for the Shopmaster RC as well as lowering the flammability rating to a 1 from a 3 for MEK.

The company was getting more products from the distributor for more trials and wanted to work on keeping the temperature in the ultrasonic steady. To help facilitate the expanded testing, the two smaller units previously provided were replaced with a larger unit (5 gal).

The company conducted follow-up testing on their own with the provided equipment. The facility supplied some additional action items that they were planning on addressing, but they

TABLE 26.2 Aluminum EHS Comparison of Current and Potential Products

Product		Current		Alt 1		Alt 2	
EH&S Data	Category	Value	Pts	Value	Pts	Value	Pts
Safety screening (conducted during Phase I)	VOC	810	0	938	0	1060	0
	GWP	0	10	0	10	0	10
	ODP	0	10	0	10	0	10
	HMIS H	3	0	0	10	2	7
	HMIS F	3		0		1	
	HMIS R	0		0		0	
	pH	NA	10	9.7	8	7	10
Safety Screening Score (best = 50)		**30**		**38**		**37**	
Full EHS Evaluation (conducted during Phase III or IV)	Acute human effects	10		4		6	
	Chronic human effects	6		2		2	
	Physical hazards	6		ND		ND	
	Aquatic hazard	4		ND		ND	
	Persistence/ bioaccumulation	6		ND		ND	
	Atmospheric hazard	2		2		2	
	Disposal hazard	10		2		2	
	Chemical hazard	10		9		4	
	Energy/resource use	4		2		2	
	Product hazard	6		2		2	
	Exposure potential	6		6		6	
EHS Score (best = 0)		**6.36**		**3.63**		**3.25**	

Note: NA, not applicable; ND, no data.

were going to move forward with the substitution of Alternative 2 for MEK. They have completed extensive in house testing in addition to the testing conducted in the laboratory setting. The company purchased a 10 gal ultrasonic unit with a bath circulator, cooler, and drying oven.

LESSONS LEARNED

By bringing in workers that would be responsible for the actual cleaning of parts, the company was able to improve the chances of successfully choosing a new cleaning product that was safer than and as effective as the current process. Without critical feedback from the workers, the follow up testing may not have been conducted which eventually allowed the company to identify the optimum cleaning conditions for the new cleaning process.

MULTIMETAL CLEANING

The company wanted to lower air emissions as the current levels were close to being in violation of the companies air permit. Secondly, the company wanted to reduce the risk of fire by finding a solvent with a lower flammability level, thus improving the work environment. This company manufactures and engineers electronic components for many types of consumer and industrial products as well as network solutions and systems for telecommunications and energy sectors.

The current process uses many materials and makes many types of parts. They make parts from many inches down to very small parts with blind holes. Metal fines can be left behind and need to be removed by the prewash, currently done using mineral spirits, at each machine station.

The machines are all newer models that are computerized and enclosed. The parts come out of an opening into a wire, mesh basket. Oils, fluids and lubricants are visible on the parts coming out of the machine. The parts in the strainer are taken to a close-by bucket filled with mineral spirits. The strainer is manually agitated in the mineral spirits, both back and forth as well as up and down. This precleaning process is followed by cleaning in an aqueous machine or in a vapor degreaser.

PHASES I, II, AND III: PRODUCT SELECTION, TEMPERATURE/ CONCENTRATION, AND MECHANICAL ENERGY STUDIES

The first objective was to evaluate selected products for the removal of the supplied cutting oil and compare to current cleaning. Sixteen alternative products were identified based on supplied client information using the CleanerSolutions online database. Two supplied products were included in testing to establish a baseline.

Copper coupons were coated with a known amount of Hangsterfer Laboratories Hard Cut 5418 cutting fluid (long chain chlorinated paraffin). Cleaning utilized immersing parts into each cleaning solution with manual raising and lowering in the cleaning solution to provide minimal mechanical agitation. There were eight products that worked very well on the cutting fluid under conditions similar to the existing cleaning process. The effective products were selected to be evaluated for cleaning performance on the second supplied soil.

Copper coupons were coated with Master Chemical Corporation Trim Sol SF (water-miscible cutting and grinding fluid containing petroleum oil, petroleum sulfonate, and chlorinated alkene polymer) and cleaned following the same dunk and swish method used during the initial contaminant cleaning. All but one product were effective on the second supplied contaminant. The most successful products on both soils were to be tested on supplied dirty parts.

PHASE IV: ACTUAL PRODUCT CLEANING STUDIES

Two groups were evaluated on the supplied parts. The first batch consisted of all the cleaners, which had a removal efficiency of more than 99% for both the Hard Cut and the Trimsol. The second set evaluated were the cleaners which had removal efficiency less than 99% but more than 80% for both the Hard Cut and the Trimsol.

All four of the first group of products removed oils from the parts, which could be seen in the cleaning vessels. Some oil or residual solvent was observed on the parts. Metal fine removal was very important at this step.

The second set of cleaners showed evidence of removing fines and cutting fluids based on visual observations of used cleaning fluids. In both cases, the parts were to be inspected using a microscope following the company's QA/QC procedures. Figure 26.9 shows two sets of parts cleaned by some of the alternative cleaning products.

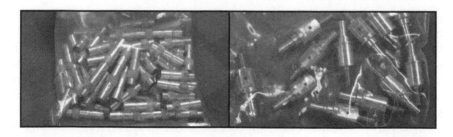

FIGURE 26.9 Brass and copper parts cleaned.

PHASE V: PILOT PLANT/SCALE-UP FEASIBILITY STUDIES

The cleaned parts were brought to the company to look at them using their QA/QC program. On initial inspection, the company executives were pleased with what they saw. They liked the fact that there was a little oil left on the parts and that there was no rusting of the parts after a couple of days of sitting after cleaning.

During the on-site evaluation, the top performing cleaning products were evaluated to determine the potential impacts to the worker and/or the environment. There were a handful of products that had safety screening values greater than the current solvent and several with equal values. A similar result was found when using the comprehensive assessment methodology. Table 26.3 lists the current solvent score and the top two alternative cleaning products.

The company wanted flammability as low as they could get and the lower the VOC level the better. Once the company finished the initial QA/QC, additional Lab cleaning of parts would be conducted using the products with lower calculated efficiencies (under 90%) if necessary. The company supplied new sets of parts to try on the other products.

Follow-Up Lab Testing

The cleaners selected for testing had lower then 100 g/L VOCs, with exception of one product. Two had flammability ratings of 2. One product had a pH of 12.6. The parts cleaned in the five cleaners were bagged and sent to the company. The products appeared to work well. The company was trying to obtain 5 gal of each of the lower VOC products to do their own testing.

TABLE 26.3 Multimetal EHS Comparison of Current and Potential Products

Product		Current Solvent		Multimetal Alt 1		Multimetal Alt 2	
EH&S Data	Category	Value	Pts	Value	Pts	Value	Pts
Safety screening	VOC	770	0	44	9	1060	0
(conducted during	GWP	0	10	0	10	0	10
Phase I)	ODP	0	10	0	10	0	10
	HMIS H	1	7	0	9	2	7
	HMIS F	2		1		1	
	HMIS R	0		0		0	
	pH	NA	10	7	10	7	10
Safety Screening Score (best = 50)		**37**		**48**		**37**	
Full EHS evaluation	Acute human effects	6		2		6	
(conducted during	Chronic human effects	2		2		2	
Phase III or IV)	Physical hazards	ND		ND		ND	
	Aquatic hazard	ND		ND		ND	
	Persistence/ bioaccumulation	ND		ND		ND	
	Atmospheric hazard	2		2		2	
	Disposal hazard	6		2		2	
	Chemical hazard	8		4		4	
	Energy/ resource use	6		4		2	
	Product hazard	6		2		2	
	Exposure potential	10		6		6	
EHS Score (best = 0)		**5.75**		**3.00**		**3.25**	

Note: NA, not applicable; ND, no data.

Adoption

The company evaluated many of the products in-house to determine which product would be the best fit for the cleaning line. They liked the first multimetal alternative but were concerned with its cost, which was two times the cost of their current mineral spirits cleaner. The Lab offered to go back to the drawing board and try to find a more economical cleaner that fit their criteria.

The company worked with another vendor to possibly formulate a similar product to the high-cost alternative; however, this vendor was unable to formulate anything that met the specifications.

The company finally implemented the switch to the second multimetal cleaner, Soy Clear 1500 (soy based cleaning solvent) without much trouble. The one issue they had was that the product biodegraded too rapidly. The rags they used at the stations with the cleaner on them were put into drums for pick up by the laundry company, but they were heating up due to the biodegradation of the product. To deal with it immediately, they put the rags in water. Then the manager of the project did some research and realized soy was a food product and with the help of their on staff chemist the realized they could add a food additive to the cleaner and retard the biodegradation that was causing the heat and acrid smell. This appeared to work and next steps were to see if the manufacturer could add this preservative to the product for them.

Lessons Learned

On-site evaluations proved to the company that the alternative cleaners would work for their cleaning needs. By piloting on a larger scale, the company identified a possible problem before making a large purchase of the selected alternative. The product's biodegradability would have been a large safety concern for the workers of the company. By working with the vendor, they were able to identify a possible additive that would allow the company to use the product safely.

Metal Cleaning Summary

In closing, tests conducted in a controlled laboratory setting need to involve all aspects of the particular cleaning process to be changed. These tests confirm that the future of metal cleaning depends upon establishing industry standards or a ranking system for the energy and water efficiency of related equipment as well as a more complete understanding of the environmental and health consequences of newly developed chemical cleaners.[9]

Conclusions and Lessons

Most cleaning projects are not solved easily. No single drop-in cleaning alternative exists. The product selection process is one that must be tailored to each situation to ensure that the chosen cleaning alternative will be the best fit for the company. Without worker's buy-in during the transition, success will be difficult to achieve. Therefore the keys to success are conducting systematic performance testing using the specific soils encountered at the facility and then to pilot the potential product on actual parts by the personnel responsible for running the cleaning process.

The lack of adoption by companies only receiving Lab testing shows the importance of providing an on-site assistance phase of the substitution process. By conducting the on-site work, questions or concerns can be met in real time, facilitating a successful adoption of safer cleaning practices.

Acknowledgments

The author wishes to express his gratitude to Heidi Wilcox, the Toxics Use Reduction Institute's Laboratory Field Implementation Technician, for her vital contributions in gather data for the field work that make up the case studies. Special thanks are also due to Dr. Carole LeBlanc for her efforts on establishing the TURI Lab's protocols during her time at the Institute.

References*

1. Nelson, W. Waste Management Resource Center (WMRC), Champaign, IL, <http://www.wmrc. uiuc.edu>. 2001.

2. Kusz, J. P. Diffusion, confusion or illusion? Measuring the success rate of alternative cleaning processes. In: *CleanTech 2002 Conference Proceedings. Cleaning Resources Conference*. Atlanta, GA, May 2002. Gardner Publications: Cincinnati, OH, 2002.

3. LeBlanc, C. The toxics use reduction Act of 1989: Lessons learned. In: *Annual Conference Proceedings: Precision Cleaning*. Cincinnati, OH, April 1997.

4. Marshall, J. P. CleanerSolutions: A tool for toxics use reduction through solvent substitution in surface cleaning applications. Dissertation, University of Massachusetts Lowell, 2008. UMI: Ann Arbor, MI, 2008.

5. Toxics Use Reduction Institute University of Massachusetts Lowell. P2OASys tool to compare materials. TURI. March 2009. <http://www.turi.org/home/hot_topics/cleaner_production/p2oasys_tool_to_compare_materials>

6. Grace Metal Working Fluids. *Aqueous Cleaning Handbook*. Lexington, MA, 1995.

7. Toxics Use Reduction Institute University of Massachusetts Lowell. CleanerSolutions on-line tool for solvent substitution in surface cleaning. TURI. July 15, 2009. <http://www.cleanersolutions. org/?action=static_page&page=about>

8. Marshall, J. P. Documentation Database of the Surface Solutions Laboratory. Toxics Use Reduction Institute's Database DocumentLab.mdb. 2009.

9. Kanegsberg, B. and LeBlanc, C. The cost of process conversion. In: *Annual Conference Proceedings: CleanTech*. Chicago, IL, May 1999.

---

* Additional details may be obtained by visiting the laboratory web sites, www.turi.org/laboratory and www.cleaner solutions.org.

27

Wax Removal in the Aerospace Industry

Bill Breault
*Petroferm, a Business of
Vantage Specialties Inc.*

Jay Soma
*Petroferm, a Business of
Vantage Specialties Inc.*

Christine Fouts
*Petroferm, a Business of
Vantage Specialties Inc.*

Introduction 429
Selective Plating and Masking 429
D1 Immersion Method
Wax Removal Technologies 430
Cabinet Ovens • Steam Stripping • Hot Water • Wax Tanks • Vapor
Degreasing • Aqueous Cleaning • Semi-Aqueous Cleaning
Aerospace Specifications/Industrial Practices 432
Methods 2 and 3
General Process Guidelines for Wax Removal 435
Step 1—Prewash (Optional) • Step 2—Wax Removal • Step 3—Water or
Alkaline Rinse • Step 4—Final Water Rinse • Step 5—Drying (Optional)
Summary 438
References 438

Introduction

From the manufacture of individual aircraft components to the maintenance, repair, and overhaul (MRO) of major assemblies, cleaning is an important practice in the aerospace industry. During manufacturing, removal of contamination is often a critical step prior to subsequent metalworking processes. During component inspection, surfaces must be clean of contamination that can mask or shield potential defects. Prior to painting, cleaning is required to remove impurities that can affect coating adhesion and functionality. In selective plating operations, masking agents must be cleaned prior to downstream fabricating processes. Other processes with critical cleaning steps include segment bonding, component disassembly, and welding. Ultimately, cleaning processes safeguard an aircraft from defective parts that inflate maintenance costs and most importantly threaten public safety.

Selective Plating and Masking

During use, aircraft engine and landing gear components experience significant wear and surface loss. Dimensional restoration rebuilds these surfaces and is common in the aerospace industry where it is costlier to replace than to repair a component. There are several processes used for rebuilding surfaces including thermal spray, weld overlay coatings, and selective plating. Selective plating is a commonly used electroplating technique that returns the substrate surface to its original condition by coating specific areas of a component. These areas must exhibit wear and corrosion resistance against extreme mechanical and environmental conditions. Typical selective plating applications include hard chrome, electroless nickel, and sulfamate nickel. Prior to selective plating, non-wear surfaces are protected from undesired build up by the use of a masking agent. The most common masking agents include hard tooling and fixturing, caps, plugs, tapes, peelable resins, lacquers, and waxes.

Lacquers and waxes have been used for years because of their resistance to elevated plating bath temperatures and their ability to coat components with complex geometries. Lacquers are often used during plating operations, such as electroless nickel plating where the bath temperature of the plating operation exceeds the melt point of the available waxes. Lacquers resist harsh plating chemicals and unlike waxes do not require heating for application onto the component surfaces. However, lacquers typically contain volatile organic compounds (VOCs) and have slow dry times.

Waxes are effective masking agents in plating operations with lower bath temperatures, such as sulfamate nickel plating. Waxes also solidify quickly, an advantage in higher throughput operations. They exhibit good chemical resistance, substrate adhesion, and pliability and, unlike lacquers, can be reclaimed for reuse after gross cleaning operations. Moreover, waxes can be easily removed with low-VOC cleaning products that have low health and safety risks.

It is common practice to apply two coats of masking wax, though more may be added during the masking process. Pratt & Whitney's SPOP (Standard Practices and Operating Procedures)-36[1] and POP (Process Operation Procedure) 1800-U[2] are operating procedures that outline the steps for wax masking prior to selective plating. In these procedures, three types of waxes are available for use based on the bath temperature of the plating operation: low-, moderate-, and high-temperature wax. The melting point ranges for these wax types are (1) low temperature 138°F–148°F (59°C–64°C), (2) moderate temperature 165°F–185°F (74°C–85°C), and (3) high temperature ≥175°F (79°C). The steps described in POP 1800-U for masking by immersion are

D1 Immersion Method

Wax with a minimum of two coats of low-temperature wax […], moderate-temperature wax […], or high-temperature wax […].

- For good adhesion, the temperature of the wax for the first coat should be 225°F±15°F (200°F±15°F for [the low-temperature wax]). Allow the parts to remain in the wax tank until the part reaches a temperature of approximately that of the wax, or preheat the part to approximately 225°F before immersing into the wax.
- Subsequent coatings of the low-temperature wax […] may be more easily obtained by using it at a temperature of approximately 150°F.
- Subsequent coatings of the moderate-temperature wax […] may be more easily obtained by using it at a temperature of approximately 180°F.
- Subsequent coatings of the high-temperature wax […] may be more easily obtained by using it at a temperature of approximately 200°F.

Wax Removal Technologies

A number of methods for wax removal are practiced in the aerospace industry. While aerospace operations typically involve cleaning other soils such as greases and lubricants apart from waxes, most facilities favor a cleaning process capable of both de-waxing and degreasing. Some of the methods currently available are only effective in removing waxes, necessitating the use of additional degreasing processes. The more commonly available de-waxing methods are described in this chapter.

Cabinet Ovens

Electrically heated cabinet ovens are one option used by some plating shops for the removal of masking wax from aerospace parts. These ovens use forced-air convection to evenly heat the parts and are typically equipped with self-draining trays for easy removal of the wax from the cabinet. This method removes the majority of the wax, but use of cabinet ovens for very large components is expensive, and the handling of the components is difficult.

Steam Stripping[3]

This is a variation of the hot oven method, wherein high pressure steam cabinets are used to remove wax from masked parts after plating operations. After plating, the parts are placed in "de-wax" cabinets equipped with numerous high-pressure steam nozzles. High-pressure steam impingement directed at parts placed in the center of the cabinet, along with high temperatures in the cabinet, work in combination to remove the wax from the parts. This process is effective in de-waxing the parts, but is not very effective for degreasing the parts.

Hot Water

Another common method to remove wax is the immersion of the parts in hot water at temperatures greater than 190°F (88°C). At those temperatures, typically 10°F–30°F higher than the typical wax melting point, most of the wax is melted away leaving just a thin layer of wax adhering to the parts. This method is mainly used to remove gross wax from the parts and is not intended as a final step. Wax removal with hot water is typically followed by either a semi-aqueous or aqueous cleaning process as the final step of cleaning to remove those last traces of wax. Like steam stripping, this method is a poor degreasing process.

Wax Tanks

Another gross wax removal technique is one in which the parts are dipped into hot molten wax tanks. This technique allows most of the wax to be removed, leaving a thin layer of wax coating. It also provides the advantage of recycling most of the wax; but care should be taken to prevent contamination of the wax baths with heavy metals or other process soils. Typically, to prevent contamination of the wax from metal salts and from soils used in electroplating, a layer of water (consisting of up to 10% of the volume of the hot wax tank) is maintained in the tank to help dissolve salts. Similar to the hot water de-waxing process, parts must still go through an additional cleaning process to completely de-wax the parts.

Vapor Degreasing

For many years, vapor degreasing has been widely used for de-waxing. Chlorinated solvents such as trichloroethylene (TCE), perchloroethylene (PCE), and 1,1,1-trichloroethane (TCA) had been the solvents of choice for de-waxing and degreasing. However, due to the environmental and toxicity concerns, many aerospace Original Equipment Manufacturers (OEMs) and Maintenance Repair Operations (MROs) have moved and continue to move away from vapor degreasing with chlorinated solvents.

The advantages of vapor degreasing include small equipment footprint, waterless process with no corrosion issues, simultaneous use for de-waxing and degreasing, effective (does not require additional cleaning steps), and relatively inexpensive. Apart from the environmental, regulatory, and toxicity issues associated with the traditional solvents used in the vapor degreasers, the disadvantages also include the need for regular clean outs to remove the wax from the degreasers, continuous monitoring to ensure the serviceability of the solvent (to assure that the solvent has not turned acid), and the expensive disposal of the used solvents.

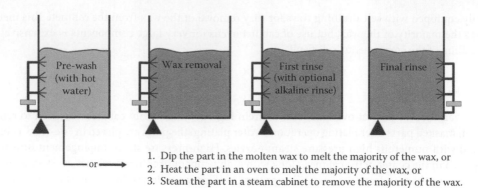

1. Dip the part in the molten wax to melt the majority of the wax, or
2. Heat the part in an oven to melt the majority of the wax, or
3. Steam the part in a steam cabinet to remove the majority of the wax.

FIGURE 27.1   Semi-aqueous cleaning process for wax removal.

Aqueous Cleaning

Aqueous cleaning using highly caustic cleaners for wax removal is in practice in some facilities, but it has limited success as a stand-alone process due to its limited ability to remove wax. It is usually preceded with hot water or steam stripping for gross wax removal. In most commercial applications, alkaline cleaning is typically used as a final process step in a semi-aqueous cleaning process so that the bath life is not quickly exhausted.

Semi-Aqueous Cleaning[4–8]

Semi-aqueous cleaning is a process by which the parts are cleaned first in an organic solvent and then rinsed with water or an aqueous cleaner. This process offers the most versatility and cost-effective solution for de-waxing and degreasing, and it is one of the most widely used processes in the industry. Figure 27.1 depicts a common semi-aqueous system designed for wax removal.

In this process, wax removal is accomplished by dissolving the wax in an organic solvent bath at a temperature usually 25°F–35°F degrees above the melting point of the wax. The organic solvents used in wax removal applications vary in solvency and can load up to 60% of the wax by weight depending on the chemistry. The heavy soil loading has the advantage of providing a long bath life. The first cleaning step can be preceded by an optional step to remove the majority of the wax with hot water, steam, a hot oven, or immersion in molten wax tanks to extend the cleaning bath life of the subsequent steps. After cleaning in the organic solvent tank, the parts are rinsed in either hot water or alkaline cleaner to remove the organic solvent from the parts. Some of the commercial organic solvent blends have built-in surfactants and wetting agents to allow rinsing in hot water alone. One to three water rinses are typical in obtaining clean parts. If an alkaline rinse step is required, it is followed by a water rinse to completely remove any residual alkaline cleaner. Typical alkaline cleaner concentrations range from 5% to 20%.

The products currently commercially available have a variety of organic solvent bases. Some of the properties of the solvents used are listed in Table 27.1.

Aerospace Specifications/Industrial Practices[9,10]

OEMs such as Pratt and Whitney, Rolls-Royce, GE Aviation, CFM International, International Aero Engines (IAE), Turbomeca, and Snecma have established multistep processes for wax removal. OEMs maintain operating manuals for cleaning processes that specify which cleaning products are approved for specific aircraft surfaces and components. Prior to approval, these products are tested in accordance

TABLE 27.1 Properties of Commonly Used Organic Solvent Bases in Semi-Aqueous Cleaning

Chemistry	Bio-Based[a]	Alkaline Rinse	Odor	Wax Loading Capacity
Mineral oil	No	Required	Mild	Medium
Hydrocarbon solvents and surfactants	No	Optional	Mild/heavy	Medium
Ester-based solvent blends	Some	Optional	Mild/ moderate	High
Terpene-based solvents	Yes	Required	Heavy	Low
Acetate ester	No	Required	Pungent	Low

[a] United States Department of Agriculture (USDA) defines bio-based products as "commercial or industrial products that are composed in whole or in significant part of biological products or renewable domestic agricultural materials or forestry materials."

with the individual OEM's cleaning specifications. The intention of these specifications is to verify the effectiveness and compatibility of the cleaning products with specific aircraft materials that include both metallic and nonmetallic substrates.

Commonly tested substrates include aluminum, stainless steel, titanium, magnesium, nickel-based alloys, acrylics, silicones, and rubbers. Compatibility is determined by testing cleaning solutions in accordance with widely accepted standard test methods. Test methods common among most aircraft OEMs include Sandwich Corrosion (ASTM F 1110), Hydrogen Embrittlement (ASTM F 519), and Stress Corrosion of Titanium (ASTM F 945). In addition to compatibility testing, many OEMs require cleaning products to meet specifications related to cleaning performance, health and safety risks, and product composition. Table 27.2 lists some of the tests typically required for OEM approval.[11,12]

In addition to these tests, many OEMs require a cleaning product to be field tested prior to approval. Testing is typically conducted at an OEM site or by an approved supplier or an MRO facility. Testing is

TABLE 27.2 Tests Typically Required for OEM Approvals

Characteristic	Test Method	Type
Total Immersion Corrosion Test for Aircraft Maintenance Chemicals	ASTM F 483	Compatibility
Corrosion of Titanium Alloys by Aircraft Engine Cleaning Materials	ASTM F 945	Compatibility
Sandwich Corrosion Test	ASTM F 1110	Compatibility
Hydrogen Embrittlement	ASTM F 519	Compatibility
Intergranular Attack	ASTM F 2111	Compatibility
Stock Loss Test Method	SAE ARP 1755	Compatibility
Copper Corrosion	ASTM D 130	Compatibility
Salt Spray Test	ASTM B 117	Compatibility
Paint Softening Test	ASTM F 502	Compatibility
Chemical Analysis	Varies by OEM	Composition
Soil Cleaning Tests	Varies by OEM	Performance
Cleaning Prior to Fluorescent Penetrant Inspection	Varies by OEM	Performance
Cleaning Prior to Bonding	Varies by OEM	Performance
Cleaning Prior Painting	Varies by OEM	Performance

often time-consuming and may require up to six months to complete. These tests are designed to verify the performance of a cleaning solution under normal operation conditions. During the field-testing period, data are collected on cleaning effectiveness, bath life, and solution makeup. Other factors in consideration may include foaming characteristics and solution stability, as well as any observable scaling of equipment and components.

Many of the wax removal processes in use today by OEMs mirror Pratt & Whitney specification SPOP 37,[13] which is a multistep process based around the semi-aqueous wax removal method. SPOP 37—Methods 2 and 3 in the Pratt & Whitney *Commercial Standard Practices Manual* defines the steps required for a semi-aqueous wax removal process as follows:

Methods 2 and 3

1. *Remove wax from part in a tank of masking wax.*
   [Additional parameters as outlined in the SPOP for this step, such as allowable time and temperature ranges are not included here or in subsequent steps.
      The bulk of the wax is removed by dipping in the hot wax tank. This serves to recover and reuse the wax, thus enabling reduced wax consumption. It also helps prolong the cleaning agent life.]
2. *Put the part fully in remover* [the cleaning agent].
   [Mechanical agitation is recommended for some cleaning solvents, while it is optional for others.]
      Depending on the solvent, the use temperature can vary from 25°F to 35°F above the melting point of the wax to a maximum temperature of 250°F (121°C). Most facilities operate the bath near 180°F (82°C) to reduce or eliminate odors and smoke formation. Typical contact times in the cleaning agent range from 5 to 20 min depending on the size and quantity. If agitation is required, spray under immersion is the most common method of bath agitation.
      Methods 2 and 3 are similar, except that Method 3 requires the use of an automatic wax-collecting equipment or overflow tank to avoid dragging of the part through the wax layer formed on the surface of some cleaning agents.
3. *Put fully in mechanically agitated hot water and then pressure spray rinse.*
   [Water should be slightly overflowing to skim removed wax residues. The temperature of the water should be 140°F–212°F (60°C–100°C) and the contact time should be 1–5 min.]
4. *If necessary, clean the part by SPOP 209.*
   [SPOP 209 is an aqueous process involving a cleaning agent sometimes used as a rinse aid after the solvent de-waxing step of the semi-aqueous process to help remove the organic solvent from the parts.]
5. *Preserve in corrosion preventative, as necessary, with SPOP 5.*

Similarly, United Technologies has patented a process that uses a combination of methods and cleaning agents in a series of baths. The process involves first removing the wax from substrates by immersing in a hot wax bath to remove and recover most of the wax. This is followed by immersion in a single or a series of hot mineral oil baths to remove any remaining wax. The oil left on the parts is removed by immersion in a semi-aqueous or organic cleaner. The parts are cleaned in an alkaline cleaner bath to remove the semi-aqueous cleaner. This is followed by a series of water rinses to remove any traces of the alkaline cleaner. The substrates are finally dried.

Semi-aqueous cleaning was also found to be critical to the successful de-waxing of a space shuttle main engine (SSME) main combustion chamber (MCC). Tests have shown that the best process for wax removal for these components uses a multistep process. The first step is to clean the part in hot oil for a gross removal of the wax. This is followed by cleaning in a semi-aqueous cleaner to remove residual wax/oil film. The parts are then cleaned in an alkaline cleaner to remove the residual semi-aqueous cleaner. Finally, the parts are rinsed in water to remove the residual alkaline cleaner before drying.

General Process Guidelines for Wax Removal

General guidelines for wax removal using a multistep process based on the semi-aqueous cleaning process are described in the following text.

Step 1—Prewash (Optional)

Gross wax removal using one of several pre-clean techniques is usually recommended. This can be done in several ways as described earlier—by dipping the parts back in the wax tank to dissolve buildup, by dipping parts in hot water >190°F (88°C), by steam stripping, or by melting in an oven. With most of these methods, wax can be captured and recycled. This step serves to reduce wax consumption and prolong cleaning agent life.

Step 2—Wax Removal

This step is the primary cleaning step and removes wax by immersing the parts in a bath containing the organic solvent. Choosing an economical cleaning agent with high solvency and good environmental, safety, and toxicity characteristics is crucial in today's business operations.

Time and Temperature

The optimal time and temperature combination will depend on variables such as the size, complexity, and thermal mass of the part being cleaned, the presence and effectiveness of any agitation, and the wax melting point. For optimal cleaning performance, the temperature of the cleaning bath should be at least 25°F–35°F above the melting point of the wax being removed. A temperature range of 180°F–185°F (82°C–85°C) is recommended to reduce or eliminate smoke and odor problems. Organic solvents with closed-cup flash points exceeding 248°F (120°C) allow for a comfortable margin of safety when operating at the typical melting points of popular waxes (less than 180°F (82°C).)

Bath Agitation

For parts with complex geometries, aggressive agitation in the bath may be needed for effective removal of the wax within a reasonable time frame. The most common method of bath agitation is spray under immersion typically achieved via a manifold system fed by a high-volume pump.

Other types of agitation include single-source agitators, such as direct-drive mixers, recirculating pumps to return fluid to the bath through a single discharge tube, platform-lift agitators, and air agitation.

Plumbing Precautions

For systems that incorporate a pump, spray under immersion manifold, and associated plumbing network, care should be taken not to cool the system below the freezing point of the solution. Typically, the freezing point of the solution increases as the cleaning agent becomes loaded with wax, with the solution no longer pourable at room temperatures for many waxes.

Installation of steam traces or heat tape around any exterior plumbing network can make startups simpler if a system is shut down and allowed to cool. An alternative method is to install valves and drain cocks at the appropriate points in the plumbing network so that draining is possible prior to cooldown and whenever the system is shut down. The bath itself is not of concern since the heating elements, once activated, will liquefy the bath.

Additionally, the viscosity of the solution increases with wax loading and plumbing requirements for the viscosity increase may need to be addressed.

Bath Life

Organic solvents used in semi-aqueous processes are robust in nature and can handle high wax loading. In a properly designed system, a cleaning bath can normally be used until it is loaded with wax to approximately 40% by weight. For some wax removal solvents, evaporative losses are minimal and the bath can be used until it is loaded. Bath lives of up to three years have been reported. Most operations routinely top off the tank with virgin cleaner to compensate for drag-out and evaporative losses, further extending bath life. Periodic change-outs are undertaken as necessitated by the process cleaning requirements.

Emissions

Some organic solvent blends used in semi-aqueous wax removal processes have vapor pressures less than 0.001 mmHg at 20°C. Very little vapor or odor is produced from the cleaner and typically minimal ventilation is required. However, many waxes produce an odor when heated, which may be detected when parts are loaded into the cleaning bath. Adequate ventilation can be achieved with lip vents around the cleaning tank or by a hood above or behind the bath among other methods.

Waste Disposal

The most common method of disposal of the wax-loaded cleaning solvent is fuel blending. Some of the solvents used are considered nonhazardous waste in the United States and can be disposed of by fuel blending as they have high BTU values, some as high as 15,000 BTU/pound.

Step 3—Water or Alkaline Rinse

The wax removal step is followed by a water or alkaline rinse step to remove the residual mixture of cleaning agent and wax. Often, two rinses following the cleaning bath are needed to keep the final water rinse contamination free. While water only may be used to rinse, an alkaline solution as a rinse aid in the first rinse step typically improves the completeness of the rinsing process and decreases rinsing time.

Time and Temperature

If water without alkaline rinse aids is used for rinsing, immersion of the part in mechanically agitated hot water for 1–5 min at temperatures of 140°F–212°F (60°C–100°C) improves rinsing. Skimmers or reverse cascading tanks are used to remove wax residues and the residual solvent that float to the top.

When an alkaline cleaner is used as a rinse aid, the operating temperature of the bath is maintained as per the manufacturer's instructions. Typical temperatures of the rinse aid bath range from 140°F to 180°F (60°C–82°C). Immersion of the part in the alkaline solution for 1–5 min with mechanical agitation is common.

Most often, the concentration of wax in the cleaning bath will affect the rinse bath temperature. As the wax load increases to fifteen weight percent or more, higher rinse temperatures may be required to effectively rinse the cleaning agent/wax mixture from the parts. Operating the rinse baths near the melting point of the wax improves and maximizes bath life.

Bath Agitation

Agitation in the rinse bath is necessary for optimal performance if no aqueous rinse aid is used. Spray under immersion is commonly used and is typically achieved via a manifold system fed by a high-volume pump.

Spray-in-air agitation used throughout the rinse cycles provides excellent results when rinse-water sprays are configured so all surfaces of the parts are contacted.

While ultrasonic agitation can be very effective, it is rarely used, as it is prohibitively expensive in applications where very large parts such as jet engine component assemblies necessitate large bath volumes.

Platform-lift agitators are also used, but are not as vigorous as the other methods discussed. These systems often require a rinse aid to yield effective rinsing.

Air agitation, while an inexpensive option, is problematic due to its lack of aggression, its tendency to cause foaming problems, and the potential for creating mists above the tank.

Bath Life

If an alkaline cleaner is used as a rinse aid, the rinse bath life is typically dependent on the level of wax loading in the solvent bath. As wax loading increases, the alkaline cleaner is expended more quickly. Typical rinse aid concentrations range from 5% to 15%; and periodic additions are required to maintain the concentration of the bath at the desired level.

Some alkaline cleaners require frequent addition of pH adjusters and makeup chemicals to prevent scale formation or precipitation of some of the cleaner's components. Choosing a stable alkaline cleaner that does not require pH adjustments or makeup chemicals can reduce the complexity of managing the rinsing step and increase process efficiency.

Emissions

In most jurisdictions, rinse baths are vented to the atmosphere since the emissions are steam. Some especially restrictive regulatory areas require de-misters in the air stream. If the rinse aid contains any organic solvents, emissions from such baths may be subject to local, state, or federal regulations. Proper ventilation may also be required to reduce odors in the workplace.

Waste Disposal

Wastewater from the process may be handled in a variety of ways. Rinse water containing low amounts of organics may be discharged into the sewer depending on the local regulations that set the maximum levels for organics and certain heavy metals. Wastewater from the rinse baths can be recycled, if desired, through membrane filtration or ion-exchange resin beds. An evaporator may also be used to concentrate the cleaning agent and wax mixture for waste disposal.

Step 4—Final Water Rinse

Time and Temperature

A final water rinse is crucial for the completeness of the cleaning process and is configured similar to the first rinse step with regard to heat and agitation. Rinse water temperatures at or above the melting point of the wax yield maximum rinse effectiveness. A small stream of fresh water can supply the rinse bath with an overflow to drain.

Bath Agitation

Final rinse stage agitation is typically either spray under immersion or spray in air. Final spray-in-air rinses have been used with excellent results for systems without aggressive rinse agitation. Handheld spray wands are fed either by fresh water or with water from the final rinse.

Water Quality

The water quality requirement of the final rinse bath is determined primarily by the level of cleanliness needed. If the wax removal process also serves as the final cleaning step, the water quality requirement for the final rinse will be higher. While tap water is used in some processes, deionized water is recommended for most operations. Rinse processes are usually configured so that water is metered into the final rinse bath at a rate sufficient to keep the bath adequately purged of contamination.

Wastewater Discharge

As in Step 3, the disposability of the final rinse water bath into the sewer is dependent on local regulations. In applications where wastewater discharge must be kept to a minimum, water from the cleaning process can be recycled through membrane filtration or ion-exchange resin beds.

Step 5—Drying (Optional)

Parts cleaned in semi-aqueous processes with final rinses in hot water usually flash dry as a result of heat retention in the part. Forced air (sometimes heated) is used in cases where drying assistance is needed. Parts can also be dried by a variety of other methods including oven, vacuum chamber, or centrifugal force depending upon the parts involved and the specific needs of the application.

Summary

In the aerospace industry, the removal of masking wax is a common and critical practice for selective plating operations. Waxes can be difficult to remove if processes are not properly configured. A variety of techniques are currently available for OEM and MRO operations as described in OEM manuals. Successful wax removal processes are often multistep and require matching the specific equipment with the selected cleaner to effectively clean within the process parameters outlined in the OEM procedures. Also, factors such as environmental impact, worker health and safety, economics, and performance must also be considered.

References

1. Plating Procedures—SPOP 36, *Pratt & Whitney Standard Practices Manual*, June 1, 2009.
2. Plating Procedures—POP 1800-U, *Pratt & Whitney Process Operation Procedure Manual*, February 14, 2008.
3. Case Study #1: Dewaxing aircraft components using steam instead of solvents. Eliminating CFC-113 and methyl chloroform in aircraft maintenance procedures. *EPA/ICOLP Aircraft Maintenance Manual*, pp. 157–158, September 1993.
4. US 5,209,785, Non-Chlorinated Solvent Dewax Process. United Technologies Corporation, 1993.
5. Steven B. Hayes, A semi-aqueous process for removing masking wax used during plating operations. *Plating and Surface Finishing*, 84 (7), 29–32, 1997.
6. C. Eden, Can a plating shop operate without vapor degreasers—Even for wax coatings? *Plating and Surface Finishing*, 87 (2), 18–19, 2000.
7. E. A. Bivins and S. B. Hayes, Wax and pitch compound removal: Refining the new, revisiting the old. *Precision Cleaning*, 5(12), 11–15, December 1997.
8. A. U. Aktapi, SSE main combustion chamber (MCC) hot oil dewaxing. NASA Contractor Report, NASA-CR-202079, 1994.
9. F. Brandisi, Whyco Finishing Technologies, LLC., personal communication.
10. Guy Boucher, Pratt & Whitney, personal communication.
11. MIL-PRF-680A Performance Specification, Degreasing Solvent, July 25, 2003.
12. PWA 36604 (Revision D, 01/26/05) Approval of Cleaners Used in Manufacture and Overhaul of Parts, January 26, 2005.
13. Plating Procedures—SPOP 37, *Pratt & Whitney Standard Practices Manual*, June 1, 2006.

28

Implementation of Environmentally Preferable Cleaning Processes for Military Applications

Overview 439
Life Cycle Cost Focus
DoD Alternatives Implementation Road Map 440
Process
Barriers to Implementation.. 442
Decision-Making Process • Evolving Regulations • Definition of
Requirements • Logistics of "Greening" the Supply Chain • Institutional
Resistance to Change
Keys to a Successful Alternative Technology Implementation Effort.... 444
Define Required Performance • Focus on Implementation, Not
Science • Engage the User • Influence the Design Process • Work the
Logistics • Grow Champions
DoD Solution 447
Proactive View of Evolving Regulatory Issues • Joint Service Solvent
Substitution Working Group • Materials Information Management for
Sustainable Chemical and Materials Management
Conclusion 451
References..451

Wayne Ziegler
Army Research Laboratory

Tom Torres
*Naval Facilities Engineering
Service Center*

Overview

Can the U.S. Department of Defense (DoD) effectively implement green cleaning alternatives with minimal impact on weapon system performance, operations productivity, or cost? Executive Orders and EPA regulations are placing increasing pressure on depot and maintenance facilities to operate in an environmentally friendly manner. Traditionally, vehicle, equipment, aircraft, and ship maintenance operations have utilized organic solvents containing VOCs and Hazardous Air Pollutant(s) (HAPs), so reducing the impact of cleaning and surface preparation operations is a significant part of any compliance scheme. As a result of real and perceived materials compatibility issues or process impacts from environmentally preferred alternatives, the DoD continues to use large quantities of organic solvent cleaners at great expense. There have been many instances where "good science" failed to make the transition to the shop floor. In some cases, these "green" initiatives have come at the sacrifice of technical performance and productivity, at an increase in the cost of operating an industrial facility, and with a negative impact on weapon system performance. Recent technical concerns are influencing a change from an ESOH (Environmental Safety and Occupational Health) compliance focus to a healthier focus on risk management by improving technical and economic performance while reducing ESOH impact.

Life Cycle Cost Focus

In the current global economy, the DoD cannot rely on traditional approaches of pollution control and special exemptions to ensure continued operations worldwide. Implementing Executive Order (EO) 13423, Strengthening Federal Environmental, Energy, and Transportation Management and preparing for the Department of European Union (EU) "REACH" initiative, Registration, Evaluation, Authorization, and Restriction of Chemical Substances have necessitated the DoD to focus on sustainable chemical and material management.[1] Compliance with the conventional end-of-pipe environmental pollution control focuses only on the emissions. However, adverse impacts on the environment occur from the other life cycle stages, such as production, distribution, operation, maintenance, and disposal. Without reducing environmental impacts from the entire life cycle of a weapon system, the environmental problems that accrue from the production, use, and disposal of the product result in a significant cost burden. In addition, the record-keeping burden necessary to demonstrate compliance with both traditional end-of-pipe regulations and recent regulations focused on risk management represents an enormous cost.

Recognizing these factors, the DoD has shifted focus from pollution control to pollution prevention and sustainability. For developmental weapon systems, this has been accomplished by efforts to implement "design for sustainability" concepts. Efforts are underway to modify the acquisition management decision-making process. The approach is to minimize the life cycle costs of ESOH impacts by considering those impacts systematically during the design process. This requires the identification of key environmental issues related to the product throughout its entire life cycle. The implementation of sustainability aims at attaining and balancing mission risk and environmental preservation. Legacy systems are a particular challenge. For such systems, the design is already in place. Of necessity, to assure reliability of such mission-critical systems, processes, including cleaning and other manufacturing processes are frozen in place. Therefore, modifying the manufacture or repair of associated instruments and components is exceedingly complex and cumbersome. There is a shift from looking at environmental regulations as being purely environmental engineering problems to that of looking at environmental issues as part of systems engineering. This shift in viewpoint reopens the design door so that the focus is on mission risk, and the regulatory driver becomes an opportunity to improve a weapon system.

There are significant barriers to the implementation of sustainability concepts and opportunities. Elucidating and overcoming these barriers will be discussed.

DoD Alternatives Implementation Road Map[2]

The first challenge to developing an effective implementation strategy in a large bureaucratic organization is to understand the review and approval process. The objective is to implement "green" cleaners/degreasers while ensuring continued, reliable operations in DoD general maintenance applications. While all parties may agree on this general goal, there are many competing and sometimes conflicting objectives.

Process

The Army Alternative Cleaner Program will be used as an example of barriers to steps along the pathway to implementation. During the initial phases of the Army Alternative Cleaner Program, the project team attempted to map out the process of finding and implementing alternative products or processes. This effort included identifying the key technical communities involved in each aspect of the process. The process flow diagram in Figure 28.1 identifies the key steps in a typical alternative process or product evaluation and implementation project and assigns each step to the principle technical community.

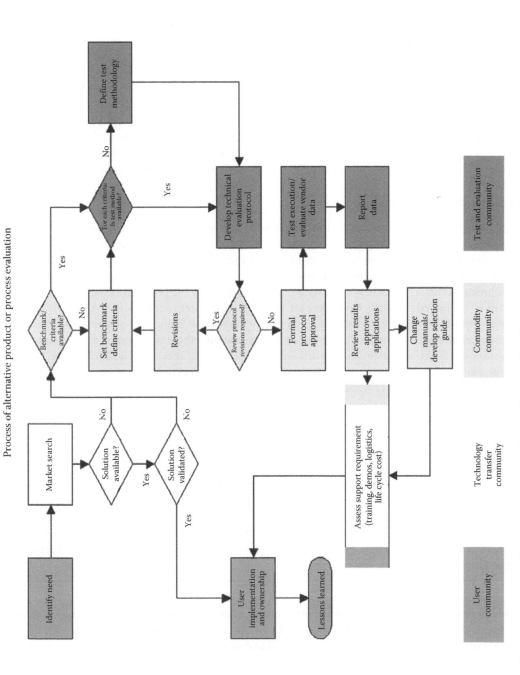

FIGURE 28.1 Key steps in a typical alternative process or product evaluation and implementation project assigning each step to the principle technical community.

Each time the responsibility moves from one technical community to another, there is a potential break-down within the process and an opportunity for miscommunication.

It was found that, as a general rule, the process breaks down in the initial and final stages: identification of requirements and the implementation of the alternative. The science of finding and evaluating alternative products is the "easy" part of the process compared to the politics and the battle to overcome inertia that characterizes the identification of the requirements and the implementation of alternatives on the shop floor.

Barriers to Implementation

There are significant barriers to the implementation of alternative cleaners and processes.[3]* Many of these barriers result from policies and procedures that have been developed over years of acquisition experience because these procedures work well in most cases to manage and to regulate a complex and critically sensitive purchasing environment. These policies, however, often have the impact of impeding or preventing the development, qualification, and implementation of alternative materials and processes.

Decision-Making Process

The greatest barrier to the implementation of alternative materials and processes is the acquisition decision-making process itself. The current acquisition policy is focused on control and is often compartmentalized. This approach impedes the implementation of alternatives in general processes and operations that cross-functional areas. Starting with system development, there is little incentive for a weapon system technical authority to incorporate any alternative approach. The program manager's success continues to be evaluated primarily based on cost, schedule, and system performance. Although there are efforts to include life cycle cost and ESOH considerations in the evaluation of a program manager's performance, this will continue to be an issue because currently there is no formal mechanism for the recognition of efforts to implement alternatives. Consideration of alternative technologies, even those that potentially improve performance, reduce cost and mitigate life cycle and maintenance burdens, represent an increased program cost and risk. In addition, the technology development cycle has historically been significantly longer than the engineering design cycle (Figure 28.2). The lack of "maturity" of many of the alternative chemicals and processes adds to the program risk. From the perspective of the program manager, the downside of utilizing a "new" approach often outweighs the potential benefits.

A second acquisition barrier to implementing alternatives is the process for approving innovative processes or technologies. Each service and each weapon system platform has its own policies and procedures for defining the success criteria and the test methodology to determine if a new process meets acceptable performance requirements. On one level, this is understandable since the cognizant authority is responsible for any technical failures. However, the net effect is to impede innovation and the implementation process. Considering the number of cleaning operations that are ubiquitous to industrial and maintenance operations, this results in a complex web of requirements that an alternative must comply with to gain wide implementation.

The fact that the costs incurred as a result of environmental compliance are typically incurred by the maintenance organization, not the weapon system manager, is a third acquisition barrier. While environmental requirements significantly increase operation and maintenance budgets for facility managers, historically these costs were not assigned back to the weapon system technical authority. In their defense, program managers are often not equipped to consider the future environmental consequences

---

* This paper was drafted based on a roundtable discussion that occurred during the final session at the *13th Annual Solvent Substitution Workshop* in Scottsdale, AZ on December 12, 2002. The results of a survey asking about barriers to adopting new technologies and recommendations for overcoming them were provided to attendees. The discussion was lead by Mr. Joe Vallone and Mr. Carl Scott of the Office of the Deputy Assistant Secretary of the Army for Environment, Safety, and Occupational Health.

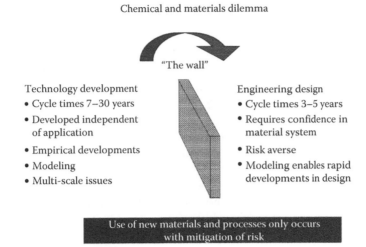

Chemical and materials dilemma

"The wall"

Technology development
- Cycle times 7–30 years
- Developed independent of application
- Empirical developments
- Modeling
- Multi-scale issues

Engineering design
- Cycle times 3–5 years
- Requires confidence in material system
- Risk averse
- Modeling enables rapid developments in design

Use of new materials and processes only occurs with mitigation of risk

FIGURE 28.2 Dilemma of introducing new materials or processes.

of current practices. The lack of systems to aggregate and disseminate information on current and projected ESOH issues and requirements hampers the ability to influence the design process, address current issues, and anticipate emerging issues. In addition, the design information typically available does not include ESOH factors.

Evolving Regulations

The regulatory environment in which the DoD operates is not consistent or constant. The DoD operates industrial facilities in every region of the United States and worldwide. These facilities focus on a vast variety of operations and maintenance activities across a wide spectrum of commodity areas, from aviation to ground vehicles to electronics and communications. The complex web of regulations leads to the necessity for alternatives to be evaluated in the context of many different compliance situations. At times, this web of regulations results in conflicting guidance related to worker safety and the protection of the environmental.

Definition of Requirements

The first step in identifying alternatives is to understand the requirements. While this may seem obvious, in practice it is often difficult. For example, if there were an effort to replace a VOC/HAP cleaner for a given maintenance operations, the first question asked of the technical authority for the weapon system and the maintenance manager would be, "What does the alternative cleaner need to do?" There are three typical responses to this question; none of them are helpful in setting up a realistic definition of requirements for alternatives.

In most cases, the response would be, "We are currently using XYZ solvent and we need something that works just like that." There is more to defining a cleaning process than identifying the existing cleaner, and since a proper definition of requirements is the key to effective implementation, this typical response constitutes an information gap.

At the other end of the spectrum is the request for an alternative to do everything. This request comes with a long list of soils and substrates, an impossibly short cleaning cycle time, and a cost range that eliminates the profit margin for any potential cleaning agent manufacturer. There is no magic, one-size-fits-all, drop-in alternative. All cleaning applications are not equivalent and an understanding of what, why, and, realistically, how fast is critical.

The third unproductive perspective is the "green only" view. This alternative request lists applicable regulatory restrictions with no consideration of mission impact and economics. "Green" is not enough. There must be a focus on the capability and the cost of an alternative in addition to the environmental impact.

Logistics of "Greening" the Supply Chain

Even after preferable alternatives have been identified and approved, actual implementation is problematic, especially at the field maintenance level. The logistics of replacing existing products and greening the supply chain are complex. Once a change has been authorized, all related documents need to be changed. The necessary changes to individual technical manuals and control documents are expensive, difficult to coordinate, and slow. Acquisition reform dictated a move from technical specifications to performance-based specifications. Performance specifications define the required performance as opposed to prescribing materials or processes. Specifying an alternative requires the presence of a technical authority to maintain a list of approved products and to monitor product integrity.

Let us say that a substitute for a widely-used VOC was found to be acceptable. In most cases, it is impossible to remove that VOC from the supply chain since there may be critical applications that still require that product.

The manner in which the DoD identifies products becomes a barrier to the deployment of an alternative. National Stock Numbers (NSNs) identify product groups that meet a set of requirements as opposed to a single product. This approach is critical in minimizing document changes by allowing changes to a single document to flow to all referencing documents. A maintenance facility can order a stock numbered item and receive a different product to what they are expecting. While this product also meets the requirements of the respective performance specification, the maintenance authority may view it as a process change. As a result, purchases with credit cards and attempts to circumvent the system by ordering product-specific NSNs are used with the unintentional result of undermining efforts to implement environmentally preferred alternatives.

Institutional Resistance to Change

The DoD, like any large bureaucratic organization, has a lot of inertia that must be overcome to effect a culture change. Influencing the acquisition decision-making design paradigm will not take place overnight. Changing the information flow between R&D technical personnel and design engineers will take time. Convincing a technician on the floor of a depot to change the way he cleans a widget will take training and engagement. There are efforts underway to influence small changes where possible and put systems in place to support significant changes in the future.

Keys to a Successful Alternative Technology Implementation Effort

Define Required Performance

Effective implementation requires an understanding of the current cleaning application. Simply identifying the current cleaner does not sufficiently define the performance and compatibility requirements for an alternative. The cleaning application must be considered as part of a process, not a stand-alone operation. The Joint Service Solvent Substitution (JS3) working group, described later in this chapter, has identified the following questions to help guide the definition of the required performance for a given cleaning application.

- What is the "regulatory driver"? What is the anticipated timeline for this regulatory driver?
- Do we need an alternative? Why is there a cleaning process? Is there a requirement for an alternative? Is there another way to achieve the desired result? Is the process redundant and is there an opportunity for process optimization?

- What is the current controlling "documentation"? Who controls the document and the process?
- What is the "contaminant"?
- What are the "substrates"? What are materials compatibility requirements? What will be cleaned? Is this a dedicated cleaning station? Or will other parts and soils be added to the mix?
- What are the process requirements"? Define the process flow boundaries: cleaning time, drying time, and residue restrictions. What happens after cleaning/degreasing? What are the possible downstream impacts?
- What are the "economic factors"? What is the ROI required for switching to an alternative? What cost drivers are in play? What is the total ownership cost for the current cleaning process, including bath life and disposal costs?
- What are the keys to "worker acceptance"? What previous efforts have been tried? With what success or failure? Have workers been engaged in the process of determining an alternative? What are the training requirements for a new process?

The challenge is to be part of the solution, not the perpetuation of the problem.

Focus on Implementation, Not Science

Within the DoD, the two groups of professionals most commonly involved in materials substitution for pollution prevention are the scientists/technologists and the compliance officers. Traditionally, these two groups of professionals have had a limited view of the process of implementing alternatives. As a result of the DoD command structure and the compartmentalization and specialization of professions, it has been unusual for one professional group to consider the impacts and constraints outside of their area of expertise. The compliance office understands the regulatory drivers but they often do not understand the practical challenges of implementation. The technologist, on the other hand, sees the regulatory drivers as performance criteria along with materials compatibility and effectiveness. The scientist faces several challenges related to their worldview. There is a natural tendency to shoot for the moon and look for the silver bullet solution that will be a drop-in replacement for currently used solvents. It is important that the scientist understand the short term realities and pursue solutions that have a reasonable likelihood of being successful. The scientist also needs to see the big picture, consider the related processes, and seek to understand user input. The scientist needs to get out of the lab and onto the shop floor.

The challenge for both of these groups is to include other significant factors that will control the successfulness of the implementation of an alternative. Both groups are focused on an aspect of the science without a proper appreciation of the economics, politics, and psychology of change.

Engage the User

There are many "users" or "customers" to be considered when developing an effective implementation plan. Each user has a unique and limited perspective of the cleaning operation, and what you do not know can hurt you. Working in technology transfer, it is important to understand these perspectives and to talk with each user group in the language that they understand. The "design engineer" has a bias toward materials and processes that he knows and understands. Like the program manager in the decision-making process, there is very little incentive to adopt alternatives. On the contrary, there is a significant risk for a design engineer in recommending alternatives. To engage design engineers, they must be provided reliable information in a useable format that they can use to make decisions about the compatibility and efficacy of alternatives and, ultimately, to defend those decisions. The "program office" is a second user group that is risk averse and skeptical of alternatives. The issues that concern the program office were discussed at length in the section above on the decision-making process. The direct user group is the "maintenance community." This end user group is concerned with productivity impacts and the life cycle costs of the alternative. This is the guy who is getting his hands dirty, who must use, dispose of, and face the consequences of impacts of the alternative cleaner or process.

Each of these user groups needs to be engaged in the process of understanding the potential weapon system or process impacts, identifying criteria, estimating life cycle costs, and qualifying alternatives.

Influence the Design Process

The ideal solution for new weapon systems is to design out the undesirable materials and processes. The design engineer uses his background and experience to design items and select materials and processes. Then the design drawing is submitted and a materials or environmental or safety person evaluates and provides feedback, which may or may not include help in identifying alternatives. Design engineers and process engineers make decisions based on the information available. It is very costly to rethink a design that was made based on incomplete information. Seventy percent of life cycle cost is locked in by the end of the concept and development stage. Therefore, a key to implementing alternatives is to make the information about environmental and safety issues readily available as well as to provide alternatives to the design engineer during the concept development phase.

Work the Logistics

Ultimately, a successful implementation has not occurred until the end user consistently uses the alternative and the use of the undesirable chemical is eliminated or controlled. Because there are critical applications where performance characteristics of traditional HAP and VOC solvents are necessary, it is impossible to completely remove a chemical from the logistics system. Therefore, it is important to put in place systems and controls that work. The Defense Logistics Agency (DLA) has the responsibility for commodity management. However, DLA does not establish specifications and has limited ability to impact buyer actions. The purchases by the end user are controlled by operating manuals, technical manuals, and specifications. End users need to be educated about the alternatives and the issues with the cleaner being replaced.

One of the difficult logistical issues in specifying environmentally preferred cleaning agents is the ability of the end user to make small purchases by credit card without engaging the logistics system. It is, therefore, critical to engage the user's cooperation. The modification of existing specifications and the development of new specifications is part of the compliance scheme as well. The end user needs guiding documents and cleaning specifications that tell them what to use and how to use it.

The cleaning specification also provides a mechanism for adapting to changing regulations. Changing a cleaning document that is referenced in maintenance documents allows a single document change to impact many operations. This was demonstrated with the modification of MIL-PRF-680, one of the most commonly used cleaning specification documents within the DoD. Since there was no language in the specification prohibiting hazardous air pollutants (HAPs), the products that were listed on the qualified products list for MIL-PRF-680 were not NESHAP (National Emission Standards for Hazardous Air Pollutant) compliant. A survey of the qualified products revealed that they were all HAP-free and a subsequent change to the document made MIL-PRF-680B compliant with the proposed Defense Land System and Miscellaneous Equipment (DLSME) NESHAP.

Grow Champions

History has demonstrated that efforts to introduce alternative chemicals and processes are much more successful when the intended user is pulling for the change and when the intended user sees the benefit of the change to his job. Working within the intended user communities and growing champions for change lead to a much better chance of implementation than attempting to force acceptance of alternatives as being the right thing to do. Within the DoD, this is accomplished by celebrating success stories and attempting to build a fan base from within. With each user group, it important to speak the language that they understand. With the design engineer, the focus is on the substitution options and their technical performance. The program office needs to be educated about the life cycle cost benefits and returns on investment. The user and the facility owner need to understand the ESOH benefits and the process impacts. The compliance office is looking for ease of documenting compliance and future potential impacts.

DoD Solution

The focus on "greening" in the DoD is based on a realization that environmental regulations impacting the industrial base are system engineering challenges and not environmental engineering problems. There is a commitment to working with regulators to obtain sustainable solutions and to obtain system engineering/pollution prevention solutions in preference to pollution control solutions. All alternatives must consider program risk (cost, schedule, performance), mission risk (readiness and capability), and war fighter impact (readiness, health, and safety) up front and often. The alternatives must be technology driven, war fighter focused.

Proactive View of Evolving Regulatory Issues

There are efforts within the DoD to adopt a proactive vs. reactive view of environmental and safety risks. ESOH risk management must be integrated into DoD risk assessment and decision-making processes as early and frequently as possible.[4] The key is to develop a process for early identification of potentially problematic materials and processes. This early identification reduces life cycle costs and expands operational options.

The Materials of Evolving Regulatory Interest Team (MERIT), a DoD group formed to focus on defining risks and finding solutions, devised a scan–watch–action process. The process was developed to scan the horizon for emerging contaminants, to assess their potential impact on DoD operations using a defined risk assessment process, and to create a risk management plan. Emerging contaminants with possible DoD impacts are moved to the Watch list. If the qualitative risk assessment identifies a probable high DoD impact, the material or chemical is moved to the Action list. For materials on the Action list, a detailed analysis of impact is executed, and risk management options are identified and proposed. This process relies on rapid collection and dissemination of reliable information.

The DoD has also worked to cooperate with regulators to define workable compliance options that reduce impacts on the environment and that at the same time reliably support both the weapon system and the defense industrial infrastructure. For example, the EPA was in the process of proposing a number of NESHAP regulations that would have a significant negative impact on DoD operations. The DoD maintenance and operations activities span many areas and are covered by many NESHAPs using the same facilities. Therefore, the record keeping to demonstrate compliance with multiple regulations for the same coating operation would have been cost prohibitive. The DoD Clean Air Act Services Committee worked with the EPA and other experts to show that this regulatory plan was not in the best interest of either the environment or the DoD. The EPA agreed and worked to develop the proposed Defense Land Systems and Miscellaneous Equipment NESHAP that covers all DoD and NASA applications that are not subject to the Aerospace or Shipbuilding NESHAPs.

Joint Service Solvent Substitution Working Group[5,6]

Joint service
Solvent substitution
Working group

In 2001, a collaborative effort was funded by the DoD Strategic Environmental Research and Development Program (SERDP) to identify R&D gaps in the solvent substitution technology area. One of the outcomes of that project was the establishment of a joint service working group to continue the collaboration initiated by the gap analysis effort.

The JS3 working group was formed to facilitate open communication, encourage collaboration, and promote a well-coordinated response to solvent issues related to regulatory requirements. The keys to successful implementation discussed previously apply to any individual effort. However, one of the key elements of widespread implementation is the cross pollination of ideas and results within a large bureaucracy like the DoD.

There are many examples of lack of communication within each of the services and obvious communication gaps between services. For instance, as part of a recent effort to qualify alternatives for a vapor degreasing operation, a study was funded to identify the process requirements and to research the performance of currently available, commercial, off-the-shelf alternatives. At the conclusion of the study, it was discovered that an identical effort had been carried out 8 years earlier and that the same conclusions had been reached. The two studies involved the same facility, the same operation, and the same conclusion, but there was no mechanism for sharing and storing the results of the two efforts. The DoD cannot afford to waste time and resources in duplicating work across services.

JS3 Mission

- Ensure coordination and facilitate solvent substitution information exchange.
- Establish a joint methodology.
- Develop and support a database to provide a mechanism for promoting and facilitating project coordination and information exchange.
- Validate & coordinate DoD solvent research, development, test and evaluation (RDT&E) efforts. They serve as subject matter experts to regulatory and decision-making agencies.

The JS3 working group has developed tools that have been endorsed by each of the services and NASA. In addition, these tools serve as primary mechanisms of coordination, collaboration, and interaction with industry.

JS3 Methodology[7]

The JS3 methodology is descriptive as opposed to prescriptive. It does not identify alternative cleaners. Instead, the methodology identifies the critical steps necessary for the acceptance of alternative cleaners for DoD industrial maintenance activities. Solvent substitution for DoD maintenance activities is a complex process entailing a significant amount of coordination and testing. A well-thought-out plan must be developed to successfully replace a current solvent with an environmentally friendly alternative cleaner. The Naval Facilities Engineering Service Center, with input from the JS3 working group, derived a methodology tool for establishing the acceptance criteria needed for the approval and successful implementation of alternative cleaners. Acceptance criteria include material compatibility; environmental, safety, and occupational health criteria; chemical properties; and performance.

The methodology was derived in part from existing information supplied by the DOD agencies who participated in the JS3 working group, the Army Research Lab (ARL), Army Research, Development and Engineering Centers (RDECs), Naval Air Systems Command (NAVAIR), Naval Sea Systems Command (NAVSEA), Naval Facilities Engineering Command (NAVFAC), Marine Coups Logistics Bases (MCLBs), and NASA. The methodology is a tool for establishing the acceptance criteria needed to obtain approval and successfully implement alternative solvents. The methodology focuses on the importance of involving stakeholders in the development of acceptance criteria, in test plans, and in the evaluation of results. By introducing and gaining the endorsement of a joint methodology, the JS3 working group has significantly increased the likelihood of leveraging resources among the services. The across-the-board endorsement has enabled the DoD to engage industry (particularly suppliers of cleaning agents) in addressing broad application challenges. The methodology directs that the project lead to engage all stakeholders and to develop an implementation strategy early in the program.

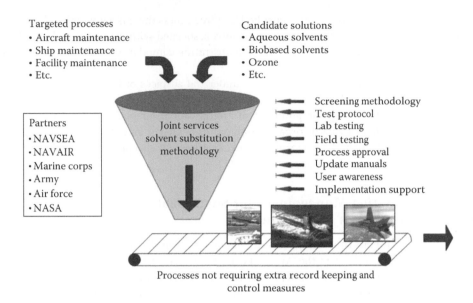

FIGURE 28.3 JS3 methodology process.

In addition, JS3 has incorporated acquisition reform efforts within the DoD that focus on the search for alternatives so as to emphasize process requirements. This supplants efforts that focus on quantifying the properties of the historical cleaner and then attempting to find alternatives that replicate those properties.

The JS3 methodology is requirement driven, which means that a maintenance process warranting an environmentally friendly solvent must be identified before executing the methodology. As one follows the methodology, each process will have its own unique set of requirements. Of course, to reduce implementation costs, some processes under investigation may have commonalities that can be shared. Figure 28.3 schematically shows the JS3 methodology process.

JS3 Project Tracking Database

It is also important to ensure that information collected and generated for alternative cleaners is stored centrally and is readily available. The JS3 database is a web-based information system that allows the exchange of historical, ongoing, and proposed solvent substitution efforts. It is not an alternative cleaner selection database. The process description and acceptance criteria entered in the database allow DoD agencies to identify leveraging opportunities and provide an opportunity for industry to see the details of processes currently under investigation. Vendors are encouraged to submit technical test data on the performance of alternative cleaners and to review the process descriptions so that they can nominate a product for consideration. The JS3 database can be accessed from the Joint Group on Pollution Prevention website, www.jgpp.com.

JS3 Data Exchanges

The JS3 working group generally meets twice a year to discuss ongoing efforts, to brainstorm opportunities for collaborative efforts, and to review changes in the regulatory environment. This meeting is often held in conjunction with pertinent meetings and conferences to maximize the benefit to participants.

General Cleaning Specification Development

One of the issues identified by the members of the JS3 working group is the significant number of general cleaning applications that are not governed by a specification. Under proposed regulations like the Defense Land System and Miscellaneous Equipment (DSLME) NESHAP, all cleaning operations that are part of the surface coating operation must be HAP free unless there is a governing document that specifies otherwise. Because many cleaning operations associated with ground vehicles and support equipment are not governed by a specification, demonstrating compliance would be very tedious and costly.

Therefore, a development team was assembled under the auspices of the JS3 to prepare a general cleaning specification that would be applicable to these operations. The scope of this specification covers general cleaners and cleaning compounds that are hazardous air pollutants (HAP) free and that contain either a low content of volatile organic compounds (VOC) or an exempt VOC, for use on military systems. The introduction of a performance specification will also provide adaptability for future regulatory changes.

Materials Information Management for Sustainable Chemical and Materials Management

One of the consistent themes throughout this discussion has been the need for information to support decision making. A gap analysis of the Army's current materials development process indicates that there is a lack of exchange of materials and process information early in the development cycle. This lack of systems to aggregate and disseminate information on the current and projected ESOH issues inhibits the ability of those charged with developing preferred, sustainable processes to influence the design process, address current issues, and anticipate emerging issues. Insufficient data coordination also results in a duplication of efforts and expensive design modifications later in the life cycle. Further, the data gap persists throughout the life cycle making the identification of implementation opportunities problematic. In addition, the design information that is available does not include ESOH factors and may not be in a format that is useful to the design engineer. The traditional stovepipe organization (Figure 28.4) and the compartmentalization and specialization of technical personnel in RDT&E impede the exchange of information. This traditional approach placed chemical and materials research very early in the acquisition life cycle (for the DoD, prior to the concept development decision milestone, and in many cases pre-acquisition life cycle). The information generated by these activities is "thrown over the wall" into the design stovepipe with little or no care about the format or usefulness of the information provided. There is limited communication between stovepipes, and the communication that does occur is usually the result of individual efforts, so the link is lost when personnel change. There is no efficient process for the collection and distribution of corporate knowledge. Eighty-five percent of the life cycle costs and environmental impact is built into a weapon system during the technology development phase. By not addressing ESOH issues in the early phases of development, the DoD is condemned to attempt to apply design improvements and to fix field problems.

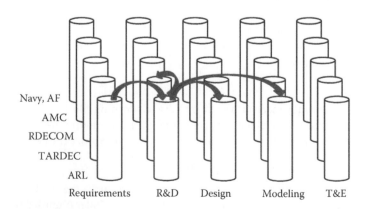

Navy, AF
AMC
RDECOM
TARDEC
ARL

Requirements   R&D   Design   Modeling   T&E

FIGURE 28.4 Stovepipe organization.

There are focused efforts within the DoD to implement ways to deliver research products better, faster, and more affordably. The objective is to move from stovepipes to a distributed information system. The DoD is teaming with NASA and industry to implement a materials and process information system for the purpose of recording and disseminating chemical and materials information to inform the decision-making process. For design engineers, this information will be available through modules in their design engineering packages that will flag ESOH issues and provide information on potential alternatives. This information system also has the potential to significantly reduce the administrative burdens associated with compliance demonstration and conformity with precautionary regulations like REACH.

Conclusion

We started this chapter asking the question, "Can the U.S. Department of Defense (DoD) effectively implement green cleaning alternatives with minimal impact on weapon system performance, operations productivity, or cost?" Looking forward, the JS3 working group has developed and gained endorsement for a common process. JS3 has encouraged the communication and cooperation that is necessary to address the implementation of alternative cleaners in an increasingly restrictive regulatory environment and in spite of decreasing RDT&E funding.

The keys to success can be adapted to many efforts involved in sustainable, environmentally preferred cleaning processes:

- Focus on the life cycle impact.
- Engage all affected groups.
- Define process requirements.
- Be realistic.
- Put the information into the right hands.

Finding alternatives is the "easy" part, however, being "green" is not sufficient. There must also be a focus on the performance and economic benefits of alternatives.

References

1. *Sustainable Chemical and Material Management: Next Steps at the Department of Defense (DoD)*, presented at NDIA Environment, Energy & Sustainability (E2S2), May 2009, Carole LeBlanc, PhD Chemical and Material Risk Management Directorate, Office of the Deputy Under Secretary of Defense, ODUSD (Installations & Environment).
2. Final Solvent Usage Technology Process Roadmaps to Address Technology Gaps Report (CDRL A012), dated March 24, 2004 (Task No. 000-08, "Sustainable Painting Operations for the Total Army (SPOTA)," approved October 1, 2002).
3. *White Paper: Implementing Environmentally Preferable Industrial Processes in DoD*. (This paper was drafted based on a roundtable discussion that occurred during the final session at the *13th Annual Solvent Substitution Workshop* in Scottsdale, AZ on December 12, 2002.)
4. *Meeting DoD's Emerging Contaminant Challenges*, presented at CSIMP, February 2007, Shannon Cunniff, Director, Emerging Contaminants, OUSD.
5. The Efforts of the DoD Services and NASA to Green Cleaning Operations, Process Cleaning Magazine, 2006.
6. Demonstration/Validation of Alterative Cleaners for DoD Applications, presented at National Manufacturing Week 2006, Wayne Ziegler, Army Research Laboratory.
7. *Technical Memorandum TM-2361-ENV; Joint Service Solvent Substitution Methodology*, prepared by Pollution Prevention Technology Development Branch Code ESC421; NFESC; May 2003.

III

Safety and Regulations

29 **Worker Protection and the Environment: Current Editorial Observations**   *Barbara Kanegsberg* 455
Introduction • Environmental Regulations versus Worker Safety • Tropospheric Ozone (VOCs) • Material Safety Data Sheet et al. • Known Hazards versus the Great Unknown • Air Monitoring, Process Monitoring, and Emergency Supplies • Your Input Is Important • Communicating with Safety Professionals and Environmental Regulators • Additional Environmental Issues • Conclusion • References

30 **Health and Safety**   *James L. Unmack* 465
Overview • Introduction • Identifying Hazards • REACH • Implementing Safer Critical Cleaning Processes • Personal Protective Equipment • Conclusion • References

31 **Critical Cleaning and Working with Regulators: From a Regulator's Viewpoint**   *Mohan Balagopalan* 483
Introduction: The Challenge • Historical Perspective • Regulations and Communication • Bridging the Communication Gap • After Receiving a Permit • Compliance Inspections • New Tools for Communications • Ongoing Communication • Reporting and Record-Keeping Is an Ongoing Activity That Has to Be Maintained • Conclusions

32 **Momentum from the Phaseout of Ozone-Depleting Solvents Drives Continuous Environmental Improvement**   *Stephen O. Andersen and Margaret Sheppard* 491
Introduction • After the Montreal Protocol Was Signed, the Industry Changed Direction • The Clean Air Act Amendments of 1990 Set Further Environmental Goals • The Phaseout of Ozone-Depleting Substances Continues • Lessons from the Montreal Protocol for Climate Change • Proactive Companies That Protected the Ozone Layer Are Now Protecting the Climate • Challenges of Sustainable Enterprise • Conclusions and Climate Leadership Opportunities • Glossary of Acronyms • References

33   Screening Techniques for Environmental Impact of Cleaning Agents
Donald J. Wuebbles 501
Introduction • Focus of This Chapter • The EPA SNAP Program • Good Ozone vs. Bad
Ozone • Air Quality and VOCs • Concerns about Stratospheric Ozone: Ozone Depletion
Potentials • Concerns about Stratospheric Ozone: Equivalent Chlorine Loading • On
the Recovery of Stratospheric Ozone • Concerns about Climate Change: Global Warming
Potentials • Summary and Conclusions • References

29

Worker Protection and the Environment: Current Editorial Observations

Introduction 455
Environmental Regulations versus Worker Safety 456
Tropospheric Ozone (VOCs) 457
Material Safety Data Sheet et al. 458
Known Hazards versus the Great Unknown 459
Air Monitoring, Process Monitoring, and Emergency Supplies 460
Your Input Is Important 460
Communicating with Safety Professionals and Environmental
Regulators 461
Additional Environmental Issues 462
Hazardous Air Pollutants
Conclusion 462
References 463

Barbara Kanegsberg
BFK Solutions

Introduction

Components and parts manufacturing takes place in a highly regulated world. Environmental regulations often provide conflicting requirements to the manufacturing community. We presented results of a study concerning U.S. regulations and comparing federal, state, and local environmental regulations (Kanegsberg and Kanegsberg, 2003). Even within the United States, and even considering environmental regulations in the absence of worker safety rules and provisos, there was substantial ambiguity. If anything, the trend toward confusing rules has increased during the past decade. Manufacturers may be left with few, if any, valid options, and many are left in a Catch-22 situation. With the globalization of manufacturing, this problem has become even more acute. Companies attempt to become more productive with fewer resources, and so do regulatory agencies. This means that agencies increasingly operate with blinders on; they address the immediate concern without understanding the impact of their rules on other requirements. The people involved may not have sufficient time to fully research the impacts of what they propose. For example, agencies involved with worker safety may impel the use of ventilation in processes where environmental regulations do not allow ventilation. Individual companies may be left with few viable options; they may be faced with using respirators, investing in engineering controls, or changing the process. We should be optimizing worker protection, neighborhood protection, and environmental protection.

Those involved in critical cleaning should also be able to optimize process performance and process efficiency. All too often, the goal of manufacturing takes a back seat to the need to satisfy the environmental crisis du jour. By the time all of the environmental controls and safety controls are in place, the efficiency of cleaning may be so compromised that orders of magnitude more cleaning agent or energy may be used. The various controls required may in fact be at odds with one another. Sometimes, manufacturing in a particular location becomes so onerous that the decision is made to ship the process or even the entire manufacturing plant to a geographically remote location. One can only imagine the environmental impact of such a decision.

At the same time, environmental crises starting with the depletion of stratospheric ozone have led to new technology in cleaning chemistries, cleaning equipment, and overall processing, which have provided valuable additional options to manufacturers. Given the increasing fragmentation and diversity of regulations, questions remain as to whether or not companies should invest in newer technologies. The solution of today can become the problem of the future.

Environmental Regulations versus Worker Safety

Manufacturers must be concerned with the environment, but they must also protect the individual worker. One point of confusion is that many engineers think of environmental regulations as being synonymous with safety regulations. In terms of impact on the manufacturing process, worker safety and environmental regulations may actually be at odds.

Let us look at a plausible, if somewhat exaggerated, scenario. Let us assume you are a manufacturing engineer. You enter an assembly area to be greeted by the breathtaking aroma of fluoro-iodo propyl daffodilic acid (a fictional strong organic acid) bubbling merrily out of the contained safety bath and onto the floor and which is currently splashing all over you. At the same time, you spot the warning light on your degreaser full of hexamethyl nitro chicken wire (a fictional volatile organic compound or VOC and a hazardous air pollutant or HAP). The light indicates that someone on the notorious third shift has managed to override all the safety switches—the room is full of those lovely vapors, as well. How do you handle this? One good way might be to sound the alarm, get yourself and everyone else out of the work area, and dive under the safety shower while stripping away your clothing in an attempt to reenact Woodstock. At the same time, someone else turns on the fans full blast to exhaust the vapors from within the room. Then the crew dons protective gear and cleans up the mess on the floor. You have protected yourself and your workers. At the same time, depending on where you work and on other environmental controls, you have perhaps increased air and water pollution.

Granted, this was an emergency.

However, it is important to be aware that policy impacting the ongoing operation of a process may be based on ongoing protection of the neighborhood environment or the global environment, not on protection of the worker. It is important to work with the appropriate health and safety professionals and to be aware of potential problems yourself. Just because a chemical is not on an environmental hit list does not mean you can automatically assume it will be safe for your workers.

Just because a product is not on a safety or environmental list does not mean it is safe. You must evaluate your own manufacturing situation carefully and thoughtfully.

The Significant New Alternatives Policy (SNAP) program, for example, is involved with determining allowable substitutes for ozone-depleting compounds (ODCs). In your particular process, the solvent that best suits your needs may be acceptable as an ODC substitute, and it may be delisted as a VOC. However, it may also have a much lower allowable worker inhalation level. Or, the chemical could itself have a very high allowable exposure limit, but the vapors may react with acids or bases used in a nearby process. Without looking at worker safety in terms of the full process and the overall manufacturing environment, you would not necessarily be adequately protecting the individual worker. If the material has a low flash point, it would be unsuitable for use in standard heated equipment. Yet, this author has observed what can best be described as doubtful proposals to use isopropyl alcohol to replace an ozone depleter or acetone to replace VOCs, without considering the engineering controls required in the process. Even cleaning agents without a flash point can oxidize during use; azeotropes (and certainly, many "near azeotropes"), if not stored appropriately can change in composition, sometimes resulting in significant amounts of flammable vapors being produced.

Tropospheric Ozone (VOCs)

VOCs are smog producers. VOCs react with nitrogen oxides in the presence of light to form photochemical smog, which contributes to poor air quality. All organic compounds (carbon-containing compounds) are VOCs. Even trees and other vegetation can produce solvent emissions that contribute to smog. However, not all VOCs are created equal; some are more reactive than others. A number of scales indicating reactivity of various organic compounds were developed (Carter, 1998, 2000). Some are based on smog chamber studies, while others are models which include actual smog episodes. Research, discussions, and refinements are ongoing.

Particularly in areas of poor air quality, manufacturers are encouraged to use cleaning agents that have low or no VOCs. In the United States, at the federal level, a "line-in-the-sand" approach is used. Certain compounds are determined to have low tropospheric reactivity relative to ethane. Compounds with lower reactivity than ethane are not treated as VOCs. Organic compounds more reactive than ethane are all considered VOCs, even though reactivity may vary by orders of magnitude. Reactivity of the solvent vapors is considered, not the speed of volatilization. The goal of the policy was to encourage these negligibly reactive compounds to be used in place of more reactive ones. A few compounds are now classified as being negligibly reactive. If a VOC exemption is applied for, usually by a producer and/or an industrial group, and if the U.S. Environmental Protection Agency (EPA) determines that it has "negligible reactivity," and if enough states agree that it is not a VOC, the chemical becomes an attractive option for industry. The process of VOC exemption is fascinating. Access the files for the exemption process for acetone; they are publicly available. I once spent a delightful weekend reading through most of the files. Acetone has a relative reactivity close to that of ethane, and the technical and political chemistry arguments presented were complex and innovative. The technical reality is that all organic compounds are reactive, and therefore they are, to some extent, VOCs. Non-VOCs are probably best thought of as diet potato chips—eat too many, and you are still pleasingly plump. Emit a very high level of VOC-exempt compounds, and air quality is still compromised. There have in fact been ongoing efforts to use relative reactivity (Carter, 2000) directly rather than the "line-in-the-sand" approach. The California Air Resources Board (CARB) appears to have taken a leading role in this effort, particularly for what is referred to as consumer goods. Many of these consumer goods are actually produced for industrial applications.

Table 29.1 indicates VOC-exempt compounds or categories of compounds that (from the viewpoint of the author) are, might be, or might have been useful in critical cleaning applications. Table 29.1 is based on the current federal definition of a VOC (U.S. EPA, 2009), current as this book goes to press. The number of VOC-exempt options that are effective cleaning agents and that are not under regulatory distress is very limited. You will notice that *t*-butyl acetate (TBAC) is in its own category. TBAC is not a VOC for the purpose of VOC emissions limitations of VOC content requirements; however, it is a VOC for the purposes of recordkeeping, emissions reporting, photochemical dispersion modeling, and inventory requirements applying to VOCs, and it has to be uniquely identified in emission reports. This separate treatment stems from toxicity concerns, and our understanding is that some of these concerns may have been raised by one or more state agencies. Further, it appears that more guidance to state agencies as to exactly how to manage TBAC is needed. In other words, the situation is as clear as mud. Our understanding is that there are attempts to resolve the situation with the EPA, but at this point, TBAC is still enmeshed in a Cinderella story. In addition, the State Implementation Plans (SIPs), notably from California, provide other approaches that are in effect feedback loops. One involves relative reactivity, and other methods of measurement are also allowed under the SIP.

Particularly in areas of poor air quality, VOCs (and, in fact, all solvents) are discouraged by some regulatory groups in favor of water-based processes. Let's suppose you avoid solvents and adopt a

TABLE 29.1 Compounds Determined by U.S. EPA to Have Negligible Photochemical Reactivity (VOC-Exempt) Compounds, with Past, Current, or Potential Utility in Critical Cleaning

Compound or Category	Utility for Critical Cleaning and Some Provisos
Methylene chloride	Hazardous air pollutant (HAP), worker safety concerns
1,1,1-Trichloroethane (methyl chloroform)	Class I ODC, no longer produced HAP
1,1,2-Trichloro-1,2,2-trifluoroethane (CFC-113)	Class I ODC, no longer produced
Parachlorobenzotrifluoride (PCBTF)	Cost, odor, worker exposure
Cyclic, branched, or linear completely methylated siloxanes	Cost, efficacy of cleaning agent, variable worker exposure profile, other environmental considerations, some have low flash point
Acetone	Strong solvent (compatibility issues), very low flash point
Perchloroethylene (tetrachlorethylene)	HAP, worker safety, and air quality issues
HCFC 225 ca and cb	Moderate solvency, solvent costs, Class II ODC's, phaseout scheduled 2015
HFCs, assorted	Low solvency, high cost, potential atmospheric lifetime issues
HFEs, assorted	Low solvency, high cost
Perfluorocarbon compounds, assorted	Very low solvency, long atmospheric lifetime
t-Butyl acetate	Separate category, confusing
VOC measured by SIP or 40 CFR Part 60 Appendix A	For determining compliance limits; may be location specific, may provide additional options
California's aerosol coatings reactivity-based regulation recognized	May provide additional options in California

water-based process. Such a process may resolve ODC, VOC, and HAP issues. However, even environmentally friendly aqueous cleaners require appropriate handling to minimize hazards to workers. This was observed with some "beyond-compliant" so-called clean air solvents (CASs). CASs are cleaning agents which have been analyzed by the South Coast Air Quality Management District (SCAQMD) in California and have been found to meet their stringent environmental requirements for VOCs, ODCs, global warming potentials (GWPs), and air toxics. The additives in the CAS, particularly when the mixture was heated, could produce skin irritation. This does not mean the products cannot be used; the CAS provides an extra measure of reassurance to the end user. It does mean that all products must be used with careful regard for the safety of the individual worker.

With organic additives, heated mists may release significant amounts of solvent vapors. In addition, the force of cleaning action can be sufficient to cause injury to workers. In some cases, cabinet washers may be so designed as to be linked to a sink on a drum cleaner for economic reasons. In such cases, it is important to restrict the maximum operating temperature of the cabinet washer, so as not to scald the worker who unknowingly begins to operate the attached sink on a drum for hand cleaning. It is also important to walk through the plant to see how the process is being used. If the worker is supplementing the aqueous process with aerosol sprays, it is important to determine the content of the sprays and to evaluate potential worker exposure, perhaps with vapor monitoring. In addition, it is important to look at the spent cleaning solution in terms of soils, and unexpected solvent additives in terms of impact on the individual employee and on the potential for water pollution.

Appropriately impervious gloves, goggles, aprons, etc. become important. In addition, there is no substitute for a well-designed process run by a well-educated workforce.

Material Safety Data Sheet et al.

The material safety data sheet (MSDS) has become a standard part of the workplace. Unfortunately, an MSDS can be difficult to read and to interpret, the information format is not standardized, and the degree of information and recommendations provided may vary from one manufacturer to the next.

I personally like to compare the MSDS from several suppliers. A bit of time spent looking over these sleep-inducing documents can save headaches in the long run.

You should beware of an overabundance of proprietary data. If you suspect a hazard, ask questions. Sign a confidentiality agreement, if necessary, but get as full a picture as possible. If you cannot get disclosive information and are suspicious, I suggest you find another supplier.

It is also important to be aware that you may have to go beyond the immediate MSDS, depending on the way the cleaning or processing agent is used. For example, an MSDS may indicate that chemical A dissolved in chemical B will release vapors of C and D. The MSDS may vaguely indicate that the vapors of C and D are harmful, but gives no further details of these latter chemicals. If you are using chemicals A and B separately and in portions of the plant which are isolated from each other, there may be no significant problems. However, if you are indeed dissolving A in B, you would be well advised to obtain safety information on chemicals C and D.

The mere presence of books of MSDSs at the work site, while helpful, is not a substitute for an educated, aware, and informed workforce. Some workers look at the MSDS as a sort of talisman. If there is a problem, they will wave the book of MSDSs in the general direction of the chemical spill or source of solvent vapors, with the expectation that the problem will be solved. Mere ownership of MSDSs is not enough. Workers must be able to read them, understand them, and apply the information to day-to-day process operation.

It is important to understand the role of the MSDS. It is mandated by the Occupational Safety and Health Administration (OSHA) as a means to warn and protect workers from materials that could be harmful to **themselves**. It does not regard any potential harmful consequences to **the product** itself. A material that might be well within acceptable limits as regarding the health of the workers may cause an undesirable modification to the surface or bulk of a component.

Known Hazards versus the Great Unknown

We sometimes fear an abundance of data without considering the alternatives. A solvent with a well-established toxicity, which has been used for decades, may have a relatively low allowable inhalation level. People may be fearful of using the chemical. Or, because it is on a list of hazardous air pollutants, ozone depleters, or VOCs, there may be regulatory restrictions or company policy restrictions on using the product.

However, suppose an alternative cleaning agent or an alternative cleaning blend has a list of ingredients indicating that no inhalation level has been set. Does this mean you should be blithely breathing great gulps of it or drinking it as a liqueur? Of course not. Not all possible hazards to the worker or the environment have been evaluated; there simply aren't the resources available to do so. It is preferable to exercise reasonable prudence.

On the other hand, the argument has also been made that some of the older, more established air toxics have well-known hazards and can therefore be used more readily. However, this does not mean that they are necessarily relatively safe. The closest analogy I can make is to a speeding train. We understand the power, we understand the utility in getting us from one place to the next. However, this does not mean we can stand on the tracks.

Older chemicals are said to be safer because there have been decades of use and decades of epidemiological data, whereas newer chemicals only have data based on animal exposure. In the first place, it is difficult to evaluate epidemiological data. In the second place, we now ought to know enough to understand that **all** chemicals should be used less emissively and with less exposure to workers than they were in the past. In the third place, let's not put our production workers in the position of unwittingly supplementing testing in lab rats. In summary, all chemicals, aqueous and solvent based, have potential safety and environmental hazards, and should be handled to minimize exposure.

Air Monitoring, Process Monitoring, and Emergency Supplies

Designing and maintaining an effective, profitable cleaning process is very site specific. In the same manner, assuring that a process is environmentally sound and minimizes employee exposure to chemical is also very site specific. A detailed treatment is beyond the scope of this book. However, a few general guidelines are appropriate.

Monitoring emissions of organic solvents to the environment is often heavily mandated and inspected. In the same manner, it is important to monitor employee exposure. This involves taking a dispassionate look at the process in terms of all of the workers in the area as well as particular workers who might have more exposure. For cleaning processes, monitoring may include the following:

- Overall air quality in the manufacturing plant
- Overall air quality in the work area
- Air quality proximal to the sample-handling area of the cleaning system
- Individual employee exposure

Monitoring may be appropriate not only at process startup, but also at regular intervals. Often, a process may be set up so as to assure that parts are dry and free of trapped solvent. Over time, day-to-day considerations of the production schedule may abbreviate process time, resulting in increased solvent loss and potential employee exposure. If the solvent has an obvious odor, such losses may be apparent. Choosing a cleaning agent because it has an additive to mask odors may be unwise. The human nose is an extremely sensitive detection instrument; if there is no emission, there will be no odor. However, workers can become desensitized to odors, so one cannot depend on odor alone for solvent detection.

Selection of the solvent detection device is also important. As with analytical techniques, one must consider the lowest detection limit and specificity for the solvent of interest. In addition to working with the safety/environmental professional, the cleaning agent supplier and safety supply distributors can be good sources of information.

With all processes, whether aqueous or solvent based, ongoing process monitoring to minimize employee exposure is crucial. In most cases, it is productive to manage this along with overall process control. Clear, unambiguous process instructions, employee education and training, and accountability are required (Petrulio and Kanegsberg, 1998). As regulations change, process instructions and educational training must be updated.

Emergency supplies also have to be considered. If your process involves chemicals with the potential for acute, extraordinary hazards, such as hydrofluoric acid, it may be wise to have specific emergency materials beyond the typical eyewashes and safety showers on hand in case of spills or employee exposure. How much to keep on hand depends on the process as well as the capabilities of outside emergency teams.

The best way to design an effective cleaning process is to involve not only the environmental staff and advisors but also health and safety professionals during process design.

Your Input Is Important

Most of us have had the experience of having company management or an environmental advisor present us with the latest environment hit list, with the orders to stop using whatever is on the list. Or they may provide a "safe" list with an indication to use ONLY chemicals or products on that list. The detrimental consequences to your production line can be economic, performance, or safety related. Rather than respond with an "Oh, no, I can't possibly do this," or "Great! More unreasonable regulations. This means career security," perhaps you should consider some proactive work.

Environmental policy may be based on research by the greatest minds in the world. However, based on my experience, I think we could set a more realistic, holistic, and environmentally sound policy if you, the manufacturing engineer, tracked environmental policy and contributed your own views.

A perusal of advisory committees to regulatory agencies from the local to the national and even the international level is likely to show a reasonable membership by employees of or advisers to various chemical companies and/or manufacturers of cleaning equipment. This includes aqueous and solvent-based processes. It is laudable and valuable that such involvement takes place. After all, the technical people who design cleaning agents or cleaning processes are often in the best position to explain their products to regulators and policymakers. You might also see participation by advisors, advocates, and lobbyists. No matter how impartial the participants, I do not think anyone can completely operate out of the context of their company. Each participant has a viewpoint that must inherently be influenced by their own training, background, and experience. Given the independent nature of funding, the more active members include individuals employed by or affiliated with particular chemical and/or equipment companies.

At a local level, one might observe that those involved in regulatory advisory committees concerned with solvent usage often include a large proportion of manufacturers of cleaning agents and cleaning equipment, often aqueous cleaning. Equipment and chemical suppliers associated with solvent cleaning are often also involved, and many of them are interested in fostering their own particular technology.

In my view, I also see the need for YOU, the manufacturing engineer to become directly involved. Even where companies are represented in such groups, very often an individual concerned with regulatory compliance rather than people concerned with the technical aspects of the manufacturing process are involved. Sometimes, trade associations will send advisors. Such views are valuable in providing input from various overall sectors. Even if there is a trade association involved, by the time issues percolate from the work area to the environmental group, to the legal advisor, to the association committees, to the regulatory advisory group, and then back down to the production area again, the message may become diluted, lost, homogenized, or otherwise compromised.

Regulatory advisory committees aside, you might find it instructive to look, listen, and observe not only the formal presentations and question-and-answer sessions, but also the informal questions and comments to representatives of regulatory agencies. You might observe representatives of particular cleaning technologies explaining how their approach is far environmentally superior to that of the competition.

I am not suggesting that you in the manufacturing world don an imported designer suit, grab a cell phone, and head for Washington, DC. You have more important things to do. However, it is desirable to be proactive. If your trade association or environmental advisor asks for your input, even though you may be preoccupied with the immediate production schedule, my advice is to help them. If you are in a small company, keep track of the regulatory issues and impending policy decisions. Take your safety officer to lunch; take your lawyer to lunch—find out what is going on. If you see a regulator at a conference, you don't need to explain details of a process, but if contemplated, impending regulations are going to leave you without reasonable options, say so *before* you are burdened with an unreasonable regulation. If you are asked to write a note, do so—working with your company management where appropriate. One to two pages is plenty; regulators have lots to read in their spare time.

Your actions just might help. The interplay of the chemistry, physics, biology, toxicology, environmental science, and politics will ultimately impact your engineering and economic choices. Further, because unintended consequences of environmental policy may be detrimental to workers or to other aspects of the environment, your input will allow policymakers to set a regulatory atmosphere that ultimately better protects the environment.

Communicating with Safety Professionals and Environmental Regulators

Having said that we should regularly communicate with regulators, this author must add the obvious: many engineers run, hide, and otherwise avoid communicating process plans with the safety and/or environmental community. Why?

Some of the fear is based on previous unpleasant, inconvenient experience. A chemical may be needed to perform the process, but the safety advisor may, for reasons of his or her own convenience (maybe it requires more paperwork), discourage its use. Recollections of regulatory problems or permitting hassles can also produce a "run for the hills" response.

In some cases, the company safety/environmental group or the company lawyer may strongly advise that technical people have no communication with the outside regulatory community. Each individual manufacturing situation is different; it is important to evaluate your own legal status. Under ordinary circumstances, however, a more open communication with the regulatory world is probably more productive. With no communication, restrictive regulations that do not allow the company to operate productively are more likely to be enacted. Companies may want to choose advisers who, while protecting company privacy and security, also provide a clear, proactive presentation to regulators.

Additional Environmental Issues

A dizzying array of environmental concerns regarding air, water, and soil contamination can potentially impact manufacturing processes. Environmental issues are complex, because the underlying science, economics, and politics are complex. As such, environmental issues are subject to change. Future regulations are likely to reflect impacts on both air and water. With this understanding, two additional pressing environmental issues, current as of the time of writing, are noted.

Hazardous Air Pollutants

HAPs are a list of nearly 200 compounds that the U.S. federal government considers damaging to air quality. Presence on the list does not mean that the compound cannot be used safely. Absence from the list does not mean that there are no potential toxic issues. One major impact on components manufacturing is that certain chlorinated solvents are regulated under the National Emissions Standard for Hazardous Air Pollutants (NESHAP) for halogenated solvents. Many of the newer cleaning systems are inspired by the requirements of this NESHAP. The Halogenated Solvent NESHAP is currently under review.

The Aerospace NESHAP impacts volatility of cleaning agents used in hand-cleaning operations.

In the future, additional NESHAPs with additional impact on industry are likely.

Conclusion

There is no such thing as free lunch. All cleaning agents and all cleaning processes inherently have some environmental and safety concerns. In this author's opinion, the situation is not likely to change. When we manufacture, we basically remove or redistribute soils. Soils are matter out of place. Oils, greases, cleaning agents all have some underlying similarity to the compounds that make us alive, that make the grass green, that form the basis of our ecosystems. Regulations still often seem to be on a compound-by-compound basis. There seems to be an attitude that if only compounds A, B, and C were not used, or if we could restrict the manufacturing community to using only processes Q, R, and S, all would be well.

During the phaseout of ozone-depleting chemicals (ODCs), components manufacturers responded to an environmental crisis with a variety of innovative and creative process changes. This author has been very much involved in the changes. However, it seems safe to say that all of these changes have had their own environmental consequences. Despite the best efforts of regulators to control adverse safety impacts, there have doubtless been impacts relating to employee exposure.

At the micro level, you, as the individual components manufacturer, have to evaluate the menu of process options available in terms of performance, economics, and employee safety. At the same time, environmental regulations are becoming an increasingly complex task. The regulators and policy makers need your input. Otherwise, we will all continuously need to reformulate and redesign, moving from one crisis du jour to the next, without necessarily seeing a net environmental improvement (Kanegsberg, 1994).

References

W. P. L. Carter, Current status of reactivity research, Presented at the *Photochemical Reactivity Workshop* Durham, NC, May 12–14, 1998.

W. P. L. Carter, Development and evaluation of an updated detailed chemical mechanism for VOC reactivity assessment, Presented at the *A&WMA 93rd Annual Conference and Exhibition*, Salt Lake City, UT, June 18–22, 2000.

B. Kanegsberg, Safety, the environment, and profitable manufacturing, *EIA/EIC Conference*, San Francisco, CA, October, 1994.

B. Kanegsberg and E. Kanegsberg, Conflicting Environmental Regulations, *2003 AESF/EPA Conference for Environmental Process Excellence*, Daytona Beach, FL, February 3–6, 2003.

R. Petrulio and B. Kanegsberg, Back to basics: The care and feeding of a vapor degreaser with new solvents, *Presentation and Proceedings, CleanTech '98*, Rosemont, IL, May 21, 1998.

U.S. EPA, 40 CFR 51.100(s)—Definition—Volatile organic compounds (VOC) March 31, 2009; www.epa.gov/ttn/naaqs/ozone/ozonetech/def_voc.htm

30

Health and Safety

Overview 465

Introduction 465

Identifying Hazards.. 466
Mechanical Hazards • Thermal Trauma • Flammable and Reactive
Materials • Noise • Chemical Hazards • Defining Hazardous
Materials • Minimizing Chemical Exposure • Quantitative Exposure
Numbers

REACH 471
What Is REACH?

Implementing Safer Critical Cleaning Processes 473
Step 1: Collect Information • Step 2: Identify Hazards • Step 3:
Assess the Risk Arising from Each Hazard Identified • Step 4: Plan
Actions to Eliminate or Reduce Risk • Step 5: Document the Risk
Assessment • Evaluating and Managing Complex and Proprietary
Blends • Process Comparison • Harmonizing Safety and Environmental
Regulations • Integration with Other Manufacturing Activities • Working
Productively and Proactively with the Safety Professional

Personal Protective Equipment 477
Contact Lenses and Safety Glasses • Respiratory Protection • Contact
Lenses with Respirators • Facial Hair • Levels of Protection • Using
Personal Protective Equipment • Selection Matrix

Conclusion 481

References...481

James L. Unmack
Unmack Corporation

Overview

With the development of cleaning and manufacturing processes comes the potential for introducing hazards that may affect the health or safety of workers. This chapter is intended to provide those involved in industrial and critical cleaning with an awareness of the various hazards, with major risks imposed by those hazards, and with approaches to recognize, manage, and minimize those risks. Given the number of variables and the site-specific requirements of critical cleaning processes, rather than attempt to provide specific "recipes" for safety, the approach is to educate and guide the user as he or she addresses the challenges of protecting the health and safety of workers who use cleaning processes.

Introduction

Safety is having knowledge of the hazards associated with the intended activities and applying that knowledge to minimize the risk of injury or damage. Safety describes bringing those resources, information, and equipment to the activity to improve the likelihood of a successful outcome and reduce the risk of failure.

The concept of safety implies having available the appropriate information to provide the basis for decisions regarding processes, materials, and equipment to accomplish the objectives. Safety is more than just risk avoidance; when decisions are optimized, job performance is enhanced significantly.

Appropriate information provides the basis for achieving the objectives. The safety information when working with hazardous materials or processes must include information about those specific materials and processes. The material safety data sheet (MSDS) provides a convenient starting point to acquire information about the materials, particularly if the material is used in the manner intended by the producer. However, critical cleaning includes both the cleaning agent and the cleaning process. Therefore, achieving a safe cleaning process also requires an understanding of the tools, equipment, and machinery. Knowledge of the hazards associated with a process and the materials used in the process helps guide decisions on the selection of processes.

In addition to the safety of the individual worker, as various technologies are evaluated to achieve the cleanliness objective, related issues must also be considered, including environmental issues (sometimes thought of as environmental safety) and product safety. The effect of the technology on the global environment, the local environment, and the people who use this technology must also be considered. For example, it is because of their effect on the global environment that chlorinated and photoreactive solvents are avoided.

The safety of the object to be cleaned and the effect of the cleaning technology on the product being manufactured are prime considerations. For example, materials compatibility concerns may impact the selection of specific chemicals and processes. Restrictions on surface residue, including cleaning agent residue, may lead to the use of relatively toxic but highly volatile solvents.

Therefore, worker safety is more complex than restricting choices in cleaning and manufacturing chemicals. Rather than prohibiting the use of a hazardous material or process, consider cost trade-offs. The more hazardous the processes and materials, the greater the degree of isolation of the process from the worker. The greater the degree of isolation, the more automated the process becomes.

Identifying Hazards

Hazards can be conveniently categorized as mechanical hazards, physical hazards, and chemical hazards. Mechanical hazards include force that, when applied, may cause injury. Examples include hitting, squeezing, cutting, puncturing, abrading, rubbing, scuffing, shaking, vibrating, and blasting. Physical hazards include electrical currents, temperature extremes, radiation in any form, and noise. Chemical hazards can be thought of as inhalation or contact hazards. A material has hazardous properties if it is toxic, flammable, corrosive or reactive.

Mechanical Hazards

Mechanical hazards are associated with tools, process equipment, and machinery. Trauma is usually the result of contact with moving parts of equipment and machinery. Standards have been developed for guarding power transmission and point of operation. As a rule of thumb, the moving parts of machinery must be guarded to the extent that not even a finger can touch a moving part of a machine that would cut, squeeze, or strike with such force as to cause injury. Moving parts of a machine are commonly isolated from the employee through the use of barrier guards, but many processes do not lend themselves to barrier guards. For these situations, other guarding methods are employed, such as two-hand controls to keep the operator's hands out of the danger zone. Barrier guards should be interlocked with the operation of the machine so that the machine cannot operate when the guard is out of place. Switches and sensors used for interlocking are becoming much more sophisticated and much harder to defeat. Depending on the degree of hazard, emergency stops may be required. The

Occupational Safety and Health Administration (OSHA) has devoted an entire subpart to machine guarding, with specified guards for various parts of machines. OSHA compliance guides are available from the National Safety Council.[1] The standards for protection against shaking, vibration, and pressure are less precise. The threshold to cause injury is less well defined. The *American Conference of Governmental Industrial Hygienists* (ACGIH) has developed guidelines to protect workers from injurious vibration. The guidelines for hand-arm (segmental) vibration are time (duration), frequency, and axis dependent and depend on the direction of the vibration relative to the hand. Separate guidelines are published for whole body vibration. Any process where workers may contact vibrating surfaces in the 1–80 Hz frequency range should be carefully evaluated against the ACGIH guidelines for vibration exposure. Vibration exposure and other guidelines are available in the ACGIH TLV booklet,[2] available for purchase at www.acgih.org.

Thermal Trauma

In terms of air temperature, workers in environments outside the normal comfort zone need to be protected against the effects of exposure. Extreme environments may require rest areas where the workers can go to recover normal body temperature. The greater the deviation from normal comfort temperatures, the greater the hazard and the more time must be spent in rest and recovery areas. Air velocity (wind) and relative humidity must also be considered.

Workers located proximal to large cleaning systems, whether or not they are directly involved in the cleaning process, may be potentially impacted by thermal trauma. For example, cleaning equipment can generate heat due to incomplete insulation. Spray-in-air systems can generate heat and humidity, not to mention vapors and mists that contain cleaning chemistries.

To protect workers from temperature extremes, the temperature of the parts that are handled during the washing, rinsing, and drying processes must be considered. The temperature comfort zone for bare-handed handling of metal objects is 12°C–43°C (55°F–110°F). Momentary contact with metal objects hotter than 54°C (130°F) may produce tissue injury. Metal objects colder than 0°C (32°F) may freeze moisture and stick to the skin. Much colder objects can freeze tissue on momentary contact and may require insulated gloves or special handling tools to eliminate the need to touch the part with bare skin.

Flammable and Reactive Materials

The definitions of flammability and reactivity are purposely broad. The intent is to prevent fire and explosion under a variety of conditions.

What Is a Flammable Material?

A material is considered flammable if at normal ambient temperatures enough vapor is given off through evaporation or sublimation to sustain a flame. The flash point is defined as the minimum temperature to evolve enough vapor to sustain flame. A gas is flammable if it will burn at some concentration in air. The minimum concentration to sustain a flame is the *lower flammability limit* (LFL), sometimes expressed as the *lower explosive limit* (LEL). For every gas, there is a concentration in air above which it will not sustain a flame. This is called the *upper flammability limit* (UFL) or *upper explosive limit* (UEL).

What Is a Reactive Material?

A reactive material is an unstable substance that in the pure state, or as produced or transported, will vigorously polymerize, decompose, condense, or will become self-reactive under conditions of shocks, pressure, or temperature. Commercial and military explosives, blasting agents, and pyrotechnics are considered reactive materials, as are other unstable substances that can violently release energy without the need for other substances to react with. Fuels are generally not considered reactive materials, because they need an oxidizer to react with. There are a few exceptional fuels that, when properly initiated, need no external air or oxidizer to explode. For more information on fire, electrical, and building safety, consult the National Fire Protection Association.[3]

Noise

Excessive noise can damage the ability to hear. Since the industrial revolution, it has been accepted that we will continually lose some hearing ability as we age. Years ago, the medical community called this *presbycusis*. We now know that we do not have to lose hearing ability as we age; the decrease in hearing ability is the result of damage to ears from loud noise. The modern term is *sociocusis*. In the United States, OSHA adopted a guideline in 1972 that would protect 75%–85% of the workers from debilitating hearing loss. OSHA permissible exposure limit (PEL) for noise is 90 dB for 8 h, with a 5 dB exchange rate. That is, the allowable exposure time is cut in half for each 5 dB increase in loudness. Newer standards adopted in Europe and elsewhere strive to protect 95% of the workers from hearing loss. The newer standard is 85 dB for 8 h with a 3 dB exchange rate. This is an equal energy exchange rate supported by the ACGIH and the National Institute for Occupational Safety and Health (NIOSH).

Every time an exposure to loud noise results in a temporary threshold shift, expect about a 98% recovery. This means that every time an exposure to loud noise results in a temporary threshold shift, there is some permanent damage to the ears. This damage is cumulative and the ability to hear slowly and imperceptibly fades away. Ears should always be protected against loud noises with hearing-protective devices, such as ear plugs, ear muffs, or canal caps.

Cleaning with high-frequency sound, including ultrasonics and megasonics, is widely and increasingly used in critical cleaning and in other manufacturing processes. While nearly all of the functional frequencies are above the hearing range of most humans, the sub-harmonics are audible. OSHA and ACGIH have developed standards and guidelines that include exposure to noise as a result of the use of ultrasonics. Sound at ultrasonic frequencies rarely presents any risk to hearing. Ultrasound propagates poorly through air, and if it does reach the ear, it is not transmitted to the inner ear to cause damage to the auditory nerves. However, sometimes sub-harmonics are created on the ear drum that may be in the audible frequency range and may cause annoyance, distraction, or stress. A full discussion of ultrasound is provided in *The Occupational Environment: Its Evaluation, Control, and Management*.[4]

Chemical Hazards

Materials that are corrosive, irritating, or toxic are hazardous. While there is no bright line between hazardous and nonhazardous chemicals, several government agencies have developed guidelines. These guidelines and definitions are inherently situational, but they can be helpful in reducing hazards to workers (see Table 30.1).

The toxicity of a material is rated on its ability to cause harm. The usual routes of exposure in the workplace are inhalation for materials that are airborne, skin contact with liquids and solids, and ingestion, transferring materials from work surfaces to hands and then to the mouth.

In the workplace, the most common route of entry is inhalation. Laboratory rats are exposed to various airborne concentrations of the material to determine the lethal concentration that statistically would kill half the rats. This value is expressed as LC_{50}. LC_{50} is an indication of acute, immediate toxicity. Long-term, multigenerational animal studies are often conducted to more completely characterize potential risks to workers.

For other materials, the most hazardous exposure is ingesting the material, either in food or water. Oral toxicities are usually measured on laboratory rats and expressed as the amount of material taken at one meal required to kill 50% of the rats. This is expressed at the *lethal dose-50*, LD_{50}. Toxicities are expressed as milligrams of the poison per kilogram of body weight of the test subject. The guiding principle is that the larger the animal, the more poison is needed to kill it.

TABLE 30.1 Comparison of Hazardous Materials Definitions

	OSHA 29 CFR 1910	EPA 40 CFR	U.S. DOT 49 CFR
Toxic	§1910.1200 LD_{50} <500 mg/kg oral, rat LD_{50} <1,000 mg/kg skin, rabbit LC_{50} <2,000 ppm inhaled, rat LC_{50} <20 mg/m³ inhaled, rat	§261.24 LD_{50} <5,000 mg/kg oral rat LD_{50} <4,300 mg/kg dermal rabbit LC_{50} <10,000 ppm LC_{50} <500 mg/L fathead minnows	§ 173.132 LD_{50} <500 mg/kg liquid LD_{50} <200 mg/kg solid LD_{50} <1,000 mg/kg dermal LC_{50} <10 mg/L dust/mist LC_{50} <5,000 mL/m³ gases and vapors
Corrosive	§1910.1200 Irreversible tissue damage <4h	§261.22 pH ≤2.0 pH ≥ 12.5 Corrodes steel > 6.35 mm/year (0.25 in./year)	§173.136 Corrodes > 6.35 mm/year (0.25 in./year) steel or aluminum Destroys skin <60 min
Flammable	§191.1200 LEL <13%, UEL—LEL > 12% FP <100°F (37.8°C) Pyrophoric	§261.21 Ignitable FP <60°C (140°F) Pyrophoric Oxidizer	§173.121 FP <60°C (140°F) §173.120 Temperature > FP LEL <13%, UEL—LEL > 12% §173.115, FP <100°F (37.8°C)
Reactive	§1910.1200 Vigorously polymerize, decompose, condense, or become self-reactive under conditions of shock, pressure or temperature.	§261.23 Unstable, undergoes violent change without detonating Water reactive Forms explosive mixtures with water Generates toxic gas Cyanide or Sulfide Detonates or explodes DOT forbidden explosive	§173.56 Explosive
Other	§1910.1200 Presents significant safety or health hazard	§261.3 Listed or not excluded by USEPA Administrator in 40 CFR 261	§171.8 Determined by the Secretary of Transportation to be capable of posing an unreasonable risk to health, safety, and property when transported in commerce, and which has been so designated. The term includes hazardous substances, hazardous wastes, marine pollutants, cryogenic and elevated temperature materials.

For materials that are hazardous on your skin, the toxicity is determined by placing a measured quantity of the material on the shaved skin of a rabbit. OSHA considers a material corrosive if the skin is destroyed after 4 h of contact. If the skin is inflamed, reddened, or exhibiting a rash, the material is considered an irritant. If the material is absorbed through the skin to produce systemic effects, the material is considered toxic.

Defining Hazardous Materials

The answer to the question of what is a hazardous material is not simple. To give a good answer, we need to know who is asking the question and why. As an example, it is illuminating to compare and contrast the approach to hazardous materials adopted by three U.S. federal agencies: OSHA, the Environmental Protection Agency (EPA), and Department of Transportation (DOT). Of course, local and state agencies in the United States may have additional related priorities, and comparing agencies at an international level presents even greater complexity.

OSHA is charged with minimizing work-related injuries, illness, and death. Environmental toxicity is an important issue, one that is distinguishable and distinct from worker safety. EPA considers environmental effects. For example, the degree of hazard of a material getting into rivers, lakes, or oceans is rated by its aquatic toxicity. EPA rates aquatic toxicity by the killing of fathead minnows, rainbow trout, golden shiners, or daphnia. Aquatic toxicity is the concentration in water that kills half of the test species in 96 h. DOT oversees transportation via highway, railroad, air, maritime, etc.

The differences in definitions outlined in Table 30.1 reflect the differing agendas of the three agencies. The mission of OSHA is to ensure the health and safety of people at work. The assumption OSHA makes is that people work for 8 h per day, 5 days per week, for a total of 40 h per week. Inherent in this assumption is that people have time to recover from the effects of the exposure each day before being exposed again. OSHA also assumes an adult of working age and a relatively healthy worker population. EPA must assume that environmental exposures are continuous, 24 h per day, without time for recovery each day. Therefore, the criteria used by EPA must be more protective. Note that a material such as table salt (sodium chloride) with an LD_{50} of 1200 mg/kg (oral, rat) would not be considered toxic by OSHA but would be considered toxic by EPA.

The hazardous materials criteria used by DOT are intended to protect workers who are involved in cleanup at the site of a transportation accident and also to protect the infrastructure, vehicles, and equipment. Note that an isotonic brine solution could be a corrosive to DOT for its corrosive effects on steel, but not a corrosive to OSHA because it is not corrosive to skin. In fact, an isotonic brine is used medicinally to protect skin from corrosive effects.

Other Risks in the Workplace

Are Hazardous Materials Really That Dangerous?

By knowing the hazards and taking reasonable precautions, most hazardous materials can be handled with comparatively little risk.

In fact, many more injuries in the workplace are caused by trauma than by exposure to hazardous substances. The leading causes of fatal injuries to workers are highway accidents, violence, and falls. The ranking of these three leading causes depends on the business sector in question. Violence is the number one cause in retail sales, while highway accidents are the number one cause among workers who travel to work sites outside their permanent work location. Falls to the same level account for 12% of fatal falls (U.S. Bureau of Labor Statistics, U.S. Department of Labor 2009).

The number and severity of slips, trips, and falls can be reduced by keeping all walking surfaces clean, dry, and clear of obstructions and debris. Wintertime ice on walking surfaces is a major hazard in colder climates and contributes to more falls than any other hazard.

Minimizing Chemical Exposure

In the workplace, most exposures to hazardous chemicals occur through inhalation of gases, vapors, or aerosols and by direct skin contact with liquids or solids on the hands. In extreme situations, respirators can be used to protect against inhalation hazards. However, respirators require special attention to ensure their adequate functioning. OSHA requires a written respiratory protection program to ensure that all critical aspects of the program are implemented.

Quantitative Exposure Numbers

Quantitative numbers associated with inhalation hazards have been developed by many agencies and professional organizations. All other things being equal, it is easier for manufacturers to use chemicals that have higher exposure numbers. For occupational exposures in the United States, the most commonly used guidelines are the OSHA PELs and the ACGIH threshold limit values® (TLV®). Other commonly cited occupational exposure limits (OELs) include NIOSH recommended exposure limits (RELs) and State of California (Cal/OSHA) PELs. The United Nations, through the World Health Organization and the Organization for Economic Cooperation and Development (OECD), has also developed occupational exposure guidelines. Other countries throughout the world have developed separate chemical exposure levels. Most guidelines or requirements specify the airborne concentration of specific chemicals.

This chemical-by-chemical approach has limitations. OSHA has published PELs for 447 substances, ACGIH has established TLVs for about 696 chemicals, in California Cal/OSHA has published 747 PELs, and the American Industrial Hygiene Association (AIHA) has published workplace environmental exposure levels (WEELs) for 113 substances. In 2000, OECD compiled a list of over 5000 high production volume chemicals, chemicals with an annual production volume of more than 1000 metric tons per year. There are many millions of chemicals in commerce, with more being added each year. To provide guidance for chemicals for which neither OSHA nor ACGIH has established guidelines, the State of California, through Cal/OSHA has independently embarked on a standards-setting process using health-based risk analysis. Cal/OSHA worker exposure numbers are legally enforceable throughout California. Many users of hazardous chemicals have developed internal OELs to protect their own employees. In addition, for new chemicals or where governmental agencies have not established standards, individual chemical producers may set OELs. To address this gap in exposure guidance, the European Community created REACH.

REACH

What Is REACH?

REACH is a new European Community Regulation[5] on chemicals and their safe use (EC 1907/2006). It deals with the registration, evaluation, authorization and restriction of chemical substances. The new law entered into force on June 1, 2007, and overhauled the former legislative framework on chemicals, which had been slow and showed limited efficiency.

The provisions of REACH will be phased in over 11 years. The first deadline under REACH is for registration of highest volume chemicals and is set for November 30, 2010. Substances known to be particularly hazardous are also included in this first deadline for registration. Manufacturers and importers have to document their management of risk of chemicals in their registration files to be able to continue their production and marketing. Only manufacturers or importers of chemicals must register; downstream users do not have to. It is estimated that about 9000 substances are covered.

REACH is a key example of striking a balance between the three pillars of sustainable development: competitiveness, social awareness, and environment. REACH ensures a high level of protection of human health and the environment, while also playing an important role in encouraging innovation, fostering competitiveness, and better enabling enterprises to meet the essential demands of consumers.

The aim of REACH is to improve the protection of human health and the environment through the better and earlier identification of the intrinsic properties of chemical substances. At the same time, innovative capability and competitiveness of the EU chemicals industry should be enhanced. The benefits of the REACH system will come gradually, as more and more substances are phased into REACH.

The REACH Regulation gives greater responsibility to industry to manage the risks from chemicals and to provide safety information on the substances. Manufacturers and importers will be required to gather information on the properties of their chemical substances, which will allow their safe handling, and to register the information in a central database run by the European Chemicals Agency (ECHA) in

Helsinki. ECHA will act as the central point in the REACH system: it will manage the databases necessary to operate the system, coordinate the in-depth evaluation of suspicious chemicals, and run a public database in which consumers and professionals can find hazard information.

The Regulation also calls for the progressive substitution of the most dangerous chemicals when suitable alternatives have been identified.

One of the main reasons for developing and adopting the REACH Regulation was that a large number of substances have been manufactured and placed on the market in Europe for many years, sometimes in very high amounts, and yet there is insufficient information on the hazards that they pose to human health and the environment. There is a need to fill these information gaps to ensure that industry is able to assess hazards and risks of the substances, and to identify and implement the risk management measures to protect humans and the environment.

It has been known and accepted since the drafting of REACH that the need to fill the data gaps would result in an increased use of laboratory animals for the next 10 years. At the same time, in order to minimize the number of animal tests, the REACH Regulation provides a number of possibilities to adapt the testing requirements and to use existing data and alternative assessment approaches instead.

Guidance for controlling human exposure is required in the dossier for registration. Annex 1 of REACH defines the derived no-effect level (DNEL)[6] as the level of exposure above which humans should not be exposed. Manufacturers and importers are required to calculate DNELs as part of their Chemical Safety Assessment (CSA) for any chemicals used in quantities of 10 ton or more per year. The DNEL is to be published in the manufacturer's Chemical Safety Report and, for hazard communication, in an extended Safety Data Sheet. The DNEL is used in the risk characterization (RC) part of the CSA as a benchmark to determine adequate control for specified exposure scenarios. Risk to humans can be considered to be adequately controlled if the exposure levels estimated do not exceed the appropriate DNEL. REACH specifies that DNELs shall reflect the likely routes, duration, and frequency of exposure. If more than one route of exposure is likely to occur (oral, dermal, or inhalation), then a DNEL must be established for each route of exposure and for the exposure from all routes combined. It may also be necessary to identify different DNELs for each relevant human population (e.g., workers, consumers and humans liable to exposure indirectly via the environment) and possibly for certain vulnerable subpopulations (e.g., children, pregnant women).

Manufacturers and importers must give a full justification for the DNEL, explaining their choice of the information used, the route of exposure, and the duration and frequency of exposure to the substance for which the DNEL is valid. They must take into account the uncertainty arising from the variability in the experimental data, intra- and interspecies variation, the nature and severity of the effect, and the sensitivity of the human (sub)population to which the quantitative and/or qualitative information on exposure applies.

The resulting value can be considered as an overall no-observed-adverse-effect level (NOAEL) for a given chemical based on an integration of all available and relevant human health hazard data. REACH then requires the manufacturer or importer to perform a RC for the leading health effect, that is, for the toxicological effect that results in the most critical DNEL. The exposure/DNEL ratio, in principle, presents a simple tool for RC, especially for downstream users who do not have the hazard data at their disposal. For any exposure scenario, the risk to humans can be considered to be adequately controlled if exposure levels do not exceed the appropriate DNEL. For some endpoints, especially mutagenicity and carcinogenicity, the available information may not enable a threshold, and therefore a DNEL, to be established. This is the case when the available data do not allow reliable identification of a threshold, or when a substance exerts its effect by a non-threshold mode of action. In such cases, it is assumed that even at very low levels of exposure, residual risks cannot be excluded.

The DNEL methodology is intended to harmonize the approach to occupational health risk assessment with those used for other types of risk. It can be compared, for example, with the predicted no-effect concentration (PNEC) used to assess environmental risks. This is important under REACH, as manufacturers must assess not only human health risks but environmental and physical safety risks as well.

DNELs are more precautionary than conventional occupational exposure limits (OELs, TLVs®, and PELs). The calculation of DNELs follows a rule-based approach in which a series of standardized assessment factors are applied to the toxicological endpoints to allow for uncertainties and inter-/intraspecies differences. This can result in a very conservative figure, perhaps one or two orders of magnitude lower than that from the traditional OEL-setting process. While the extra margin of safety in DNELs will be welcomed by some, it raises the question of practicability. How would industry handle, say, a 10-fold reduction in exposure limits?

Some people hope and expect that DNELs will eventually replace OELs because they are more stringent. What will happen to the traditional system of OELs used by hygienists? The European process for setting Indicative Occupational Exposure Limit Values (IOELVs) is well established, involving experts from member countries on a scientific committee (SCOEL) and providing an opportunity for stakeholders in industry and government to comment on the proposals. In contrast, these democratic safeguards will not be there under REACH. DNELs calculated by individual manufacturers and importers are not subject to any requirement for consultation or opportunity for input by interested parties.

Debate is also underway to decide what to do when a DNEL cannot be established. Is it possible to set a Derived Minimum Effect Level (DMEL) based on some concept of acceptable or negligible risk (such as the "Threshold of Toxicological Concern"), or should such materials automatically be banned because they cannot be adequately controlled? The methodologies, tools, and technical guidance needed for REACH are currently being developed by the European Chemicals Bureau through a series of REACH Implementation Projects (RIPs). RIP 3.2 is focused on technical guidance for preparing the chemical safety report and has a subgroup tasked with looking at the derivation of DNELs. Further information can be found on the ECB Web site at www.ecb.jrc.it/reach.

The British Occupational Hygiene Society (BOHS)[7] REACH steering group will continue to monitor the progress on DNELs and would welcome input or comment via the REACH discussion point on the BOHS Web site.

There is often wide disagreement between the exposure guidelines set by the different agencies. For example, the OSHA PEL for acetone is 1000 ppm, the ACGIH TLV is 500 ppm, the NIOSH REL is 250 ppm, in Japan the MAC is 200 ppm, and in Poland the OEL is 84 ppm. With all chemical exposures, the principle of controlling exposures to the "lowest feasible concentration" (LFC, NIOSH) or "as low as reasonably achievable" (ALARA, Nuclear Regulatory Commission) should be the guiding rule.

Implementing Safer Critical Cleaning Processes

Implementing a safer critical cleaning process begins with an occupational risk assessment. The main aim of an occupational risk assessment is to protect workers' health and safety. Risk assessment helps to minimize the possibility of the workers or the environment being harmed due to a critical cleaning process. A thorough risk assessment typically leads to a more efficient and cost-effective process and therefore also helps to keep the business competitive and effective. Risk assessment is part of an injury and illness prevention program (sometimes called an accident prevention program). Such programs are required in the European Union, California, and by many government contracts.

Step 1: Collect Information

A risk assessment begins by identifying the processes used in the critical cleaning, the equipment, and the materials and substances.

TABLE 30.2 Evaluation of Health and Safety Risk

Severity	High	3	4	5
	Moderate	2	3	4e
	Low	1	2	3
		Low	Moderate	High
			Probability	

Step 2: Identify Hazards

The next step is to identify the hazards associated with each machine, tool, and instrument. Hazards may include lifting and manual materials handling, pinch points, vibration, noise, hot or cold surfaces, sharp edges or points, power transmission systems, point of operation hazards, etc. This is only a partial list, to give an idea of the range of hazards that must be considered.

Step 3: Assess the Risk Arising from Each Hazard Identified

Estimate the probability and severity of the consequences for each hazard. A useful tool for deciding whether the risk is acceptable or unacceptable is to rate the probability and severity as low, moderate, or high for each hazard and then plot in a table. Table 30.2 may be used to assign a priority for addressing the risks, where 5 is the highest priority and 1 is the lowest priority. A risk with priority 5 should receive immediate attention.

Step 4: Plan Actions to Eliminate or Reduce Risk

Priority 4 and priority 3 should be considered unacceptable. For each hazard for which the risk is unacceptable, action is needed to manage the hazard to bring the risk into the acceptable range. Examples of actions include material substitution, guarding to reduce likelihood of contact, and enclosure or local exhaust ventilation to control the release of hazardous air contaminants.

Step 5: Document the Risk Assessment

At each step of the risk assessment, the findings should be committed to writing. The completed document will be a rational plan to address the risks that are deemed unacceptable.

What Are Considered Reasonable Precautions?

Reasonable precautions are determined by the risk assessment. An inventory is developed to include all the hazardous materials in the workplace. For each material on this list, the opportunities for exposure are identified. Where exposure may result in injury or illness, means of preventing the exposure must be identified. The preferred way of preventing exposure is through what are known as engineering controls. Engineering controls include process or material substitution, process enclosure, process automation, local exhaust ventilation, and general dilution ventilation. It is often said that with an unlimited budget (and unlimited time) any process can be accomplished with no measurable risk. The aim is not to reduce risk of injury to zero, but to produce a product at a competitive cost. This leads back to the risk assessment and determining when a risk is unacceptable. The management of that risk then comes down to selecting engineering controls, such as process or material change, or personal protective equipment (PPE).

Sometimes a risk can be made acceptable through the use of work practice controls, that is, training the workers on techniques and behaviors to minimize exposure while performing the process. While work practice controls were the preferred risk management philosophy at the beginning of the industrial revolution 200 years ago, work practice control and administrative controls are distinctly out of favor today. Instead, risk can be eliminated or reduced through appropriate guards, shields, ventilation, or other practical engineering controls.

If all practical engineering controls have been applied and still there exists a significant possibility of exposure, PPE is needed. PPE includes devices and clothing that are worn to protect and prevent direct contact. Some of the more common items of PPE include gloves, aprons, face shields, hearing protectors, hardhats, and respirators. Many others types of PPE have been developed for specific hazards.

Evaluating and Managing Complex and Proprietary Blends

Many cleaning agents are blends of chemicals. Blending is used to achieve a desired set of performance, cost, safety, and/or environmental properties. In many instances, the chemicals in the blend have been evaluated individually for their hazards of toxicity, corrosivity, and flammability (flash point, etc.). However, few if any of the blends have been evaluated for the toxicity of the blend. Therefore, in general, blends should be handled with a degree of prudence and caution.

The TLV Committee of the ACGIH recommends a reciprocal calculation procedure to develop a group guidance value (GGV) for the blend when each member of the blend has a similar biological effect on the same target organ system. This method of evaluating a blend was specifically developed for mixtures of aliphatic (alkane), cycloaliphatic (cycloalkanes), and aromatic hydrocarbons ranging from 5 to 15 carbons.

$$GGV_{mixture} = \frac{1}{F_a/GGV_a + \cdots + F_n/GGV_n}$$

where

$GGV_{mixture}$ = the calculated 8-h TWA OEL for the mixture

GGV_a = the guidance value (or TLV®) for the group (or component) a

GGV_n = the guidance value (or TLV®) for the nth group (or component n)

F_a = the liquid mass fraction of group (or component) a in the hydrocarbon mixture, with a value between 0 and 1

F_n = the liquid mass fraction of the nth group (or component n) a in the hydrocarbon mixture, with a value between 0 and 1

The calculated GGV of the mixture should be considered a rough estimate of a health-based risk assessment with a precision of approximately 25%, and should be rounded off accordingly. While this reciprocal calculation procedure has been applied to solvent blends other than middle petroleum distillates, the resulting GGV may differ widely from true health-based risk assessments because the pharmacokinetics of each component of the blend may vary widely. The GGV should not be applied to blends that have one or more components with unique toxicological properties, such as benzene, n-hexane, or methylnaphthalene. If a solvent blend has any components with unique toxicological properties, exposure to those components should be evaluated individually.

Azeotropic blends are often used because an aggressive solvent with a low flash point can be flash point inerted by an inert chemical. While azeotropes should not be left in open containers, true azeotropes, with reasonable process control, tend to maintain their relative composition during normal use. However, unless the blend is a true azeotrope, its physical properties, such as boiling point and flash point, may vary as the blend ages and different components evaporate at different rates. Experienced industrial hygienists have observed that so-called safety solvents or "near-azeotropes" may become flammable as the flame-suppressing component evaporates faster than the hydrocarbon base, leaving a residue with a low flash point.

Process Comparison

The risk assessment outlined in the section "Implementing Safer Critical Cleaning Processes," provides a handy and convenient way to compare cleaning processes. A process with information gaps and an incomplete risk analysis should be deferred for further study when compared to a process with a complete hazard analysis and risk assessment.

Harmonizing Safety and Environmental Regulations

Since any chemical is potentially hazardous, many of the chemicals used in cleaning have come to the attention of environmental protection organizations. Photoreactive chemicals, volatile organic compounds (VOCs), have been heavily restricted in areas of poor air quality. Most environmental protection organizations will only permit the use of VOCs in closed systems with strict emission limits. Going a step further, chemicals that can migrate to the upper atmosphere and react with ozone are facing a global ban. The chemicals, called ozone-depleting chemicals (ODCs) or ozone-depleting substances (ODSs), are becoming increasingly difficult to procure on the open market and/or to use legally.

Environmental regulations must deal with both point source and fugitive emissions, whereas occupational safety and health regulations aim to protect the worker from harmful concentrations of airborne contaminants. Occupationally, it used to be the practice to exhaust harmful contaminants to the outdoors and away from the worker. Now, consideration must be given to the external environment around the industrial operation. Exhaust streams must now be controlled to a level that meets annual emission limits. Such limits are often very different than the concentration of a given chemical to which a worker may be exposed.

Integration with Other Manufacturing Activities

Cleaning is often integrated with quality assurance inspection. This implies that cleaning is a process that is separate and distinct from other manufacturing processes. Sometimes cleaning can be integrated with other processes through the selection of process chemicals to minimize residue, thus minimizing the amount of cleaning agent and restricting the frequency or even the occurrence of the cleaning process.

In addition, risks introduced by the cleaning process cannot be fully evaluated without consideration of other proximal processes. For example, if a low- flash-point solvent is to be used, all ignition sources must be determined, not just ignition sources from the cleaning equipment.

Working Productively and Proactively with the Safety Professional

The environmental health and safety professional is a valuable resource for information about chemical and physical hazards. Too often, safety professionals are avoided as naysayers and obstacles, when in fact they can be the key to overcoming obstacles. Safety professionals have a broad view of all the activities and operations of the plant. This gives them access to information across the organization, a real advantage for cross-fertilization of ideas. But, to gain access to this rich source of information, the safety professional should be brought into the project during the design phase.

In problem solving, a dialog works best. When a safety professional says: "You can't do that," it is time to stop and explain, "This is what we are trying to accomplish." Then ask, "How can we do it?" A "can do" attitude is needed by every member of the team. The "can do" is developed by establishing a dialog, discussing the obstacles, and exploring the paths around the obstacles. In occupational safety and health, our vision is not one of accepting roadblocks, but one of how to navigate through the obstacles: "Yes, we can do it and this is how we can do it."

Because the safety professional is usually spread very thin, he or she likely will not know enough about your operation to understand the interactions between the materials and processes. This lack of knowledge may lead to erroneous assumptions and may get in the way of making an accurate assessment of the hazards and risks associated with your operation. Therefore, take the time to go through the process and explain the operation of the equipment. The safety professional may have to do some research on the materials you are using. Material safety data sheets (MSDS) are necessary but are only a beginning. Often the chemistry is complex, and interactions need to be charted to identify intermediate and side products. The operation of the machines can be very esoteric, and critical points of operation are sometimes not fully explained.

Going into a new operation, expect that the safety professional may not even ask the right questions. Take the time to establish a dialog. The rewards can be very satisfying.

Personal Protective Equipment

PPE describes items that are worn to protect against a defined hazard. Examples of PPE, starting at the top, include hard hats, ear plugs, safety glasses, goggles, respirators, coveralls, aprons, gloves, cups, knee pads, shin guards, metatarsal guards, steel-toe work boots, and other articles. PPE does not include clothing, hats, or shoes that would be considered ordinary street garb.

Starting with the premise that no one wears PPE because it makes them look sexy or is more comfortable than ordinary clothing, the need for PPE must be justified by the risks to health that it is designed to protect against. It is generally preferable to eliminate or reduce the risks so that PPE is not needed. This is not always possible, feasible, or practical. Situations arise where the most practical way to manage a risk is the use of PPE.

Contact Lenses and Safety Glasses

There is much controversy regarding the wearing of contact lenses in areas with the potential for exposure to hazardous concentrations of airborne contaminants. Contact lenses are clearly inappropriate in dusty atmospheres unless protected with a cover goggle. Exposure to hazardous vapors and gases is less clear-cut. Hydrophilic gases, such as ammonia, formaldehyde, and hydrogen chloride may cause more harm to contact lens wearers than non-wearers, and, therefore, contact lenses are not recommended where these gases may reach hazardous concentrations.

Further, standard eyeglasses may not provide sufficient coverage to prevent chemical hazards or other hazards specific to the workplace, such as from exposure to high-intensity light of certain wavelengths. The need for safety goggles should be carefully evaluated and an appropriate program instituted.

The selection of PPE depends on the results of the risk assessment. The risk assessment provides characterization and analysis of the hazardous materials, job tasks, physical hazards of the work location, intended use, and duration of potential employee exposure. When the risk assessment identifies hazards that remain after reasonable, feasible, or practical controls have been applied, then PPE must be provided. The appropriate PPE must be selected, and a program written. A minimum program has the following key elements:

1. The hazard and the controls used to minimize the risk are identified.
2. The PPE that will be used to manage the remaining risk after the controls have been applied is selected.
3. The person who will supervise the PPE program to ensure every element of the program is appropriately implemented is identified, by name or by position.
4. How the user will obtain the PPE is explained.
5. Some types of PPE, especially respirators, require a medical surveillance program. Certain hazardous materials, including asbestos and lead, are addressed in regulations that define in detail the scope of the annual medical examination.

6. Fit testing is required for respiratory protection to ensure maximum protection from the respirator. A number of fit test protocols are available and are described in detail in OSHA regulations requiring respiratory protection. A fit test can only be accomplished if there is no hair where the respirator seals against the face.

7. Describe the training and how it will be provided to the user, so that the user will understand the need for the PPE, the rationale for the selection or choices (where choices are provided), how to put on (don) and take off (doff) the PPE, and cleaning and maintenance procedures for the PPE.

8. How will inspection, maintenance, and repair of the PPE be accomplished? Some types of PPE, such as spectacles and respirators, must be cleaned daily. Some respirators, ear plugs, gloves, and coveralls are intended for single use. Arrangement for proper collection and disposal of single-use items will ease the burden of housekeeping.

9. Whether the PPE is single use, limited multiple use, or intended for daily use, a change out schedule should be developed. Respirator cartridges, in particular, require a change out schedule to ensure their continued effectiveness. For most PPE, a visual inspection is sufficient to determine when to change. However, the degree of contamination of a respirator cartridge is not obvious to visual inspection, and few have an end-of-service-life indicator. Therefore, OSHA requires that the end of service life be conservatively calculated rather than wait for respirator cartridge breakthrough.

10. The success of a PPE program is often determined by the storage of the PPE. PPE must be stored out of harm's way, kept clean and dry. Too often, this puts the PPE out of reach for the user, who then decides to skip the PPE when he or she should be using it. The more conveniently located the storage locker, the more likely that workers will use the PPE when needed. When the storage location is not convenient, either the PPE will not be worn when needed or the PPE will not be stored in the storage location but in toolboxes, on worktables, or hung from a nail, collecting dirt and getting stretched out of shape.

11. The PPE program should be reviewed periodically by a certified environmental health and safety professional to ensure all elements are being properly implemented.

Maintenance and storage of PPE, decontamination, donning and doffing procedures, inspection and monitoring of effectiveness, and limitations are outlined in this section.

Respiratory Protection

All employees whose jobs may require the use of respiratory protection should be certified medically fit to use a respirator before being fit tested and issued a respirator. Respirators put an extra load on the heart and lungs, which is not well tolerated by some people. All employees whose jobs may require the use of respiratory protection should be certified annually as medically fit to use a respirator. Only employees who have successfully completed respiratory protection training should be allowed to use respiratory protection. Respiratory protection training includes how to wear and maintain respirators properly, the proper use and limitations of respirators, and familiarization with respirators to be used on the job.

Employees should be fit tested using a fit testing protocol recognized by OSHA. Qualitative and quantitative fit testing protocols are well described in the appendices of many OSHA standards, such as the lead standard, 29 CFR 1910.1025, Appendix D. The objective of the fit testing is to determine that the protection factor exceeds a minimum value. The protection factor is defined as the ratio of the concentration of the contaminant outside the respirator to the concentration inside. The fit testing should be conducted before issuing a respirator to ensure the selected respirator will provide an adequate seal on the employee's face. The fit test should be repeated annually to ensure that a good seal is maintained. The contours of the face change with age and loss or gain of weight.

If an employee has difficulty in breathing during the fit test or during use, the employee should be evaluated medically to determine if he or she can wear a respirator safely while performing assigned tasks.

No employee should be assigned to tasks requiring the use of respirators if, based upon the most recent examination, a physician determines that the health or safety of the employee will be impaired by respirator use.

Contact Lenses with Respirators

Contact lenses are well protected by full face respirators. OSHA has reversed a long-held position and now permits contact lenses with full face respirators. However, contact lenses should not be worn with half-face respirators.

Facial Hair

Facial hair that might interfere with a good facepiece seal or proper operation of the respirator is prohibited. The seal that a respirator makes when seated on facial hair is unreliable. A favorable level of protection in a fit test is not a good indication of seal performance during actual work conditions. Facial hair that might interfere with a good facepiece seal or proper operation of the respirator must automatically disqualify respirator use. Only a respirator for which an employee has been successfully fit tested should be made available to the employee.

To ensure an adequate level of protection against airborne contaminants, the selected respirator should be approved by NIOSH for protection against the identified hazards. Many respirator manufacturers have a variety of styles and sizes. When the selection is based on cost, life-cycle costing techniques should be used. The cost of wearing a respirator includes the purchase price of the respirator plus all the accessories purchased over the life of the respirator and the cost of labor to maintain the respirator.

The respirators must be properly cleaned, maintained, and stored when not in use, to preserve the level of protection provided by the respirator.

Air-purifying respirators (APR) should not be used in heavily contaminated atmospheres where the protection factor is likely to be exceeded. When the fit test is performed with qualitative methods, the maximum protection factor is 10. The nature and concentration range of the contamination must be known before an APR may be selected for use.

There is no clear guidance on how often the air-purifying cartridges should be replaced. Cartridges must be replaced when breakthrough or increased breathing resistance is detected. However, many users recommend replacing the air-purifying cartridges at the end of each shift. Few cartridges have an end-of-life indicator to show when the cartridge needs to be replaced. Powered air-purifying respirator (PAPR) cartridges will be changed when flow through the cartridge falls below 4 cfm.

Positive and negative pressure tests must be performed each time the respirator is donned.

Air-supplied respirators should be assembled according to the manufacturer's specifications. Hose length, couplings, valves, regulators, manifolds, and all accessories should meet American National Standards Institute (ANSI) and the manufacturer's requirements.

Respirators should be cleaned and sanitized daily after use.

Respirators should be inspected during cleaning. Warn or deteriorated parts should be replaced.

Respirators maintained on site for emergency response should be inspected monthly by a trained technician.

The person responsible for the respiratory protection program should check the users periodically to ensure that they are wearing and maintaining their respirators properly and that the respiratory protection is protecting them adequately.

Regulations advise that a health and safety professional, such as an industrial hygienist, should evaluate the respiratory protection program annually to ensure its continuing effectiveness.

TABLE 30.3 PPE Selection Matrix

EPA Protection Level	Airborne Contaminant Level	PPE
D	<PEL	Work clothes
C	<10× PEL and <1000 ppm	Air-purifying respirator
B	<1000× PEL and <10% LEL	Supplied air respirator
A	<10% LEL	Total encapsulating, gas-tight suit

Levels of Protection

The U.S. EPA has defined levels of protection based on the protection afforded. See Table 30.3.

Level A Protection

Level A protection is a gas-tight suit. Level A provides protection to the skin and lungs against gases, vapors, dusts, and mists. Level A is often called the "moon suit," and is the ultimate in protection.

Level B Protection

Level B protection provides full protection to the lungs from gases, vapors, dusts, and mists, but only provides splash protection for the skin. Level B is defined by the supplied air respirator, such as a self-contained breathing apparatus (SCBA) or an airline respirator.

Level C Protection

Level C protection provides limited protection for the lungs against gases, vapors, dust, or mists. Level C is defined by the air-purifying respirator. The selection of the respirator depends on the airborne contaminant. Air purifying filters, cartridges, and canisters are generally specific to a particular airborne contaminant and can be overwhelmed by high concentrations. Most gases and vapors are limited to 1000 ppm or 10 times the OSHA permissible exposure limit, whichever is lower, when air-purifying respirators are used.

Level D Protection

Level D protection is defined by ordinary work clothes appropriate to the job. Level D protection may include coveralls, gloves, or an apron to protect the skin from contact with liquid or solid hazardous materials.

Using Personal Protective Equipment

All persons entering the work area where hazardous materials are used should put on the required PPE according to established procedures, to minimize exposure potential. When leaving the work area, PPE should be removed according to these established procedures to minimize the spread of contamination.

Donning Procedures

To minimize the spread of contamination, the following donning procedures have been developed and will serve as a good model:

1. Remove street clothes and store in a clean location.
2. Put on underwear and coveralls. Underwear should be absorbent and may be cotton or disposable.
3. Put on boots and boot covers. Secure boot covers to coveralls with tape. If boot covers are loose, adjust fit with wraps of tape.
4. Put on under gloves.
5. Put on chemical and abrasion resistant outer gloves. Secure gloves to coveralls with tape.
6. Don respirator and check for secure fit.

7. Put hood or head covering over the respirator.
8. Put on remaining protective equipment, such as hard hat, safety glasses, etc.

One person should remain outside the work area to check that each person entering has the proper protective equipment. No persons should be allowed to enter an exclusion improperly attired.

Doffing Procedures

When the potential for contamination with hazardous materials is severe, it is essential to have a good procedure to exit the work area and leave the hazardous materials behind. For highly contaminating work areas, the following procedure will provide an adequate level of protection. Whenever a person leaves the work zone, the following proper decontamination sequence will be followed:

1. Upon entering the Contamination Reduction Zone, rinse contaminated mud and debris from boot covers. Then remove boot covers.
2. Clean reusable protective equipment.
3. Remove protective garments and equipment, except respirator, on the contaminated side of the shower. All disposable clothing should be placed in plastic bags and labeled as contaminated waste.
4. Take shower and begin washing hair, neck, and face.
5. Remove respirator in shower and finish washing.
6. Dry with fresh towel.
7. Proceed to the clean area and dress.
8. Clean respirator and prepare for next use.
9. Proceed to the sign out point.

All disposable equipment, garments, and PPE should be bagged in two 6 mil plastic bags, properly labeled, and disposed.

Selection Matrix

The level of personal protection can be based on measurements of the work environment, when such measurements can be made in real time. When the assessment of the work environment depends on laboratory analysis of samples collected, then the selection of PPE must be made on professional judgment of possible or expected exposures.

Conclusion

It is possible to work with hazardous materials and processes by recognizing the hazards and taking steps to mitigate the risks to health and safety. Knowledge is an important key to safety. Knowing the properties of the materials and how they interact is vital to controlling the hazards. Controls have been developed for most hazards. Tapping into this body of knowledge allows effective and efficient controls to be utilized for the hazards associated with the cleaning processes.

The most productive approach is to involve the safety professional early on in the process design and development. If you are told "no," or if you are provided with an unacceptable list of options, explain the situation clearly. Most chemicals can be used safely, if appropriate controls are implemented.

References

1. The National Safety Council has over 450 publications on specific topics of occupational safety and health, including 15 OSHA compliance guides in the OSHA Regulatory Training Series. A good general reference is the Accident Prevention Manual. National Safety Council, 1121 Spring Lake Drive,

Itasca, IL 60143-3201, 630–285–1121, www.nsc.org. Accident Prevention Manual for Business and Industry: Engineering & Technology, ed., Philip Hagan, John Montgomery, and James O'Reilly, NSC ISBN 0-87912-213-7.

2. Guidelines to assist in the control of health hazards are published in the TLVs and BEIs booklet of the American Conference of Governmental Hygienists, which can be purchased from ACGIH, 1330 Kemper Meadow Drive, Cincinnati, OH 45240–1634, www.acgih.org. Threshold Limit Values for Chemical Substances and Physical Agents & Biological Exposure Indices, ACGIH, ISBN 978-1-882417-79-7.

3. The National Fire Protection Association is the authority on fire, electrical, and building safety. The NFPA publishes a series of codes and guidance manuals for the application of their codes. A general reference is the Fire Protection Handbook. National Fire Protection Association, 1 Batterymarch Park, Quincy, MA 02269 617-770-3000, www.nfpa.org. Fire Protection Handbook, NFPA.

4. A wealth of information is available for occupational safety and health professionals in the "White Book," published by the American Industrial Hygiene Association, 2700 Prosperity Avenue, Suite 250, Fairfax, VA 22031, 703-849-8888, www.aiha.org. The Occupational Environment: Its Evaluation, Control, and Management, ed., Salvatore DiNardi, AIHA, ISBN 1-931504-43-1.

5 European Commission, REACH: http://ec.europa.eu/environment/chemicals/reach/reach_intro.htm

6. Information about REACH and derived no-effect levels (DNELs) is available from the European Commission at http://ec.europa.eu/environment/chemicals/reach/reach_intro.htm

7 British Occupational Hygiene Society, REACH: http://www.bohs.org/resources/res.aspx/Resource/lename/777/06_REACH_Feb_07_newsletter_article.pdf

31

Critical Cleaning and Working with Regulators: From a Regulator's Viewpoint

Introduction: The Challenge 483
Historical Perspective 484
Montreal Protocol • Federal Actions, Stratospheric Ozone
Protection • The 1990s: VOCs and HAPs • Current VOC Restrictions
Regulations and Communication 485
Web Sites • Workshops
Bridging the Communication Gap 487
After Receiving a Permit...487
Compliance Inspections 488
New Tools for Communications 488
Ongoing Communication 488
Reporting and Record-Keeping Is an Ongoing Activity That
Has to Be Maintained 488
Conclusions..489

Mohan Balagopalan
South Coast Air Quality
Management District

Introduction: The Challenge

Communication is the one of the biggest challenges faced by all parties concerned with critical cleaning applications. In the past, when the use of chlorofluorocarbons (CFCs) and 1,1,1-trichloroethane (TCA) as cleaning solvents was ubiquitous and their usage was encouraged by the air agencies, the regulatory issues were much simpler to communicate. These solvents were considered unreactive organic gases, and were considered not responsible in the formation of photochemical smog (ozone in the troposphere) in the air basins. Therefore, they were exempt from most of the regulations. Permits for degreasing equipment using these solvents were issued by the air agencies, and there were no restrictions on usage, especially for CFC–113. TCA typically contained a co-solvent, 1–4 dioxane, approximately 5% by weight, and this co-solvent was considered photochemically reactive. The reactive hydrocarbon emissions calculated from a degreaser using TCA were based on a 1–4 dioxane weight fraction and TCA usage. The physical chemistries of these solvents include low boiling point, low toxicity, nonflammability, and fast drying time; this resulted in large consumption of the solvents in the cleaning industry. The air permits for degreasers using CFCs and TCA were fairly standard, and issued without much difficulty.

Historical Perspective

Montreal Protocol

However, in 1974, two scientists presented a paper on the role of CFCs in the depletion of the ozone layer, and this launched an international debate on the destruction of the ozone layer along with further studies and experiments. The Food and Drug Administration (FDA) was the first U.S. agency to act, and on October 15, 1976, announced a proposal to phase out all nonessential uses of CFC propellants in food, drug, and cosmetic products. The following year, the FDA, the U.S. Environmental Protection Agency (EPA), and the U.S. Consumer Product Safety Commission (CPSC) announced that CFCs used in spray cans would be phased out and banned within two years, except for some limited use, such as in asthma medication and a few other essential products. This ban did not have much effect on industries using CFCs for solvent degreasing.

Eventually, as more data were gathered, further impacts of the CFCs in causing global warming were presented. To prevent further deterioration of the ozone protective layer, representatives from 27 countries, including the United States, signed an agreement in 1987, pledging to reduce and eventually stop the production of solvents that deplete the ozone layer. This global agreement, which was quite unprecedented and which involved the participation of many industrialized nations and developing countries, was known as the Montreal Protocol as the conference was held in Montreal, Canada.

Federal Actions, Stratospheric Ozone Protection

Responding to the data on the destructive nature of CFCs on the protective ozone layer in the stratosphere, in 1990 the U.S. Congress also directed EPA to protect the ozone layer through several regulatory and voluntary programs in Title VI of the Amendments to the Clean Air Act of 1990. In addition, under Title VI, Section 612 of the Clean Air Act, EPA's Significant New Alternatives Policy (SNAP) program was established. SNAP's mandate is to identify alternatives to ozone-depleting substances (ODSs) and to publish lists of acceptable and unacceptable substitutes. These substitutes are reviewed on the basis of ozone depletion potential, global warming potential, toxicity, flammability, and exposure potential. The SNAP program makes decisions on a particular substitute in a particular end-use within the nine sectors identified by EPA: refrigeration and air-conditioning, foam blowing, solvent cleaning, fire suppression and explosion protection, aerosols, sterilants, tobacco expansion, pesticides and adhesive coatings, and inks. What is unique about the SNAP program, in my opinion, is the consideration of all the effects of the substitute solvent, and not the piecemeal approach that the air agencies have been accused of. The approval or disapproval of a substitute solvent through the SNAP program can be lengthy, allowing the vendor of the substitute solvent to market the solvent if no decision is made within 90 days after the petition for acceptance, with the proper documentation, is made to EPA.

Unfortunately, due to this interim legality to market, in some cases solvents have been marketed as "EPA approved" even though they were only pending approval, and were eventually disapproved by the EPA. This causes more problems for industry as they have to start anew looking for a substitute solvent if they had started using the one that was disapproved. This is another example of miscommunication where all the facts about the SNAP program were not communicated or understood.

In the United States, the production phaseouts of CFCs, halons, carbon tetrachloride, and methyl chloroform (TCA) were accelerated and were to be phased out by December 31, 1995. With the accelerated phaseout of the CFCs and TCA, the prices of these solvents, still legal to use but no longer produced, started to rise, and companies were forced to scramble to find cheaper alternatives that met regulatory requirements.

The 1990s: VOCs and HAPs

Traditionally, before the 1990 amendments to the Clean Air Act, the emphasis of air quality regulations had been on the control of the "criteria" pollutants that included reactive hydrocarbons, oxides of nitrogen, oxides of sulfur, carbon monoxide, particulate matter less than 10 microns, ozone, and seven hazardous air pollutants (HAPs). The reactive organic solvents used in the cleaning industry were thus regulated. This included perchloroethylene (PCE), trichloroethylene, and acetone. Methylene chloride (MC) and TCA were also used as degreasing solvents, but since these were unreactive organic solvents, their use was actually encouraged until much later, when MC was identified as a toxic air contaminant by the California Air Resources Board and a HAP by EPA, and TCA was phased out as an ODS. Ironically, PCE and acetone, which were regulated as reactive hydrocarbons, are now classified as unreactive hydrocarbons, and PCE use as a degreasing solvent increased in the 1990s. Acetone is used in cleaning solvent formulations to reduce the overall volatile organic compound (VOC) content of mixtures, but because of its high flammability properties, its use is limited, unlike PCE. PCE use in solvent cleaning is regulated by the National Emission Standard for Hazardous Air Pollutant for Solvent Cleaning (NESHAP). In the NESHAP for vapor degreasers, there are a number of options to comply with the equipment standards (increased freeboard height, superheated system, increased dwell time, freeboard refrigeration, etc.).

In the South Coast Air Quality Management District (SCAQMD) in southern California, PCE was regulated as a toxic air contaminant as of September 8, 1998. This drastically limits its emissions, practically making it impossible to use except in an air tight or airless cleaning system and after a health risk assessment is conducted, so its use as a degreasing solvent has decreased in the past decade. MC and trichloroethylene were listed as toxic air contaminants in June 1990, and their use has been restricted since then.

Current VOC Restrictions

The SCAQMD's regulation for the control of VOC emissions from solvent cleaning operations (Rule 1122) has been modified, and now has some of the most stringent requirements anywhere in the country. Cleaning processes using solvent containing VOC levels greater than 50 g/L used in a degreaser have to be conducted either in an airless or airtight system for cold cleaning. Rule 1171, covering vapor degreasing operations, has similar provisos.

With the multitude of restrictions on the use of solvents for critical cleaning operations, it becomes even more important that proper communication occurs between the industry, the regulatory agencies, the environmental organization, and the public.

Regulations and Communication

Web Sites

So how can we improve communication? The Internet is the new communication medium, and there are a prolific number of Web sites that have information on solvent cleaning. The EPA and most air quality management districts have their own Web sites, but oftentimes the information is scattered in these sites and there is no central repository for specific information such as "What are the permitting requirements for a vapor degreasing in (*place*)?" The degreaser equipment manufacturers and solvent manufacturers have their own Web and portal sites with information on the criteria for the selection of their equipment or their solvent properties and chemistries; they also offer complete solutions for critical cleaning. However, very few Web sites are available as yet that make it easy for someone to determine the regulatory and fiscal impact of a solvent selection at different agencies.

Workshops

Another useful venue for obtaining information is by attending trade shows, seminars, and workshops where specific information can be obtained. At these events, vendors, manufacturers, and regulators meet and discuss issues and share information.

While all these portals of information are good sources, frequently it is best to go to the source for the regulations: the regulators. There is some reluctance to do so, because the agencies are perceived to be bureaucratic and it is sometimes difficult to find the "right person" to talk to. So how do you contact a regulator, approach the agency for information, and get connected with the person or persons most knowledgeable on the cleaning processes and regulations? Asking the wrong questions or talking to the wrong persons may lead to inaccurate or misleading information that could prove costly in the long run.

Case Situations

Methylene Chloride

In the SCAQMD, the use of MC was encouraged, by permitting engineers, as a replacement for CFCs and 1-1-1 TCA, as it is an unreactive organic solvent and it did not contribute to the formation of photochemical smog. Its use was later restricted (in June 1990) when MC was listed as a toxic air contaminant in EPA's new source rule for air toxics, and the cancer risk from its usage was considered high even for a moderate user of the solvent. Permits that were issued prior to the effective date that MC was listed as a toxic air contaminant were considered to be "grandfathered," and were not affected. It only affected the permits of new cleaning machines or those increasing their usage. In this situation, permits were issued for the use of a solvent that was later determined not to be an ideal choice.

Perchloroethylene

An opposite situation to this occurred with the use of PCE as a cleaning solvent. PCE was considered a reactive hydrocarbon, and its use was regulated as such, whereby emission offsets and technology standards and practices were required. It was identified as a hazardous air pollutant by the Clean Air Act of 1990, and as a probable human carcinogen. In 1991, the California Air Resources Board listed PCE as a toxic air contaminant pursuant to its Air Toxics Program (AB 1807) and Proposition 65. At the local permitting agency level, PCE was not listed as a toxic air contaminant subject to health risk assessment until September 1998. In 1994, EPA adopted the National Emission Standards for Hazardous Air Pollutants for Solvent Cleaning. This regulated the use of PCE and established emissions standards, equipment design, and work practices.

In 1996, EPA delisted PCE as a VOC, and the SCAQMD followed suit in June 1997. When this delisting occurred at the SCAQMD, industries looking for a suitable substitute rushed to PCE, as there were no regulations limiting its use in the interim, until it was added to the toxic list in September 1998. Industries that switched to it were looking for a drop in replacement without any obstacles in obtaining an air permit.

In this situation, permits were issued because the regulations during this period (June 1997–September 1998) allowed it, even though it was known that PCE was listed a toxic air contaminant.

In the first situation, when the use of MC was encouraged, its toxicity and cancer-causing potential were relatively unknown to the permitting staff. In the second situation, the potency of PCE was known, and it was hopefully communicated to the permit applicant that there were other ramifications to its use, such as the NESHAP and OSHA permissible exposure limit.

Role of the Regulator

What role can the regulator play in the process of solvent and equipment selection? Should the regulator be consulted and made a partner early in the selection process? Can the regulator be trusted to be partial and forthcoming in providing information? This is one area where, due to poor communication, the regulator is oftentimes ignored or considered to be an adversary, and is therefore not consulted early in the selection process. The regulator frequently becomes involved only when he or she has to work on the permit application for approval of the permit with the chosen solvent. The operator may have chosen the solvent based on its solvency, its cost, the capital cost of a new equipment if one is needed, the ongoing maintenance and repair cost, cost of insurance, cost of employee training, and customer satisfaction and with some assurance from the chemical and equipment vendors that the permit would be approved.

A permitting engineer responsible for issuing permits may have limited knowledge of the cleaning processes or the reasons behind the selection of the cleaning solvent. However, he or she is normally familiar with the rules and regulations, and evaluates the permit application accordingly. During a normal site inspection, a compliance inspector is concerned mainly with verifying compliance with the permit conditions and whether the requirements for record keeping are met.

Typically, unless asked, the engineer does not volunteer information, and there is no expectation of the regulator to offer advice or provide information unless specifically requested. Alas, in many cases, the requests are not made. Similarly, due to lack of communication and trust, the learning and the gathering of knowledge by the permitting engineer or compliance inspector regarding the various cleaning processes from the regulated source are not maximized. Thus, the opportunity for both sides to benefit from each other regarding the processes and regulations is lost.

Bridging the Communication Gap

To bridge the communication gap, both parties have to trust each other and take the initial steps in getting together to meet and confer on the issues. Industry should ask the regulator for participation and assistance, and to be a partner in the selection of a suitable solvent. The regulator should learn from industry about the process and the different items that have to be considered in the selection of a solvent.

From the regulators, there must be better customer relationship management and more willingness to impart information and share knowledge. From industry, there should be more trust and openness to ask questions and to solicit information. Quite often, it appears to the regulator evaluating a permit application that the decision to select a cleaning system or a solvent is made strictly for economic reasons. In reality, numerous other considerations such as the health and safety of the workers, insurance, as well as cost may have been factored into the decision to select a cleaning process or solvent.

Some forums where informal communications occur are at seminars, workshops, conferences, and association meetings. A strong association with active members in the regulatory arena that liaises between the regulated community and the regulators is an effective way to communicate the concerns and needs of the cleaning industry.

It is possible for misinformation and misleading information to be provided by the regulatory agency unintentionally, because of lack of knowledge or misunderstanding of the questions posed. Sometimes, there is the "shopping around" for answers from different regulators, a consultant, or a company's representative until they get the information they want to hear. This can create problems and confusion down the road.

After Receiving a Permit

When you receive your permit, do not file or post it until you have read and verified the information listed on the permit and the permit conditions. Verify the equipment description on the permit for accuracy relative to the physical equipment on site. Inform the regulator as soon as possible, in writing, if there are any discrepancies. As it sometimes happens, the cleaning equipment/degreaser purchased and installed may be different from what was submitted in the permit application. It is important to correct any discrepancies between the equipment installed and operated and the equipment description on the permit. It is also critical to verify that the conditions listed on the permit are acceptable and can be complied with.

Typically, a source (e.g., the manufacturing facility) has 10 days after a permit is received, to request any corrections. After this period, the facility may have to appeal to a Hearing Board at the agency to change the conditions or submit another application to modify the equipment description or permit conditions. It is not uncommon to find that a source files away a new permit without reading it and without being aware of the need to keep records on solvent usage, conduct periodic maintenance, or other requirements listed on the permit conditions.

Compliance Inspections

Insufficient understanding of the information in a new permit may become an issue during the compliance inspection, as the inspector may issue a notice to comply or a notice of violation to correct discrepancies. The notice could be a minor issue, such as a need to correct the model number of the equipment, or the serial number or the manufacturer of the equipment. In other cases, there could be a more severe violation if the solvent used is different than what the equipment was permitted for. This is considered an alteration/modification of the system. Often, what the compliance inspector has is a permit that shows discrepancies between what is on the permit and what is actually there in terms of equipment and its operation and maintenance. There are typically two standard conditions listed on all permits that address this situation.

1. Operation of this equipment shall be conducted in accordance with all data and specification submitted with the application under which this permit is issued unless otherwise noted as follows.
2. This equipment shall be properly maintained and kept in good operating condition at all times.

The conditions on the permit may be listed for compliance with rule requirements for a source-specific rule for the degreasing operation; a NESHAP rule requirement; monitoring, record-keeping, and reporting provisions of applicable local, state, and federal agencies; Best Available Control Technology (BACT) requirements; and emission limits.

New Tools for Communications

Agencies are now realizing the great potential to communicate with interested parties, with tools such as Wiki, Tweet, Facebook, MySpace, YouTube, podcasts, video, and interactive Web sites. This will help in further keeping all concerned parties informed about changes to rules and regulations that affect the permits.

Ongoing Communication

Unless there are changes to the equipment or process change, the permit does not have to be modified. However, the permit should be kept active and renewed annually. In most cases, this would involve paying the annual operating fees; the permit is renewed without a reissue. This might be different if the facility is a Title V permit, in which case the permit may be issued if other equipment listed in the permit undergoes permit modification. The Title V permit is renewed every 5 years. It is the responsibility of the facility to keep track of when the fees are due and to pay them on time. The agency should be notified in writing if there are any changes to the mailing address, contact person, or responsible official. The consequence in not renewing the permit can result in the expiration of the permit. If the source is unaware that the permit has expired and continues to operate the equipment, it could lead to a violation notice if the facility is inspected. In addition, once a permit has expired, the source has to reapply for the permit, and it would be treated as a new source, subject to new source regulations, that may subject the equipment to current BACT requirements or limits on usage of solvent, due to risk assessment.

Reporting and Record-Keeping Is an Ongoing Activity That Has to Be Maintained

Sources have to report usage and emissions of solvents oftentimes to local, state, and federal agencies if usage exceeds a certain threshold or if the solvent is listed as a toxic air contaminant subject to reporting. Sources must, however, bear in mind that most of the data gathered by the agencies are accessible to the community. It is therefore critical that care is taken in preparing these reports and it is ensured that the reported emissions are indicative of actual emissions at the facility. For example, facilities reporting their annual Toxics Release Inventory (TRI) data to EPA should be aware that the reports are provided to the public and give the communities access to information on their exposures to toxic chemicals. This may cause undue alarm to the public if the data reported were not verified and if they were overestimated.

It is also critical that if changes are going to be made to the process or equipment, the permitting staff is contacted to determine if any modification would be needed. Often, the source-specific rule that applies to the degreasing operation may be amended and requirements added that would apply for the existing operation. However, the permit is often not updated when the rule is amended, and it is up to the source to keep abreast of the changes and make any changes needed to demonstrate compliance with the applicable rules and regulations.

If there is any accidental release of the solvent due to an accident or poor maintenance, it is necessary to call the agency to report a breakdown. The facility can apply for a variance to operate while the problem is being fixed. It pays to be actively in communication with the permitting engineer and keep the channel open so that there is no hesitancy in picking up the phone and calling the person to ask for input or report a problem.

Conclusions

Facilities that integrate environmental compliance as a strategic component of business have a better advantage over those who do not, since the former would have consulted with the regulatory agency in a proactive manner to determine the necessary steps to ensure continued compliance. This could be in the form of self-audit programs, participation in International Organization for Standardization (ISO) 14000, no-fault compliance inspections that are available from some agencies, or periodic meetings and communications with the regulators. Improving industry–government relations by having more open communication, sharing solutions, and working together to solve problems fosters development of sound regulations that facilitates permit issuance and compliance.

32

Momentum from the Phaseout of Ozone-Depleting Solvents Drives Continuous Environmental Improvement*

Introduction 491
After the Montreal Protocol Was Signed, the Industry
Changed Direction 492
News-Making Corporate Leadership Left No Doubt That Ozone-Depleting
Solvents Were Obsolete and Unwanted • Military Leadership Was
Important • Electronics Companies, Associations, and Committees
Became Environmental Leaders • No-Clean Soldering Was a
Remarkable Team Success
The Clean Air Act Amendments of 1990 Set Further
Environmental Goals 495
The Phaseout of Ozone-Depleting Substances Continues.....................496
Lessons from the Montreal Protocol for Climate Change 496
Proactive Companies That Protected the Ozone Layer Are
Now Protecting the Climate...496
Challenges of Sustainable Enterprise..497
Conclusions and Climate Leadership Opportunities 497
Watch List
Glossary of Acronyms...498
References..498

Stephen O.
Andersen
*Institute for Governance &
Sustainable Development*

Margaret Sheppard
*U.S. Environmental
Protection Agency*

Introduction

Experts from industry and government are very proud of progress in critical cleaning that is safe for workers, satisfies technical performance criteria, and is environmentally sustainable. This chapter takes a quick look at the extraordinary environmental innovation occurring from the early 1970s until the

---

* This publication has not been subjected to EPA review and therefore does not necessarily reflect the views of the EPA, and no official endorsement should be inferred. Mention of trade names or commercial products does not constitute endorsement or recommendation for use.

present time. It finds the roots of current innovation in the Montreal Protocol and then looks to the future to see how lessons from the past can guide the industry in the daunting task of climate protection and sustainable enterprise.

Consider the extraordinary change for the better. The cleaning industry of the 1960s and 1970s was a dangerous place to work and an expensive place to cleanup. Workers were exposed to toxic carbon tetrachloride and other chlorinated solvents, groundwater was contaminated, and drinking water was compromised. Private and public litigation reached Superfund proportions.

In the 1980s, the Montreal Protocol went after less toxic chlorofluorocarbon (CFC) and methyl chloroform solvents as ozone-depleting substances (ODSs). The annual disclosure of the company name, facility location, and quantity of emissions of chemicals listed on Toxic Release Inventory (TRI) of the U.S. Environmental Protection Agency (EPA) empowered citizens and communities to demand safe manufacturing. Cleaning experts with green credentials prospered, and saved their clients operating, litigation, and cleanup expenses. Often, companies learned how to reduce or eliminate chemical use while improving product performance and durability.

The 1987 Montreal Protocol is the most successful environmental treaty to date. It is the only treaty with every country as a member and with both developed and developing countries in almost total compliance with phaseout schedules. The ODSs are phased out and the replacement products and processes are technically and environmentally superior. The ozone-depleting solvents were primarily replaced with "not-in-kind" technology, including no-clean or low-clean, aqueous—with less than 20% of previous applications using hydrochlorofluorocarbon (HCFC), hydrofluorocarbon (HFC), hydrocarbon (HC), or other chemical solvents. HCFC solvents are already being phased out, and HFC solvents are proposed for phasedown under the Montreal Protocol, are agreed for emissions limits under the Kyoto Protocol, and would be subject to phasedown under legislation pending before Congress.

Looking forward from the 1987 signing of the Montreal Protocol was frightening for aerospace, electronics, and metal cleaning companies and their civilian and military customers. This chapter describes how industry genius and government-industry partnerships innovated to protect the stratospheric ozone layer and how that momentum now supports continuous innovation in cleaning. This chapter also outlines how the lessons from the Montreal Protocol can be applied to the challenge of protecting the fragile climate under the Kyoto and Copenhagen Protocols.

In the 1990s, cleaning experts working to eliminate toxic and ozone-depleting solvents became part of a wider campaign to completely transform manufacturing to a more profitable and greener footprint.

After the Montreal Protocol Was Signed, the Industry Changed Direction

In the early 1980s, the industry initially opposed ozone layer protection in much the way some industries oppose climate protection today. During the debate before the signing of the Montreal Protocol, manufacturers of chlorofluorocarbons (CFCs) and many of their industrial customers fought aggressively against regulations. Industry argued that scientists had not yet proven that CFCs destroyed stratospheric ozone, that the products made with or containing ODSs were absolutely vital to society, that there were no safe substitutes, and that potential substitutes would be ineffective and costly.

Fortunately, industry then began listening more carefully to the scientific findings linking ozone depletion to CFCs, halons, and the other potent ODSs. In late 1987 and early 1988, leading multinational companies began announcing corporate goals to halt the production and use of ODSs. They mobilized their technical experts to seek solutions, motivated suppliers to offer alternatives, and commercialized and implemented new technology that eliminated the need for ODSs. Corporate support and the availability of technical solutions enabled the parties to the Montreal Protocol to make strong political decisions to expand the list of controlled substances and to accelerate the phaseout schedule (Andersen and Sarma, 2002).

News-Making Corporate Leadership Left No Doubt That Ozone-Depleting Solvents Were Obsolete and Unwanted

- In 1986, DuPont made an about-face, by embracing scientific evidence as conclusive and advising customers to seek alternatives.
- In 1988, AT&T broke ranks with industry associations and announced that a new solvent made from oranges cleaned as effectively as CFC-113.
- In 1988 and 1989, Nortel/Northern Telecom and Seiko Epson pledged to phase out CFCs faster than required by international treaty or domestic regulation, and frequently updated those pledges with faster action on a growing list of solvents.
- In 1989, EPA helped to organize the Industry Cooperative for Ozone Layer Protection (ICOLP) and the Halons Alternatives Research Corporation (HARC) under the Cooperative Research Act, allowing collaboration to protect the ozone layer with protection from antitrust concerns. Japanese associations combined their considerable strength in the Japan Industrial Conference on Ozone Layer Protection (JICOP).

Military Leadership Was Important

The U.S. military also surprised skeptics by spearheading technical innovation, green procurement, and market influence. U.S. Department of Defense (DoD) manufacturing standards requiring the use of CFC-113 had contributed to a global dependence on this potent ozone-depleting solvent. Many companies manufactured all their sophisticated, high-reliability products with CFCs in order to qualify for military sales. In response to this manufacturing customer demand, suppliers to electronics manufacturing perfected components, equipment, flux, solder, and other products to optimize performance. CFC-113 manufacturers offered valuable training and trouble-shooting services to their customers. Because the DoD specifications became globally respected for quality assurance, many companies and some countries cited these specifications in manufacturing products for civilian markets. In effect, the DoD standard was a de facto global standard in 1987 when the Montreal Protocol was agreed.

The DoD joined with EPA and the Institute for Interconnecting and Packaging of Printed Circuits (IPC—now called the Association of Electronic Interconnecting Industries) in organizing the Ad Hoc Solvent Working Group to advise the official DoD soldering technical committee. That team changed the *prescriptive standard* requiring CFCs to a *performance standard* that would allow and ultimately encourage CFC-free assembly. In less than two years, the working group created a test board and procedure for measuring cleaning potential of solvent alternatives, bench-marked cleaning with CFC-113, established a test verification team to observe the cleaning tests, persuaded DoD to accept solvents that cleaned as well as or better than the CFC benchmark, and verified the first solvents to pass the demanding test. As this project was proceeding, the working group experts concluded that solvent-free assembly with conductive adhesives and "no-clean" flux could be an additional alternative to CFC solvents. By the time the first solvents were passing the test, DoD had revised the standard to encourage CFC-free assembly and had written the performance standard to accommodate innovative solvent-free technology. Later, DoD prohibited cleaning with CFCs unless high-level DoD exceptions were granted.

DoD was so technically confident of the CFC phaseout that they cosponsored technical seminars in cooperation with the North Atlantic Treaty Organization (NATO). By 1992, leading military experts had recognized ozone layer depletion as a national security threat requiring the strongest actions. NATO took the unprecedented step of writing directly to the Executive Director of the United Nations Environment Program (UNEP), advocating a complete phaseout of substances that deplete the ozone layer (Andersen, Sarma and Taddonio, 2007).

Electronics Companies, Associations, and Committees Became Environmental Leaders

Nortel/Northern Telecom and Seiko Epson were technically optimistic in 1988 when they became the first electronics and precision manufacturers to pledge a phaseout of CFC solvents, but they soon realized that they needed the creative force of the entire industry to achieve their goals. AT&T, Nortel, and the U.S. EPA conceived the Industry Cooperative for Ozone Layer Protection (ICOLP) to encourage competing companies to cooperate on the development and implementation of environmentally protective industrial technologies. The founding members included respected multinational companies that together exercised extraordinary market clout: AT&T, The Boeing Company, British Aerospace, Compaq, Digital Equipment Corporation (DEC), Ford, Hitachi, Honeywell, Hughes Aircraft Company, IBM, Lockheed-Martin, Matsushita Electric, Mitsubishi Electric, Motorola, Nortel/Northern Telecom, Ontario Hydro, Seiko Epson, Texas Instruments, and Toshiba. ICOLP helped to fast-track implementation of innovative technologies by fostering a spirit of collaboration rather than competition among industry rivals and then transferred their successes by sharing their findings with the electronics industry worldwide.

DEC and Nortel/Northern Telecom were the first companies to donate patented technology to the public domain in order to speed ozone layer protection. Dozens of companies opened new production facilities that were ODS-free. Public tours and technical cooperation projects helped promote use in all countries.

Industry participation was also key to the success of the UNEP Solvents, Coatings and Adhesives Technical Options Committee (STOC) which advised the parties to the Montreal Protocol on alternatives to replace CFC-113 and 1,1,1-trichloroethane (methyl chloroform). The Solvents TOC created a forum for identifying and documenting promising technology, in addition to increasing global awareness of the role of the electronics industry in ozone layer protection. Their contributions have been critical in the development of domestic ozone protection regulations by the governments of many countries.

No-Clean Soldering Was a Remarkable Team Success

The STOC advising the parties to the Montreal Protocol exemplifies the power of proactive thinking and cooperation. In 1989, the German company SEHO demonstrated a controlled atmosphere soldering technology to the Solvents TOC, which immediately recognized its potential, but realized that the equipment developers had only basic soldering skills. Upon close examination, the TOC realized that the component parts were oxidized, the flux was mismatched, and the soldering wave was poorly formed. However, the promise of the technology was so compelling that TOC members from AT&T, Ford, and Nortel/Northern Telecom persuaded their companies to experiment with the technology. Soon other companies joined the ad hoc team of inventors in a "skunk works without walls."

Some experts concentrated on the flux composition, while others believed that the flux was chemically suitable and worked ways to apply it to the board more precisely. Still others concentrated on the mixture of gases in the soldering chamber. Several months into development, one company discovered that gas monitoring and control calibrations were critical. Better calibrations dramatically improved the soldering quality. The team consulted with flux suppliers who grasped the opportunity for developing and commercializing new products and intensified development in cooperation with the electronics manufacturers. EPA encouraged the work by using ICOLP status under the Cooperative Research Act to relieve the companies of antitrust concern, and by documenting and publishing the global environmental advantages that such a technology could provide.

Meanwhile, engineers at AT&T Bell Laboratories were developing state-of-the-art spray fluxing machines to apply precisely the optimal amount of flux to the locations in the printed circuit boards where necessary. Nortel/Northern Telecom was developing equipment to verify flux concentrations on production boards, and Motorola was experimenting with soldering ultra-miniaturized circuits with hybrid components, including optical devices and flexible connectors. One by one, companies satisfied internal quality controls and moved from lab scale to pilot and finally to full implementation. During implementation, experts from the intercompany team continued to cooperate to debug operations and optimize performance.

It is impossible to say just when the engineering team realized that their no-clean technology would revolutionize electronics assembly. Engineers, who had cautiously reported as-good performance, began to report improved performance. Line managers cautiously increased the speed of soldering to rates never achieved with conventional soldering and found that in some cases defect rates actually decreased. EPA in cooperation with ICOLP published the first no-clean handbook in order to make the expertise and technology available worldwide.

The Clean Air Act Amendments of 1990
Set Further Environmental Goals

Congress's amendments to the Clean Air Act (CAA) in 1990 renewed the local, state, and federal commitment to improving local air quality. The federal government sets local air quality standards to protect human health, particularly the National Ambient Air Quality Standards. State governments can then be delegated authority to regulate and enforce the particular way to meet those national standards by creating and receiving EPA approval of State Implementation Plans. By the early 1990s, most states in the eastern half of the country and many states with large cities still had counties that did not meet the national ambient air quality standard for ground-level ozone (smog). The CAA Amendments set new deadlines for state governments to attain existing national air quality goals. EPA also determined that a stricter standard was need for ground-level ozone to protect public health. These factors resulted in waves of changes to federal and state regulations, and particularly those related to reducing ground-level ozone.

One approach that states took to reduce smog was to reduce the volatile organic compounds (VOCs) that, together with oxides of nitrogen from combustion and sunlight, create ground-level ozone. Many cities or states started setting limits on the VOC content of cleaners and aerosol products. Formulators and end users paid close attention to substances that EPA exempted from being VOCs (found at 40 CFR 51.100(s), and online at http://www.gpoaccess.gov/ecfr). These are substances that EPA found react little when exposed to sunlight, and therefore should not add significantly to smog. Examples include acetone, a number of fluorinated compounds, perchloroethylene, methylene chloride, complete methylated volatile siloxanes, and methyl acetate.

The CAA Amendments also required EPA to regulate hazardous air pollutants. Congress listed specific compounds (CAA Section 112(b)) and allowed for petitions to add or take off compounds from the list. Hazardous air pollutants are of particular concern because of serious health effects such as cancer, damage to the immune system, as well as neurological, reproductive, developmental, respiratory and other health problems. Examples of solvents on the list of hazardous air pollutants include benzene, carbon tetrachloride, chloroform, hexane, methanol, chlorinated solvents such as methylene chloride, perchloroethylene, trichloroethylene, toluene, and xylene. The complete list of hazardous air pollutants that EPA regulates is available at http://www.epa.gov/ttn/atw/188polls.html.

EPA put out standards for different hazardous air pollutants by industry category (found at 40 CFR Part 63). These standards affected formulators and users of coatings and solvent wipes, the aerospace industry, the boat building industry, the semi-conductor industry, users of paint strippers, and users of chlorinated solvents in vapor degreasing equipment, among others.

The National Emission Standard for Hazardous Air Pollutants (NESHAP) for halogenated solvent cleaning was first issued in 1994 (40 CFR Part 63, subpart T) and has been revised since then. (For details, see http://www.epa.gov/ttn/atw/degrea/halopg.html.) The halogenated solvent cleaning NESHAP controls emissions of methylene chloride, perchloroethylene, trichloroethylene, 1,1,1-trichloroethane, carbon tetrachloride, or chloroform when used in batch vapor, in-line vapor, in-line cold, or batch cold solvent cleaning machines. As a result of these rules, vapor degreasers built after the mid-1990s were built with features that reduced emissions, lowered worker exposure to toxic substances, and saved solvent. For example, vapor degreasers needed to have larger spaces above the cleaning regions ("freeboard") and one or more sets of cooling coils.

A new section of the CAA, Title VI, developed national policies and requirements to implement and to complement the Montreal Protocol for protection of the stratospheric ozone layer. One section of the CAA prohibited the sale of products containing ODSs for nonessential uses, including most uses of CFCs, 1,1,1-trichloroethane, and HCFCs in aerosol cans. EPA regulations provide an exception that allows use of ODSs in aerosols for cleaning of electronics and aviation or aerospace equipment. Other sections of the CAA require labeling of manufactured goods containing or made with ODSs. Another section requires EPA to evaluate the health and environmental impacts of substitutes for ODSs, and to make lists of alternatives that EPA finds are safe (or not). This is the basis of the Significant New Alternatives Policy (SNAP) program.

Industry needed to comply with requirements for VOCs, HAPs, and ODSs at the same time. Users evaluated alternatives for these environmental factors, as well as considering cleaning performance, cost effectiveness, and workplace safety considerations. Where possible, many users chose to switch to aqueous cleaning. Others tried techniques that minimized the use of solvent or even removed the need to clean altogether ("no-clean"). It was a time of challenge and industry innovation.

The Phaseout of Ozone-Depleting Substances Continues

After remarkable success with eliminating CFCs and 1,1,1-trichloroethane, the cleaning industry is facing another challenge with transitioning away from ozone-depleting HCFCs. In 2003, EPA regulations prohibited the production and import of HCFC-141b (1,1-dichloro-1-fluoroethane), a compound used as a substitute for CFC-113 as an aerosol solvent for cleaning electronics and aerospace equipment. The CAA prohibits the sale and use of HCFCs (and not just production or importation) as of 2015, except for recycled HCFCs, refrigerant used in existing equipment, and HCFCs used up in chemical reactions (CAA Section 605(a)). Thus, users should expect that they can no longer buy cleaning products containing HCFC-141b or HCFC-225ca/cb (e.g., AK-225) by 2015, and plan accordingly.

Lessons from the Montreal Protocol for Climate Change

This experience with the Montreal Protocol holds several lessons for the world as it confronts the problem of climatic change (Velders et al., 2007):

- Science-backed stringent regulation motivates industry.
- Corporate environmental goals focus company priorities and motivate suppliers—companies are beginning to set such goals for climate protection.
- Partnerships and associations between industry and government attract companies taking leadership.
- Information is critical.

Proactive Companies That Protected the Ozone Layer Are Now Protecting the Climate

A number of chemical companies are investigating hydrofluoroolefins (HFOs) as substitutes for greenhouse gases. HFOs are a subset of non-ozone depleting HFCs. HFOs contain an unstable double bond that causes a short lifetime in the atmosphere and a low global warming potential (GWP). For example, DuPont and Honeywell have developed refrigerant HFO-1234yf to replace HFC-134a in motor vehicle air conditioners. HFO-1234yf has a remarkably low GWP = 4 (Nielsen et al., 2007) (compared to the GWP = 1430 (IPCC, 2007) for HFC-134a). Encouraged by the market potential of HFO-1234yf, Asahi Glass and Arkema have patented their own process for manufacture. A closely related chemical from Honeywell, HFO-1234ze, with a GWP of 6 (Søndergaard et al., 2007) is under consideration for use as a heat transfer agent, aerosol propellant, and foam blowing agent.

3M developed new fluorochemicals that could replace ODSs and perfluorocarbons (PFCs). PFCs had become popular in the semiconductor manufacturing industry and had a role in cleaning. 3M created hydrofluoroethers (HFEs), such as those in HFE-7100 and HFE-7200 (GWPs of 297 and 59, respectively, IPCC, 2007), that can replace perfluorohexane (C_6F_{14}, with GWP of 9300, IPCC, 2007) for cleaning and heat transfer. Another class of fluorochemicals that 3M is developing, fluoroketones, show promise for heat transfer and fire suppression applications. For example, a fluoroketone with 6 carbons, FK-5-1-12mmy2, is a heat transfer agent and fire suppressant with GWP less than 2 (EPA, 2004); for comparison, halon 1301 has an ozone depletion potential of 10 and a GWP of 7140 and HFC-227ea has a GWP of 3220. A dozen more new low-GWP chemicals have been patented or otherwise disclosed by Asahi Glass, Arkema, Mexichem Fluor, Dow, DuPont, Honeywell, and others.

The World Semiconductor Council, representing over 90% of global semiconductor manufacturing, has reduced PFC emissions dramatically, even as semiconductor output will increase substantially.

Challenges of Sustainable Enterprise

Despite the extraordinary environmental progress, the Earth is more in peril than ever. Scientists warn that we must cut greenhouse gas emissions significantly within decades to avoid or slow predicted average temperature rises of 1.1°C–6.4°C (IPCC, 2007) by the end of this century.*

The challenge is to welcome new technology and to reward innovation. In many cases, command and control regulation may be necessary to motivate change and to level the competitive playing field if some firms resist investment in the technology that is necessary for the prosperity of all.

As we write this, national and world leaders are discussing how to address climate change. The signatories to the United Nations Framework Convention on Climate Change (FCCC) and the parties to the Kyoto Protocol on global climate change will be discussing where international efforts should go to protect climate, with the Kyoto Protocol set to expire at the end of 2012. The parties to the Montreal Protocol will be discussing a proposal to address HFCs under the Montreal Protocol (rather than under the Kyoto Protocol). The U.S. House of Representatives has passed the American Clean Energy and Security Act of 2009, with the Senate still debating similar issues. It seems possible that in this decade, there will be agreement on addressing climate change, energy security, and sustainability. Industry may again have the challenge and the opportunity to create better products using environmentally preferable approaches.

Conclusions and Climate Leadership Opportunities

Technical experts and their companies who have protected the ozone layer can build on their global reputation by contributing to climate protection. Electronics are the key to improvements in manufacturing process efficiency, key to products that save energy, and key to efficient organization to conserve resources. Furthermore, business can help government with market transformation to more sustainable commerce.

---

* Scenarios range from a lowest likely range of 1.1°C–2.9°C to a highest likely range of 2.4°C–6.4°C for surface warming. All the scenarios discussed assumed higher emissions than in 2000.

Watch List

A "watch list" of technologies important to the environment and the bottom line includes:

1. New low-GWP HFE and HFC (HFO) solvents—safe for the ozone layer, safe for climate, performing like historic solvents, and with superior "Life-Cycle Climate Performance—LCCP" incorporating direct chemical emissions and energy use at every stage.
2. Design for Environment with green sourcing, reuse, and full recycling.
3. E-commerce, telecommuting, and information systems that reduce travel, improve productivity, and enhance quality of life.

Remember, technologies alone cannot build the momentum necessary to protect the climate. People, as always, are the force of decision, persuasion, innovation, and implementation. People protected the ozone layer, people can protect the climate, and people can make production and consumption sustainable. Engineers and managers are invited to spearhead that noble effort.

Glossary of Acronyms

CAA	Clean Air Act
CFC	Chlorofluorocarbon
CFR	Code of Federal Regulations
DoD	Department of Defense
EPA	Environmental Protection Agency
FCCC	Framework Convention on Climate Change
GWP	Global Warming Potential
HAP	Hazardous Air Pollutant
HARC	Halon Alternatives Research Corporation
HC	Hydrocarbon
HCFC	Hydrochlorofluorocarbon
HFC	Hydrofluorocarbon
HFE	Hydrofluoroether
HFO	Hydrofluoroolefin
ICOLP	Industry Cooperative for Ozone Layer Protection
IPCC	Intergovernmental Panel on Climate Change
JICOP	Japan Industrial Conference on Ozone Layer Protection
LCCP	Life-Cycle Climate Performance
NATO	North Atlantic Treaty Organization
NESHAP	National Emission Standard for Hazardous Air Pollutants
ODS	Ozone-depleting substance
PFC	Perfluorocarbon
SNAP	Significant New Alternatives Policy (Program)
STOC	Solvents, Coatings and Adhesives Technical Options Committee
TEAP	Technology and Economic Assessment Panel
TRI	Toxic Release Inventory
VOC	Volatile organic compound

References

Andersen, S. O. and K. Madhava Sarma, 2002. *Protecting the Ozone Layer: The United Nations History*, Earthscan Press, London, U.K. (Official publication of the United Nations Environment Programme).

Andersen, S. O., K. Madhava Sarma, and K. N. Taddonio, 2007. *Technology Transfer for the Ozone Layer: Lessons for Climate Change*, Earthscan Press, London, U.K. (Official publication of the Global Environment Facility (GEF) and the United Nations Environment Programme).

EPA, 2004. Protection of stratospheric ozone: Notice 19 for significant new alternatives policy program. *Federal Register*, 69, 58903, October 1, 2004. Available online at http://www.gpo.gov/fdsys/pkg/FR-2004-10-01/pdf/04-21928.pdf

IPCC, 2007. Intergovernmental panel on climate change, Fourth assessment report (AR4), in *Climate Change 2007: The Physical Science Basis*, Solomon, S., Qin, D., Manning, M., Chen, Z., Marquis, M., Averyt, K. B., Tignor, M., and Miller, H. L., eds., Cambridge University Press, Cambridge, U.K., 996 pages. Available online at http://www.ipcc.ch/publications_and_data/publications_ipcc_fourth_assessment_report_wg1_report_the_physical_science_basis.htm

Nielsen, O. J., M. S. Javadi, M. P. Sulbaek Andersen, M. D. Hurley, T. J. Wallington, and R. Singh, 2007. Atmospheric chemistry of $CF_3CF=CH_2$: Kinetics and mechanisms of gas-phase reactions with Cl atoms, OH radicals, and O_3. *Chem. Phys. Lett.*, 439(2007), 18–22. Available online at www.sciencedirect.com

Søndergaard, R., O. J. Neilsen, M. D. Hurley, T. J. Wallington, and R. Singh, 2007. Atmospheric chemistry of trans-$CF_3CH=CHF$: Kinetics of the gas-phase reactions with Cl atoms, OH radicals, and O_3. *Chem. Phys. Lett.*, 443(2007), 199–204. Available online at www.sciencedirect.com

Velders, G. J. M., S. O. Andersen, J. S. Daniel, D. W. Fahey, and M. McFarland, 2007. The importance of the Montreal Protocol in protecting the climate. *Proc. Natl. Acad. Sci.*, 104, 4814–4819.

33

Screening Techniques for Environmental Impact of Cleaning Agents

Introduction 501
Focus of This Chapter 502
The EPA SNAP Program 503
Good Ozone versus Bad Ozone..505
Air Quality and VOCs 507
Concerns about Stratospheric Ozone: Ozone Depletion Potentials......508
Concerns about Stratospheric Ozone: Equivalent Chlorine Loading.....512
On the Recovery of Stratospheric Ozone..514
Concerns about Climate Change: Global Warming Potentials 514
Summary and Conclusions 517
References..517

Donald J. Wuebbles
University of Illinois

Introduction

The development of meaningful environmental policies for cleaning agents and other compounds depends on many factors, including concerns about the toxicity of the compound, its potential as a fire hazard, its potential effects on air quality, its potential effects on stratospheric ozone, and its potential effects on climate, as either a greenhouse gas or as a precursor to particle formation. Determining the environmental impact of a given compound, weighing the relative risks to human health and to community health, and then deciding on the appropriate regulatory status of that compound is therefore a monumental task. Scientists in specialized diverse fields must evaluate the compound in question. However, science is not the only factor considered in policy decisions. Certainly, economics and, unfortunately, politics can also play a role. Policymakers often desire simple ways of representing the science, and thus various metrics and concepts such as ozone depletion potentials (ODPs) and global warming potentials (GWPs) have arisen to meet that need. As an example, scientists determine the ODP and impact by the best available modeling and experimental techniques. Subsequently, the United States Congress and an international consortium of scientists, policy makers, and representatives of chemical manufacturers in United Nations policy committees worldwide, ultimately determine what ODP level will be significant. However, it must be emphasized that scientific input has a very strong influence on environmental policy.

Scientists and policymakers use a variety of approaches and techniques to analyze the potential environmental effects of chemicals, including cleaning agents. A number of factors affect the screening approaches that need to be applied. These factors then drive a series of questions that need to be addressed. Some of these questions are as follows:

- Could the agent potentially affect stratospheric ozone? Does it contain chlorine, bromine, or iodine?
- Is the substance toxic? To what degree, and what is the nature of the toxicity?
- What is the vapor pressure of the agent? That is, will it go to a gas at standard temperature and pressure)?
- What is the atmospheric lifetime of the agent?
- Will the agent be a volatile organic compound (VOC) and thus potentially affect urban/regional ozone?
- Is the agent a greenhouse gas? That is, does it have absorption features at infrared wavelengths? Could the agent potentially affect climate?
- In what processes will the agent be used?
- Will the agent be emitted into the atmosphere, or can it be contained for repeated use?
- If introduced for commercial use, is the agent likely to replace agents with a more favorable environmental profile?
- What is the likely production level of the agent?

The primary goal of this chapter is to focus on just a few of the issues related to cleaning agents, particularly those now being used or being considered for use as substitutes for class I ozone-depleting substances (ODSs). Also discussed are some of the newer issues, such as the potential for policies relating to concerns about climate change. The depth of understanding and the requirements to evaluate, meld, and determine the relative importance of various factors should provide you with a flavor of the complexity involved in determining environmental policy. Further, determining the likely risk to the world of a given compound requires some degree of looking into the future, opening up uncertainties about the factors affecting the environment. However, scientific concerns about effects of human activities on stratospheric ozone and on climate are sufficiently well understood (e.g., WMO, 2007; IPCC, 2007; USGCRP, 2009) that a strong sense of confidence can be provided about the use of the metrics discussed here.

As a scientist who studies these issues and not someone from a regulating agency, my purpose is to discuss the state of the understanding of these issues from a science perspective. Our goal is to make sure that we provide the most complete and accurate information possible to the regulating agencies. Accurate policies require a comprehensive analysis of the science. We do not want to harm the environment, but we also want to make sure that the cleaning agent industry is not unnecessarily restricted. It is therefore of value to those in the cleaning agents community, including manufacturers, to understand the science behind the regulations so that they can make appropriate input to the regulatory process. While the purpose here is to not provide all of the information available to policymakers, it is useful to summarize the current processes used.

Focus of This Chapter

This chapter focuses primarily on some of the areas of concern to policymakers in evaluating replacement compounds: concerns about toxicity, effects on stratospheric ozone, and effects on climate, with effects on urban and regional ozone being covered in a separate chapter (Andersen and Sheppard, 2011). Within the United States and its Environmental Protection Agency (EPA), the Significant New Alternatives Policy (SNAP) program is EPA's program to evaluate and regulate substitutes for the ozone-depleting chemicals that are being phased out under the stratospheric ozone protection provisions of the Clean Air Act (CAA).

Scientific analyses of current and possible future changes in ozone are largely based on complex, three-dimensional (3D) numerical models of the chemical and physical processes controlling the atmosphere. In the past, ozone studies were based on zonally averaged two-dimensional (2D) models of the atmosphere, but these are not as generally accepted for journal publications and therefore are not used much any longer in such analyses, although they may still have value for some studies. Similarly, analyses of possible future changes in climate largely depend on complex, numerical, 3D models of the global climate system (atmosphere, oceans, and land surface) that attempt to represent the many processes controlling climate. These models are computationally expensive and, given current computer capabilities, severely limited in the number of calculations that can be and have been examined. As a result, several additional approaches have been developed to examine and evaluate gases for their potential effects on global atmospheric ozone and climate. The concepts of ODPs and GWPs are discussed; both provide a means for comparing the potential environmental effects of various compounds.

The EPA SNAP Program

As solvents or solvent blends that are composed of or contain ODSs are phased out under the CAA, it is critical that replacements be found. These replacement solvents must have greatly reduced or zero ODPs and must not pose other unacceptable risks to human health or the environment. In particular, CAA Section 612(c) directs EPA to prohibit users of chlorofluorocarbon (CFC)-containing solvents from replacing ODSs with any substance that EPA has determined may present adverse effects to human health and the environment. These replacement substances, or "substitutes," are defined as any chemical, product substitute, or alternative manufacturing process, existing or new, that could replace a substance that has the potential to deplete the stratospheric ozone layer.

In evaluating the acceptability of substitutes, EPA must review the overall risk to both the environment and to human health resulting from exposure of consumers, workers, and the general population to any proposed substitute. The important factors considered by EPA in these evaluations are (1) the ability of the chemical to deplete the stratospheric ozone layer and/or to contribute to global warming; (2) the environmental fate and transport of substitutes, including information on bioaccumulation, biodegradation, adsorption, volatility, and transformation; and (3) whether the substitute is a VOC and thus a potential contributor to local/regional smog.

To evaluate these risks, EPA established the SNAP program to ensure that substitutes for ODSs pose lower aggregate levels of risk than the substances that they replace. To accomplish this goal, EPA conducts a screening-level assessment of the health and environmental risks of the ODSs for which substitution is proposed. These risk screenings combine conservative assumptions about worker/consumer/general population exposure levels and the substance's toxicity to estimate human health risks resulting from use of the substitute in a particular application, such as solvent cleaning.

The guiding principles of the SNAP program (http://www.epa.gov/ozone/snap/about.html) are

- *Evaluate substitutes within a comparative risk framework*. The program evaluates the risk of alternative compounds compared to those of ozone-depleting compounds and the available alternatives. The environmental risk factors that are considered include ODP, flammability, toxicity, occupational health and safety, as well as contributions to global warming and other environmental factors. There are also risk factors associated with quality of information, uncertainty of data, and economics factors, including feasibility and availability.
- *Do not require that substitutes be risk free to be found acceptable*. EPA finds substitutes to be either acceptable, acceptable with restrictions, or unacceptable. For substitutes to be found acceptable, they must have a reduced risk (which is not necessarily risk free) compared to the ozone-depleting compounds of primary concern (i.e., the chlorine- and bromine-containing halocarbons already being controlled under the Montreal Protocol).
- *Restrict only those substitutes that are significantly worse*. EPA does not intend to restrict a substitute if it has only marginally greater risk than another substitute. Drawing fine distinctions concerning the acceptability of substitutes would be extremely difficult, given the variability in

how each substitute can be used within a specific application. and the resulting uncertainties surrounding potential health and environmental effects. EPA also does not want to intercede in the market's choice of available substitutes, unless a substitute that has been proposed or is being used is clearly more harmful to human health and the environment than other alternatives.

- *Evaluate risks by use.* CAA Section 612 requires that substitutes be evaluated by use. Environmental and human health exposures can vary significantly, depending on the particular application of a substitute. Thus, the risk characterizations must be designed to represent differences in the environmental and human health effects associated with diverse uses. This approach cannot, however, imply fundamental trade-offs with respect to different types of risk to either the environment or to human health.

- *Provide the regulated community with information as soon as possible.* EPA recognizes the need to provide the regulated community with information on the acceptability of various substitutes as soon as possible. Future determinations on the acceptability of new substitutes are published several times a year.

- *Do not endorse products manufactured by specific companies.* EPA does not issue company-specific product endorsements. In many cases, EPA may base its analysis on data received on individual products, but the addition of a substitute to the acceptable list based on that analysis does not represent endorsement of that company's products. Generally, placement on the list merely constitutes an acknowledgement that a particular product made by a company has been found to be acceptable under SNAP.

- *Defer to other environmental regulations when warranted.* In some cases, EPA and other federal agencies have developed extensive regulations under other statutes or other sections of the CAA that address any potential cross- or inter-media transfers that result from the alternatives to class I and class II substances. The SNAP program will take existing regulations into account. For example, EPA considers the additional safety added by existing environmental regulations on hazardous air pollutants or hazardous waste, or by existing regulations for occupational safety and health.

The future atmospheric impacts of substances with reduced or no ODP are assessed by EPA relative to a "no substitution" baseline that assumes continued use of the ODS. These atmospheric impacts are modeled at EPA using their Atmospheric Health Effects Framework (AHEF), which estimates changes in stratospheric ozone concentrations based on estimated future ODS emissions. This simple model is then used to project changes in future skin cancer and cataract incidence and skin cancer mortality based on increased ultraviolet (UV) exposure. Because substitutes usually have less ODP than the materials they replace, the atmospheric impacts of these substitutes are generally found to be less than the continued use of the ODS.

Many proposed substitutes have zero ODP but may lead to other adverse health effects. To assess the overall risks of exposure of workers, consumers, and the general population to these substitutes, EPA considers a wide range of factors. Information from a variety of animal models is required on the acute and chronic toxicity of a substitute chemical and its potential impurities. EPA requests a minimum submission of the following mammalian tests: a range-finding study conducted using the appropriate exposure pathway for the specific end use (e.g., inhalation or oral), and a 90 day sub-chronic repeated dose study in an appropriate rodent species (e.g., rats or mice). For some substitutes, a cardiotoxicity study, usually measuring cardiotoxic effects in dogs, is also required. EPA may request additional toxicity tests on a case-by-case basis, depending on the particular substitute and application being evaluated. In addition to evaluation of primary literature, EPA also reviews various occupational and consumer exposure guidelines, deriving missing parameters where possible. Typically, the following parameters are reviewed for each chemical: short- and long-term occupational exposure limit, cardiotoxic lowest-observed adverse effect level (LOAEL) and no observed adverse effect level (NOAEL), reference concentration, and cancer slope factor.

Exposure assessments are used to estimate concentration levels of substitutes to which workers, consumers, the general population, and environmental receptors may be exposed over a determined period of time. These assessments are based on personal monitoring data or area sampling data, if available. Exposure assessments may be conducted for many types of releases including

- Releases in the workplace and in homes
- Releases to ambient air and surface water
- Releases from the management of solid wastes

Toxicity data is used to assess the possible health and environmental effects for exposure to substitutes. The Occupational Safety and Health Administration (OSHA) or EPA approves wide health-based criteria that are available for a substitute such as

- Permissible Exposure Limits (PELs for occupational exposure)
- Inhalation reference concentrations (RfCs for noncarcinogenic effects on the general population)
- Cancer slope factors (for carcinogenic risk to members of the general population)

If OSHA has not issued a PEL for a compound, EPA also considers workplace environmental exposure limits set by the American Industrial Hygiene Association or threshold limit values set by the American Conference of Governmental Industrial Hygienists. If limits for occupational exposure or exposure to the general population are not already established, then EPA derives these values following the Agency's peer-reviewed guidelines.

Flammability is examined as a safety concern for workers and consumers. EPA assesses flammability risk using data on

- Flash point and flammability limits (e.g., OSHA flammability/ combustibility classifications)
- Data on testing of blends with flammable components
- Test data on flammability in consumer applications conducted by independent laboratories
- Information on flammability risk minimization techniques

The SNAP program also examines other potential environmental impacts such as ecotoxicity and local air quality impacts. A compound that is likely to be discharged to water may be evaluated for impacts on aquatic life. Some substitutes are VOCs, which are chemicals that increase tropospheric air pollution by contributing to ground-level ozone formation. In addition, EPA notes whenever a potential substitute is considered a hazardous air pollutant or hazardous waste.

If the initial risk screening indicates a potential health or environmental risk, in-depth evaluations are conducted to ascertain whether the risk was accurately estimated and if management controls could reduce the risk to acceptable levels. In cases where a substitute poses a significant risk that does not seem likely to be abated by alternative methods, EPA can examine the risk screening analyses and decide to classify the substitute as unacceptable in certain end uses.

Good Ozone versus Bad Ozone

Ozone, O_3, is composed of three oxygen atoms and is a gas at atmospheric pressures and temperatures. Most of the ozone exists in the stratosphere (approximately 90%), the layer of the atmosphere about 10–50 km above the Earth's surface. The remaining ozone is in the troposphere, the lower region of the atmosphere extending from the Earth's surface up to approximately 10 km at midlatitudes and 16 km in the tropics. Ozone in the stratosphere is largely formed naturally following the dissociation of oxygen molecules by sunlight. In the lower atmosphere, ozone is also a major component of photochemical smog in urban areas.

While the ozone in the troposphere and stratosphere is chemically identical, it has very different effects on life on the Earth depending on its location. Stratospheric ozone, the "good" ozone, plays a beneficial role by absorbing solar ultraviolet radiation (UV-B), thereby preventing biologically harmful levels of UV radiation from reaching the Earth's surface (UNEP, 1989, 1991; SCOPE, 1992). It is the absorption of solar radiation by this ozone that explains the increase in temperature with altitude in the stratosphere. It is also the concerns about increased UV-B from the decreasing levels of ozone that have been the driver for policy actions to protect the ozone layer.

Closer to the Earth's surface, ozone displays its destructive side. Ozone is a strong oxidizer; hence, direct exposure to high levels of ozone has toxic effects on human health and plant viability (SCOPE, 1992). Thus, the "bad" ozone formed in urban areas is also of significant concern to policymakers.

In recent years, it has become progressively clearer that human activities have been affecting the amount of ozone in the global atmosphere, including the formation of the ozone "hole," a significant decrease in the amount of ozone roughly over Antarctica from August through October. Atmospheric measurements indicate that the amount of ozone in the global stratosphere is decreasing, while that in the global troposphere appears to have increased in the past, but is relatively unchanged in the last few decades (WMO, 1992, 1995, 1999, 2003, 2007). Although ozone formed in urban smog is not thought to be playing a significant role in the global distribution of ozone, it may be affecting the troposphere at larger scales than the urban regions it is being formed in. Overall, the vertically integrated ozone column has been decreasing globally; however, because of the effects of international policy to control the production and emissions of compounds that are effecting ozone, the maximum decrease may now have been reached (WMO, 2007; Yang et al., In press). Understanding the changes occurring to ozone and determining the appropriate societal response have presented important challenges to scientists and to policymakers. Atmospheric measurements and associated analyses have clearly implicated chlorine and bromine from chlorofluorocarbons (CFCs), halons, and other compounds as the primary cause of the decreases in stratospheric ozone that have occurred over recent decades.

Although chlorine does exist in the stratosphere as a result of natural sources, the source of this chlorine, predominantly from methyl chloride (CH_3Cl), can explain only about 0.6 ppb (parts per billion, by volume) of the 3.6 ppb of reactive chlorine in the current stratosphere. Although many additional natural sources of chlorine exist on Earth (including sea salt and chlorine in volcanic emissions and from swimming pools), this chlorine is water–soluble, and most of it is washed out of the atmosphere by precipitation processes before reaching the stratosphere.

On the other hand, CFCs are very long-lived, with atmospheric lifetimes of 50 years or more, implying they are essentially inert and their concentrations well mixed in the troposphere. Once they are transported into the stratosphere, however, the presence of high-energy UV light allows photolysis to destroy them, thus releasing the chlorine they contain. It is this active chlorine atom that takes part in the catalytic reactions responsible for destroying ozone. The chlorine catalytic cycle can occur thousands of times before the catalyst is converted to a less reactive form. Because of this cycle, relatively small concentrations of reactive chlorine can have significant impact on the amount and distribution of ozone in the stratosphere. In the lower stratosphere, atmospheric and laboratory measurements indicate that heterogeneous chemistry on sulfate particles is leading to enhanced effects on ozone from chlorine by converting less reactive chlorine compounds to the reactive forms, such as Cl and ClO (WMO, 1995, 2003, 2007). These heterogeneous reactions are a significant factor in explaining the enhanced ozone destruction in the lower stratosphere. The other major factor affecting ozone depletion in the lower stratosphere comes from the effects of bromine chemistry and human sources of bromine compounds that reach the stratosphere.

Bromine is potentially far more destructive to stratospheric ozone than is chlorine. The bromine catalytic cycle is about 60 times more efficient than the chlorine catalytic mechanism at destroying ozone (WMO, 2007). However, since the emissions and amounts of brominated compounds in the atmosphere are much smaller than those of the chlorinated compounds, the impact from bromine on the current atmosphere is smaller, though not negligible, than the effects from increasing chlorine. CFC-11 and CFC-12 are of primary concern as they account for roughly half of the organic chlorine loading of the current atmosphere. Brominated compounds of importance include methyl bromide, CH_3Br, and several halons (brominated halocarbons) such as H-1211 (CF_2ClBr) and H-1301 (CF_3Br).

Another halogen, iodine, is also very reactive with ozone, but current natural and human-related sources are too small for iodine to have any significant effect on stratospheric ozone. On the other hand, iodine compounds are under consideration for some uses, and are discussed in the section on ODPs. The fluorine found in many of the halocarbons is not important to ozone, as it quickly converts to hydrofluoric acid (HF).

The total atmospheric burden of a halocarbon is determined both by its release rate and the atmospheric lifetime. The long atmospheric lifetimes and past increases in the rate of use have contributed to a sustained trend of increases in the concentration of CFCs. The recognition of the deleterious effects of chlorine and bromine on ozone spawned international action to restrict the production and use of CFCs and halons, and protect stratospheric ozone, such as the 1987 Montreal Protocol on Substances that Deplete the Ozone Layer (UN, 1987, 1997) and the subsequent 1990 London Amendment (UNEP, 1990) and 1992 Copenhagen Amendment (UNEP, 1992). These agreements called for elimination of CFC consumption in developed countries by the end of the decade. A November 1992 meeting of the United Nations Environment Program held in Copenhagen resulted in substantial modifications to the existing protocols because of the large observed decrease in ozone and called for the phaseout of CFCs, carbon tetrachloride (CCl_4), and methyl chloroform (CH_3CCl_3)—often referred to in industry as 1,1,1-trichloroethane—by 1996 in developed countries. Production of these compounds was totally phased out in developing countries in 2005. Production of halons was stopped in 1994 in developed countries. Human-related production and emissions of methyl bromide are not to increase after 1994. Phaseout of methyl bromide occurred in developed countries in 2005, and is set to occur in developing countries in 2015. The Beijing Amendment (1999) included tightened controls on the production and trade of hydrochlorofluorocarbons (HCFCs). Bromochloromethane was also added to the list of controlled substances, with phaseout targeted for 2004. At the Meeting of the Parties in Montreal on September 17–21, 2007, the parties agreed to more aggressively phase out HCFCs in both developed and developing countries.

Air Quality and VOCs

In general, VOCs are defined as any compound containing carbon, excluding carbon monoxide, carbon dioxide (CO_2), carbonic acid, metallic carbides or carbonates, and ammonium carbonate, which participates in atmospheric photochemical reactions. However, the real concern with VOCs occurs if the emission of a given VOC can contribute to the formation of ground-level ozone. VOCs can affect production of ozone through their chemistry converting nitric oxide to nitrogen dioxide, which then photolyzes, leading to ozone production. For that reason, emissions of certain organic compounds, termed VOCs in the regulatory process, have been subject to controls for a number of years. The emissions of certain VOCs are also controlled or banned because they are toxic, can deplete the ozone layer, can contribute to formation of particulate matter, or have other impacts on the environment.

The impact of a VOC on formation of ozone or other measures of air quality is often referred to as its atmospheric "reactivity." VOCs vary by large factors in their ability to affect ozone. Because some VOC compounds have almost no effect in producing ozone, EPA has had a policy of exempting some compounds from VOC regulations. However, the vast majority of compounds have been regulated as if they essentially had the same effect on ozone. Because of the recognized inadequacies of this "either/or" approach, EPA has been working to define areas of research to assist policymakers in developing VOC emission control strategies based on the reactivity of individual species of VOC. An interim policy in 2005 provides examples of innovative use of reactivity information in developing VOC control measures, and clarifies the relationship between innovative reactivity-based policies and EPA's current definition of VOC. The notice states that as "States develop their 8 h ozone SIPs, the EPA encourages them to consider how they may incorporate VOC reactivity information to make their future VOC control measures more effective and efficient." EPA notes that although most existing VOC control programs do not discriminate between individual VOCs based on reactivity, "they continue to provide significant ozone reduction benefits and will remain in place unless and until they are replaced by programs that achieve the same or greater benefits." EPA "will continue its policy of granting VOC exemptions for compounds that are negligibly reactive" and "will continue to evaluate new scientific information regarding VOC reactivity and will update this interim guidance as appropriate."

Concerns about Stratospheric Ozone: Ozone Depletion Potentials

The cleaning agents most likely to have a potential effect on stratospheric ozone are those containing chlorine, bromine, or iodine. These atoms react extremely efficiently with ozone in catalytical processes. Thus, chlorine, bromine, or iodine released in or transported to the stratosphere can destroy thousands of ozone molecules. The concept of ODPs arose as a means of determining the relative ability of a chemical to destroy ozone. Thus, a key screening approach relative to stratospheric ozone is to ask whether the chemical being considered contains chlorine, bromine, or iodine. If it does, then the ODP needs to be determined.

If a halogenated compound does not contain chlorine, bromine, or iodine, then it most likely will not affect the stratosphere (other long-lived gases such as nitrous oxide or methane definitely affect stratospheric ozone). Exceptions may occur if the compound could achieve such significant use that it could become a source of stratospheric nitrogen oxides, hydrogen oxides, or sulfuric particles. Such exceptions are unlikely in the case of solvents unless the chemical also gained extremely wide use for other applications.

The concept of ODPs (Wuebbles, 1981, 1983; Wuebbles et al., 1999a,b; WMO, 1992, 1995, 1999, 2003, 2007) provides a relative cumulative measure of the expected effects on ozone of the emissions of a gas relative to CFC-11, one of the gases of most concern to ozone change. The ODP of a gas is defined as the integrated change in total ozone per unit mass emission of the gas, relative to the change in total ozone per unit mass emission of CFC-11. Alternatively, the ODP can be derived by using a constant emission calculated to steady-state relative to the same for CFC-11. Numerically, the two approaches are equivalent. As a relative measure, ODPs are subject to fewer uncertainties than estimates of the absolute percentage of ozone depletion caused by different gases. ODPs are an integral part of national and international considerations on ozone protection policy, including the Montreal Protocol and its Amendments and the United States CAA. ODPs provide an important means for analyzing the potential for a new chemical to affect ozone relative to CFCs, halons, and other replacement compounds.

ODPs are currently determined by two different means: calculations from models, primarily from models of the global atmosphere (WMO, 1995, 1999, 2003, 2007), and the semiempirical approach developed by Solomon et al. (1992). The calculation from models is now generally from 3D models of atmospheric chemistry and physics, although a few studies are still being done with zonally averaged (average around longitude) 2D models. The simpler 2D models—that represent the atmospheric physics and chemistry temporally, but spatially with only latitude and altitudewhile being averaged over longitude—were used for many years in the analyses used for policy development, but generally are not as accepted by the scientific community today (they are still used, but not generally as accepted for peer-reviewed publication). The current scientific literature focuses on results largely from 3D models that represent the atmosphere in all three spatial directions as well as temporally. ODPs and other analyses of ozone effects related to cleaning agents are now being done with 3D models. These models are much more computationally intensive compared to the 2D models so, at this time, only a limited number of studies are being done to look at replacement compounds.

The two approaches give similar results. Note, however that the empirical approach is not useful for short-lived gases that are not well mixed in the troposphere. The numerical models attempt to account for all of the known chemical and physical processes affecting chemical species in the troposphere and stratosphere. The compounds are assumed to enter the atmosphere at ground level, be transported in the atmosphere by dynamical processes, and react by a variety of pathways, depending on their molecular structure. The compounds may undergo photolytic breakdown by UV or near-UV light, react with hydroxyl (OH) in the troposphere and stratosphere, or react with atomic oxygen. Products resulting from these reactions, such as atomic chlorine and bromine, can react in the modeled atmosphere, which in turn may affect the calculated distribution of ozone. A major uncertainty in the models is the amount of tropospheric OH. Since few reliable measurements of OH are available, the global distribution has not been directly measured. As a result, given the importance of the atmospheric lifetime in determining the ODP for a substance, those gases where reaction with tropospheric OH is the primary loss mechanism, as it is for many of the replacement compounds, a scaling to the partial lifetime of CH_3CCl_3 due to its reaction with tropospheric OH is used. Because CH_3CCl_3 has been well measured and its emissions are considered to be well understood, it provides a useful measure of the amount of atmospheric OH in the troposphere.

The semiempirical approach (Solomon et al., 1992; WMO, 1995, 2007) for determining ODPs is based on direct measurements of select halocarbons and other trace species in the stratosphere. The observed fractional dissociation is used to determine the amount of chlorine and bromine released, and is then compared with the observationally derived ozone loss distribution, with the assumption that the ozone loss results only from halogen chemistry. The correlation between different compounds is determined on the basis of their relative reactivity in the troposphere and stratosphere. This semiempirical method avoids some of the demanding requirements of accurate numerical simulation of source gas distributions (i.e., of the CFCs, HCFCs, and other compounds) and of the resulting ozone destruction. However, the semiempirical approach also depends on the accuracy of the measurements used with this approach. As mentioned earlier, the semiempirical derivation results agree well with the ODPs derived in current state-of-the-art models.

Table 33.1 shows a partial list of derived ODPs for several compounds, including HCFCs and other gases where reaction with tropospheric OH is the primary loss mechanism. These ODPs are derived from various sources based on different scaling relative to the partial tropospheric OH lifetime of CH_3CCl_3. Table 33.1 does not contain any of the HFCs, as they are thought to have insignificant effects on ozone (WMO, 1995, 1999, 2003, 2007). The ODPs from the World Meteorological Organization (WMO) assessment (WMO, 2007) are shown in the column after the list of gases. Other more recent ODP studies are discussed below.

Since ODPs are defined in terms of the steady-state ozone change (or alternatively as the integrated cumulative effect on ozone), they are not representative of the relative, transient effects expected for short-lived compounds during the early years of emission. Time-dependent ODPs that provide information on the shorter time scale effects of a compound on ozone can also be defined. However, the steady values generally are preferred and are used in regulatory considerations. By definition, the ODP for CFC-11 is 1.0. At this point, the calculated ODPs for other CFCs being banned are all greater than 0.4. This is policy, and it is always subject to modification. The ODPs for halons are all extremely larger, much greater than 1.0, reflecting the reactivity of bromine with ozone. The ODPs for the HCFCs being used or considered as replacements are generally quite small relative to CFCs. The effect on ozone from a unit mass emission of one of these HCFCs would correspondingly be less than a hundredth of the effect on ozone than the CFC or halon they would replace. The ODPs for all the HFCs, PFCs, and for sulfur hexafluoride are near zero, owing to the low reactivity of their dissociation products with ozone.

The ODP of HCFC-123 has recently been updated with 3D model results from Wuebbles and Patten (2009). They get the same atmospheric lifetime of 1.3 years as was found in prior 2D modeling studies, but get an ODP slightly smaller at 0.0098, compared to 0.014 in previous 2D modeling studies and 0.02 from the semiempirical approach, as found in Table 33.1.

The past evaluations of ODP were generally conducted for chemicals with atmospheric lifetime sufficiently long (more than approximately 1 year) that they are well mixed throughout the troposphere

TABLE 33.1 Steady-State ODPs Based on the International Ozone Assessments

Gas	Formula	ODP
Chlorofluorocarbons		
CFC-11	CCl_3F	1.0
CFC-12	CCl_2F_2	1.0
CFC-113	CCl_2FCClF_2	1.0
CFC-114	$CClF_2CClF_2$	1.0
CFC-115	CF_3CClF_2	0.44
Bromocarbons		
Methyl bromide	CH_3Br	0.351
Halon-1301	CF_3Br	16.0
Halon-1202	CBr_2F_2	1.7
Halon-1211	CF_2ClBr	71
Halon-2402	$CBrF_2CBrF_2$	11.5
HCFCs		
HCFC 22	$CHClF_2$	0.05
HCFC-123	CF_3CHCl_2	0.02
HCFC-124	CF_3CHClF	0.02
HCFC-141b	CH_3CCl_2F	0.12
HCFC-142b	CH_3CClF_2	0.07
HCFC-225ca	$CF_3CF_2CHCl_2$	0.02
HCFC-225cb	$CClF_2CF_2CHClF$	0.03
Others		
Carbon tetrachloride	CCl_4	0.73
Methyl chloroform	CH_3CCl_3	0.12
Methyl chloride	CH_3Cl	0.02

Source: World Meteorological Organization (WMO), *Scientific Assessment of Ozone Depletion: 2006.* Global Ozone Research and Monitoring Project. Report 50, Geneva, 2007.

Note: These are based on the semi-empirical approach to analyzing ODPs and don not necessarily match the values being assumed in the Montreal Protocol. Additional compounds are discussed in the text.

after surface release, and a large portion of surface emissions can reach the stratosphere. However, many of the compounds being considered either for new applications or as replacements for substances controlled under the Montreal Protocol are now designed to be very (or even extremely) short-lived, on the order of days to a few months, so as to reduce the impacts on the total ozone column (both tropospheric and stratospheric ozone) and climate. Many of these very short-lived (VSL) compounds still contain halogens, including chlorine, bromine, and iodine. The VSL replacement gases still can be vertically transported into the lower stratosphere by intense convection in the tropical troposphere, which is relevant to possible effects on ozone from these compounds. Thus the ODP of a VSL species depends upon its distribution in the atmosphere and the location of its source (Wuebbles and Ko, 1999; Wuebbles et al., 2001; Ko and Poulet, 2003). Unfortunately, the determination of ODPs for these gases is not straightforward, because their short atmospheric lifetimes leave them poorly mixed in the troposphere (Wuebbles and Ko, 1999; Wuebbles et al., 2001; Ko and Poulet, 2003; Butler et al., 2007). The traditional 2D model analysis of ODPs is not sufficiently accurate for calculating the integrated amount of the halogenated VSL source and reaction product gases in the troposphere that enter the stratosphere (Wuebbles et al., 2001).

Three-dimensional models of atmospheric chemistry and physics fully representing the complete troposphere and stratosphere are necessary for evaluating the halogen loading and ozone depletion in the stratosphere for VSL species. Thus the definition of ODPs has been revised for VSL compounds (Wuebbles et al., 2001; WMO, 2003). The new ODP definition for VSL compounds accounts for the variation that can occur in the ODP as a function of where and when the compound is used and emitted. The most important factor in evaluating the ODP of VLS compounds is shown to be geographical distribution, or latitude, of the surface emissions, because gases emitted at higher latitudes take longer to reach the stratosphere than gases emitted in the tropics (Bridgeman et al., 2000; Olsen et al., 2000; Wuebbles et al., 2001). Seasonal variations in the emissions can also be important, if that is a relevant criterion in the atmospheric emissions of the compound.

The only VSL compound currently evaluated using state-of-the-art 3D models of the troposphere and stratosphere is n-propyl bromide (nPB, 1-C_3H_7Br), and only limited studies and a range of estimated ODPs are available for it (Wuebbles et al., 2001; WMO, 2003, 2007). The past studies of the ODPs for nPB have been only partially based on 3D model results (see Ko and Poulet, (2003) for a more complete summary). Three-dimensional models, which include a much more comprehensive treatment of transport than do 2D models, are now much more preferable for calculations of ODPs, both for the longer-lived gases and for the very short-lived candidate replacement compounds. These models cannot only determine the amount of the substance reaching the stratosphere directly but can also follow the processes affecting the reaction products. Studies suggest that the vast majority of chlorine or bromine reaching the stratosphere from VSL substances is transported there in reaction products (Wuebbles et al., 2001; WMO, 2003, 2007).

The first study of the ODPs for nPB fully derived using a 3D model representing all relevant chemistry and physics was recently done by Wuebbles et al. (2009). For midlatitude emissions (land surfaces from 30°N to 60°N), the atmospheric chemical lifetime obtained for nPB is 24.7 days and the ODP is 0.0049. The ODP from the 3D modeling study is smaller than the values found in the earlier studies that are shown in Table 33.2. Wuebbles et al. (2009) also calculated the atmospheric lifetimes and ODPs for the existing solvents trichloroethylene (TCE) and perchloroethylene (PCE). For midlatitude emissions, the derived atmospheric lifetime for TCE is 13.0 days and for PCE it is 111 days. The corresponding ODPs are 0.00035 and 0.0060, respectively.

TABLE 33.2 Estimated ODPs for nPB as a Function of Location of Emissions Based on Wuebbles et al. (2001) and Updated in WMO (2003)

Location of Emissions	Estimated ODP
Global land masses (60°S–70°N)	0.027–0.038
North America (to 70°N)	0.015–0.017
North America, Europe, China, Japan (to 70°N)	0.017–0.026
India, Southeast Asia, and Indonesia	0.071–0.10

Sources: Wuebbles, D.J. et al., *J. Geophys. Res.*, 2001, 106, 14551; World Meteorological Organization (WMO), *Scientific Assessment of Ozone Depletion: 2001.* Global Ozone Research and Monitoring Project. Report 47, Geneva, 2003.

Note: Estimates are based on combined results from analyses with the UIUC 2D model and the MOZART2 3D model. Emissions were evenly distributed over the landmasses based on their representation in the 3D model. The primary values are based on scaling using the tropopause burdens between the two models, while the range also includes an alternative approach using stratospheric burden scaling. See the text for discussion of the newest finding.

Two other VSL iodine compounds were recently investigated by Youn et al. (2010). Iodotrifluoromethane (CF_3I) is a gaseous fire suppression flooding agent for in-flight aircraft and electronic equipment fires, and a candidate replacement for trifluoroiodomethane (Halon-1301, CF_3Br). Methyl iodine (CH_3I), a potential replacement fumigant for methyl bromide (CH_3Br), has been of interest as a natural source of atmospheric iodine that is primarily emitted from the oceans, and as a potentially useful tracer of marine convection and the subsequent role of iodine in atmospheric photochemistry (Vogt et al., 1999; Bell et al., 2002; Cohan et al., 2003). CF_3I is rapidly photo-dissociated even under visible light conditions, and CH_3I is also photolyzed, as well reacting with the OH radical. For both species, the initial reaction rapidly results in the availability of atmospheric iodine for chemical reactions with ozone and other gases. Although current atmospheric concentrations of iodine-containing gases is very small, CF_3I and CH_3I are both a potentially significant source of atmospheric iodine, despite being very short-lived and are of interest in the upper troposphere and lower stratosphere as a source of iodine radicals for ozone destruction (WMO, 2007). Ozone destruction in the lower stratosphere due to catalytic cycles involving iodine was not well understood, but recent laboratory studies have improved the understanding of atmospheric chemical processes of iodine species, including their photochemistry (Bösch et al., 2003; Li et al., 2006; WMO, 2007). Like other chemicals with extremely short lifetimes, the stratospheric halogen loading and resulting ozone effects from these compounds are strongly dependent on the location of emissions. For CF_3I, a possible candidate replacement for bromotrifluoromethane (CF_3Br), ODPs derived by the 3D model are 0.008 with a chemical lifetime of 5.06 days, and 0.016 with an atmospheric lifetime of 1.13 days for emissions assumed to be evenly distributed over land surfaces at midlatitudes and the tropics, respectively. While this is the first time the ODPs have been evaluated with a 3D model, these values are in good agreement with those derived previously. The model calculations suggest that tropical convection could deliver a larger portion of the gas and their breakdown products to the upper troposphere and lower stratosphere if the emission source is located in the tropics. The resulting ODP for CH_3I emitted from midlatitudes is 0.017 with an atmospheric lifetime of 13.59 days. Also, the model derived distribution of background CH_3I compares well with available observations.

Concerns about Stratospheric Ozone: Equivalent Chlorine Loading

Although ODPs provide a useful guide to the relative effects on ozone from different gases, other analyses are also useful in fully evaluating the potential effects of a chemical on ozone. For example, it is useful to consider how much of an ozone decrease would really be expected to occur over the coming decades from the emissions of a new chemical like nPB. The concept of Equivalent Chlorine Loading (ECL) has been used as a straightforward means of evaluating these effects (Wuebbles et al., 1995, 1997; WMO, 1992, 1995; Prather et al., 1996). ECL can be used as a proxy for the approximate amount of ozone depletion resulting from a given set of emissions of chlorine and bromine containing gases at ground level. With the ECL concept, one can examine the effects of different assumptions about changing emissions of these gases on the ozone layer over the coming decades. The concept of ECL provides a measure of the total amount of chlorine and bromine reaching the stratosphere that is available to affect ozone. This concept has proven to be a useful tool in evaluating the total potential effects on ozone from the use of CFCs and halons and their replacement compounds.

However, ECL also has major limitations that have been recognized in the last few years. For example, it does not account for the effects of climate change on the stratosphere, including the effects of increasing CO_2 concentrations on stratospheric temperatures. While increasing amounts of CO_2 have a warming effect, there is a cooling effect in the stratosphere, leading to an increase in ozone. Also, the effects of increasing concentrations of nitrous oxide (N_2O) and methane (CH_4) on the ozone distribution in the troposphere and stratosphere are not accounted for in ECL.

The concept of ECL assumes that changes in stratospheric ozone are largely dominated by the changes occurring in the concentrations of chlorine and bromine. It is aimed at combining the knowledge gained from chlorine and bromine transferred from the troposphere to the stratosphere, where these halogens can react with ozone. ECL is directly proportional to the surface emissions of the halocarbons, their reactivity as reflected in their atmospheric lifetimes, and the number of chlorine and bromine atoms released per molecule. Model calculations and laboratory measurements indicate that bromine is much more reactive with ozone than chlorine. In order to represent bromine loading "equivalent" to chlorine loading, ECL includes a multiplicative factor, α, on the amount of bromine released into the stratosphere to account for its larger reactivity with ozone. The ECL values presented here assume that α has a value of 55, to account for the greater reactivity of each bromine emitted with stratospheric ozone compared to chlorine (WMO, 2007) suggests that 60 might be a better value for α, but 55 is well within the uncertainty range). This value is based on integration of the relative effects of chlorine and bromine on ozone from chemical-radiative-dynamical models of the global atmosphere (Wuebbles et al., 1991, 1999a,b,c; Kinnison et al., 1994; WMO, 2007). Although the new value of 60 for α has only a small effect on the derived ECL, it does indicate that gases containing bromine are now thought to have a larger effect on atmospheric ozone than thought previously.

Another form of ECL is referred to as equivalent effective stratospheric chlorine (EESC). EESC accounts for the chlorine and bromine released in the lower stratosphere where a majority of the ozone destruction has occurred over recent decades. Figure 33.1 shows EESC due to past emissions. The impact of CFCs and other halocarbons on EESC are shown individually in Figure 33.1, with the topmost curve representing the sum of all of the individual components. Similar results are also presented in the international assessments of stratospheric ozone (WMO, 1999, 2003, 2007). Another version of EESC also accounts for the ozone losses that have occurred at high latitudes (Newman et al., 2006).

Figure 33.1 also shows the EESC for future emissions assuming global emissions corresponding to the Montreal Protocol and its modifications under the latest amendments. From Figure 33.1, *EESC should have reached its peak value by 1998 (at a value of about 3.4 ppb of* air by volume or ppbv) and should slowly recover to levels observed before the extensive human-related emissions in recent decades. The smaller effect on projected ECL from HCFCs is related both to their much shorter atmospheric lifetimes and to the halt in their production by 2030 in support of the regulatory actions called for by the current

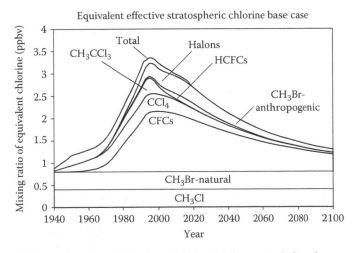

FIGURE 33.1 EESC for the baseline case following the historical changes in halocarbon concentrations and the projected emissions following the Montreal Protocol. The topmost curve shows the combined effect of all of the contributions of chlorine and bromine to EESC. Based on model of Jain and Wuebbles similar to that used in WMO (1999, 2003, 2007).

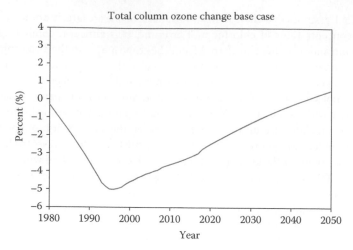

FIGURE 33.2 Derived changes in globally averaged ozone (in percent, relative to 1979) based on the EESC evaluation in Figure 33.1 relative to observed trends in total ozone.

amendments under the Montreal Protocol. Several studies (e.g., Wuebbles and Calm, 1997; Wuebbles and Patten, 2009) indicate that there are a few situations where this halt in production, such as for the use of HCFC-123 in large chiller units, may not necessarily be in the best interest of environmental policymaking.

On the Recovery of Stratospheric Ozone

Based on the measurements from the Nimbus 7 TOMS instruments (WMO, 1999, 2007), the depletion of global mean total column ozone between 1979 and 2005 was about 5%. Assuming a linear relationship between EESC and the total column ozone, Figure 33.2 shows the change in total ozone corresponding to the scenario evaluated in Figure 33.1. The highest and lowest values for ECL and ozone reductions were reached in about 1998, reversing thereafter, although it is not until about 2040 that the 1979 level will again be obtained. However, as noted in Newman et al. (2006), the recovery of high latitude ozone is unlikely before about 2065. One can then evaluate other scenarios, for example that of the assumed global emissions of a new solvent, to then evaluate its potential effects on ozone relative to this baseline scenario. WMO (2007) discusses various potential policy options using EESC. However, these studies do not account for other effects on ozone, such as the increasing concentrations of CO_2, N_2O, and CH_4 as discussed earlier.

Concerns about Climate Change: Global Warming Potentials

Greenhouse gases and other radiatively active substances in the atmosphere are important influences on climate, the aggregation of the weather generally expressed in terms of averages and variances of temperature, precipitation, and other physical properties. Greenhouse gases in the atmosphere absorb infrared radiation emitted by the Earth that would otherwise escape to space. This trapped radiation warms the atmosphere, creating a positive forcing on climate, called radiative forcing, which in turn warms the Earth's surface. The concept of radiative forcing provides an estimate of the potential effect on climate from greenhouse gases (the importance of this so-called greenhouse effect has been known about for over 180 years). For the given concentration of a gas, the radiative forcing depends primarily on the infrared absorption capabilities of the gas. Once the infrared absorption cross sections have been measured in a chemical laboratory, the radiative forcing can be determined by using computer models of the physics affecting radiative transfer in the Earth's atmosphere.

All the halocarbons are greenhouse gases. In fact, any gas containing a C–F or C–H bond will be a greenhouse gas. The effects of such gases on climate will depend on their radiative absorption properties and the amount of the compound in the atmosphere.

GWPs provide a means for comparing the relative effects on climate expected from various greenhouse gases. The concept of GWPs is used to estimate the relative impact of emission of a fixed amount of one greenhouse gas compared to another for globally averaged radiative forcing of the climate system over a specified time scale. GWPs are a better measure of the relative climatic impacts than radiative forcing alone because they also account for the atmospheric lifetime, and thus the change in concentration for a given emission, of the gases. GWPs have been evaluated for a number of replacement compounds, and have been reported in the United Nations Intergovernmental Panel on Climate Change (IPCC) and WMO assessments. The GWP concept is also being used in policymaking considerations associated with concerns about global warming from greenhouse gases. While GWPs are evaluated for a range of integration time periods, policymakers have generally chosen to use the GWP values associated with a 100 year integration.

GWPs are expressed as the time-integrated radiative forcing from the instantaneous release of a kilogram of a gas expressed relative to that of a kilogram of the reference gas, CO_2,

$$\text{GWP}_X(t') = \frac{\int_0^{t'} F_X \exp(-t/\tau_X) \, dt}{\int_0^{t'} F_{CO_2} R(t) \, dt}$$

where the integral of F_X (the radiative forcing per unit mass of species X) times the removal rate for X is divided by a similar integral for CO_2. The atmospheric lifetime of the gas X is represented as τ_x, whereas $R(t)$ represents the complex removal function for CO_2, which does not follow a simple exponential decay like most gases. The numerator and the denominator represent the absolute GWP (AGWP) of species X and CO_2 respectively. The atmospheric lifetime of a gas is the time at which the atmospheric concentration after the original emission is reduced by 63.2% (equivalent to $1/e$, where $e = 2.71828$).

IPCC (2007) reported the GWPs for a number of replacement compounds evaluated relative to CO_2. In the latest assessment, WMO (2007) reported the GWPs for the previously evaluated species and expanded the list to add some newer compounds. Shown in Table 33.3 is the most up-to-date evaluation of GWPs from IPCC (2007). For HCFC-123, Wuebbles and Patten (2009) derive a slightly smaller 100 year GWP of 77. Most compounds that absorb in the infrared will have GWPs larger than CO_2 if their atmospheric lifetimes are more than a few days; this is because there is already enough CO_2 in the atmosphere that the center of its main absorption features is already saturated and its absorption is primarily occurring in the tails of the absorption lines, whereas the other gases do not have this issue.

One additional concept needs to be discussed, namely Indirect GWPs (Daniel et al., 1995; WMO, 2003, 2007). Indirect GWPs arose as a means of accounting for the ozone effects on climate resulting from emissions of chlorinated and brominated halocarbons. For some of the halons (e.g., CF_3Br) especially, it has been thought that the decrease in ozone might fully compensate for the climate effect implied by the direct GWP for the halocarbon. However, Youn et al. (2009) have now used 3D atmospheric model results to show that the ozone effect is much smaller and such cancellation is not as important as previously believed, thus reducing the likelihood for potential use of Indirect GWPs in policy considerations.

TABLE 33.3 Evaluated 100 Year Integrated GWPs

Gas	Time Horizon in Years				
	20	100			500
	This Study	This Study	WMO (Granier et al., 1999)	% Difference	This Study
CH_4	72	28	24	14	9
N_2O	296	340	360	6	188
Chlorofluorocarbons (CFCs)					
CFC-11	6,100	4,700	4,600	−2	1,700
CFC-12	9,800	10,600	10,600	0	5,200
CFC-13	10,000	14,600	14,000	−4	17,000
CFC-113	5,800	6,000	6,000	0	2,700
CFC-114	7,100	9,700	9,800	1	8,700
CFC-115	6,000	9,100	10,300	13	12,700
Chlorocarbons (CCs)					
CCl_4	2,700	1,800	1,400	−22	600
CH_3CCl_3	500	160	140	−12	50
Hydrochlorofluorocarbons (HCFCs)					
HCFC-22	4,900	1,900	1,900	0	590
HCFC-123	280	90	120	33	28
HCFC-124	1,800	590	620	5	180
HCFC-141b	2,000	690	700	1	220
HCFC-142b	4,300	2,000	2,300	15	640
HCFC-225ca	460	140	180	28	45
HCFC-225cb	1,600	500	620	24	160
Hydrofluorocarbons HFCs					
HFC-23	15,000	19,600	14,800	−24	15,900
HFC-32	3,500	1,100	880	−20	350
HFC-125	6,700	4,300	3,800	−12	1,400
HFC-134	3,400	1,200	1,200	0	390
HFC-134a	4,400	1,800	1,600	−11	560
HFC-143	1,100	350	370	6	110
HFC-143a	6,900	5,800	5,400	−7	2,100
HFC-152a	480	150	190	26	46
HFC-161	25	8	-	-	2
HFC-227ea	6,300	4,400	3,800	−14	1,500
HFC-236fa	7,200	9,500	9,400	−1	7,400
HFC-245ca	2,600	870	720	−17	270
Perfluorocarbons (PFCs)					
SF_6	14,600	22,500	22,200	−1	33,200
CF_4	4,400	6800	5,700	−16	10,600
Bromocarbons (BCs)					
H-1211	2,900	1,100	1,300	18	340
H-1301	6,800	6,300	6,900	10	2,500
CH_3Br	11	4	5	25	1

TABLE 33.3 (continued) Evaluated 100 Year Integrated GWPs

	Time Horizon in Years				
	20	100			500
Gas	This Study	This Study	WMO (Granier et al., 1999)	% Difference	This Study
CH$_2$Br$_2$		3	1	−67	1
CH$_2$F$_2$Br	1,200	390	470	20	120
Iodocarbons (ICs)					
CF$_3$I	1	<1	<1	0	<1
CF$_3$CF$_2$I	1	<1		—	<1

Source: Based on IPCC (2007).

Summary and Conclusions

This paper has summarized many of the environmental issues of concern in considering replacement solvents, focusing particularly on the concerns about global changes in ozone and climate. The metrics discussed, ODPs, GWPs, and ECL, are being used in national and international policy considerations related to the concerns about stratospheric ozone and about climate as ways to relate different gases to each other. As new compounds are developed and considered for use, including those being considered as cleaning agents, the U.S. EPA and similar agencies around the world are requiring that they be subject to a series of tests such as those discussed for the SNAP program, including the evaluation of their ODPs and GWPs if it looks like the new compound may potentially affect stratospheric ozone and/or climate.

References

Andersen, S.O. and Sheppard, M. 2011. Momentum from the phaseout of ozone-depleting solvents drives continuous environmental improvement, In: *Handbook for Critical Cleaning: Applications, Processes, and Controls*, Taylor & Francis, Boca Raton, FL.

Bell, N., L. Hsu, D. J. Jacob, M. G. Schultz, D. R. Blake, J. H. Butler, D. B. King, J. M. Lobert, and E. Maier-Reimer, 2002. Methyl iodide: Atmospheric budget and use as a tracer of marine convection in global models. *J. Geophys. Res.*, 107, 4340, DOI: 10.1029/2001JD001151.

Bridgeman, C. H., J. A. Pyle, and D. E. Shallcross, 2000. A three-dimensional model calculation of the ozone depletion potential of 1-bromopropane (1-C3H7Br). *J. Geophys. Res.*, 105, 26,493-26, 502.

Bösch, H., C. Camy-Peyret, M. P. Chipperfield, R. Fitzenberger, H. Harder, U. Platt, and K. Pfeilsticker, 2003. Upper limits of stratospheric IO and OIO inferred from center-to-limb-darkening-corrected balloon-borne solar occultation visible spectra: Implications for total gaseous iodine and stratospheric ozone. *J. Geophys. Res.*, 108, D15, 4455, DOI: 10.1029/2002JD003078.

Butler, J. H., D. B. King, J. M. Lobert, S. A. Montzka, S. A. Yvon-Lewis, B. D. Hall, N. J. Warwick, D. J. Mondeel, M. Aydin, and J. W. Elkins, 2007. Oceanic distributions and emissions of short-lived halocarbons. *Global Biogeochem. Cycles*, 21, GB1023, DOI: 10.1029/2006GB002732.

Cohan, D. S., G. A. Sturrock, A. P. Biazar, and P. J. Fraser, 2003. Atmospheric methyl iodide at Cape Grim, Tasmania, from AGAGE Observations. *J. Atmos. Chem.*, 44(2), 131–150.

Daniel, J. S., S. Solomon, and D. L. Albritton, 1995. On the evaluation of halocarbon radiative forcing and Global Warming Potentials. *J. Geophys. Res.*, 100, 1271–1285.

Granier C., K. P. Shine, J. S. Daniel, J. E. Hansen, S. Lal, and F. Stordal, 1999. Climate effects of ozone and halocarbon changes. Chapter 10 in *Scientific Assessment of Ozone Depletion: 1998*, WMO Global Ozone Research and Monitoring Project, Rep. No. 44, Geneva, Switzerland.

IPCC (Intergovernmental Panel on Climate Change), 1995. *Climate Change 1994: Radiative Forcing of Climate Change and an Evaluation of the IPCC IS92 Emissions Scenarios*, Houghton, J. T., Filho, L. G. M., Bruce, J., Lee, H., Callander, B. A., Haites, E., Harris, N., and Maskell, K., eds., Cambridge University Press, Cambridge, U.K.

IPCC (Intergovernmental Panel on Climate Change), 1996. *Climate Change 1995: The Science of Climate Change. Contribution of Working Group I to the Second Assessment Report of the Intergovernmental Panel on Climate Change*, Houghton, J. T., Meira Filho, L. G., Callander, B. A., Harriss, N., Kattenberg, A., and Maskell, K., eds., Cambridge University Press, Cambridge, U.K.

IPCC (Intergovernmental Panel on Climate Change). *Climate Change 2007: The Physical Science Basis. Contributing of Working Group I to the Fourth Assessment Report of the Intergovernmental Panel on Climate Change*, Solomon, S., Qin, D., Manning, M., Chen, Z., Marquis, M., Avery, K. B., Tignor, M., and Miller, H. L., eds., Cambridge University Press, Cambridge, U.K.

Kinnison, D. E., K. E. Grant, P. S. Connell, D. A. Rotman, and D. J. Wuebbles, 1994. The chemical and radiative effects of the Mount Pinatubo eruption. *J. Geophys. Res.*, 99, 25705–25731.

Ko, M. K. W. and G. Poulet, 2003. Very short-lived halogen and sulfur substances. In: *Scientific Assessment of Ozone Depletion: 2002*, Ennis, C. A., ed., World Meteorol. Org., Geneva, Switzerland, Chapter 2, pp. 2.1–2.57.

Li, Y., K. O. Patten, D. Youn, and D. J. Wuebbles, 2006. Potential impacts of CF_3I on ozone as a replacement for CF_3Br in aircraft applications. *Atmos. Chem. Phys.*, 6, 4559–4568.

Newman, P. A., E. R. Nash, S. R. Kawa, S. A. Montzka, and S. M. Schauffler, 2006. When will the Antarctic ozone hole recover? *Geophys. Res. Lett.*, 33, L12814, DOI: 10.1029/2005GL025232.

Olsen, S. C., B. J. Hannegan, X. Zhu, and M. J. Prather, 2000. Evaluating ozone depletion from very short-lived halocarbons. *Geophys. Res. Lett.*, 27, 1475–1478.

Prather, M., P. Midgley, F. S. Rowland, and R. Stolarski, 1996. The ozone layer—The road not taken. *Nature*, 381, 551–555.

SCOPE (Scientific Committee on Problems of the Environment), 1992. *Effects of Increased Ultraviolet Radiation on Biological Systems*. SCOPE Secretariat, Paris, France.

Solomon, S., M. J. Mills, L. E. Meiht, W. H. Pollack, and A. F. Tuck, 1992. On the evaluation of ozone depletion potentials. *J. Geophys. Res.*, 97, 825–842.

UN (United Nations), 1987. *Montreal Protocol on Substances that Deplete the Ozone Layer*. New York.

UN (United Nations), 1997. *Kyoto Protocol to The United Nations Framework Convention on Climate Change*. United Nations, New York.

UNEP (United Nations Environment Programme), 1989. *Environmental Effects Panel Report*. Nairobi, Kenya.

UNEP (United Nations Environment Programme), 1990. *London Amendments to The Montreal Protocol*. Nairobi, Kenya.

UNEP (United Nations Environment Programme), 1991. *Environmental Effects of Ozone Depletion: 1991 Update*. Nairobi, Kenya.

UNEP (United Nations Environment Programme), 1992. *Report of the Fourth Meeting of the Parties to the Montreal Protocol on Substances that Deplete the Ozone Layer*, Copenhagen, November 23–25, 1992, Nairobi, Kenya.

USGCRP (U.S. Global Change Research Program), 2009. *Global Climate Change Impacts in the United States*, Karl, T. R., Melillo, J. M., and Peterson, T. C., eds., Cambridge University Press, New York.

Vogt, R., R. Sander, R. Von Glasow, and P. Crutzen, 1999. Iodine chemistry and its role in halogen activation and ozone loss in the marine boundary layer: A model study. *J. Atmos. Chem.*, 32, 375–395.

World Meteorological Organization (WMO), 1992. *Scientific Assessment of Ozone Depletion: 1991*. Global Ozone Research and Monitoring Project Report No. 25, WMO, Geneva, Switzerland.

World Meteorological Organization (WMO), 1995. *Scientific Assessment of Ozone Depletion: 1994*. Global Ozone Research and Monitoring Project. Report 37, WMO, Geneva, Switzerland.

World Meteorological Organization (WMO), 1999. *Scientific Assessment of Ozone Depletion: 1998.* Global Ozone Research and Monitoring Project. Report 44, WMO, Geneva, Switzerland.

World Meteorological Organization (WMO), 2003. *Scientific Assessment of Ozone Depletion: 2001.* Global Ozone Research and Monitoring Project. Report 47, WMO, Geneva, Switzerland.

World Meteorological Organization (WMO), 2007. *Scientific Assessment of Ozone Depletion: 2006.* Global Ozone Research and Monitoring Project. Report 50, WMO, Geneva, Switzerland.

Wuebbles, D. J., 1981. *The Relative Efficiency of a Number of Halocarbons for Destroying Stratospheric ozone.* Lawrence Livermore National Laboratory Report UCID-18924, Livermore, CA.

Wuebbles, D. J., 1983. Chlorocarbon emission scenarios: Potential impact on stratospheric ozone. *J. Geophys. Res.,* 88, 1433–1443.

Wuebbles, D. J., 1995. Weighing functions for ozone depletion and greenhouse gas effects on climate. *Ann. Rev. Energy Environ.,* 20, 45–70.

Wuebbles, D. J. and J. M. Calm, 1997. An environmental rationale for retention of endangered chemical species. *Science,* 278, 1090–1091.

Wuebbles, D. J. and M. K. W. Ko, 1999. *Summary of EPA/NASA Workshop on the Stratospheric Impacts of Short-Lived Gases.* March 30–31, Washington, DC.

Wuebbles, D. J. and K. O. Patten, 2009. Three-Dimensional modeling of HCFC-123 in the atmosphere: Assessing its potential environmental impacts and rationale for continued use. *Environ. Sci. Technol.,* 43 (9), 3208–3213, DOI: 10.1021/es802308m.

Wuebbles, D. J., Kinnison, D. E., Grant, K. E., and Lean, J., 1991. The effect of solar flux variations and trace gas emissions on recent trends in stratospheric ozone and temperature. *J. Geomag. Geoelectr.,* 43, 709–718.

Wuebbles, D. J., A. Jain, K. Patten, and P. Connell, 1997. Evaluation of ozone depletion potentials for chlorobromomethane (CH_2ClBr) and 1-Bromo-Propane (C_3H_7Br). *Atmos. Environ.,* 32, 107–114.

Wuebbles, D. J., A. K. Jain, R. Kotamarthi, V. Naik, and K. O. Patten, 1999a. Replacements for CFCs and Halons and their effects on stratospheric ozone. In: *Recent Advances in Stratospheric Processes,* T. R. Nathan and E. Cordero, Research Signpost, Kerala, India.

Wuebbles, D. J., R. Kotamarthi, and K. O. Patten, 1999b. Updated evaluation of ozone depletion potentials for chlorobromomethane (CH_2ClBr) and 1-bromo-propane ($CH_2BrCH_2CH_3$). *Atmos. Environ.,* 33, 1641–1643.

Wuebbles, D. J., K. O. Patten, M. T. Johnson, and R. Kotamarthi, 2001. The new methodology for ozone depletion potentials of short-lived compounds: n-propyl bromide as an example. *J. Geophys. Res.,* 106, 14,551–14,571.

Wuebbles, D. J., K. O. Patten, D. Wang, D. Youn, M. Martinez-Avilés, and J. S. Francisco, 2010. Three-dimensional model evaluation of the ozone depletion potentials for n-propyl bromide, trichloroethylene and perchloroethylene. *Atmos. Phys. Chem.,* in press.

Yang, S.-K., C. S. Long, A. J. Miller, X. He, Y. Yang, D. J. Wuebbles, and G. Tiao, 2009. The modulation of natural variability on the trend analysis of the updated cohesive SBUV(/2) total ozone. *Int. J. Remote Sens.,* 30(15), 3975–3986.

Youn, D., K. O. Patten, J.-T. Lin, and D. J. Wuebbles, 2009. Explicit calculation of indirect global warming potentials for halons using atmospheric models. *Atmos. Phys. Chem.,* 9, 8719–8733. *Atmos. Chem. Phys. Disc.,* 9, 15511–15540.

Youn, D., K. O. Patten, D. J. Wuebbles, H. Lee, and C.-W. So, 2010. Potential impact of iodinated replacement compounds CF_3I and CH_3I on atmospheric ozone: A three-dimensional modeling study. *Atmos. Phys. Chem.,* in press.

Glossary of Terms and Acronyms

Note: This is intended as an explanation of some of the more commonly used terms and acronyms related to cleaning chemicals and processes; it is not a complete list of all technical terms or abbreviations used in this book. In general, the included terms are referred to in more than one chapter. These definitions are to be considered to be descriptive rather than necessarily to be formal definitions.

AAMI: Association for the Advancement of Medical Instrumentation, an organization that sets standards to increase the understanding, safety, and efficacy of medical instrumentation.

Abrasive media: Materials used to remove soil via the momentum of impact.

ACGIH: American Conference of Governmental Industrial Hygienists; sets TLVs.

Airless: A description of an enclosed cleaning system that is sealed to contain either full vacuum (~1 mmHg) or a pressure significantly elevated above ambient (~800–10,000 mmHg).

Airtight: A description of an enclosed cleaning system that is sealed to contain a light pressure above ambient, typically about 0.5 psig.

Aqueous cleaner or process: Water based, may contain significant levels of organic and/or inorganic compounds.

ASTM: American Society for Testing and Materials; a group that establishes testing standards.

Atmospheric lifetime: The length of time a chemical may persist in the atmosphere before breaking down to other compounds; a measure of the potential for climate change.

Azeotrope: A solvent blend that, over a limited range of temperatures, maintains the same relative concentrations as the mixture components evaporate.

Benchtop cleaning: Generally referred to as a small-volume, labor-intensive, nonautomated cleaning process performed in the open rather than in specially designed cleaning tanks; examples include overhaul and repair and spot cleaning.

Bio-based cleaning agent: Uses animal or more typically plant materials, favored because ingredients are renewable resources.

Bi-solvent: Patented sequential cleaning agent process.

CAA: Clean Air Act; the U.S. legislation that regulates air quality standards, including the phase out of ODCs.

Cal/OSHA: California Occupational Safety and Health Administration; sets standards that are legally enforceable in California.

CAS: Clean air solvent; cleaning agents that have been analyzed by South Coast Air Quality Management District (SCAQMD) in California and found to meet their stringent environmental requirements for VOCs, ODCs, GWPs, and air toxics.

CAS number: Chemical abstracts service numbers, unique chemical identifiers.

Cavitation: Vacuum "bubbles" created by negative pressures in ultrasonic and megasonic processes.

CFC: Chlorofluorocarbon.

Cold cleaning: A cleaning process in which the cleaning solvent is below its boiling point (as distinguished from vapor degreasing).

Contaminant: Material that has the potential to degrade the appearance or performance of a part, component, or assembly.

Co-solvent: A sequential process using a different solvent for a rinse; or two cleaning agents in the same tank.

D-Limonene: A citrus-derived organic cleaning solvent.

DMSO: Dimethyl sulfoxide; a cleaning solvent.

Dragin: Material (cleaning chemicals and contaminants) brought in from a previous cleaning step.

Dragout: Material (cleaning chemicals and contaminants) carried over to a subsequent cleaning step.

EPA: Environmental Protection Agency; the U.S. government agency responsible for setting and administering air and water standards.

ESCA: Electron spectroscopy chemical analysis; an analytic technique for determining surface contamination; also known as XPS.

Flammable: Used to describe a combustible material that ignites very easily, burns intensely, or has a rapid rate of heat spread.

Flash point: The lowest temperature of a flammable liquid at which vapors are given off to form a flammable mixture with air, near the surface of the liquid or within the container; test methods differ by geographic locality.

Freeboard: A term used in vapor degreasers defined as the distance from the point where the boiling solvent vapor idles to the top of the machine opening.

Freon: A trade name (DuPont) for CFC-113; sometimes applied generically to CFCs.

FTIR: Fourier transform infrared spectroscopy; a surface analytic technique utilizing reflected infrared light to identify types of surface contaminants.

Greenhouse gas: A gas that persists in the stratosphere and acts to trap re-radiated heat from the earth's surface.

GWP: Global warming potential; a relative measure of a material's heat trapping ability as a greenhouse gas.

HAP: Hazardous air pollutant, a U.S. EPA classification of under 200 compounds.

HEPA: High Efficiency Particulate Air, a filter with specific properties used to minimize particulates in cleanrooms or controlled environments as well as to control hazardous emissions.

HFC (or HCFC): Hydrofluorocarbon; a class of chemicals developed as ODC replacements.

HFE: Hydrofluoroether; a class of chemicals developed as ODC replacements.

Hydrophilic: Water soluble.

Hydrophobic: Water insoluble; usually soluble in organic solvents.

IPA: Isopropyl alcohol, a common organic solvent.

IPC: Association connecting electronics industries; a leading source for industry standards in the electronics sector.

ISO standard: Standards adopted by the International Organization for Standardization, a Geneva-based organization that promulgates worldwide proprietary industrial and commercial standards.

KB: Kauri-butanol; a number used to compare the solubility of heavy oils in a particular solvent. It is the volume of solvent required to produce a defined degree of turbidity when added to standard solutions of Kauri resin in *n*-butyl alcohol.

Kyoto protocol: International agreement to limit emissions of greenhouse gases responsible for global warming.

LEL: Lower explosion level; the lowest concentration at which a mixture can explode.

MC (Meth): Methylene chloride.

Megasonics: A cleaning technique utilizing sound waves at frequencies higher than those for ultrasonics, from 500 kHz to 2 MHz.

Montreal protocol: International agreement to limit or eliminate production of ozone-depleting compounds (ODCs).

MSDS: Material safety data sheet.

Neat: A term meaning pure or undiluted.

NESHAP: National Emission Standards for Hazardous Air Pollutants; a series of U.S. federal regulations involving chemicals that can cause air pollution.

NMP: *N*-methyl pyrillodone; a cleaning solvent.

NPB (or *n*PB): *n*-propyl bromide; a cleaning solvent.

NVR: Nonvolatile residue; solid material left behind when a solvent evaporates.

ODC: Ozone-depleting compound, known to persist in the stratosphere and cause depletion of the ozone layer.

ODP: Ozone-depletion potential; a relative cumulative measure of the expected effects on ozone of the emissions of a gas relative to CFC-11.

OSEE: Optically stimulated electron emission; a surface analytic technique that measures the degree (but not the nature) of contamination by using UV light to stimulate the surface to emit electrons.

OSHA: Occupational Safety and Health Agency; the U.S. government agency responsible for setting and administering worker safety standards.

Particulates: Contaminate material with observable length, width, and thickness. In practice, an observable size will be about 0.1 microns or larger.

PCE (Perc): Perchloroethylene.

PEL: Permissible exposure limit. These are exposure guidelines for workers using the given chemical. PELs may be set by EPA or OSHA or by state agencies like Cal/OSHA.

PFC: Perfluorinated compounds containing fluorine and carbon but not chlorine or bromine.

POTW: Publicly owned treatment works; a local water treatment facility.

RCRA: Resource Conservation Recovery Act. Defines hazardous wastes and how to manage them.

REACH: Registration, Evaluation, Authorisation and Restriction of Chemical substances; a new European Community Regulation on chemicals and their safe use.

RO: Reverse osmosis; a filtering mechanism through a semipermeable membrane.

ROHS: Restriction of hazardous substances directive; an EU directive that restricts the use of six hazardous materials in the manufacture of various types of electronic and electrical equipment.

Saponification: The reaction between any organic oil containing reactive fatty acids with free alkalies to form soaps.

SARA: Superfund Amendments and Re-authorization Act. This act requires reporting of inventories and emissions of listed chemicals and groups.

SCAQMD: South Coast Air Quality Management District; the air quality regulating agency in southern California.

SEM: Scanning electron microscopy; a surface analytic technique involving imaging a surface by means of an electron beam.

Semi-aqueous: A sequential process using both organic solvent and water rinse.

SIMS: Secondary ion mass spectroscopy; a surface analytic technique using atoms ejected from a surface to identify contaminants.

SNAP: Significant new alternatives policy; an EPA effort to identify CFC replacement chemicals.

Soil: Matter out of place, contamination.

Solvent: Organic (carbon-containing) liquids; usually distinguished from aqueous.

STEL: Short-term exposure limit; a 15 min TWA exposure that should not be exceeded at any time during the day (see *TWA*).

STOC: Solvent and Adhesives Technical Options Committee; a United Nations UNEP committee that provides a great deal of input into worldwide environmental policy on cleaning.

Stoddard solvent: A common hydrocarbon blend used for cleaning oils.

Stratosphere: The atmospheric layer above the troposphere; considered to be above about 7 mi.

Surfactant: A material added to water or a solvent in order to increase wettability.

TCA: 1,1,1-trichloroethane (also called methyl chloroform).

TCE (TRI): Trichloroethylene.

TDS: Total dissolved solids; a measure of concentration of dissolved contaminants.

TLV: Threshold limit value; a concentration level above which there may be adverse health risks on exposure; usually set by the American Conference of Governmental Industrial Hygienists (ACGIH).

TOC: Total organic carbon; a measure of concentration of organic matter in water.

Troposphere: The lower layer of the atmosphere.

TWA: Time weighted average; an employee's permissible average exposure in any 8 h work shift of a 40 h week (see *STEL*).

UEL: Upper explosion level; the maximum concentration at which a mixture can explode.

Ultrasonics: A cleaning technique utilizing sound waves from 20 kHz to over 100 kHz.

UNEP: United Nations Environment Programme; a United Nations group that includes the STOC.

Vapor degreasing: A cleaning process in which, at least for the final cleaning, the part is suspended above a boiling solvent and is cleaned by the condensate of freshly distilled solvent vapor.

VMS: Volatile methyl siloxane; a silicon-based cleaning solvent.

VOC: Volatile organic compound; responsible for smog formation in the troposphere.

XPS: X-ray photoemission spectroscopy (see *ESCA*).

Index

A

ACPH, *see* Air changes per hour
Active pharmaceutical ingredient (API), 64
Aerospace, cleaning
  airless/vacuum system, 208–213
  applications, 206–207
  communication and testing, 207–208
  conservatism, 208
  inertial navigation systems
    clean room conditions, 213–214
    co-solvent process, 215
    gyroscopes/accelerometers, 213
    process modification, 214
    VG cleaning agent, 215
  optics
    additional issues and suggestions, 219–221
    blocking compounds, 216
    chemicals and process, 217
    eyeglasses and contact lenses, 216
  optics deblocking agents, 217–219
  process change and improvement, 208
  product reliability, 207
Airborne molecular contamination (AMC)
  detection, 132
  monitoring techniques, 133
  sources, 132
  types, 126
Air changes per hour (ACPH), 69
Air cleaning equipment
  batch cleaning machine design, 303–304
  dry cycle, 306
  general guidelines, 306–307
  rinse cycle, 305–306
  wash cycle, 304–305
Aluminum cleaning
  adoption
    EHS comparison, 423
    follow-up testing, 422–423
    Shopmaster RC, 422
  cotton tip swabs, 421
  coupons, un-cleaned/cleaned, 421
  demasking process, 420

  lacquer removal process, 420–421
  mechanical energy, 421
  pilot plant/scale-up feasibility, 422
  product selection, temperature and concentration
    dilution, DI water, 420
    Microshield Red Stop-Off lacquer, 421
American Society for Testing and Material (ASTM)
    methods, 19
Analytical and monitoring techniques
  AMC
    detection, 132
    monitoring techniques, 133
    sources, 132
  and controls, 131
  definitive analysis, 130
  description, 129
  in situ *vs.* extractive analysis, 131–132
  standards, 130–131
  visual, 129–130
Analyzing surfaces practical aspects
  applicability
    FTIR spectroscopy, 141
    sophistication, 140
    surface analysis techniques, 140–141
  characteristics, 135–136
  haze identification, 142–143
  layer characterization, 141–142
  methods
    Auger analysis and XPS, 137
    involve direct analysis, 139
    perturbation, 139–140
  micron-sized particulate, 137
  particle analysis, 142
  preliminary evaluation, 137
  relative thickness, 136
  sample handling and packaging issues
    protective covering, 138
    XPS and Auger analysis, 137–138
  sample surfaces, 137
  SEM/EDX analysis, 137
Anterior lumbar interbody fusion (ALIF) spinal
    surgical instrument set
  interpretation, 244

product description, 242
purpose, 241
scope, 241
test procedures, 242–244
API, *see* Active pharmaceutical ingredient
Aqueous inline cleaning equipment
    air management, 310
    automated monitoring and control, 311
    drying, 311
    fluid
        control, 310–311
        delivery, 309
        management, 309–310
        storage, 309
    machine design, 307
    process monitoring, 311
    selection, 312
    structure, 308
    wash bath life, 309
Army Alternative cleaner program, 440
ASTM methods, *see* American Society for Testing and
        Material
Auger electron spectroscopy (AES)
    depth profiling, 142
    etching and analysis, 142
    haze identification, 142
    secondary and backscatter electrons, 140
    vs. XPS, 137

B

Back-end-of-line (BEOL) cleaning processes
    bulk material removing, 372
    EBR process, 372
Bath quality monitoring
    aqueous cleaner, 46
    solvent
        particulate removal efficiency, 46
        prevention, acid formation, 45
        vapor degreasing and soil loading, 45
Biobased cleaning agents
    castor oil, 21
    cleaning application, 20–21
    political chemistry
        handling method, 21
        regulatory standards and metrics, 22
        solvents, 21
Biochemical oxygen demand (BOD), 57–58
Biomedical applications
    cleaning and manufacturing, 221
    FDA guidelines, 221
    surface quality, 222
British Occupational Hygiene Society (BOHS), 473
Brush cleaning
    adhesion moment, 351
    chemical-mechanical polishing, 352
    particle-removal efficiency, 353

C

Chemical mechanical planarization
    post-CMP clean
        alumina particle, brush loading, 381
        contaminants, 380–381
        copper and waste water treatment, 382
        single wafer megasonic tools, 381
    primer
        aluminum interconnect applications,
            379–380
        copper, 380
        damascene technique, 380
        large-scale integration (LSI), 379
Chlorofluorocarbons (CFCs)
    cleaning applications, 483
    destructive nature, 484
    elimination, 507
    Montreal Protocol, 492
    solvents, 493
    trouble-shooting services, 493
Cleaning activities
    bath quality monitoring
        aqueous cleaner, 46
        solvent, 45–46
    housekeeping and loss prevention
        practices, 47
        spills and leaks, 46–47
    material substitution, 47, 48
    part drainage improvement/drag-out elimination
        configuration effect and air knifes, 43
        effect, withdrawal time, 42
        guidelines, 43, 44
        Kushner model, 41
        manual operations, 42
        rinsing and liquid film thickness, 40
        withdrawal losses, 41–42
    performance monitoring
        paint adhesion method, 44
        removal, water-soluble ionic contaminants,
            44–45
        surface analysis techniques, 43
        water spray atomization and water break, 44
    process improvements
        bath, 47, 48
        demineralized water use, makeup, 48–49
        soil removing, bath, 50
        tank agitation, 49
        two-stage cleaning, 50
        water conservation measures, 51
Cleaning agent balancing act
    attribution, technical/performance, 15
    biobased
        castor oil, 21
        cleaning application, 20–21
        political chemistry, 21–22
    factors, cleaning agent development, 15

flash point solvent and flammability
    ASTM methods, 19
    flash point, defined, 19
green cleaning
    aspects, 27
    CFC-113 and TCA, 22
    cleaning process, 24–26
    definition, 22
    environmental and worker safety regulations,
       23
    industrial and precision, 22
    manufacturing process, 27
    Montreal Protocol, 22
    ozone layer protection, 22
    regulation changing, 26
    toxicity, 24
    VOCs, 23–24
physical/chemical properties
    flash point and evaporation rate, 18
    halogen-free, 18
    selection, 17–18
solvent development
    niche market, 20
    steps, 19–20
wish list, 16–17
Cleaning agents
    ambitious adaptations, 9
    aqueous and terpenes, 7, 9
    checklist, 8–9
    chlorinated solvents, 7
    hydrocarbon, 7
    nonlinear alcohol, 7
    VOC content, 10
Cleaning equipment
    checklist, 11–12
    optimal cleaning process, 10
    sample handling, 10
Cleaning practices and pollution prevention
    activities
       good operating practices, 40–47
       process improvements, 47–51
    hierarchy, 39–40
    Pollution Prevention Act of 1990, 39
    rinsate treatment
       good operating practice, 51–53
       incidental pollutants tracking, 58–59
       preliminary treatment, 54–55
       technologies, 55–58
Cleaning process
    budget, workspace and utility constraints, 6
    chemistry and equipment vendors
       interactive communication, 3–4
       preliminary considerations, 5
       record keeping and containment, 4
       steps, 4
    cleaning agents, 7–10
    company requirements, 6–7

customer requirements and requests, 6
decision making, 13
drying, 10
equipment, 10
greener soils and cleaning agents
    biobased, 25
    fabrication process, 25
in-house
    lean cleaning, 26
    safety and environmental preferability, 26
    water conservation, 26
    worker safety and environmental concerns,
       25–26
organization and coordination
    checklist, cleaning equipment, 11–12
    impending regulatory changes, 12
product design
    performance factor, 24
    water-based cleaning agents, 24
refine and test
    equipment installations, 13
    process estimation, 12–13
    suppliers cleaning evaluation, 13
steps, 4
supply chain, 25
Cleanroom garments
    ASTM and AATCC test methods, 108–109
    body box testing, 109
    classifications, 108
    gamma radiation sterilization, 110–111
    gamma subcontractor qualification, 111
    garment system evaluation, 109–110
    seams and components, 109
    system supplier, 110
Cleanroom management
    behavior
       behavioral requirements, 107
       meticulous hiring practices, 107
       testing and certification, 107
       training program, 107
    contamination hazards, 104
    facility design
       air changes, 105
       certified levels, 105
       contamination sources, 104
       parameters, 106
       structures and furnishings, 104
       tests, 105
    gowning
       doffing, 112–114
       donning, 112
    housekeeping
       checklist, 115
       selection and validation, 114
       top-to-bottom motion, 114
    human contamination
       hiring and training, 107

uniform program, 106
viable and nonviable particles, 106
implementation, 104
ongoing assessments
environmental monitoring, 115–116
robust and reproducible, 116
requirements, 104
sterilization process routine monitoring, 111
Cleanroom validation and monitoring
design
clients and services, 89–90
principles, 90
process and supporting rooms, 90
medical device and pharmaceutical industries, 90
phases and occupancy states, 91
standard and guidance documents, 91
tests
airborne particle count, 93, 94
airflow volume and velocity, 91–92
air movement visualization and room recovery, 92
HEPA/ULPA filter installation leak testing, 92
IS EN ISO 14644-1, 91
microbial testing and report, 95
pharmaceutical facilities classification, 93–94
room pressurization, 92–93
temperature, relative humidity and sound measuring, 94
validation *vs.* monitoring
environmental program, 95–100
routine particulate, 96
Critical cleaning
aerospace
airless/vacuum system, 208–213
applications, 206–207
communication and testing, 207–208
conservatism, 208
inertial navigation systems, 213–216
optics, 216–221
process change and improvement, 208
product reliability, 207
biomedical applications
cleaning and manufacturing, 221
FDA guidelines, 221
surface quality, 222
definition, 198
description, 198
electronics cleaning/defluxing
description, 201–202
design requirements, 202–203
lead-free solder and regulatory issues, 205–206
process choosing, 206
product performance requirements, 205
research in motion, 200–201
soils, 203–205
industrial metal, 226
nanotechnology, 226

options
avoiding cleaning, 199
redesigning processes, 199–200
selection and improvement, 226–227
steps, 198

D

Defense Land System and Miscellaneous Equipment (DSLME), 449–450
Defense logistics agency (DLA), 446
De-varnishing
Carbopol/Armeen 2C solvent gel, 407–408
natural resin, 406
organic chemistry, 406
organic solvents advantage, 406
oxidized natural resin varnish removal, 407
solvents, 406–407
varnish description, 405
water pH, surfactant and chelating agents, 407
Disasters and mistakes, cleaning
charred solvent, 34
facilities manager
assumptions, 31
dropped ball, 30
growth spurt
inline solvent unit, 33
solvent leak and fixation, 33–34
Kanegsberg case
in-line aqueous process, 34
resolution, 35
machine
door dimensions, 32
requirements, 36
specifying engineer and customer, 37
transportation, 32
manufacturer and equipment vendor, 32
operator and maintenance training, 31
physics 101
air pressure, 37
remedial physics, 37
PSPS and metallurgy, 35
purchase order and cleaning system cost, 36
right chemistry
solvents, 32–33
technician, 33

E

Electrical turnover package
checklists, 83
distribution system, 81
documentation, 84
inspection reports
cable termination, 82
equipment, 81
motor checklist, 82

test reports
    emergency generator, 83–84
    GFI and emergency power load, 83
    MCC, 82–83
    switch gear, transformer and breaker, 82
    transformer and panel load, 84
Electronic assemblies, contamination-induced failure
    antenna control units, 343
    big copper connector, 343
    climatic reliability, 334
    component packing density, 333
    critical applications
        environmental testing, 335
        functional reliability, 334
    failure mechanisms, 334
    field sensors, 333
    humidity and pollution effects
        corrosion-induced leakage, 339–340
        electrochemical migration, 336–338
        environments, 336
        lead free/silver-based solders, 342
        preconditions, 336
    malfunctions and failures, 342
    signal-integrity disruption
        adequate climatic operating conditions, 341
        application, 340–341
        HDI assemblies, 341
        HF circuits, 342
        malfunctions, 341
        performance parameters, 342
Electronic assembly cleaning process
    building blocks
        decoupling/emulsification, 292
        defoaming, 292
        description, 289
        nonreactive additives, 291
        protect metallic alloys, 291–292
        reactants, 290–291
        solvency, 289–290
        surface tension, 291
        wetting, 291
    design, 286
    electrochemical migration risk
        challenges, 285–286
        flux residues, 284–285
        metal conductors, 284
        technology-based market pressures, 285
    flux residues
        lead-free soldering and miniaturization, 295
        process cleaning rate theorem, 295
        thermal phases, 295
    lead-free drive solder flux requirements
        flux types, 288–289
        soldering process issues, 287–288
    materials compatibility
        equipment and agent selection, 297
        validation, 298

miniaturization, 286
    process validation
        best-fit cleaning agent, 312
        cleaning equipment decision, 312
        flux soil characterization, 312–313
        modified honeywell cleaning process, 313
        qualification methodology, 313–314
    solubility properties, parameters, 296–297
    solvent vapor cleaning
        aqueous, 294
        mechanisms, 293
        semi-aqueous, 293–294
    thermodynamics *vs.* kinetics
        air cleaning equipment, 303–307
        aqueous inline cleaning equipment, 307–312
        effects, 299
        semi-aqueous cleaning equipment, 300–303
        vapor phase cleaning equipment, 300
Electronic device fabrication cleaning
    BEOL cleaning, 382
    CMP, *see* Chemical mechanical planarization
    copper interconnects
        CMP, 382
        damascene process, 383–384
        device parameter targets, 382
        inadequate and adequate cleaning, 385
        organic and inorganic residue removal, 384
        post via etch, residue, 384
        pre-/post-CMP clean, 383
        trench etching, 382–383
    mobile particles
        description, 391–392
        electrolyzed water, SC1 and SC2, 394–395
        laser-induced shockwaves, 393–394
        simple blow-off, inert gas, 392–393
        supercritical CO_2 process, 394
    particle
        defect control, 387–388
        electronic device fabrication, 388
        removal, surface depends, 388
        van der Waals force, 388
    and rinsing
        immersion/dipping process, 395
        spray process, 395–396
    stationary particles
        description, 388
        strongly bonded, removal methods, 389–390
        weakly bonded, 390–391
    wafer backside and bevel
        chemical, 387
        edge exclusion, 385–387
Electronics cleaning/defluxing
    description, 201–202
    design requirements, 202–203
    lead-free solder and regulatory issues, 205–206
    process choosing, 206
    product performance requirements, 205

research in motion, 200–201
soils, 203–205
Environmental impact, cleaning agents
  air quality and VOCs
    control programs, 508
    definition, 507
    impact, ozone formation, 507–508
  3D numerical models, 502–503
  EPA SNAP program
    atmospheric impacts, ODP, 504
    environmental and health risks, 503
    exposure assessments, 505
    flammability, 505
    health effects, 504
    principles, 503–504
    risk screening, 505
    solvents/solvent blends, 503
    toxicity data, 505
  equivalent chlorine loading (ECL)
    chlorine and bromine, 513
    concept, 512
    EESC, 513
    HCFCs, 513–514
    limitations, 512
  good *vs.* bad ozone
    bromine, 506–507
    chlorine, 506
    halocarbon, 507
    holes, 506
    iodine, 507
    troposphere and stratosphere, 505–506
  GPWs
    CO_2, 515
    concept, 515
    greenhouse gases, 514–515
    halocarbons, 515
    indirect and direct, 515
    100 year integrated, 516–517
  ODPs
    concept, 508
    determination, 508
    and GWPs, 501
    HCFC-123, 509
    Iodotrifluoromethane (CF_3I) and methyl bromide (CH_3Br), 512
    n-propyl bromide, 511
    numerical models, 508–509
    semiempirical approach, 509
    steady-state, International Ozone Assessments, 509, 510
    three-dimensional models, 511
    traditional 2D model analysis, 510–511
    very short-lived (VSL) compounds, 510
  ODSs, 502
  recovery, stratospheric ozone, 514
  replacement compounds, 502
  screening approaches, 501–502

Environmentally preferable cleaning processes implementation
  alternative technology effort
    and compatibility requirements, 444–445
    design process influence, 446
    grow champions, 446–447
    logistics, 446
    performance and compatibility requirements, 444–445
    scientists/technologists and compliance officers, 445
    users/customers, 445–446
  barriers
    decision-making process, 442–443
    institutional resistance, change, 444
    logistics of greening, 444
    regulations, 443
    requirements, 443–444
  DoD
    alternatives implementation road map, 440–442
    organic solvent cleaners, 439–440
    solution, 447–451
  life cycle cost focus, 440
Environmental monitoring program
  alert and action limits, 99
  guidance documents, 96–97
  map and frequency, sampling, 97
  microbial testing, 97–98
  microflora, 99
  sampling types and locations, 97
  sterilization/disinfection, 98
  trending and documenting, results
    library, organisms, 99–100
    microbiology testing, 100
Equivalent effective stratospheric chlorine (EESC), 513
European Chemicals Agency (ECHA), 471–472
Extraction techniques, cleaning validations
  contaminants identification
    families, 165
    primary and secondary, 164
    reaction products, 164–165
    volatile, 165
  defined, 161–162
  design qualification, 162
  device families, 163
  effectiveness, 170
  exhaustive extraction, 170
  extractables *vs.* leachables, 163
  parameters setting
    pool multiple devices, 169
    technique, 168
    temperature, 168
    time, 169
  purpose and goals, 162
  reproducibility and effectiveness, 163
  residue limits
    biocompatibility, 170–171

comparison, 171
risk-based assessment, 171–172
revalidation, 164
solvent selection
basic solubility, 167–168
device compatibility, 168
multiple, 168
spike recovery, 169
surface, defined, 163–164
test method selection
chromatography methods, 167
gravimetric analysis, 166
particulate analysis, 167
total organic carbon (TOC) analysis, 166
ultraviolet/visible spectroscopy, 166–167

F

Facility file table of contents, 65
FEOL cleaning process, *see* Front-end-of-line cleaning
process
Food processing industry
brewery soils
carbohydrate components, 276
chlorinated alkaline cleaners, 276
clean equipment and environment, 275
cleaning temperatures, 276
characteristics, food soils, 272
chlorinated alkaline cleaners, 272–273
dairy soils
chlorinated alkaline cleaners, 275
components, 274
heat effect, 274–275
processing temperature, 274
fatty soils, 273
films on equipment, 278
mineral salts, 273
petroleum-based soils, 273
proteins, 271
soils
alkaline cleaners, 277
attachment, 273
cleaner choosing, 278
cleaning standpoint, 277
egg products, 277
equipment, 278
soluble proteins, 277
sugar caramelization, 277–278
standards of cleanliness
biofilms, 280
bioluminescence technology, 281
clean surfaces and surrounding
environments, 281
enumerating organisms, 280–281
optical methods, 280
radioactive labels, 280

residual soil, 279
visual examination, 279
Front-end-of-line (FEOL) cleaning process
ozone-water mixtures, 371–372
quaternary ammonium hydroxides, 370–371
sulfuric-peroxide chemistry
ammonium persulfate, 370
oxidizing agents, 369
RCA, 370
silicon wafers, 369
TMAH chemistry, 371

G

Global warming potentials (GPWs)
CO_2, 515
concept, 515
greenhouse gases, 514–515
halocarbons, 515
indirect and direct, 515
100 year integrated, 516–517
Good manufacturing practices (GMP)
areas, 65
facility and prefabricated panels, 70
requirements, 72
Ground fault (GFI) test report, 83

H

Hamaker constant, 346
Hazards identification
chemical, 468–469
flammable materials, 467
material, 469–470
mechanical, 466–467
minimization, chemical exposure, 470
noise, 468
quantitative exposure numbers, 470–471
reactive materials, 467–468
thermal trauma, 467
Health and safety
cleaning technology, 466
description, 465–466
hazards identification, 467–471
MSDS, 466
personal protective equipment
contact lenses and safety glasses, 477–478
doffing procedures, 481
donning procedures, 480–481
facial hair, 479–480
protection levels, 480
respirators and contact lenses, 479
respiratory protection, 478–479
selection matrix, 481
quantitative exposure numbers
chemical-by-chemical approach limitation, 471
inhalation hazards, 470–471

REACH
    aim, 471
    BOHS, 473
    derived no-effect level (DNEL), 472–473
    description, 471
    ECHA, 471–472
    provisions, 471
    safer critical cleaning process implementation
        blends, complex and proprietary, 475
        comparison, 476
        documentation, risk assessment, 474–475
        harmonizing safety and environmental
            regulations, 476
        hazard identification, 474
        information collection, 473–474
        manufacturing activities, 476
        plan actions, 474
        professional and working, 476–477
        risk assessment, 473
Heating, ventilating and air-Conditioning (HVAC)
    ACPH, 69
    balance sheet analysis, 62
    documentation, 78, 80
    duct routing, 62–63
    equipment, 68
    MUA pretreatment, 68–69
Hydrodynamic cleaning
    adhesion force, 348
    rotating disk, 349

I

ISO standards and cleanroom classification
    aseptic processing, 67–68
    Federal Standard 209E, 67
    ISO 14644 and 14698, 68

J

Joint service solvent substitution (JS3) working group
    data exchanges, 449
    methodology
        DOD agencies, 448
        process, 449
        tool, 448
    mission, 448
    project tracking database, 449

L

Life science cleanroom facility design
    architectural issues
        gypboard *vs.* prefabricated panels, 69–70
        "plan and spec" projects, 70
        rounded corners and smooth surfaces, 69
    central glycol system *vs.* air cooled direct expansion
        units, 63

clinical production, 87–88
commissioning
    direct impact systems, 85, 86
    plan elements, 85, 86
contamination control criteria
    CFM per filter, 66–67
    ISO standards and classification, 67–68
coordination and communication, 61–62
criteria and containment technologies, 66
FDA warning letter, 86–87
HEPA leak testing criteria, 87
HVAC, 62–63, 68–69
isolation technology
    described, 63
    isolator-based fill machines and line, 64
    modern barrier technology integration, cGMP,
        63–64
    product protection, 64
plan, 62
preconstruction review
    bulk API production, 64
    facility file document, 64–65
preparation
    BOD elements, 74
    FDA reviews, 74–75
    GMP requirements, 72
    modular construction advantage, 72–73
    operator intrusions and equipment, 73
    product and component flows, 73
project planning and delivery
    design/build approach, 72
    IQ/OQ protocols, 71
    modular/pre-engineered technology, 70
    preventive maintenance rules, 70–71
    virtual design–build company, 71
    walkable ceiling benefits, 70
TOP, 75–85
value engineering, 62
walkable ceilings, 63

M

Material compatibility
    alter hardware performance, 185
    changing parts, 188
    cleaning
        fluid changing, 187–188
        process changing, 187
        solvents, 186
    definition, 184
    entrapment fade, 186
    environmental requirements, 183
    finding data, 189
    interpreting data, 189
    optimum materials and processes, 184
    preventing damage to hardware, 184–185
    processing equipment protection, 185–186

risks, 186
testing
    Clevis McKleen's degreasing system, 192
    coatings, 190
    elastomer, 190
    immersion, 189, 191
    liquid mixing, 190
    material, 191
    quantify degradation, 189
    service environment, 189–190
Material safety data sheet (MSDS), 458–459, 466
Materials of evolving regulatory interest team
        (MERIT), 447
Measuring surface cleanliness
    acceptable level
        controlled experiment, 155–156
        production testing, 156
    analytical methods, 150
    cost impact, 154–155
    definitions, 146
    direct methods, 150
    indirect methods, 149–150
    microbial contaminants, 146–147
    monitoring/measurement method
        accuracy and precision, 149
        acquisition and operating costs, 149
        features, 149
        skill level, 149
        speed, 149
        substrate being checked, 148–149
        type of contaminants, 148
    operating parameters, 148
    optimum cleanliness level, 156
    particle contaminants, 146
    precision cleaning, 148
    principles of operation
        analytical methods, 153–154
        direct methods, 151–152
        indirect methods, 150–151
    process variables, 148
    thin-film contaminants, 146
Medical applications
    biomedical instrumentation, 222
    bonding problems, 223
    cleanliness standards, 223–224
    company cleaning requirements, 224
    device manufacture
        combination devices, 225–226
        paralysis through analysis, 226
        prototype to type, 225
        regulatory constraints, 224–225
        ultrasonic's and flammables, 224
        validation and pre-validation, 225
    plastic and metal components, 223
Medical device cleanliness
    analytical testing
        hardness testing, 178

re-sterilized medical devices, 177
        tensile test, 177
    biological safety assessment
        aim, 180
        ISO 10993, 181
        materials chemical characterization, 180
        toxicological risks, 179
    chromatographic analysis, 178
    cleaning and sterilization residues
        chemical methods, 176
        conventional cleaning process, 176
        ISO EO residue requirements, 177
        organic solvent usage, 176
        toxic residues, 177
    critical device, 174
    infrared (IR) instruments, 179
    noncritical device
        description, 174
        design features, 175
        reusable devices, 175
        sterilant-tolerant materials, 176
        sterilization efficacy testing, 176
    semicritical device, 174
    spectrophotometric analysis, 179
    total organic carbon (TOC) analysis, 179
    ultraviolet/visible spectroscopy, 179
Megasonic cleaning
    acoustic streaming, 350
    cleaning time and temperature, 351, 353
    input power, 353
    removal efficiency, 352
    transducer frequency, 350
Mobile particles cleaning
    description, 391–392
    electrolyzed water
        hydrogen and oxygen generation, 394
        ionization, 394
        SC1 and SC2, 394–395
    laser-induced shockwaves
        adhesion forces, 393
        low gap distance, 394
        supercritical carbon dioxide (SC-CO$_2$) process,
            394
        velocity, 393
    simple blow-off, inert gas
        argon gas, 392
        cleaning solution, 393
        size, 393–394
Montreal Protocol
    CFCs and ODSs, 492
    climate change, 496
    description, 492
    electronics companies, associations and
            committees, 494
    military leadership, 493
    news-making corporate leadership, 493
    no-clean soldering, 494–495

Moore's law, 286
Motor control center (MCC) test report, 82–83

N

Nanoparticles removal
    semiconductor manufacturing, 355
    silicon-cap films, 355–356
    single-wafer megasonic, 356
National Emission Standard for Hazardous Air
        Pollutants (NESHAP), 495–496

O

Occupational safety and health administration
        (OSHA), 459
Optics
    additional issues and suggestions
        cleaning *vs.* surface modification, 219
        process maintaining and improving, 221
        troubleshooting, 220
    blocking compounds, 216
    chemicals and process, 217
    deblocking agents
        acetone, alcohols, and low-flash-point
            blends, 217
        aqueous/surfactant, 217–219
        chlorinated and brominated solvents, 217
        co-solvent and bi-solvent processes, 219
        hot watercold shock, 217
    eyeglasses and contact lenses, 216
Ozone-depleting solvents
    Clean Air Act (CAA)
        air pollutants regulation, 495
        NESHAP, 495–496
        Title VI, 496
        VOCs reduction, 495
    Montreal Protocol
        CFCs and ODSs, 492
        climate change, 496
        description, 492
        electronics companies, associations and
            committees, 494
        military leadership, 493
        news-making corporate leadership, 493
        no-clean soldering, 494–495
    phaseout of, 496
    proactive companies, climate
        hydrofluoroethers (HFEs), 497
        hydrofluoroolefins (HFOs), 496–497
    sustainable enterprise, 497
Ozone-depleting substances (ODSs), 476
Ozone depletion potentials (ODPs)
    concept of, 508
    determination, 508
    and GWPs, 501
    HCFC-123, 509

Iodotrifluoromethane (CF_3I) and methyl bromide
        (CH_3Br), 512
    n-propyl bromide, 511
    numerical models, 508–509
    semiempirical approach, 509
    steady-state, International Ozone Assessments,
        509, 510
    three-dimensional models, 511
    traditional 2D model analysis, 510–511
    very short-lived (VSL) compounds, 510

P

Paintings, cleaning
    aqueous
        anti-biologicals, 403
        chelating materials, 402–403
        critical factors, fine art surfaces, 400
        enzymes, 403
        ionic strength, 401–402
        materials, 401
        nonvolatile compounds clearing, 403–404
        oxidants, 403
        solution pH, 401
        surfactants, 402
        viscosity-modifying materials, 403
    conservators, 399
    contaminant accumulation, 400
    cotton swabs, 400
    efficacy, 400
    "ethical" concerns, 400
    overpaint/retouch removal
        conservator, 408–409
        old, 409
    problem, 399
    soils, 399–400
    surface
        conservators, 404
        description, 404
        grime, aged layer removal, 404–405
        rinse solution, 405
        saliva, cotton swabs, 404
    varnish removal, *see* De-varnishing
Particle-removal mechanism
    lifting, 354
    rolling, 354
    sliding, 354
Perfect sample part syndrome (PSPS), 35
Personal protective equipment (PPE)
    contact lenses and safety glasses
        maintenance and storage, 478
        program, key elements, 477–478
        selection, 477
    facial hair
        air-purifying respirators (APR), 479
        positive and negative pressure tests, 479

procedures
    doffing, 481
    donning, 480–481
    protection levels, 480
respirators and contact lenses, 479
respiratory protection
    fit test, 478–479
    respirators, 479
selection matrix, 481
Pharmaceutical applications, cleaning
agents
    selection, 256–258
    suppliers, 258–259
analytical detection methods
    types, 264
    validation, 264–265
biological contamination
    containers and closures, 268
    equipment, 268
    test methods, 268
change control, 267–268
critical parameter, 259–260
description, 255–256
equipment design considerations, 260–261
method design, 261–262
product and equipment grouping, 262–263
QbD and PAT, 269
recovery studies, 263–264
regulations, 256
residue limits and acceptance
    acceptable level, 265
    build-up and distribution, 266
    limit of detection (LOD), 265
    swab sampling, 266
sample site selection, 263
sampling methods, 263
standard operating procedures, 267
validation master plan and protocol design, 266–267
Photoresist performance factors
adhesion capability, 366
exposure speed and sensitivity
    polymerization, 366
    positive and negative photoresists, 367
    structure, 366
resolution, 365–366
selection, 365
Pollution Prevention Act of 1990, 39
Pollution prevention, cleaning product selection
adoption, 426
aluminum, *see* Aluminum cleaning
chemicals, TURI Lab, 412
goals, 412
health and safety screening
    CleanerSolutions database, 414
    score, 415–416
    surface, environmental indicators, 415

multimetal
    brass and copper parts, 424
    EHS comparison, 425
    fines and cutting fluids, 424
    parts, 423–424
    pilot plant/scale-up feasibility, 425
    removal efficiency, 424
    strainer, 424
    temperature, concentration and mechanical energy, 424
on-site pilot plant and scale-up feasibility
    batch cleaning testing, 416
    flammability, 425
    follow-up lab testing, 425
    potential impacts, worker/environment, 425
performance-based
    aqueous-based surface cleaning test, 414
    CleanerSolutions database, 414, 415
    friendlier aqueous detergents, 413
    solvent, 412–413
P2OASys, health and safety, 416
steel
    adoption, 419–420
    EHS comparison, 419
    mechanical energy studies, 417–418
    pilot plant/scale-up feasibility, 418–419
    safety, 420
    soiled, 418
    temperature and concentration, 417
Polystyrene latex (PSL), 347
Precision cleaning in electronics industry
aqueous cleaning products
    attributes, 325–326
    encountered residues and associated risks, 329
    process integration, 327–328
critical cleaning
    chemical isolation, 322
    environmental, 324–325
    requirements and qualification, 322
    solvency, 323
    temperature, 324
ownership cost
    agent technology, 329–330
    bath monitoring, 330
    challenges, 329
    vapor recovery, 330
surfactant-free aqueous cleaning products
    builders and conditioners, 326
    encountered residues and associated risks, 329
    exposure times, 326
    material compatibility considerations, 326
    monitoring techniques, 328
    pH level, 326
    refractive index, 327–328
    residue, 326
    temperature and concentration effects, 325–326
    titration, 327

surfactants
    classification, 320–322
    description, 319–320
    structure, 320
Process guidelines, wax removal
    bath
        agitation, 435
        life, 436
    drying, 438
    emissions and waste disposal, 436
    final water rinse
        bath agitation, 437
        time and temperature, 437
        wastewater discharge, 438
        water quality, 437–438
    plumbing precautions, 435
    prewash, 435
    time and temperature, 435
    water/alkaline rinse
        bath agitation, 436–437
        bath life, 437
        emissions, 437
        time and temperature, 436
        waste disposal, 437
Process operation procedure(POP), 430
PSPS, *see* Perfect sample part syndrome

R

Registration, Evaluation, Authorisation and
        Restriction of Chemical substances
        (REACH)
    aim, 471
    BOHS, 473
    derived no-effect level (DNEL), 472–473
    description, 471
    ECHA, 471–472
    provisions, 471
Regulators
    chlorofluorocarbons (CFCs) and TCA, 483
    communication
        gap bridging, 487
        ongoing, 488
        tools, 488
    compliance inspections, 488
    Montreal Protocol, 484
    permit
        corrections period, 487–488
        equipment verification, 487
    regulations and communication
        methylene chloride, 486
        perchloroethylene, 486
        role, 486–487
        web sites, 485
        workshops, 485–486
    reporting and record-keeping, 488–489

restrictions, VOCs, 485
stratospheric ozone protection, 484
VOCs and HAPs
    air quality regulations, 484–485
    organic solvents, 485
    PCE, 485
Reusable medical devices
    ALIF spinal surgical instrument set
        interpretation, 244
        product description, 242
        purpose, 241
        scope, 241
        test procedures, 242–244
    compendium
        critical items, 232–233
        device categories, 232
        non-critical items, 233
        semi-critical items, 233
    description, 230
    designing, testing and labeling, 231
    guidance, 230–231
    issues and test programs
        design considerations, 233–236
        endotoxins and exotoxins, 236
        sterilant-tolerant materials, 233
        transmissible spongiform encephalopathy,
            236–237
    sterilization, 232
    testing methodology and validation
        BCA protein assay, 239
        contamination type, 238
        microorganisms, 237
        phenol-sulfuric acid test, 238–239
        replication, 238
        verification, 238
        viable count determination, 238
Rinsate treatment
    BOD reduction, 57–58
    definition, pollution prevention, 51
    good operating practice
        event of upset, 52–53
        importance, 51–52
        knowledge, prohibitions and limits, 52
        proactive approach, 52
        setting standards, reuse, 53
    heavy metal precipitation, 56–57
    incidental pollutants tracking, 58–59
    manufacturing facilities, 55–56
    pH adjustment, 56
    physical oil and grease separation, 57
    preliminary
        flow equalization, 54–55
        screening, solids, 54
        stream segregation, 54
    suspended solid removal, 57
    toxic organic pretreatment, 58

S

Safer critical cleaning process
  blends, complex and proprietary, 475
  comparison, 476
  documentation, risk assessment
    engineering controls, 475
    precautions, 474–475
    work practice controls, 474–475
  harmonizing safety and environmental
      regulations, 476
  hazard identification, 474
  information collection, 473–474
  manufacturing activities, 476
  risk assessment, 473
  safety professionals, 476–477
Semi-aqueous cleaning equipment
  air batch systems, 303
  immersion/ultrasonic systems, 302–303
  inline semi-aqueous equipment, 301–302
  solvent washing/water rinsing, 300
Semiconductor wafer manufacturing cleaning
      processes
  BEOL
    bulk material removing, 372
    EBR process, 372
  challenges
    copper interconnects, 375–376
    low-k dielectric material, 376–377
  fabrication
    description, 360
    layering, 360
    negative-acting photoresists, 363–364
    patterning, 360
    photolithography and masking, 361
    photoresist chemistry, 361
    photoresist performance factors, 365–367
    positive-acting photoresists, 364–365
    positive *vs.* negative resists, 363
    radiation-sensitive polymers, 361–363
    resist layer, 361
  FEOL
    ozone-water mixtures, 371–372
    quaternary ammonium hydroxides,
        370–371
    sulfuric-peroxide chemistry, 369–370
    TMAH chemistry, 371
  negative photoresist strippers, 373
  positive photoresist strippers
    NMP-based, 373
    non-NMP-based, 373
  post-plasma etch polymer removers
    HA/Amine chemistry, 374–375
    HF/Glycol chemistry, 375
    oxygen plasma ashers, 374
    sidewall polymers, 373

  techniques
    FEOL, 368
    optimization, 368
    temperature effect, 368–369
    ultrasonic and megasonic effect, 369
Significant new alternatives policy (SNAP) program,
      456–457, 484
Solvent development
  animal testing, 20
  cleaning niche, 20
  niche market
    industrial cleaning, 20
    parachlorobenzotrifluoride, 20
    small blip, 20
  screening survey, 19
  toxicity testing, 19
Standard operating procedures (SOPs), 119
Standard practices and operating procedures
      (SPOP), 430
Stationary particles cleaning
  description, 388
  strongly bonded, removal methods
    chemical procedures, 389–390
    solution effect, optical waveguide, 390
    undercut process, 389, 390
  weakly bonded
    electro-repulsion force, 390
    glass wafers, solution effect, 392
    pH rinse, commercial solution, 392
    wet clean process, 390
    zeta potential cleaning process, 391
Steel cleaning
  adoption, 419–420
  EHS comparison, 419
  mechanical energy
    ChemSearch Aerolex Plus Moly dry film
        lubricant, 418
    ultrasonic cleaning, 417–418
  parts, 417
  pilot plant/scale-up feasibility
    Daraclean 282 GF, 418
    five-indicator screening process,
        418–419
  product selection, temperature and
      concentration, 417
  safety, 420
  soiled, 418
Stratospheric ozone
  ECL, 512–514
  ODPs
    concept of, 508
    determination, 508
    and GWPs, 501
    HCFC-123, 509
    Iodotrifluoromethane (CF_3I) and methyl
        bromide (CH_3Br), 512

n-propyl bromide, 511
numerical models, 508–509
semiempirical approach, 509
steady-state, International Ozone
     Assessments, 509, 510
three-dimensional models, 511
traditional 2D model analysis, 510–511
very short-lived (VSL) compounds, 510
recovery, 514
Surface cleaning
adhesion forces
     capillary force components, 347
     classification, 346
     electrons, 347
     particles and substrate material, 345–346
     relative humidity, 347, 348
chemistry-enhanced, 355
description, 345
nanoparticles removal, 355–356
particle removal
     brush cleaning, 351–353
     hydrodynamic, 347–349
     mechanism, 354
     megasonic, 350–351
     ultrasonic, 349–350
zeta potential, 356

T

Threshold limit value (TLV), 215
Total organic carbon (TOC), 264
Toxics Use Reduction Act (TURA), 411
1,1,1-Trichloroethane (TCA), 22
Turnover package (TOP)
architectural
     finishes, furnishings and specialty, 76, 77
     guidelines, drawings and specifications,
         76–77
controls and software, 84–85
described, 75
design/build team, 76
documentation, 75
electrical
     distribution system, 81
     documentations, 84
     inspection and test reports, 81–84
HVAC, 78, 80
packaged equipment
     definition and basic specifications, 78
     documentation, 78, 80
     testing, 78, 79
piping
     documentation, 78–79, 81
     non-sanitary systems, 81
     sanitary systems, 78
systems, 75, 76

U

United Nations Framework Convention on Climate
     Change (FCCC), 497
U.S. Consumer Product Safety Commission
     (CPSC), 484
U.S. Department of Defense (DoD)
alternatives implementation road map
     objective, 440
     process, 440–442
     steps, 441
chemical and materials management
     Army's current materials development
         process, 450
     ESOH issues, 450
     NASA, 451
     stovepipe organization, 450
cleaning specification development
     DSLME NESHAP, 449–450
     scope, 450
joint service solvent substitution working group
     communication gap, 448
     JS3 working group, 448
     SERDP, 447–448
JS3
     data exchanges, 449
     methodology, 448–449
     mission, 448
     project tracking database, 449
organic solvent cleaners, 439–440
regulatory issues
     MERIT, 447
     NESHAP, 447
U.S. Environmental Protection Agency's Toxic Release
     Inventory (TRI), 411

V

Vapor phase cleaning equipment, 300
Volatile organic compounds (VOCs)
acetone, 23
control programs, 508
definition, 507
impact, ozone formation, 507–508
reduction
     inventory use, 23
     relative reactivity, 24
worker protection
     cleaning compounds, 458
     "line-in-the-sand" approach, 457
     organic additives, 458
     water-based processes, 457–458

W

Wafer backside and bevel
chemical, 387

edge exclusion
    barrier layer and seed copper layer, 385–386
    clean copper zone, 386
    residues, absence, 386–387
Wax removal, aerospace industry
    D1 immersion method, 430
    process guidelines
        bath agitation, 435
        bath life, 436
        drying, 438
        emissions and waste disposal, 436
        final water rinse, 437–438
        plumbing precautions, 435
        prewash, 435
        time and temperature, 435
        water/alkaline rinse, 436–437
    selective plating and masking
        applications, 429–430
        dimensional restoration, 429
        lacquers and waxes, 430
        SPOP and POP, 430
    specifications/industrial practices
        compatibility, 433
        OEMs, 432–433
        test methods, 433–434
    technologies
        aqueous cleaning, 432
        cabinet ovens, 430–431
        hot water, 431
        semi-aqueous cleaning, 432, 433
        steam stripping, 431
        vapor degreasing, 431
        wax tanks, 431
Wiping and cleaning validation principles
    cleaning validation
        aerospace manufacturing, 126
        pharmaceutical manufacturing, 125
        process equipment surfaces, 124
        semiconductors, 125
        visual examination, 125
    cleanroom/mini-environment
        bucketless system, 122

        cleaning weapons, 120
        disinfectants use, 120
        isolator cleaning tool, 123
        mops, 122
        readily accessible surfaces, 121
        relevant site protocol, 122
        vertical surfaces, 121
        wipers, 120
    contamination, 119–120
    protocols and SOPs, 123
    swabs, 124
    total organic carbon (TOC), 125
    training, 124
Worker protection
    communication
        environmental regulators, 461–462
        safety professionals, 461–462
    emergency supplies, air and process monitoring, 460
    environmental
        issues, 462
        policy, 460
        regulations *vs.* safety, 456–457
    hazards, 459
    manufacturing goals, 455–456
    material safety data sheet (MSDS)
        formation and recommendations, 458–459
        OSHA, 459
        proprietary, 459
        work site, 459
    regulatory advisory committees, 461
    safety rules and provisos, 455
    stratospheric ozone depletion, 456
    VOCs
        cleaning compounds, 458
        "line-in-the-sand" approach, 457
        organic additives, 458
        water-based processes, 457–458

X

X-ray photoelectron spectroscopy (XPS), 137, 138